DEVELOPMENT AND EVOLUTION OF BEHAVIOR; essays in memory of T. C. Schneirla, ed. by Lester R. Aronson and others. Freeman, 1970. 656p il tab bibl 76-84600. 12.00. ISBN 0-7167-0921-X

CHOICE   MAR. '71

Biology

T. C. Schneirla was one of the few pre-1950 zoologists to show much of an interest in animal behavior, and his influence both on ethology and comparative psychology has been considerable. This volume of essays in his honor is, like most such memorial volumes, written chiefly by his surviving friends, colleagues and students. The book covers a wide range of topics, from "Functional evolution of the forebrain in lower vertebrates" to "The concept of development and evolution in psychoanalysis." Some of the papers are good some are much better than that. Almost all of them show Schneirla's influence. The coverage is too broad and unsystematic to make this a useful undergraduate text — for that, see Maier and Maier's *Comparative animal behavior* (CHOICE, Oct 1970) or R. A. Hinde's *Animal behaviour* (CHOICE, Jan 1967). Or, for the matter, refer undergraduates to the last edition of Maier and Schneirla's classic *Principles of animal psychology* (CHOICE, April 1966).

DEVELOPMENT AND EVOLUTION OF BEHAVIOR

Essays in memory of T. C. Schneirla

EDITED BY

LESTER R. ARONSON   ETHEL TOBACH
Department of Animal Behavior
The American Museum of Natural History, New York

DANIEL S. LEHRMAN   JAY S. ROSENBLATT
Department of Animal Behavior
Rutgers University, Newark

W. H. FREEMAN AND COMPANY
San Francisco

A Series of Books in Psychology

EDITOR: Stanley Coopersmith

Library of Congress Catalog Card Number: 76-84600
International Standard Book Number: 0-7167-0921-X
Printed in the United States of America

1 2 3 4 5 6 7 8 9

Contents

PART IV   SOCIAL BEHAVIOR

PART V   HUMAN BEHAVIOR

Foreword

I first met Dr. Theodore C. Schneirla in the late 1930s when I was a graduate student at the University of Michigan. He had just completed a book, *Principles of Animal Psychology* with Professor N. R. F. Maier of the University, and had come from New York to discuss changes to be made in subsequent editions of the book. He presented a seminar for the graduate students and spoke of the broad new approach to the field called "animal behavior." In his discussion he emphasized the need for a comparative phylogenetic approach in the science of animal behavior rather than the study of different psychological problems without serious consideration of the basis for choosing a particular species for investigation. I remember his fascinating and remarkably vivid illustrations of the conceptual developments that result from comparative methods of studying the adjustments of animals to their environments. Freshly returned from a trip to Panama, he underscored his points with examples of his field studies there of the army ant. His presentation motivated many students of biology to explore the new field of animal behavior.

In the years since that seminar took place I had the good fortune to become better acquainted with T. C. Schneirla, professionally and personally, with mounting respect, admiration, and affection. He was a man of widely varied interests, and he had a capacity to stimulate the intellectual curiosity of all who came within his sphere. His primary concern was his consuming interest in the behavior of all kinds of animals, but especially ants and man; and he showed how much the study of the behavior of lower animals could teach us about human behavior. His published studies covered investigations on specific types of behavior observed in certain forms, as well as on broader perspectives and principles, such as instinctive behavior and learning processes. He devoted his energies, selflessly, to the discipline that he helped create and establish, but he never lost his empathy for his fellow man, his enthusiasm for working with students, or his warm humanity. He contributed in a personal and individual sense to the growth and evolution of the scientific eminence of The American Museum of Natural History. Although to some his work might seem esoteric or appropriate to an ivory tower, he was really a vital, participating, active force in the scientific councils of the Museum. He earned recognition and respect as an outstanding member of the Museum staff who was keenly aware of the public responsibility of the Museum and was always ready to share his knowledge with others through any appropriate channel of communication.

Ted's outstanding sense of integrity, one of his foremost attributes, characterized his every action. His integrity, for example, simply would not permit him to accept the attractive explanation that behavioral phenomena are either instinctive or learned. Rather, he emphasized that all behavior stems from highly complex developmental processes. This same sense of integrity sometimes caused dismay to editors who could not persuade him to use only ten words in a sentence if he felt twenty were necessary to remove all ambiguity. He wanted to be sure that there was absolutely no possibility of anyone's misunderstanding what he was saying. This volume is a fitting tribute to the man, Theodore C. Schneirla, and to his integrity.

June 23, 1969

James A. Oliver
Director
The American Museum
of Natural History

ETHEL TOBACH and LESTER R. ARONSON
Department of Animal Behavior
The American Museum of Natural History
New York

T. C. Schneirla
A Biographical Note

The interest in animal behavior and its relevance to human societal problems has focused attention on the various scientific approaches to the subject and the differences among these approaches. Such inquiry has had many effects, one of which has been the crystallization of what some people call "Schneirla's Theory." Dr. Schneirla, of course, never referred to his work in this way. Rather, he was concerned with the formulation of concepts that would make it possible by laboratory experimentation and field observation to explain behavioral phenomena in evolutionary and developmental terms, without recourse to vitalistic or static assumptions. He formulated his concepts inductively, as in the example of his unique analysis of the behavior of army ants and their social organization. The inductive procedures that he used along with the complexity of the problems to which he addressed himself resulted in a style of writing that does not conform with the usual presentations found in the biological and psychological literature. As often happens in such cases, his publications have been overlooked by some and misunderstood by others. Nevertheless, his work and ideas have affected his students and colleagues profoundly and have led to a growing number of experimental reports and theoretical articles by others, based to a greater or lesser extent on his concepts. This body of literature does not always spell out its relationship to Schneirla's theoretical formulations, but its character continues Schneirla's challenge to old ways of thinking about behavior in general and about the evolution and development of behavior in particular.

Accordingly, because of the need to bring together the material that would exemplify new ways of thinking about old problems, and to honor him for his protean contributions to the science of behavior, the editors initiated the publication of two books—this book of essays and a book of selected writings of T. C. Schneirla. Unfortunately personal difficulties slowed the writing of many of the manuscripts and this led to serious delays. One paper that was submitted early had to be published in a professional journal and is reprinted here. Late in August of 1968, the originally-planned Festschrift became a memorial volume, upon Schneirla's unexpected illness and untimely death. Because of the nature of the book of Schneirla's selected writings, it had been necessary at a late stage in the planning to inform him of our intentions, but he never knew the details concerning the participants and their contributions. There may be some consolation in the fact that he was pleased and gratified. In his typically considerate fashion, he never expressed more than a gentle wonder about the progress of the project.

In any cooperative enterprise, the aims and goals of the organizers are more or less at variance with the outcome. It is satisfying to note that in this case the variance is minimal. For the most part, the papers attest to their debt to Schneirla by their emphasis on the inductive method rather than the deductive method of science, by stressing the importance of fundamental theoretical concepts of behavioral process as contrasted with the more popular "atheoretical" stance in behavioral research, and by expressing the need for a holistic, integrative approach based on both analysis and synthesis instead of the more prevalent one-sided practice of reductionism.

Dr. Theodore C. Schneirla was born on July 23, 1902, in modest economic circumstances. His father had a celery farm in Michigan, and Theodore and his brothers worked hard after school and during summers, doing farmwork and selling and delivering celery in town. While in high school, he learned to play the trumpet, and later, when he attended the University of Michigan, he joined the University band. He played in the band throughout his undergraduate and graduate career, eventually becoming its president. He also studied typing and shorthand which he realized very early in life were valuable tools for an academic and scientific career. He made good use of the shorthand in the field, at meetings and conferences, and in all his writings, leaving many of his notes in that form. He was an excellent typist to the amazement of many a secretary who could not equal him in efficiency and accuracy.

At 23 years of age, he began his experiments on maze learning in ants at the University of Michigan under the direction of Dr. John F. Shepard, with whom he had studied comparative psychology. The notebook from that course was preserved by him as a precious memory of his early training. It is a leatherbound book of some 400 pages, with many original, beautifully drawn illustrations. The title page carries the "A" given by Shepard. This notebook bears many similarities to the book he eventually wrote with Dr. Norman Maier. Later, Schneirla required his students to hand in similar notebooks in his graduate comparative psychology course at New York University, but it is doubtful that many students, if any, have produced the equivalent in quality and precision.

In 1925, Dr. Schneirla met Leone Warner who was also a student at

Drawing made by T. C. Schneirla as a heading for the chapter "Ant Work" in his notebook for the comparative psychology course taught by Professor John F. Shepard at the University of Michigan.

Female

Worker

Deälated females and workers of Formica subsericea (2x) (from Wheeler)

Michigan. They were married about a year later when he became a teaching fellow. In 1927, they spent the summer in Oklahoma, where he taught at the University of Oklahoma. In the fall Schneirla left Ann Arbor to teach at New York University, where he finished writing his Ph.D. thesis, "Learning and Orientation in Ants." After three years of his teaching at New York University, Schneirla and his wife went to Chicago in 1930 where he spent a year in Lashley's laboratory as a National Research Council Fellow. During their stay in Chicago, their daughter Janet was born; their son Donn was born in New York in 1933. Also while in Chicago, the Schneirlas met and made lifelong friends with another young couple, Norman and Ayesha Maier. From this friendship came the classic book *Principles of Animal Psychology,* which was published in 1935 and was reissued in paperback form in 1964 with additional material. Dr. Schneirla returned to New York University in 1931 and he remained on the faculty of this institution until his death.

In 1932, Dr. Schneirla made his first field trip to the Canal Zone Biological Area on Barro Colorado Island in Gatun Lake, Panama. Before this first trip was over, he had developed the hypothesis that the migratory and nonmigratory episodes of the army ant colonies were related to the condition of the brood. He concluded that the high degree of stimulation by the active larval brood arouses the workers to large raids, ending in emigrations. When the larvae become enclosed as pupae, and the rest of the brood consists of nonmotile eggs, stimulation of the workers is reduced, the raids are smaller and do not end in emigrations. In this way he demonstrated that the changes in behavior of these ants are intimately related to changes in morphology and in physiological processes. This radical hypothesis stood in sharp contrast to the then prevalent and static explanations of the migratory phenomenon, which either relegated the phenomenon to the category of innate behavior controlled by the genes (and hence requiring no further investigation) or related it to a decline in the abundance of food in the raiding areas.

Schneirla's return from the second trip to Barro Colorado Island in 1933 was the setting for one of his favorite stories. He grew a beard while he was away (a ritual that he probably acquired from his growing association with field biologists) and he often told with amusement of his infant daughter's greeting upon his return—"Go 'way, man; daddy come soon."

While at New York University, he was not only an excellent teacher who inspired his students, but he also found time to do original research, go on field trips, and involve himself in the problems of the professional psychologist as well. A resolution signed by Schneirla, along with Ward Halstead, David Krech, Theodore Newcomb, Ross Stagner, Gordon Allport, N. R. F. Maier, Edward C. Tolman, Karl F. Muenzing, Goodwin Watson and others, was presented in 1935 to the American Psychological Association to "approach the Federal government immediately with a request that money be made available from the works-relief fund" for the employment of a large number of unemployed psychologists so that they might set up clinics, carry out research, and practice their profession. For the students of today it must be difficult indeed to adequately comprehend the situation during the depression when so many competent psychologists could not find professional employment or even menial work for minimal wages. From the activities of this group there arose the Society for the Psychological Study of Social Issues, which is now a division of the American Psychological Association.

Dr. Schneirla also participated in the organization of the faculty of New York University and in various activities in support of the legitimate government of Spain against the Falangists. This awareness of the relationship between his scientific and professional work and the social welfare of his fellow men was one of Schneirla's outstanding characteristics.

In 1943, Dr. Frank Beach, Chairman of the Department of Animal Behavior of The American Museum of Natural History, invited Schneirla to become an Associate Curator there in recognition of his fine field work and ability as an experimentalist and theoretician. He remained as an Adjunct Professor at New York University where he continued to give courses until 1962 in Comparative Psychology, Thinking, and Development of Behavior. Through the years he directed the research of doctoral candidates not only from New York University, but from several other universities in the metropolitan area.

In September 1944, Dr. Schneirla was awarded a John Simon Guggenheim Fellowship which enabled him to spend 6 months in the Tehuantepec region of Mexico where he studied the social behavior of army ants under dry season conditions. It was after this trip that he developed a continuous tinnitus, apparently as a side effect of an anti-malarial drug. This condition never left him and grew progressively worse. Two other unfortunate events which occurred in the midst of Schneirla's flourishing career exaggerated the tinnitus. In 1954, shortly after the classic Conference on Orientation that he organized in Washington, D.C. under Office of Naval Research sponsorship, he contracted the Guillain-Barré syndrome, probably as a result of overwork. This disease, a viral infection of the spinal cord, almost completely paralyzed his body and legs. Despite the affliction, he never lost his spirit or will to recover and to continue his life's endeavors. Even while stretched out on the hospital bed he did a considerable amount of reading and writing and made plans for future research. When he had recuperated sufficiently to return home, he gave his course in "Thinking" while sitting in a living room chair. With great fortitude and patience he methodically performed his difficult orthopedic exercises and, to the amazement of all, his recovery was complete in just about a year. He was then in good physical shape and in 1955 he returned once again to Barro Colorado Island for his strenuous field work.

In 1960, misfortune struck again. His only son, Donn, was killed in a hotel fire in San Francisco. Despite this tragedy and the now much more acute tinnitus, he continued his research, writing and teaching. At the same time he undertook the arduous task of planning, designing and executing the Behavioral Alcove of the Museum's new Hall of Invertebrates. When this work was completed in 1967 he turned to writing a book on army ants which was nearly complete at the time of his death.

The forty years of research and writing have given us a rich inheritance of papers from which the books of selected writings was compiled. A comprehensive book on the army ant will also be published posthumously, as will several research papers on the behavior of the ant, chick, rat and cat. Among his unpublished papers are preliminary sketches for books on thinking, on development, and on primate behavior.

The progress of Schneirla's theoretical position during these productive years is not only a reflection of his personal experience, but is part of the growth of behavioral science. The ground substance and the emergent qualities are clearly evident. Fundamental was the emphasis on the use of the inductive method in the attainment and organization of knowledge. He was guided overall by the law of parsimony, by Morgan's canon, and above all, by the need to avoid the dangers and fallacies of anthropomorphism and zoomorphism.[1] His work reflects the meaningful application of a truly

---

[1] Zoomorphism, as used here, denotes giving animal-like qualities to human beings. This meaning is derived from the classical definition "the representation of God, or gods, in the form, or with the attributes of the lower animals."

comparative approach to behavior which stresses the differences as well as the similarities of various species.

He viewed the concept of levels as a fundamental tenet in all comparative and ontogenetic studies as well as in the analysis of processes within the individual organism. For example, he maintained that there is little scientific value in viewing the feeding behavior of all animals from protozoa to man as a common process, just because it results in the intake of food. There are major qualitative differences in morphology, physiology, neural, and psychological processes at various phyletic levels and these must be considered in the analysis of food consumption.

Dr. Schneirla contributed creatively to the understanding of three basic behavioral processes: motivation, social organization, and development. At first, he conceptualized the biphasic processes of approach and withdrawal as related to the state of the organism and the intensity of the effective stimuli, and as critical in the initiation and maintenance of activity. This concept of the biphasic approach-withdrawal process was expanded later by considering it to be a fundamental feature both in the behavior of simple animals such as protozoa and in early stages of behavioral development of the more complex animal forms.

Another of Schneirla's early contributions was a significant broadening of Wheeler's original concept of trophallaxis (exchange of nutriments) by formulating the new concept of reciprocal stimulation which includes the exchange of stimuli in addition to nutritive substances.

He explained socialization phenomena by integrating the process of reciprocal stimulation with the concepts of phyletic and organismic levels of organization. By doing so, he clarified the basis for the varieties of social behavior in the animal world as well as the changes in such behavior during the history of the individual.

The developmental process was conceived as a fusion of growth, experience, and maturation. Experience was redefined as the total history of all stimulation to which the organism is exposed during all stages of development. This formulation resolved the apparent dilemma of heredity versus environment, of nature versus nurture, and of instinct versus learning. Long before he published his first paper on behavioral development he had been concerned with the unscientific concept of "instinct" and its wide acceptance by behavioral scientists who came from diverse fields such as anthropology, sociology, psychology, psychiatry, biology, and animal breeding. He was especially opposed to such ethological terms as innate releasing mechanism, vacuum and displacement reactions, fixed action potential and action specific energy, which he considered to be reifications deduced from the basic assumption of the existence of instincts. Even when the terms "vacuum and displacement reactions" are used in a strictly descriptive fashion, they are applied to behavior in widely disparate groups of animals in which the structural and functional bases of behavior are very different.

Although Dr. Schneirla appreciated the observations and data gathered through the field and experimental observations of the classical ethologists, he questioned the premises and assumptions underlying their research and the interpretations of the results of their investigations. In the earliest stages of his critical analysis of "natively determined behavior," he pointed out the need to understand and take into consideration the developmental history of the individual, as in the discussions of pecking behavior in his and Maier's book, *Principles of Animal Psychology* (1935). His emphasis on behavorial development led also to a revitalization of the concept of levels of organization by the inclusion of development and change as important processes inherent in the concept. This new combination of a levels and developmental approach provides, thereby, the alternative to theories of behavior that are based on concepts of instinct. In essence, at a given phyletic level and at a given stage of ontogeny, the total behavior of the individual will be determined by the prior and continuous interaction and fusions of growth and maturational processes with the genome as a derivative of the individual's continuous experience. At higher phyletic levels and in late stages of ontogeny, specific learning processes are involved in the maturational processes. Moreover, learning, itself, is not a unitary process, but occurs in various levels of complexity. Only at the highest phyletic level—man, behaviorally speaking—and even then, at a late stage of child development, does the individual begin to use the most complex mental processes. This view of the development and evolution of behavior is a far cry from the traditional and still popular but ideologically weak dichotomy between "the innate and the acquired."

Dr. Schneirla saw the fallacies of the concept of instinct not only in the confines of behavioral research and theory, but in its application to man's behavior. Among his latest writings was a sharp denunciation of such theorizing under the rubric of "aggression" and the application of ethological theory to the complex human social behavioral patterns such as international politics, war, social conflict within a nation, and interindividual hostility. He pointed out that the uncritical extrapolation of incompletely supported conclusions from research with infrahuman species to specious solutions of the problems of humanity was the result of anthropomorphism, zoomorphism, and a disregard of the law of parsimony and of Morgan's canon. The pessimistic implications of such fallacious thinking for human society were countered by Schneirla's reliance on the highly evolved plasticity of human behavior for solutions to the problems of human aggression and violence.

This last printed encounter with those who support vitalistic and static theories of behavior as predetermined by heredity, exemplified much about

Schneirla. He could not compromise the principles of scientific thinking although he was always ready to examine theoretical explanations and test their validity. He set the highest ethical standards in his personal life, in his relationships with people, including his peers, colleagues and students, and for society at large. The frustrations of seeing mankind debase its potential by engaging in antisocial behavior were tempered for him by his great wit and delight in words and their meaning, his joy in his work and his love of nature in its varied forms. He derived daily pleasure from the esthetic aspects of his work, collecting beautiful photographs and paintings which resulted from his experiences in the field and laboratory. He continued to enjoy music and learned to play the recorder and flute during his last years.

Dr. Schneirla's sudden death in 1968 left a vacuum in the progress of behavioral theory that will not be readily filled. He united in his person and in his behavior a profound comprehension of the interrelatedness of phenomena together with an appreciation for precision and regard for the uniqueness of any one phenomenon; an integration of theory and activity which was constantly being developed in the course of new experience; and a concern not only for man's progress as a controller of his environment, but for his welfare and for the preservation of his human quality.

GERARD PIEL
Publisher, Scientific American
New York

The Comparative Psychology of
T. C. Schneirla

*"The process that I want to call scientific is a
process that involves the continual apprehension of
meaning, the constant appraisal of significance, accompanied by a running act of checking to be sure that I am
doing what I want to do, and of judging correctness or in-
correctness. This checking and judging and accepting, that
together constitute understanding, are done by me and can
be done for me by no one else. They are as private as my
toothache, and without them science is dead."*

P. W. Bridgman (1950)

Among the original contributions of T. C. Schneirla to comparative
psychology must be counted his studies of the behavior of students of
animal behavior. He was impelled to this uninviting task by his concerned
realization that behavior is everyman's science. From Aesop and Solomon
to Pliny and Aquinas and, in recent times, from Fabre, Maeterlinck, and
Seton to Lorenz (1966), Ardrey (1961), and D. Morris (1967), authors
have edified their fellow men through animal stories—drawing lessons for
human conduct from them. It is now possible to point to some seven
decades "of real progress in [the] materialistic and empirical study of
animal capacities" (Schneirla, 1949). At the same time, however, this
work has not much raised the quality of popular discourse.

Anthropomorphism no longer supplies the ready-made rationale, but
the contemporary fallacy of attributing the faculty of the brute to man is
no less pathetic. The mainstream of comparative psychology has supplied
cant along with certification to recent popular literature. Terms such
as "territoriality," "pecking order," "aggression," "imprinting," and "reinforcement" mark the trail from the newsstand back to the journal of
primary publication. The rise in popular interest in comparative psychology

reflects the ascendance of the behavioral sciences in scale and scope of activity and in the expectations of society. As a primary discipline of comparative psychology the comparative study of animal behavior has had need of a discerning and responsible critic. Schneirla filled that need.

Two centuries ago, Reaumur could declare with satisfaction: "What the erudite of former times seriously proclaimed to other savants would today scarcely be recounted by credulous nurses to their nurselings" (Schneirla, 1958). Yet, as recently as 1949, Schneirla found it necessary to remark on the propensity of colleagues to "generously endow" lower animals with capacities "not readily acquired by man." The journals of behavior still carried echoes of the quaint conflict between sentimental vitalists and vulgar mechanists that had died out in other branches of biology a half century earlier. On the one hand, Schneirla cited a fully articled contemporary who held that psychology was a science of "psychic realities," which man first discovers "by introspection" in himself and then studies in lower animals through "sympathetic intuition"; on the other, he had to confront the author of the empty declaration that "the raw basis from which [a man's interests and desires] are developed is found in the phenomena of living matter." On their divergent courses, these two writers converge in the same self-defeating "reductionism." Each reduces the psychological capacities of animals to a single level, and therefore, apart from other flaws in their viewpoints, they miss the true center of interest in comparative psychology, which is, in Schneirla's words (1959b), "the nature of the differences among levels of behavioral capacity."

One may ask why issues so sterile and works of such little merit should claim the attention of a competent scientist with his own work to do. Schneirla made the same use of this material that medicine makes of gross pathology: exposure and examination of overt symptoms sharpen the diagnosis of the covert symptoms of a disease. Animal psychology has not outgrown naive anthropomorphism when serious investigators can blur the distinction between adaptive and purposive behavior. Schneirla's principal predecessor in field studies of the army ant had, for example, explained the migratory habits of these animals as a response to the depletion of food in the nesting area (Schneirla, 1957b). "But there is absolutely no factual justification," Schneirla (1940) took great pains to show, "for the conclusion that the adaptiveness of such group behavior accounts in any way for the events in individual behavior through which it appears." Evidence of more subtle contamination of the thinking process along the same lines caused Schneirla (1943) to urge students of invertebrate learning to recognize that their publications showed "a strong tendency to impose the mammalian pattern deductively upon the performances of lower organisms and to interpret them as simpler editions of that pattern. . . . Apparently, goal-striving is seen as dominating the special process of learning in maze-adjustment on all levels."

The "spirit of the hive," evoked by Maeterlinck to explain the behavior in the social bee, finds reincarnation in the concept of the "super-organism," a notion that has bemused a number of the ablest field-working psychologists of this century. The protagonists of the superorganism concept invariably draw it forth from their astonishing cataloguing of parallels they have perceived between the anatomy and physiology of multicelled organisms and the organization and functioning of societies, both insect and human. Here, for example, the most distinguished recent advocate (Emerson, 1942) of the superorganism finds convergent evolution at work: "The common characteristics of organisms and population systems are the result of natural selection of each unit through its internal and external relations." Here, he finds deep principles to be discovered: "Significant parallels between insect and human societies have a common causation." Here, important moral lessons: "Just as the internal controlled environment of the organism is exploited by parasites, social parasites have risen to exploit the social environment."

In pinning down this specimen in his collection, Schneirla observed, as to the integrity of the concept, ". . . individual [organism] and social group are entities in very different senses of the word, differently subject to selection in evolution." As to the method, he said: "While . . . analogy has an important place in scientific theory, its usefulness must be considered introductory to a comparative study in which differences may well be discovered which require a reinterpretation of the similarities first noted" (Schneirla, 1946).

Schneirla (1949) could be moved to use a sharper pen. On the employment of "analogy at the expense of analysis," he wrote: "The term 'mind' suggests close qualitative similarity wherever it is used, evading explanation but at the same time implying it with such propagandistic success that the issue seems fairly settled." On the superorganismic notion of "crowd mind," he wrote: "Such concepts implying a social psychic agency either are used frankly as literary conceits, or as fraudulent devices substituted for any real effort to study the factors underlying a given level of group performance." On the associated practice of anecdotalism: "One who sets out to demonstrate that protozoan organisms or any others have the mental characteristics of man may convince himself at least, provided he singles out opportunely the brief episodes which seem describable as instances of perception of danger, of reasoning, or what not. By the same method, the absence of reasoning in man can be proved with ease."

In what other field of science can one today find serious workers arguing from analogy, relying upon anecdote, or pursuing the circular error of nominalism? Such weakness of intellection underlies the recurring "heredity-environment" ("nature-nurture," "instinct-learning," "heredity-environment," "innate-acquired," "preformistic-epigenetic") disputation. Nothing more insightful or penetrating distinguishes the approaches of the two schools that held the center of the stage in animal psychology during Schneirla's lifetime: the operant-behaviorists identified with B. F. Skinner and the ethologists identified with Konrad Lorenz.

The central concept of ethology is, of course, the "ethogram" or innate behavioral trait, which is programmed in the genes, faithfully unfolded by growth, and triggered into function upon presentation of its appropriate "releaser." Schneirla (1957a) argued against this notion, using evidence from physiological genetics: "Genes are complex biochemical systems, integrated from the beginning of ontogeny into processes of increasing complexity and scope, the ensuing progressive processes always intimately influenced by forces acting from the developmental context. . . . It is undesirable to confuse this natural state of affairs with the impression of heredity directly determining development." Further, he argued (1966), "There seem to be no hard and fast rules for distinguishing hypothetically innate behavior from other kinds." Pointing out that the ethologists' pre-occupation with the innate-acquired dichotomy caused them to focus on part-processes of behavior in the adult animal, Schneirla asked (1959a), "In what particular stage of development is 'original nature' to be identified?"

Schneirla commented on the "releaser," citing a well-known experiment of N. Tinbergen (1939). An airplane-shaped model of a "bird of prey" was said to release the appropriate alarm and escape behavior from young nestling birds when it was moved over them with the "shortneck" end-first but not when moved "tail-first." Schneirla (1954) proposed a simpler explanation for this result—the initial effect on the birds might be "perceptually non-qualitative, perhaps a shock reaction produced by a sufficiently abrupt stimulus." He was able to cite an experiment by another worker that showed the speed of the moving effigy to be the critical factor in eliciting the response of the nestling. "The artifact method," he concluded, "requires careful control against subjective impressions. . . . 'Shortneckedness' in the above situation is a cue to the human observer . . . but not necessarily to the bird."

Schneirla found the conceptual poverty of the notion of the innate behavioral trait plainly exhibited in the kind of experimentation it inspires. By way of extreme example, he pointed to efforts by various investigators to demonstrate "information transfer" by the injection or feeding of DNA and RNA extracts from animals trained to perform specific tasks to the eggs, embryos, and mature specimens of untrained individuals (Corning and Ratner, 1967). The ultimate enterprise in this direction is represented by the planting of electrodes in the presumed "pleasure center" of the brains of catatonic patients in a Louisiana state mental hospital (Heath, 1964).

The operant-behaviorists approach their subjects from a quite different quarter. Skinner (1957) declared that he and his colleagues had been able to demonstrate "surprisingly similar performances, particularly under complex schedules in organisms as diverse as the pigeon, mouse, rat, dog, cat and monkey." This work, said Schneirla (1966), "is directed at just one set of abilities in the area of learning: those having to do with the establishment, weakening by extinction, re-establishment, and so on, of continuity-learning (i.e.: Pavlovian) patterns." It is so contrived "as to largely exclude or obscure the animal's potentialities for behavioral organization." Schneirla noted that these behaviorists "have not claimed to offer a comprehensive psychological theory," and he pointed to the hazards "in the uncritical extrapolation of [their] results to the broader problems of comparative behavior." In another connection Schneirla (1949) identified one such hazard: "When an educational program is unscientific in that it emphasizes a mass acquisition of predigested categorical results and discourages questioning and search for better answers to old questions, the conditioned-response pattern of learning obviously is in the ascendance and the intelligence potential of the given society is broadly wasted."

Schneirla wryly took note of a conference of ethologists and operationists called to explore the possibilities of collaboration from their respective sides of the heredity-environment dichotomy. They found some grounds of common interest but believed the possibilities of collaboration to be very limited. Schneirla (1966) thought this conclusion was unduly negative. The two parties "may be more closely related in their basic rationales than this conclusion might suggest," he observed, in that both groups rely on "implicit or outright assumptions of innate behavior as distinct from learned behavior." Schneirla then added: "Behavior develops and changes—on higher integrative levels it is not often rigid except when pathological. Yet ideologies directed at interpreting it are too often cast in rigid forms."

It is a fact deserving a moment of reflection that the instincts and releasers, the stimuli, reinforcements and responses, the superorganisms and the very animals themselves featured in the contemporary literature of comparative psychology hold less inherent interest than, say, the time, space, particles, waves, forces and quanta of modern physics. Yet physics stands at the "disorganized simplicity" bottom of the hierarchy of the sciences postulated by Warren Weaver (1948), and behavior surely belongs to the "organized complexity" category at the top. In responding to the question why behavioral science should be so much less prepossessing than its subject matter, a behavioral scientist is obliged to give some credence to surveys that show physics has been attracting more able intellects, as identified by various tests. There is another answer, perhaps kinder to the practitioners. Most of them have other primary interests. They are taxonomists. They are concerned with range and wildlife management. Or they select animals as models for study, in other contexts, of perceptual and learning capacity or of straightforward neural physiology. Not many have been concerned with the work of comparative psychology as such. Not one in recent years has done more to define and to keep forward the significant questions in this field than Schneirla.

"Theories concerning how behavior and behavior patterns arise," Schneirla (1954) wrote, ". . . must do more than bridge initial and terminal stages with hypothetical shielded intra-organismic determiners." The gap "between genes and behavior" is not to be bridged; it is the territory to be explored. Schneirla remembered always that life is history and that ontogeny is therefore the proper study of comparative psychology. "Processes of behavioral organization and motivation cannot be dated from any one stage, including birth, as each stage of ontogeny constitutes the animal's 'nature' at that juncture and is essential for the changing and expanding accomplishments of succeeding phases" (Schneirla, 1957a).

The animal of Schneirla's studies is thus a productively different kind of animal from the ethologist's wind-up toy, awaiting its releaser, or from the operant-behaviorist's feathered and furry computers, so susceptible to programming and read-out.

Behavior, or for that matter, development, can be seen at any given stage to reflect factors that are internal and external to the animal. This useful preliminary analysis invariably hardens into the familiar dichotomy of heredity and environment. On more than one occasion Schneirla urged his colleagues to extricate themselves from the "heredity-environment" trap and tried to show them how. Starting with the semantically most neutral pair of terms corresponding to the preliminary analysis, Schneirla (1966) suggested that " 'maturation' be redefined to refer to the contributions to development from growth and tissue differentiation, together with their organic and functional trace effects surviving from earlier development" and that " 'experience' be defined as the contributions to development of the effects of stimulation from all available sources (external and internal) including their functional trace effects surviving from earlier development." These definitions, it should be observed, make a necessary correction of the simplistic distinction drawn between the external and the internal. Influences, both external and internal, become incorporated in the trace effects of progressive development; through feedback linkage, the output of earlier stages of development becomes input to the later stages. In consequence, Schneirla concludes: "The developmental contributions of the two complexes, maturation and experience, must be viewed as fused (i.e.: as inseparably coalesced) at all stages in the ontogenesis of any organism."

From this operation, the heredity-environment dilemma departs without its horns. Consider, for example, the familiar "isolation" experiment designed to establish the "innateness" of some behavioral trait; "proof" is furnished if the subject exhibits the trait upon release from isolation. In Schneirla's words (1954): "Isolation experiments do not tell us what is native in the normal patterns; for, if the animal survives, the atypical situation also must have contributed to the development of some adaptive pattern. Techniques of the 'isolation' type thus concern relative-abnormality-of-setting rather than isolation in the full sense, and help tell us how far extrinsic conditions may be changed at a particular developmental stage without preventing or altering further development based on the gain from preceding stages."

Study of the ontogeny of behavior sets the only secure foundation for comparative psychology—for "comparing the respectively different adaptive patterns attained by animals and for better understanding of the animal series inclusively" (Schneirla, 1957a). Quoting Z.-Y. Kuo, Schneirla pointed out that such study answers not only the question "What happens?" but "How?" Without an adequate longitudinal perspective on the developmental process, the investigator is prone to error when he compares analogous or homologous patterns of behavior observed in different phyla. The literature of animal behavior is rich in misconception, as Schneirla's periodical surveys and critiques have shown. Misconception is reinforced by cross-sectional or time-limited studies that often "turn out to be demonstrations of the investigator's initial assumptions rather than critical tests of the theory that inspired the work" (Schneirla, 1966).

Investigation of traits, habits, or modes of behavior isolated from the context of the subject's development is inappropriate for comparative studies because behavioral evolution cannot be viewed "as a process of accretion" with new mechanisms added to old. "Homologous mechanisms are transformed functionally . . . as through ontogeny a characteristically total behavior pattern arises in a different total context" (Schneirla, 1957a).

This observation has particular relevance at the present time. The public is being told, through popularizations of the scientific literature, that the stress of history is uncovering man's dark genesis and stripping the veneer of culture from the naked brute. It is not the brute, Schneirla (1946) insisted, but a pathological human personality that is beheld on such occasions: "The view that man's 'higher psychological processes' constitute a single agency or unity which is capable of being sloughed off under hypothetically extreme provocation is a naive outcome of the mind-body conception of man's nature."

~ As a common denominator for comparing behavioral capacities attained at successive levels of evolutionary development, Schneirla (1959a) pro-~posed a general theory of "biphasic processes underlying approach and ~ withdrawal." He submitted, with italics, the statement that: "*Approach* ~and *withdrawal* are the *only* empirical, objective terms applicable to *all* motivated behavior in all animals." He was careful to specify that "an animal may be said to *approach* a stimulus source when it responds by coming nearer to that source, to *withdraw* when it increases its distance from the source."

Yet he had to admit: "This point is not sufficiently elementary to escape confusion." Some people find it easier to oppose "approach" with "avoid," the source of their confusion being, of course, their inexplicit identification of adaptation with purpose. With teleology set aside, how-ever, it is plain to see that approach and withdrawal behavior must be highly adaptive in all surviving animals, for "beginning in the primitive scintilla many millions of years ago . . . the haunts and typical niche of any organism must depend on what conditions it approaches and what it moves away from." Of the biphasic response it may be said, in general, that "low intensities of stimulation tend to evoke approach reactions; high intensities, withdrawal reactions."

Compared to the imaginative hypotheses that inspire so much current investigation in animal behavior, the "approach-withdrawal response" ap-pears as a pallid truism. This quality nicely served the purpose of the Schneirla program: to develop in an unambiguous way the differences in the ontogeny of a simple, universal response, from species to species.

In the amoeba, approach or withdrawal is energized by "protoplasmic processes set off directly by the stimulus"—the sol reaction in response to a weak source of light, with streaming in the direction of the source, and the gel reaction in response to a strong source, with streaming in the op-posite direction. In the rat, the response is mediated by "specialized higher-level processes not indicated in the protozoan"—including extensor muscle dominance with weak stimulation and flexor dominance with strong (as demonstrated by Sherrington, 1923)—and is also freed by ontogeny from the sway of simple stimulative intensity. The human infant, Schneirla (1959a) observed, "specializes perceptually in reaching and smiling before he avoids and sulks discriminately."

The Schneirla program for comparative study of the approach-with-drawal response poses questions in a fruitful way for investigations yet to be mounted. It also raises serious questions about the kinds of answers that have come from past studies—answers that have given the sanction of science to the popular hankering for final causes. One set of finalisms is represented by the various constellations of emotions and drives with which human beings are supposed to be endowed at birth. J. B. Watson, the founder of the American school of behaviorism, postulated a trinity of innate emotions—"love," "fear," and "rage"—from his observations of the human neonate (Watson, 1949). As Schneirla suggested, further study will doubtless show that, in the psychosocially barren dawn of human existence, these come down to a simple approach response for the first and withdrawal for the second two. Watson himself regretted that he had attached emotion-laden words to the responses he observed and expressed the wish that he had designated them "x," "y," and "z" respectively. In that case, Schneirla said, "x" would reflect extensor dominance and "y-z" tensor dominance. As for the pleasure-pain dichotomy, now ratified by the identification of the "pleasure center" in the brain, it may yet be shown that the implanted electrodes "cut-in on A- or W-type patterns, tapping critical way-stations in circuits of distinctly different arousal thresholds" (Schneirla, 1959a).

If it can be said that the science of behavior has approached complete elucidation of the behavioral repertory of any animal, this surely must be said of Schneirla's work on the army ant. The questions that remain after Schneirla, that indeed were posed by Schneirla, must be answered through the methods of other disciplines: neurophysiology and hormonal physiology, to begin with. In this work, he not only set a model for his science— of a system of behavior mapped and structured in minute detail—but he also showed that behavioral science can develop methods and concepts worthy of its subject matter.

Schneirla (1940) chose as the centerpiece of his life work "the most complex instance of organized mass behavior occurring regularly outside the homesite in any insect or, for that matter any infra-human animal." This is a sober understatement of a phenomenon that had caused earlier students to abandon discretion. Describing an ant army on the march, Paul Griswold Howes in his treatise *Insect Behavior* (1919) has the queen "hidden from the common horde, attended by her special ladies-in-waiting"; eggs, larvae, and pupae "guarded and kept warm, lest injury result and the future of the tribe be endangered"; "lieutenants keeping order or searching for members out of step that might hinder the march"; "scouts searching out the ground to be hunted or travelled next"; and the run-of-the-mill ants obeying "commands" and evincing a "wonderful sense of duty." A still earlier observer, T. Belt (1874), noted that it is the "light-colored officers" that keep the "common dark-colored workers" in line.

Schneirla's (1944) observations and his experiments with individual colony members showed that the army ant is a highly limited animal, with less capacity for maze adjustment than the common trail-running *Formica*.

Plainly, the organization of colony behavior "does not pre-exist in any one type of individual workers, brood or queen—nor is it additive from these alone." As Schneirla showed, the system arises from the intricate, inter-dependent interaction of a large number of diverse factors. They may be conveniently classified under the headings of individual colony-member capacities, group functions, reproduction, the development of the brood (eggs, larvae, pupae, and callows), the natural environment, and the system of organized behavior itself. There are many cross-links between behavioral sub-routines; a hierarchy of interlocked feedback loops integrates the total performance. Thus, for example, the mass of living workers in the colony bivouac ensures that "a stable microclimate is so regularly established for the successive broods that the periodicity of the developmental stages is highly predictable in brood after brood" (Schneirla, 1954). But the developmental cycle of the brood plays its part in turn: "Reciprocally, through indirectly contributing to the existence of rythmic colony behavior, a normally developing brood makes the bivouacs possible" (ibid.).

The biphasic approach-withdrawal response is to be seen in action at the head of a raiding swarm or column, with individuals venturing for-ward, beyond the "trophallactic" thrall of the characteristic colony odor, and rebounding back into the mass when that stimulus has been weakened by distance. In the laboratory and in the field, Schneirla (1944) recorded episodes showing that the stereotyped army ant workers, when they are placed in neutral surroundings, such as a table top, a sidewalk or a flat rock, inexorably form up in a circular mill and march themselves to death. It follows that "heterogeneity in the operating terrain is indispensable for execution of the typical raiding-emigration sequence" (ibid.).

Considering the entire picture in all its complexity, it is tempting to attribute an endogenous control of the cycle to the queen "who becomes physogastric and delivers a new brood approximately midway in each statory phase." Yet, on closer inspection, Schneirla found that the queen's periodicity is regulated in turn by the maturation of the larval brood. As the brood approaches pupation and reduces its feeding demands, the surplus of food and worker activity is diverted to the queen. The colony is meanwhile terminating the nomadic phase of its migratory cycle, and the queen, feeding voraciously, enters her next ovulation cycle. "The cyclic pattern thus is self re-aroused in a feedback fashion, the product of a reciprocal relationship between queen and colony functions, not of a timing mechanism endogenous to the queen" (Schneirla, 1957b).

This cyclic drama, which has been reenacted in the tropical rain forests of the middle latitudes of the planet week in and week out for tens of millions of years, is not to be reduced to any single internal or external cause. Nor is it to be explained by any governing or energizing principle extrinsic to its own dynamic organization. Schneirla's careful analysis and reconstruction show the army ant behavioral system to be *sui generis* and as splendid, in its fashion, as a supernova.

Schneirla (1950) was intensely aware of the liabilities that an investigator takes with him into the field or laboratory. "Man goes to nature," he said, "to learn what nature is but, in so doing, he introduces possibilities of distortion through his own presence." He warned of a second hazard: "When initial concepts are strong and vivid and metaphorically very appealing, they may be carried through an entire study without ever being examined effectively." With reference to the special hazards attending the study of behavior, he declared, "Our attitudes toward the nature, origin and relationship of competition, cooperation and natural selection processes exert subtle influences not readily controlled in the planning and prosecution of investigations." Whether in field or laboratory, "the experimenter's responsibility for control involves the regulation of such matters, and not only the manipulation of objective factors in the phenomenon under study."

Befitting the role of the behavioral scientist, Schneirla (1959a) set out his position not only in precept but also by example: "It is the methods of objective science that must be directed at the question," he said, "and not the forensic arts." He schooled himself to keep detailed, written records of what he observed while he was observing it. With a command of shorthand and a practiced ability to write in the dark, he was able to take notes "by tactual control" while maintaining "visual touch" with events. He also practiced estimating distances, counting animals, and judging their spatial arrangement; he worked out codes and symbols to record the interaction of his subjects. Frequently his notes included on-the-spot sketches, diagrams, graphs, and photographs. Schneirla was careful to record the usual as well as the unusual. At the end of each day's work it was his habit to review and summarize the day's record. At intervals of a few days he would review the accumulated record "for discrepancies, omissions and latent meanings." With conclusions and hypotheses spelled out from earlier observations, he would repeat his observations systematically on the next round of his subject's behavioral cycle.

Many of Schneirla's papers report experiments devised in the field. Thus, by removing the larval brood from a colony, he tested his hypothesis that the squirming larvae stimulate the more energetic raiding of the nomadic phase in the cycle; raiding promptly declined and nomadism soon ceased. He later supplemented this field experiment with laboratory tests of the relative responsiveness of individual workers to active larvae and quiescent pupae (Schneirla, 1950).

By such stratagems Schneirla sought to bring to his field work the rigor that is associated with the laboratory. In his view, there was no inherent virtue that made experiment more reliable than observation. The study of animal behavior is necessarily an observational science. Knowledge from the field is essential, at the very least, to correct for "the influence of the laboratory environment, to cage." The ideal program for the future study of animal behavior, he concluded, "would involve a coordination of field and laboratory investigation. . . . Field investigation offers an opportunity to work with the animal's full pattern of activities. . . . In the laboratory one may focus on specialized problems such as sensory discrimination, motivation, learning and higher processes, pursuing them in detail and under conditions involving refined controls" (Schneirla, 1950).

Schneirla was a scientist to whom a wise student would report as an apprentice. He was a philosopher of science who practiced what he preached.

REFERENCES

Ardrey, R. 1961. *African genesis.* New York: Atheneum.
Belt, T. 1874. *The naturalist in Nicaragua.* London: John Murray.
Bridgman, P. W. 1950. Science, public or private? *Reflections of a physicist,* New York: Philosophical Library, Inc.
Corning, W. C., and S. C. Ratner. 1967. *Chemistry of learning.* New York: Plenum Press.
Emerson, A. E. 1942. Basic comparisons of human and insect societies. *Biological Symposia* 8.
Heath, R. 1964. Pleasure response of human subject to direct stimulation of the brain, physiologic and psychodynamic considerations. *The role of pleasure in behavior.* New York: Hoeber Medical Division, Harper & Row.
Howes, P. G. 1919. *Insect behavior.* Boston: Gorham Press.
Lorenz, K. 1966. *On aggression.* London: Methuen.
Morris, D. 1967. *The naked ape.* London: Jonathan Cape.
Schneirla, T. C. 1940. Further studies on the army-ant behavior pattern. *J. Comp. Psychol.* 29 (3).
Schneirla, T. C. 1943. The nature of ant learning. *J. Comp. Psychol.* 35 (2).
Schneirla, T. C. 1944. A unique case of circular milling in ants, considered in relation to trail following and the general problem of orientation. *Am. Mus. Novitates,* No. 1253.
Schneirla, T. C. 1946. Problems in the biopsychology of social organization. *J. Abnorm. Soc. Psychol.* 41 (4).
Schneirla, T. C. 1949. Levels in the psychological capacities of animals. *Philosophy for the Future.* New York: Macmillan.

Schneirla, T. C. 1950. The relationship between observation and experimentation in the field study of behavior. *Ann. N.Y. Acad. Sci.* 51 (6).

Schneirla, T. C. 1954. Interrelationships of the "innate" and the "acquired" in Instinctive Behavior. In P.-P. Grassé, ed., *L'Instinct dans le comportement des animaux et de l'homme*. Paris: Masson.

Schneirla, T. C. 1957a. The concept of development in comparative psychology. In D. B. Harris, ed., *The concept of development*. Minneapolis: Univ. Minnesota Press.

Schneirla, T. C. 1957b. Theoretical consideration of cyclic processes In Doryline ants. *Proc. Am. Phil. Soc.* 101 (1).

Schneirla, T. C. 1958. The study of animal behavior: Its history and relation to the museum. *Curator* 1 (4).

Schneirla, T. C. 1959a. An evolutionary and developmental theory of biphasic processes underlying approach and withdrawal. In M. R. Jones, ed., *Nebraska symposium on motivation*. Lincoln, Nebraska: Univ. Nebraska Press.

Schneirla, T. C. 1959b. Comparative psychology. *Encyclopaedia Britannica*. Chicago: Encyclopaedia Britannica.

Schneirla, T. C. 1959c. The study of animal behavior: Its history and relation to the Museum. *Curator* 2 (1).

Schneirla, T. C. 1966. Behavioral development and comparative psychology. *Quart. Rev. Biol.* 41 (3).

Sherrington, C. S. 1923. *The integrative action of the nervous system*. New Haven: Yale Univ. Press.

Skinner, B. F., and C. B. Ferster. 1957. *Schedules of reinforcement*. New York: Appleton, Century-Crofts.

Tinbergen, N. 1939. Uber die auslosenden und die richtunggebenden Reizsituationen der Sperrbewegung von jungen Drosseln. *Z. Tierpsychol.* 3.

Watson, J. B. 1949. *Behavior: An introduction to comparative psychology*. New York: Holt, Rinehart and Winston.

Weaver, W. 1948. Science and complexity. *Am. Scientist* 36 (4).

I

EVOLUTION OF BEHAVIOR

DANIEL S. LEHRMAN
Institute of Animal Behavior
Rutgers University
Newark, New Jersey

Semantic and Conceptual Issues
in the Nature-Nurture Problem

The question of what is called "innate" and what is called "acquired" in the behavior of animals, including man, is one that appears regularly and persistently, as a problem and as a source of controversy among students of animal behavior and psychology. This is true in all areas studied by students of behavior. The question of how to formulate the roles of "heredity" and "environment" in the determination of behavior characteristics has agitated students of human intelligence (Anastasi and Foley, 1948), of the abilities for visual perception in higher animals (Hochberg, 1963), of species-typical ("instinctive") behavior in animals (Schneirla, 1956), and virtually every other area of interest to such students.

In the present essay, I propose to discuss the role of the concepts of "innate" and "acquired" in a number of discussions of animal behavior that have occurred in recent years.

Starting in the early 1930's Konrad Lorenz and his students and colleagues developed a conception of the mechanism of "instinctive" behavior which was and is very influential and which has, for a considerable period of time, stimulated a large amount of interesting and creative research. Lorenz's theories (Lorenz, 1937, 1950), developed from work by zoologists on the behavior of lower animals (mostly birds, fish, and insects), formed the basis for a new and flourishing school of animal behavior studies, for which the name "ethology" was adopted.

The term "ethologist" is difficult to define. Insofar as it has a formal definition, it means approximately a scientist who studies the species-typical behavior patterns which constitute part of the animal's biological adaptation to its environment. During the 1940's and 1950's, the term tended to carry the additional informal connotation that a person designated by it was guided in his work primarily by Lorenz's theories of behavior. In recent years, however, it has become clear that the scientists working on the problems, and with the methods, characterized as "ethological," by no means constitute a monolithic body of opinion linked by fidelity to a particular set of theories. The term "ethologist" must therefore refer to a group of people characterized by a common interest in understanding the behavior of animals in relation to their natural environment (including fellow members of the species).

Lorenz's thinking has, from the start, depended very heavily upon the idea that it is always possible and profitable to distinguish "innate" from "acquired" elements of behavior. Lorenz (1965) agrees that "it would be hard to exaggerate the importance attributed by ethologists to the distinction between the innate and the learned." About fifteen years ago, three North American psychologists (Hebb, 1953; Lehrman, 1953; Schneirla, 1956) published discussions which, each in its own way, implied skepticism about both the heuristic value of the traditional distinction between "innate" and "acquired," and the reality of those two concepts as classes into which any given element of behavior could unambiguously be placed. Over the years since then there has been a considerable amount of discussion centering around these critiques (Eibl-Eibesfeldt, 1961; Lorenz, 1961; Thorpe, 1963a, 1963b; Tinbergen, 1963; Schneirla, 1966), culminating in Lorenz's recent (1965) book, *Evolution and Modification of Behavior.*

Since I was the author of one of the papers giving rise to this series of discussions (Lehrman, 1953) and since I have not published any further direct contribution to the discussion during the intervening fifteen years (although I naturally regard my work during that time as an illustration of my point of view), I would like to take advantage of the opportunity offered by this memorial volume, dedicated to the late T. C. Schneirla, to comment on the present status of this discussion, with particular reference to Lorenz's recent (1965) attempt at a definitive resolution. This is especially appropriate, since I regard Schneirla, who was my teacher, as the most creative, the most articulate, and the most consistent modern spokesman for the point of view that the use of dichotomies such as "innate" and "acquired" is restrictive, rather than instructive, in its effects on the analysis of behavior.

SEMANTICS, CONCEPTS, AND FACTS

When opposing groups of intelligent, highly educated, competent scientists continue over many years to disagree, and even to wrangle bitterly, about an issue which they regard as important, it must sooner or later become obvious that the disagreement is not a factual one, and that it cannot be resolved by calling to the attention of the members of one group (or even of the other!) the existence of new data which will make them see the light. Further, it becomes increasingly obvious that there are no possible crucial experiments that would cause one group of antagonists to abandon their point of view in favor of that of the other group. If this is, as I believe, the case, we ought to consider the roles played in this disagreement by semantic difficulties arising from concealed differences in the way different people use the same words, or in the way the same people use the same words at different times; by differences in the concepts used by different workers (i.e., in the ways in which they divide up facts into categories); and by differences in their conception of what is an important problem and what is a trivial one, or rather what is an interesting problem and what is an uninteresting one.

The Critiques of the Heredity-Environment Dichotomy

INTERACTION OF HEREDITY AND ENVIRONMENT

Hebb (1953) asserted that to make a dichotomy between "innate" and "environmentally-determined" behavior patterns, with the intention of assigning each element of behavior to one or the other of these classes, is misleading because the influences of heredity and of environment are not exerted upon different parts of the behavior (or of the organism), but are effective, in different ways, on the development of the *same* elements. This goes fairly directly to the heart of Lorenz's earlier use of the term "innate," since he has always implied that, if only behavior could be broken up into appropriately defined elements, it should be unequivocally possible to state which ones were wholly innate and which ones were influenced by "learning." In fact, Lorenz's characteristic method of dealing with the role of learning in behavioral development has been to conceive of an interlacement of innate and learned elements, making a chain, or of situations in which a clearly defined aspect of the behavior (e.g., its form) was innate, while another equally clearly defined aspect (e.g., its orientation to the environment) could be described as learned.

This argument of Hebb's is referred to by Lorenz (1965)[1] as the "first behavioristic argument."

ROLE OF ENVIRONMENT IN THE DEVELOPMENT
OF SPECIES-TYPICAL BEHAVIOR

Lehrman (1953) and Schneirla (1956) emphasized a somewhat different argument, which was also mentioned by Hebb. They point out that

---

[1] Except as otherwise indicated, all further references to Lorenz are to Lorenz (1965).

the ontogenetic development of species-specific behavior patterns may often depend upon influences from the environment, which interact with processes internal to the organism at all stages of development, in such a way that it is misleading to label those behavior patterns that seem to depend upon ordinary learning, and those that do not, as "learned" and "innate," with the implication that they have dichotomously different developmental origins. Schneirla (1966), in particular, has used the concept of "experience" to mean all kinds of stimulative effects from the environment, ranging from stimulus-involved biochemical and biological processes (having effects on the developing nervous system) to what we ordinarily call conditioning and learning. He speaks of maturation as "the contributions to development of growth and of tissue differentiation, together with organic and functional trace effects surviving from earlier development." Earlier I had said (Lehrman, 1953):

> The "instinct" is obviously not present in the zygote. Just as obviously it is present in the behavior of the animal after the appropriate age. The problem for the investigator who wishes to make a closer analysis of behavior is: how did this behavior come about? The use of "explanatory" categories such as "innate" and "genetically fixed" obscures the necessity of investigating developmental *processes* in order to gain insight into the actual mechanisms of behavior and their interrelations. The problem of development is the problem of the development of new *structures* and activity *patterns* from the resolution of the interaction of existing ones, within the organism and its internal environment, and between the organism and its outer environment. At any stage of development, the new features emerge from the interactions within the *current* stage and between the *current* stage and the environment. The interaction out of which the organism develops is *not* one, as is so often said, between heredity and environment. It is between *organism* and environment! And the organism is different at each different stage of its development.

The section of my paper in which I made these remarks was called "Heredity-vs.-Environment, or Development?"

It should be obvious from these quotations that what Schneirla and I (and Hebb in a slightly different way) intended was *not* to say that learning was all-important, while accepting the traditional dichotomy that maturation is an unfolding of gene-determined anatomical, physiological, and behavioral patterns (Schneirla, 1966), and that influence from the environment consists solely of conditioning or trial-and-error learning; rather, we were questioning the value of the dichotomy itself, *not* stressing one side or the other of it.

Lorenz calls this type of discussion the "second behavioristic argument."

Some Problems of Definition

My first serious task is to examine some semantic problems: i.e., those arising from the use of words as labels, which may compound the actual problems arising from the conceptions to which the words were intended to apply.

WHO IS AN ETHOLOGIST?

What is the significance of the fact that Lorenz labels the principal considerations introduced by Hebb, Schneirla, and myself as "behavioristic" arguments? The background for this labeling lies in the repeated assertions made by Lorenz and other writers to the effect that these criticisms arise from the fact that the people who wrote them are psychologists, and are therefore incapable of understanding biological problems. Lorenz refers to "American psychologists" as the source of criticism of his ideas, and repeatedly implies that these critics impute to biologists ideas which they do not hold, the implication being that as psychologists, they are not sensitive to the considerations that are important to biologists.

This type of labeling is, for several reasons, not a very constructive contribution to a discussion of the problems of heredity and development. For one thing, the views of biologists like Hinde (1966) and Tinbergen (1963) are very much like those expressed by these "American psychologists," and the implication that they have been unduly influenced by alien intrusions into their field of work is less than respectful.[2]

For another thing, this kind of labeling tends to arouse (or to reveal) a prejudice against the person being labeled which prevents (intentionally or unintentionally) a full appreciation of his contribution to the discussion. It is too easy to close one's mind to an argument by simply deciding that the source of the argument is an outsider.

Finally, the term "behaviorist" is an affront to the memory of T. C. Schneirla, whose lifelong work was a thoughtful, penetrating and broadly based analysis of the role of physiological, social, and ecological processes, and of the integrations among them, in the development and regulation of the behavior patterns by which the army ants (*Eciton*) are adapted to their environment; of the ways in which different species of these ants differ from each other; of the mechanisms that give rise to these species differences; and of evolution in this group. His work was not remotely related to the tradition of American "rat psychology" to which Lorenz refers by the term "behaviorism." As for me, the reason I chose to study

---

[2] Hinde (1968), in a paper published while this essay was being set in type, has expressed ideas quite similar to those presented here.

with Schneirla was the same as the reason I chose to become a student of animal behavior: I was interested in understanding the behavior of birds as I had observed it in nature in my youth. If Lorenz intended to be tactful by pointing out that Schneirla and I should not be regarded as biologists, then the intention failed. I would much rather be called stupid!

I should not point out irrational, emotion-laden elements in Lorenz's reaction to criticism without acknowledging that, when I look over my 1953 critique of his theory, I perceive elements of hostility to which its target would have been bound to react. My critique does not now read to me like an analysis of a scientific problem, with an evaluation of the contribution of a particular point of view, but rather like an assault upon a theoretical point of view, the writer of which assault was not interested in pointing out what positive contributions that point of view had made. It does fail to express what, even at that time, I regarded as Lorenz's enormous contribution to the formulation of the problems of evolution and function of behavior, and his accomplishment in creating a school based upon the conception of species-specific behavior as a part of the animal's adaptation to its natural environment. (This would be an appropriate point for me to remark that I do not now disagree with any of the basic ideas expressed in my critique!)

THE MEANINGS OF "INNATE" AND THE MEANINGS OF "INHERITED"

The terms "innate" and "inherited" both have, in different contexts, at least two different meanings, which do not refer to the same processes, which are not arrived at by the same operations, and which have entirely different kinds of reference to the problem of development, but which are often confused with each other.

When a geneticist speaks of a character as inherited, what he means is that he is able to predict the distribution of the character in an offspring population from his knowledge of the distribution of the character in the parent population and of the mating patterns in that population. He is *not* necessarily making any inferences whatever about the developmental processes involved in the ontogeny of the character, or even the extent to which it is subject to change under environmental influence. Another way of saying this is that a character may be said to be "inherited" or "heritable" or "hereditary," if the variation of this character from individual to individual can be shown to arise from differences in the genetic constitution, or genome, of the different individuals, rather than from differences in the kind of environment in which they have been reared, or in the way in which they have been treated. Now, the fact that selective breeding (i.e., arranging for the offspring generations to consist only of individuals resulting from the matings of members of the parent generation which have been selected for the presence or absence of some specific characteristic or characteristics) can result in striking changes in the characteristics of the group of organisms certainly means that the characteristic is hereditary, but it by no means demonstrates that the *same* characteristic cannot be influenced by the environment. A genome arrived at by selective breeding in one environment may have quite different phenotypic characteristics in a second environment, while an environmental change that has great influence upon the phenotypic development of one genome may have no effect upon that of another (Haldane, 1946).

Geneticists have dealt with this problem by restricting the concepts of "heritability" and "environmental influence" with respect to any given character to an estimate of the amount of the actually observed variability in that character that can be attributed to variations among the different genomes *actually tested,* and to the amount that can be attributed to the variety of environments in which organisms with those genomes have *actually been raised.* They thus do not preclude the possibility that other genes than the ones tested might have an effect upon the character, or that environments other than the one tested may cause unpredictable changes in the phenotypic appearance of a given genome. In genetic usage, therefore, the fact that a character can, in a given environment (for example, the "normal" environment) be strikingly affected by selective breeding (as, for example, by hybridization experiments) does not directly deal with the question of whether variations in the environment during the development of the organism would or would not have an effect upon the manner of development of the adult phenotype.

There is, however, a *second* meaning which, implicitly or explicitly, is often attributed to the words "innate" or "hereditary" or "inherited" by students of animal behavior and by nongeneticists generally. This meaning is that of *developmental fixity,* i.e., that the organism is impervious to environmental effects during development, and so it *must* develop characteristics that are preorganized "in the genes," regardless of the environment in which it is reared. Now, I am *not,* at this point (but see below, pp. 28–30), attempting to discuss the question of whether there is such a thing as developmental fixity, or whether the term "innate," meaning unavailable to environmental influence, is a useful or meaningful term to apply to behavior characteristics. I am merely trying to point out that the concept of "innateness" as referring to developmental fixity is a *different* concept, and one that exists in a different, and not parallel, dimension, from the concept of "alterable by selective breeding," which is the same as "achievable by natural selection."

But it must be obvious to every candid observer of the literature of animal behavior that these two different, and incommensurate, concepts are very often implicitly mixed into one use of the term "innate." This is, for example, what Lorenz does when, in a discussion about the legitimacy of the use of the term "innate," he introduces as evidence both the fact that the offspring of a hybrid shares behavioral characteristics of both parent species and the fact that learning does not influence the development of the behavior patterns concerned.

LORENZ'S USE OF "INNATE"

Lorenz did not identify the distinction outlined in the preceding section, or the problems raised by it, because his recent discussion was couched in somewhat different terms. He has, however, proposed a very interesting resolution of the problem as he perceives it. He states that the term "innate" should "never, on principle, be applied to organs or behavior patterns, even if their modifiability should be negligible." He does, however, think it proper to describe as "innate" a distinctive *property* of a neural structure, such as its ability to select, from the range of available possible stimuli, the one which specifically elicits its activity, and thus the response seen by the observer. Presumably, the property of the neural structure in giving rise to a particular movement pattern would also be a property of this kind, which Lorenz refers to as a "character." He is thus making a distinction between organs, structures, and behavior patterns, which he says should not be called "innate," and special properties of these organs, structures, and behavior patterns, by which they fit into the appropriate environment. His conception is that, even if, for example, an animal can see nothing without previous visual experience at the appropriate time in its development, it might still be that if it had appropriate experience (i.e., experience of light, or of contours, etc.), some of its specific responses could be linked to specific visual stimulus configurations, *without* the animal having had any *specific* visual experience that would account for its reaction to those visual stimuli, rather than to others. In this context, he asserts that the "information" that the following response of a given species of fish is elicited by the characteristic color of the mother fish of that species, rather than by other colors, may legitimately be called "innate," if it can be demonstrated that a fish that can see will prefer to follow this color, without any previous experience of the color, even though it may be true that the development of visual abilities in the first place required experience of light. In this situation, he would not call the visual capacity as such innate, but only a specific property of the fish's visual system: that it was capable of selectively responding to the appropriate color when the animal was in a mood to follow.

This distinction is made in the service of a more fundamental distinction, which is the principal argument of Lorenz's book and the principal basis for his insistence that the concept of "innateness" is an objective and necessary one. Briefly, Lorenz states that, when a behavior pattern is adapted to a given aspect of its environment, the "information" which defines the properties by which the behavior pattern is adapted to the environment can have been incorporated into the behavior pattern of the animal either by "the adaptive processes of evolution" or by "individual acquisition of information." By "adaptive processes of evolution," Lorenz refers to the creation of the genome by natural selection, so that the gene-complex characteristic of an individual is the result of a history of selection for those genes that, in the natural environment, give rise to characteristics that are adaptive to that environment, including behavior characteristics. This is a sort of historical "trial and error" process, in which mutations that lead to useful (i.e., adaptive) results are retained, while those that lead to harmful (i.e., nonadaptive) ones are eliminated. By "individual acquisition of information," Lorenz means individual learning.

Lorenz's present argument is that behavior characteristics that are adaptive to particular points of the environment must be considered "innate" if their source, or provenance, is through the incorporation of genes into the genome through natural selection, and must be considered "learned" if their source is a change in the behavior of the animal as a result of its individual experience of that environment.

Up to this point, my purpose has been to outline the type of behavior about which opposing views have been expressed, to sketch, however briefly, the nature of the disagreements, and to point to some problems involved in defining the terms about which the disagreements have flowered. I should now like to turn to a consideration of some conceptual problems, with a view to pointing out some ways in which different workers disagree in their evaluation of the importance of various questions, even when they agree about the facts concerned.

VARIATION AND DEVELOPMENT

Genetic Variation Versus Developmental Process

If we rear two animals in the same environment, and they develop different behavior patterns, it is perfectly clear that the difference in the behavior patterns depends upon differences in the genetic constitution of the animals, and not upon differences in the environment. One might refer to these *differences* as "innate," meaning only that they depended upon differences in the genome. This use of the term "innate" would be meaningful and useful, provided it was recognized that the observations that justified the conclusion that the difference between the two animals was a genetic difference do not necessarily imply anything, one way *or* the other, about the extent to which the development of the character concerned is, in either animal, influencible by changes in the environment.

This is the situation in which we find ourselves when we compare different species living in similar environments, and showing different behavior. The differences between the species obviously arise from genetic differences, and it is perfectly appropriate to use genetic terms, and terms deriving from considerations of evolutionary adaptation, in analyzing these differences and their evolutionary relationships, and the evolutionary origins of particular behavior patterns. None of these considerations, however, really bears on the question of ontogenetic origin, which is, to some degree, a question of a different kind.

To take a rather simple example, let us consider one aspect of the problem of the role of behavior patterns in the formation of species. An existing species becomes divided into two descendant species when two parts of the original population become geographically separated from each other and when, under conditions of geographical separation (i.e., when no genes are being exchanged between one population and the other), they become so different from each other genetically that they would not interbreed if the geographical barrier were removed. This may happen because, while they were in geographical isolation, they evolved in slightly different directions as a result of adaptation to slightly different environments, or it might occur because the populations were small enough so that rarely occurring mutations occurred, by chance, in different frequencies in the two populations, or it might come from some combination of these and similar factors. Whatever the reason, each of the two populations may eventually consist, more or less homogeneously, of individuals that will be sufficiently different from those of the other population so that they will no longer be as ready to mate with them as they would be with members of their own population. This may be because the members of the two populations prefer different habitats and thus do not meet each other, because the courtship behavior of a member of one population is no longer adequately stimulating to a member of the other, or for a variety of other reasons, behavioral, ecological, or morphological. Now, this "reproductive isolation" may not be complete at the time the geographical barrier between the two populations is removed. It may be that, at the time it becomes geographically possible for the two populations once more to become continuous, there has developed only a relative isolation, defined as a quantitative preference for mating with a member of the animal's own population, rather than a member of the other. In that case, a number of things may happen, the extremes of which might be defined by two alternative outcomes: (a) if the two populations are similar enough to each other so that hybrids between them can survive in the existing range of environments as well as can members of either population, the two populations may simply merge, as the genes of one are spread through the other through the hybrid matings at the points where the populations meet each other; and (b) the two populations may have become so different from each other, and adapted to such different environments, that hybrids between them will not be as well adapted for survival in *any* environment as each of the populations is to *some* environment. These hybrids will be at a selective disadvantage, compared with the offspring of within-population members of either population. In this case, the hybrids will be eliminated by natural selection, which means that, at least in the zone of overlap between the two populations, only those genomes will survive that ensure, in that environment, a preference against mating with a member of the other population (Mayr, 1942).

This process can be reproduced experimentally in the laboratory. For example, Koopman (1950) allowed individuals of two species of *Drosophila* to select mates in a mixed population. There was a moderate degree of preference for mating within the species, as could be seen by the characteristics of the larvae (the animals carried a marker gene, which made it possible to distinguish a hybrid animal from pure-bred animals in the larval stage). He removed all the hybrid larvae, and then allowed the remaining members of the offspring generation to choose mates again from a mixed population, now consisting only of the offspring of animals that had selected members of their own species for breeding. He repeated the process for a number of generations, thus tending to eliminate those genes that made possible the selection of a mate from the "wrong" species. The result was that the degree of reproductive isolation (i.e., the strength of the tendency to select a mate from within one's own species) was gradually increased.

In all these examples, the characteristics that insure that the animals will mate only with members of their own population have been arrived at entirely by selection directed against those genes that made possible a maladaptive mating choice. That is, they have been arrived at by the adaptive processes of evolution. This process is very common in nature, and may probably be assumed to have taken place in almost every case where two closely related species breed in the same area. If, however, an animal is reared in association with members of another, closely related, species, it soon becomes clear that in some cases, such as the cowbird, the rearing conditions have no effect upon the mating preferences, while in other cases, such as some species of doves (Whitman, 1919), the mating preferences may be strongly affected, being shifted sharply in the direction of a willingness to mate with a member of the "foster" species (Mayr and Dobzhansky, 1945)! In both cases, the features of the animal that are, in normal circumstances, responsible for its absolute preference for mating with a member of its own species, have been incorporated by natural selection, by selective breeding, by "the adaptive processes of evolution." But in one case, these features include a role of learning, in the other case not! Nature selects for *outcomes:* it does not care whether this outcome is arrived at through the development of features of the animal that make it impossible for it to respond to stimuli offered by members of the other species, or whether it is arrived at through the development of features that make the animal prefer to mate with a member of the species that it experienced in its early life!

Now, I would not dream of implying that Lorenz, the discoverer of imprinting, does not know all that I have just said, and I am aware that his way of dealing with these facts would be to call the preference of the cowbird innate and that of the dove learned. I am merely pointing out that if a scientist is not overwhelmingly convinced that characteristics incorporated into the species by the actions of natural selection are, *by that fact,* demonstrated to be impervious to individual experience, he is not necessarily guilty of "a very deep misunderstanding of biological ways of thinking" or "a lack of acquaintance with phylogenetic and genetic thought" (Lorenz, 1965).

What I intend to indicate by the example that I have just given is that the clearest possible genetic evidence that a characteristic of an animal is genetically determined in the sense that it has been arrived at through the operation of natural selection does not settle any questions at all about the developmental processes by which the phenotypic characteristic is achieved during ontogeny.

Genetics and Developmental Fixity

In the crustacean *Gammarus,* the difference between the normal red eye-color and a mutant with chocolate eyes depends upon a single mutation which affects the rate at which an eye pigment is deposited during a certain stage of development. If the mutant is reared below a given threshold temperature, the eyes will develop red, and at intermediate temperatures there will be intermediate eye colors (Ford and Huxley, 1927).

Variations in the wing structure of *Drosophila melanogaster* may be affected by a wide variety of genetic mutations (Morgan, Sturtevant, Muller, and Bridges, 1923). These mutations also have an effect upon the ability of the animal to fly. Different flies of a single genotype may be able to fly normally, weakly, or not at all, depending upon the temperature at which they are raised (Harnly, 1941).

These two examples (which, let me hasten to add, will not surprise any biologist) show that the same genes may lead to different phenotypic outcomes when the animal is subjected to different environmental influences during development. Suppose, however, that *Gammarus* or *Drosophila* were, for reasons having nothing to do with the mechanisms of eye development or wing development, unable to survive at any temperature outside the range 24°–26° C. In that case, we would have to say that the character was uninfluenced by the environmental temperature. But would this mean that the mechanism of eye development, or of wing development, was any different? I think not. Further, the situation would be similar if, instead of being unable to survive temperature variations, the animals possessed regulative mechanisms that maintained the temperature environment of the eye (or of the wing) constant in spite of variations in environmental temperature. In both of these cases the outcome of the experiments would be that we had failed to show any effect of environmental temperature upon the development of eye color, or of flying ability. But this would not *necessarily* mean that considerations of temperature were irrelevant to the development of these characters. Further, it should be clear that the failure to show the effects of a particular environmental variable does not say anything positive about the processes involved in the development of any character.

There is a fundamental question of logic involved here. I am sure that Lorenz and his colleagues perceive those of us who are oriented by Schneirla's teaching as constantly engaged in an eager search for any little snippets of evidence that learning has any effect, however small, on the development of a behavior pattern, to the exclusion of any attention to the broad problems of adaptation, and that we exaggerate the relative importance of learning influences in the service of a need to see learning everywhere and heredity nowhere. Given the role played by the phenomena of adaptation, and by the concept of the "normal environment" in Lorenz's thinking, I can certainly understand how this impression could arise. But that is not at all our conception of our situation. It seems to us that an experimental manipulation that causes a change in the behavioral outcome has thrown some light on the process by which the behavior develops, while an experimental manipulation that *fails* to cause any change in the outcome has *failed* to throw light upon the nature of the processes leading to the outcome. To Lorenz, the failure of an experimental treatment to cause any change in outcome seems just as illuminating as does the success of an experimental treatment in affecting the outcome. He makes this quite explicit when he says that he disagrees with the formulation that "it is not characters but differences between characters which may be described as innate," and says that "the opposite formulation is at least as workable: calling innate the similarities of characters developing under dissimilar rearing conditions." To an *experimental* scientist, the insight gained by observing that a variety of treatments all failed to have any effect is not at all equivalent to that gained by observing that some treatments have effects, while others do not. Indeed, it is of the essence of the experimental method that an experiment cannot be regarded as making a contribution to the understanding of any problem unless the experimenter has succeeded in finding alternate treatments that have different effects upon the outcome. It is for this reason, and not because I think that any *particular* kind of developmental influence is all-important, that I regard an experiment that shows an effect, during development, of any treatment, as a contribution to the illumination of a process of development, while a study which succeeds only in showing that some types of manipulation have no effect upon the outcome seems only like a challenge to follow the problem to an earlier stage of development, or to a more intricate level of physiological analysis.

The criterion of developmental fixity is thus a negative one in the sense that it is based upon the *lack* of effect of experimental treatments. If a class of behavior patterns is defined in large part by such a criterion (lack of effect of treatment), as in Lorenz's classification of the instinctive act (*Erbkoordination*), the assumption that all members of the class have *developmental* or *physiological* features in common is not necessarily valid.

LEARNING, EXPERIENCE, AND DEVELOPMENT

Learning and Experience

One persistent difficulty is that Lorenz, and a number of other writers, use the term "learning" to refer solely to the kind of conditioning and associative learning that are traditionally described as the learning capacities of adult animals (Kimble, 1967), and they have made no effort to incorporate into their thinking Schneirla's concept of "experience," which refers to a wide range of processes, of which learning is only a relatively small part.

Let me repeat Schneirla's definition of "experience": the contribution to development of the effects of stimulation from all available sources (external and internal), including their functional trace effects surviving from earlier development (Schneirla 1957, 1966). Contrast this with Lorenz's statement: "Not being experimental embryologists but students of behavior, we begin our query, not at the beginning of the growth, but at the beginning of the function of such innate mechanisms." By this statement, Lorenz is asserting that he is simply not interested in the type of question to which Schneirla's conception addresses itself. Now, it is not at all necessary that the problem of development should be a central problem for every scientist interested in behavior, or that all students of behavior who are interested in problems of development should be interested in the development of the same types of behavior, or should be primarily concerned with the same stages of development. I think it is important, however, to recognize that there are differences of attitude involved in these disagreements, which do not have to do with factual matters, but with what each of the parties considers to be an interesting problem, or a heuristically significant question. As Lorenz states, he is really not interested in the origins of behavior patterns at those stages of development before they begin to exist as modes of adaptation to the environment. Since his interest starts at that point, it is quite understandable that the only kind of experience that would seem theoretically relevant for him would be the kind of conditioning and associative learning characteristic of animals whose behavioral organization is already ontogenetically well formed. In effect, Lorenz would like to consider the problems of experience solely in terms of the role of conditioning and associative learning in the behavior of animals, starting at stages of development when their species-specific behavior patterns are already functional, so that the problem becomes merely one of whether an animal can learn to use a nesting material other than the ones for which its normal movements are adapted, or whether it needs to have seen a red object in order to prefer to attract a fellow-member of the species with the red belly, or whether it will respond appropriately to the sound of a young animal of its species without ever having heard one before, etc. Problems of the sort referred to by Schneirla would then be left to "experimental embryologists."

This feeling on Lorenz's part is consistent with his assertion that the innate is what must be there before learning begins. However, there is already evidence that the development of organisms is not divided into such convenient chapters, corresponding to the divisions among the professional specialties of biologists. Conditioning can occur very early in life, even prenatally in mammals (Spelt, 1948; Prechtl, 1965), or pre-hatching in birds (Gos, 1935; Hunt, 1949; Sedláček, 1962, 1964; Gottlieb, 1968). For a scientist who is *primarily* interested in the analysis of development, the existence of such early conditioning abilities cannot seem irrelevant to the problem of the ontogenetic origin of later-appearing behavior patterns.

Further, nonbehavioral physiological regulations and those that are of interest to a student of animal behavior cannot be sharply separated. Physiological events that are not normally or conventionally regarded as a part of "behavior," such as changes in body temperature, changes in bladder activity, changes in kidney activity, changes in tension in the mammary gland, dilation and constriction of the blood vessels, are all to some degree under neural control, and can be conditioned, both by Pavlovian techniques (Bykov, 1957) and by those of operant conditioning (Miller, 1969; Miller and di Cara, 1967). In addition, the conditioned stimuli may be either external stimuli or stimuli arising inside the body, including stimuli arising from changes in tension elsewhere in the body, which may themselves be conditioned (Razran, 1961). This means that the distinction between "animal behavior" and other kinds of physiological regulation are not as absolute in the organization of the animal's physiological mechanisms as they usually are in the perception of the student of animal behavior, and that it may be necessary for scientists interested primarily in animal behavior to pay attention to a great many things that are primarily of interest to other kinds of scientists (such as experimental embryologists, or even psychologists) in order to achieve a broadly based understanding of the origins and organization of the phenomena that attract their primary interest.

The separation of problems into those visible after the adaptive behavior patterns begin to function, and those which are relevant to early development (the former being the province of the student of animal behavior, while the latter is assumed to be of interest only to experimental embryologists), presupposes that it is possible to make a sharp distinction between learning and other contributions of experience to development, and that there are no intermediates. As Schneirla has repeatedly pointed out, however, sharp lines cannot be drawn, in early stages of development, between: the effects on neural development of nonbiological conditions (temperature, light, chemical conditions in the environment); nonspecific effects of gross stimulus input; the developmental effects of practice passively forced during ontogeny; the developmental effects of practice resulting from spontaneous activity of the nervous system; links and integrations between behavioral elements, resulting from early, nonfunctional partial performances; interoceptive conditioning resulting from inevitable tissue changes and metabolic activities; simple conditioning to stimulation resulting from spontaneous movements; and simple instances of conventional conditioning and learning.

Now, the introduction of the concept of "experience," in the sense described here, into the discussion of development is by no means equivalent to saying that all behavior patterns derive from learning. This point has been repeatedly made by Schneirla over many years, and I was quite aware of it when I wrote my first contribution to this discussion fifteen years ago. To quote a characteristic remark from that paper (Lehrman, 1953):

> . . . Analysis of the developmental process involved shows that the behavior patterns concerned are not unitary, autonomously developing things, but rather that they emerge ontogenetically in complex ways from the previously developed organization of the organism in a given setting. . . . The post-hatching improvement in pecking ability of chicks is very probably due in part to an increase in strength of leg muscles and to an increase in balance and stability of the standing chick, which results partly from this strengthening of the legs and partly from the development of equilibrium responses . . . Now, isolation or prevention-of-practice ex-

periments would lead to the conclusion that this part of the improvement was due to "maturation." Of course it is partly due to growth processes, *but what is growing is not pecking ability or anything isomorphic with it.* The use of the categories "maturation-vs.-learning" as explanatory aids usually gives a false impression of unity and directedness in the growth of the behavior pattern, when actually the behavior pattern is not primarily unitary, nor does development proceed in a straight line toward the completion of the pattern.

As I reread that paper, it seems clear to me that, even at that early stage, I was not insisting that all behavior is learned, but that the distinction between "innate" and "acquired" is an inadequate set of concepts for analyzing development, and that the development of behavior patterns could not be analyzed by assuming autonomously developing specific substrates for each behavior pattern, isomorphic with the behavior.

These remarks are my reaction to a recent rereading of my 1953 paper. In the intervening years, I have heard it so often said that I believed that all species-specific behavior develops through individual learning that I almost came to believe that I *had* said it! I remember reading a discussion in which I. Eibl-Eibesfeldt and W. H. Thorpe apparently succeeded in convincing an initially incredulous Donald Hebb that I had insisted that all behavior is learned through individual experience (Eibl-Eibesfeldt, 1961)!

I believe this difficulty arose from the fact that many workers in the field of animal behavior had such a firmly fixed opinion that every element of behavior ought, on *logical* grounds, to be clearly classifiable as "innate" or as "learned," that any discussion that cast doubt upon the usefulness of the concept of "innate" must inevitably have seemed like an insistence that all behavior must belong to the other category! As I know from my own experience, this could be so even when the discussion in question was addressed, not to the thesis that all behavior is learned, but to the thesis that the dichotomy *itself* does not adequately express the necessities for developmental analysis of behavior.

Is Development Necessary?

THE IDEA OF A "GENETIC BLUEPRINT"

The idea of a genome as a "blueprint," contained in the fertilized egg and representing a plan for the construction of an adult organism, is a very attractive one. Lorenz says that "what rules ontogeny . . . is obviously the hereditary blueprint contained in the genome and not the environmental circumstances indispensable to its realization. It is not the bricks and the mortar which rule the building of a cathedral but a plan which has been conceived by an architect. . . ." Further, he says that ". . . our first question concerning the ontogeny of an organism and its behavior is: 'What is blueprinted in its genome?' "

Now, it may be comforting, in the sense that it gives us the feeling that we have increased our understanding of the problem, to say that a behavior pattern (or a structure) is innate if it is "blueprinted in the genome" or, in a more modern vernacular, "encoded in the DNA." There are, of course, contexts in which such expressions are meaningful, but I believe that the comfort and satisfaction gained from disposing of the problems of ontogenetic development by the use of such concepts are misleading, and are based upon the evasion or dismissal of the most difficult and interesting problems of development.

It seems to me that there is a fundamental fallacy in the use of the analogy of the relationship between a blueprint and the structure represented by it to represent the relationship between the genome at the zygote stage and the phenotypic adult. A blueprint is isomorphic with the structure that it represents. The ratios of lengths and widths in the blueprint are the same as those in the structure; the topographical relationships among the parts of the structure are the same as those among the corresponding parts of the blueprint; each part of the structure is represented by a separate part of the blueprint, and each part of the blueprint refers only to a specific part or parts of the structure. It will be immediately obvious that this is profoundly different from the relationship between the genome and the phenotype of a higher animal. It is *not* true that each structure and character in the phenotype is "represented" in a single gene or well defined group of genes; it is *not* the case that each gene refers solely, or even primarily, to a single structure or character; and it is *not* the case that the topographical or topological relationships among the genes are isomorphic with the structural or topographical relationships among phenotypic structures to which the genes refer. It is, of course, a commonplace of modern biology to say that each gene is responsible for the production of a single enzyme. This formulation reflects the truly enormous advances that have been made in recent years in understanding the structure of the genes, primarily on the basis of research on the biochemical actions of genes in one-celled organisms. The problems of ontogeny and differentiation of structures in complex organisms, however, have hardly been touched as yet by the recent massive advances in molecular biology. A facile description of the genome as a "blueprint" gives a misleading impression of understanding a problem that is regarded by modern geneticists as one of the major unsolved problems of biology, and which ought to be regarded as a truly difficult problem by *any* biologist, even a student of animal behavior who is prepared to leave the problem to the experimental embryologist.

Another problem with the conception of the genome as a "blueprint" is, of course, that, while it poses as a contribution to the understanding of ontogeny, it is actually *irrelevant* to the question of individual development. As Lorenz himself says, ". . . it is perfectly possible that a particular motor sequence may owe to phylogenetic processes all the information on environment underlying its adaptiveness and yet be almost wholly dependent upon individual learning for the 'decoding' of this information." But it should be perfectly clear that, if a character "encoded" in the genes may or may not require individual experience for its development, then a scientist who is interested in the causal analysis of development is not helped very much by statements about the "encoding" or "blueprinting" of complex characters in higher animals. Here again, I repeat that the concept of "innate" in the sense of determination by the genome, and the concept of "innate" in the sense of imperviousness to individual experience, refer to different problems and relationships, which cut across each other, rather than making a single conceptual whole. And here again Lorenz, by inconspicuously merging these two conceptions into a single usage, is led to speak of patterns as being blueprinted in the genome, *as opposed to* being based upon experience, while simultaneously acknowledging that patterns "blueprinted" in the genome may or may not develop through individual experience.

It seems to me, then, that although the idea that behavior patterns are "blueprinted" or "encoded" in the genome is a perfectly appropriate and instructive way of talking about certain problems of genetics and of evolution, it does not in any way deal with the kind of questions about behavioral *development* to which it is so often applied.

PROVENANCE AND ONTOGENY

In my 1953 critique, I referred to the work of Kuo (1932a, b, c, d), who made detailed observations of the behavioral development of the domestic chick embryo within the egg. On the basis of these observations, Kuo suggested that the pecking behavior that can be seen in chicks immediately after hatching develops through a series of stages in which the neck is first (early in embryonic life) passively bent when the heartbeat causes the head (which rests on the thorax) to rise and fall, with active bending of the head occurring later, at first in response to tactual stimulation. Kuo also suggested that the opening and closing of the bill (associated with pecking in the post-hatching animal) first occur when the bird's head is nodding during the embryonic period, apparently through nervous excitation furnished by the head movements through irradiation in the still-incomplete nervous system, while the opening and closing of the bill become independent of head activity only somewhat later. Kuo noted that fluid forced into the throat by movements of the bill and head apparently causes swallowing, beginning at a characteristic time during embryonic development. Kuo's suggestion was that the movements forced by the timing and order of development of the various structures, neural and motor, and by the conformation of the bird's body enforced by its position within the egg, provided an experiential contribution to the development of the integration of the head, bill, and throat components of the food-pecking lunge, which is already to be seen (although in incompletely integrated form) by the time of hatching. It has recently become clear that some aspects of the development of motility in the chick embryo do not depend upon sensory input (Hamburger, 1963; Hamburger, Wenger, and Oppenheim, 1966), and caution is required in interpreting Kuo's data, which have not yet been subjected to direct experimental test. However, the existence of conditioned responses several days before hatching is very well established in these birds (Gottlieb, 1968; Sedláček, 1962, 1964), and the nature of behavioral and neural development during embryonic life in birds, and the problem of the role of experience at this stage, are being actively investigated in several different laboratories (see Gottlieb, 1968, for review).

Of this discussion, Lorenz says "If Lehrman (1953) gives serious consideration to the assumption that a chick could learn, within the egg, considerable portions of the pecking behavior by having its head moved rhythmically up and down through the beating of its own heart, he totally fails to explain why the motor pattern thus individually acquired should fit the requirement of eating in an environmental situation which demands adaptedness to innumerable single givens. . . ." [3]

Here I must repeat something I said earlier in this essay. Nature selects for *outcomes*. Natural selection acts to select genomes that, in a normal environment, will guide development into organisms with the relevant adaptive characteristics. But the path of development from the zygote stage to the phenotypic adult is devious, and includes many developmental processes, including, in some cases, various aspects of experience. This is clear from many considerations, and is acknowledged by Lorenz himself. What then is the difficulty about assuming that, *whatever* the characteristics

---

[3] This is Lorenz's oversimplified version of my own description, which was a very cautious description of what I imagined to be a very complex series of events, in connection with which I did not use the word "learning." I did refer, at one point in a 500-word description of Kuo's observations, to a "process of development, which involved conditioning at a very early age. . . ." Even fifteen years ago, under Schneirla's influence, I was trying to convey the idea of continuity and interpenetration between the processes of growth and those of the influence of environment, and to express a feeling of tentativeness and ambiguity about the distinction between the effects of experience on a developing organism and the effects of experience in a mature nervous system. And even today Lorenz is perfectly confident of the sharpest distinction between "morphological ontogeny producing structure" and "trial-and-error behavior" producing learning, with no sense of difficult intermediates or unsolved conceptual problems.

of the developing nervous system, they must be such as to give rise to the adaptive form of pecking which is seen after the bird hatches? The relationships described by Kuo, involving certain putative effects of experience that might be inevitable in the context of the developing structures in the egg, are no more mysterious a product of embyronic development than any other characteristics of the developing nervous system. It does not matter to the process of natural selection whether what is being selected for is a genome that gives rise to adaptive pecking at food through a developmental process that does not involve experience, or whether it is a genome that gives rise to adaptive pecking behavior via a course of development that *does* include effects of experience. This is another case in which the statement that a characteristic has been arrived at through selective breeding (i.e., in this case, through natural selection) says nothing at all about whether its development does or does not include an effect of experience. Natural selection can select for specific ways of being sensitive to experience, or for phenotypic structures that make experience possible, just as readily as it can for any other characteristics.[4]

In the same context, Lorenz speaks of "some American psychologists" as "trying to avoid, at all costs, the concepts of survival value and phylogenetic adaptation for no other reason than that they regarded them as 'finalistic.' " As I hope the discussion so far has made clear, however, I have not been trying to avoid the concepts of survival value and phylogenetic adaptation, but only to prevent them from being merged with the concepts of the *causal* analysis of *development,* in order that the understanding of ontogeny should not be confused by merging two different meanings of the term "innate," which are to some extent irrelevant to each other.

Lorenz's objection to the formulation that "it is not characters but differences between characters which may be described as innate" is not as clear as he implies. The concept of evolutionary adaptation is not arrived at, and is not maintained in the minds of observers, by perceiving one animal or one species in its adaptation to the environment. The concept of adaptation, both historically and in its everyday application, depends upon the fact that we observe *different* species to show elaborate adaptations to *different* environmental requirements. The adaptive elegance of the way in which a newly hatched pheasant pecks at food on the ground is fully apparent only to the observer who is on some level aware, while he

---

[4] Hamburger's recent work (1970), makes it clear that Kuo's conceptions of the sources of early behavioral organization in the chick embryo are not tenable, and are based on incorrect assumptions about the embryology of the chick's nervous system. Lorenz's feeling and mine was wrong. I would not now use Kuo's work as an example of the study of behavioral development. I have, however, retained the present discussion of Lorenz's reaction to it because it still illustrates the conceptual and methodological problem I am discussing here.

watches the pheasant, that a newly hatched thrush would not peck at the ground, but would gape (beg) from its parent, who would be willing to feed it in a way of which the parent pheasant is incapable. Lorenz makes this quite explicit when he says (of Kuo): "It also remains unexplained why only certain birds peck after hatching, while others gape like passerines, dabble like ducks, or shove their bills into the corner of the mouth. . . ." Here Lorenz clearly, if inadvertently, acknowledges that it is the *differences* among the behavior patterns of different species living in similar environments that give rise to the sharp feeling, which I share, that the species have different genomes. It remains true, of course, that differences in the genome may give rise to differences between animals, at a very early stage of development, which *consist* of differences in the extent to which they are able to take advantage of information offered by the environment, or differences in what they will pay attention to in the environment. Therefore, although animals reared in the same environment that behave differently must have started with different genomes, this does not in any way tell us whether or to what extent differences in experience might have played a role in the development of the phenotypic differences between them as adults.

NORMAL AND ABNORMAL

The differences in attitude and interests between scientists whose primary interest is in evolution and adaptation and those whose primary interest is in the causal analysis of development are fairly well demonstrated by Lorenz's reaction to a paper by Donald Jensen (1961). Jensen had suggested that many operations other than genetic selection or training could produce differences in behavior between animals. These operations include nutritional variations, alterations of the nervous system, hormone treatments, etc. Jensen suggested that studying the effects of a wide range of differential treatments upon the development of behavior differences, with the intention of inductively integrating the information thus acquired, would be a more fruitful and less controversial way of coping with problems of ontogeny and of causality than the prevalent attitude of treating the question "innate or learned?" as a primary and ultimate question on which all others must hinge.

This modest suggestion has aroused a special ire in Lorenz, which is noticeable even against the background of the generally indignant tone of his book. This is because, in Lorenz's opinion, the investigation of the effects of a very wide variety of treatments which can alter the behavior runs directly counter to the main task of the biologist: to understand how the genome that is arrived at by natural selection gives rise, in the *normal* environment, to the *normal* behavior pattern adapted for that environment.

Lorenz's concern that the introduction of "abnormalities" will distort and misrepresent the study of adaptive characters is shown by the following selection from his remarks:[5]

> Non-adaptive differences in structure and behavior are of but secondary interest to the biologist, while they are the primary concern of the pathologist. . . . As students of behavior, we are not interested in ascertaining at random the innumerable factors that might lead to minute, just bearable differences of behavior bordering on the pathological. What we want to elucidate are the amazing facts of adaptedness. . . . We need not bother about the innumerable factors which may cause "differences" in behavior as long as we are quite sure that they cannot possibly relay to the organism that particular information which we want to investigate.

Now, Lorenz is quite right to point out that experimental treatments cannot be selected at random; they must be chosen with some intuitive feeling for their relevance to the normal phenomenon, the development of which we wish to understand. I am not persuaded, however, that the distinction between "pathological" and "normal" is a very useful guide for understanding the causes of development; and I am not convinced that a biologist interested in understanding *development* is obligated to recoil from any treatment that disturbs the "very complicated and very finely balanced system" (Lorenz, 1965) which is the living organism. Indeed, a very good case can be made for the proposition that it is precisely by interfering with normal development and noting in what way the resulting abnormalities develop that we gain the most illuminating insights into the normal processes of development.

Experience and "Normality"

Lorenz's tendency to regard conventional learning paradigms as the only method for defining environmental influences that are of any interest to a student of behavior expresses itself in a tendency to regard any other developmental effect of experience as simply a pathological effect of "bad rearing." This distinction is very clear, for example, in his statement that "we try to produce an individual whose genetical blueprints have been realized unscathed in the course of healthy phenogeny. Should we fail in this, we would incur the danger of mistaking some effects in our subject's behavior for the consequences of information withheld, while they really are the pathological results of stunted growth."

---

[5] These sentences are not consecutive in Lorenz's text but occur, with intervening text, in the space of a page or so.

This is, of course, a logical extension of Lorenz's position that, with the exception of trial-and-error learning or classical conditioning in the fully developed nervous system, the effects of stimulation from the environment during development are matters of interest, not to the student of animal behavior, but to the experimental embryologist or the pathologist.

I think, however, that it is not so easy as Lorenz implies to make sharp distinctions between "learning" and "mere" pathologies of development. Let us look at some examples of the effects of rearing in abnormal conditions upon behavioral adaptations:

1. In many species of birds, the young characteristically follow their parents about within a very short time after emerging from the egg. It has been clear since the early work of Lorenz (1935) that, in many ducks, geese, and other species of precocial birds, the ability of the young to follow selectively adults of their own species is dependent upon a very quick learning process which occurs during a restricted period very early in life; the birds thus learn through this "imprinting" experience to follow the models that they experience immediately after emerging from the egg. These will, in nature, usually be the birds' parents. If newly hatched ducklings are exposed to adult ducks of another species than their own, they may later prefer to mate with the members of the species on which they were imprinted, rather than with members of their own species (Schutz, 1965). This may happen either through long-term effects of the early experience, or through intervening (adolescent) experience with birds with which they associate because of the earlier experience (Hinde, 1962; Bateson, 1966). Now, a bird which, because of this early experience, wants to mate only with a member of another species, which refuses to mate with it, has certainly had its development altered in an abnormal direction; it is pathological, since the abnormal conditions of its development have led to an adult condition in which it is no longer adapted to its environment. This treatment, which is widely and correctly regarded as demonstrating, for the student of development, a form of learning, must also be regarded, for the student of evolutionary adaptation, as an example of pathological interference with an evolutionary adaptation through rearing in an abnormal environment.

2. When Harlow reared rhesus monkeys without giving them any opportunity to interact with age-mates, they developed striking and pervasive abnormalities of behavior: as adults, they were not able to maintain social contact with other monkeys, their sexual behavior was so drastically interfered with that most of them were totally unable to copulate, and the balance between the role of fear and aggression and the role of more positive social responses in the social relationships of these monkeys was

severely distorted. The deprived monkeys, in general, were incapable of normal integration into a group of monkeys. This distortion of the normal early experience apparently has widespread effects upon the emotional responsiveness of these monkeys, which are reflected both in specific distortions of particular behavior patterns and in more general interferences with a wide range of behavior patterns (Harlow and Harlow, 1965).

3. The rat shows a characteristic response of fear and anxiety to a strange environment. The level of this response and many details of it can be altered by selective breeding, and are therefore heritable in the geneticist's use of the term (Fuller and Thompson, 1960). The rat also shows a characteristic tendency to be curious about a novel environment, and to explore it (Berlyne, 1960). The tendency to explore a novel environment (and thus to find food and a nesting place) and the tendency to be fearful of it (and thus to avoid precipitate entry into new areas where predators might be lurking) are both adaptive, and the balance between these two tendencies is undoubtedly arrived at through natural selection. This balance is *also* arrived at through early experience, however, and the amount of fearfulness shown by adult rats introduced into a strange environment can be substantially influenced by early weaning, by handling during early life, or by preweaning experience with different types of mother (Beach and Jaynes, 1954; Levine, 1962; Denenberg and Whimbey, 1963). The "normal" amount of fearfulness shown by a rat in a strange environment is therefore in part a function of the way in which its mother treated it during its infancy. It is impossible to say that the rat has "learned" anything about the characteristics of the environment (including the predators) to which it will later respond; it is equally impossible to deny that the response to the strange environment is in part an effect of experience.

4. The visual cortex of the cat contains cells that fire in response to the movement of a contour (a dividing line between a light and dark area) across an appropriate area of the retina, and which are differentially sensitive to contours in different orientation (Hubel and Wiesel, 1959, 1962). These units, which were first discovered in the cortex of the adult cat, are found in newborn kittens, and are already differentiated in their function at or shortly after birth, even in the absence of patterned visual experience, although the orientation of the receptive fields is not so clear-cut as in adult cats (Hubel and Wiesel, 1963). If the kittens are reared to the age of two or three months with one eye deprived of pattern vision, contours moved across the deprived retina will not activate the cortical cells (Wiesel and Hubel, 1963). Further, this deprivation causes a partial failure of normal cell growth at a lower level in the visual system (lateral geniculate nucleus) (Wiesel and Hubel, 1963a). These observations on the electrophysiological effects of visual deprivation are compatible with the results of

behavioral studies, which show that some mammals reared without patterned visual experience are deficient in the ability to learn visual discrimination habits (Riesen, 1960) and in the ability to transfer visual pattern discriminations learned through one eye to the other eye (Riesen, Kurke, and Mellinger, 1953). The performance of visual discrimination behavior requires not only experience in the sense of visual patterns reaching the eye, but also some experience of the coordination between motor activities and the visual consequences of these activities (Riesen and Aarons, 1959; Held and Hein, 1963). This suggests that the effects of visual experience on behavior are not limited to the development of the electrophysiological mechanisms described by Wiesel and Hubel, but include developmental effects upon wider areas of behavior. Different kinds of experiential effects upon the development of behavior range from the degeneration of an already developed neural mechanism, as shown by Hubel and Wiesel,[6] through more and more specific effects, some of which must be interpreted as conventional learning (Riesen, 1961). Of this range of effects, Lorenz says:

> It is a matter of taste whether or not one choses (sic) to call it learning when an activity is necessary to prevent atrophy and disintegration of a physiological mechanism, but it can be regarded as adaptive modification and it may well involve ontogenetic acquisition of information. . . . These modifications . . . must, therefore, never be forgotten or overlooked in our attempts to analyze this function. On principle, however, they are no obstacle to the solution of our fundamental question concerning the provenience of the information underlying each point of adaptedness in behavior. . . . But there is little danger, with circumspect experimentation . . . and with an experimenter knowing its pitfalls, that any process of true learning, particularly classical conditioning, might pass unnoticed.

Here again, Lorenz indicates his opinion that "true learning, particularly classical conditioning" is the only kind of effect of experience on the development of animal behavior that is of serious interest to a student of such behavior.

5. The structure of the joints in the foot and leg of the domestic chick,

---

[6] I am embarrassed to recall that, in a review of the first edition of the book by Thorpe (1963a), I said of his attitude to this problem: "When experiment shows that some 'instinctive' act does not develop when practice is prevented, Thorpe speaks of the 'regression' of the instinct through non-use, thus preserving its 'innateness' in the face of the most direct possible evidence to the contrary" (Lehrman, 1957). The work of Hubel and Wiesel shows that I was much too abrupt in my reaction to this comment of Thorpe's. As I hope will be clear from the present discussion, at that time Thorpe and I both used the term "innate" without differentiating between different meanings that the term could have, or between the different conceptions that scientists interested in different kinds of problems could have.

and of the articulating surfaces of the foot and leg bones, depends upon the movement of these bones during their period of embryonic development. If the muscles of the embryonic limb are paralyzed either by interrupting the nerve supply or by treatment with pharmacological blocking agents, striking abnormalities develop, including complete lack of movement of the joints. The abnormalities include failure of the joint cavities to develop, distortion of the articulating surfaces, and failure of development of the cartilages and ligaments which surround the joint and bind the bones (Drachman and Sokoloff, 1966). It is thus clear that skeletal muscle contractions are essential for normal formation of the structural prerequisites for walking in these animals. Movements of embryonic muscles, which are to some extent the result of spontaneous activity of the central nervous system (Hamburger and Balaban, 1963; Hamburger, Balaban, Oppenheim, and Wenger, 1965), may also be affected during later stages by external stimuli, including stimuli arising from spontaneous movements (Carmichael, 1946). This is a borderline case in which it is not at all clear to what extent afferent inflow plays a role in determining the amount or direction of the relevant movement, but it is clear that the participation of the nervous system in the development of the normal morphological prerequisites for locomotor behavior are only illuminated by a treatment which produces a striking abnormality.

6. If embryos of the fish *Fundulus heteroclitus* are kept in magnesium chloride solutions, a small percentage of them will develop into hatchlings with only one centrally-located eye (Stockard, 1909), and these fish will apparently be able to see (Rogers, 1957). This sort of treatment, and this result, would seem to me to be the quintessence of the production of a pathology by abnormal rearing, of the kind which Lorenz asserts is of no interest to the student of adaptation to the environment. These fish do, however, suggest a couple of questions, which might provide food for thought: first, are the number and location of the eyes an adaptive character; and second, is the information about the number and location of the eyes located in the genome or in the relationship between the genome and the chemical environment? If we say that the information about the structure and location of the eyes is contained in the genome, rather than in the relationship between genome and environment, then we must be referring to the fact that different kinds of animals reared in the same environment will develop eyes of different structure and location. This again, however, means appealing to the fact that there are *differences* between animals reared in the same environment as proof that there are *differences* in the genomes. I do not see any way out of this apparent paradox except to acknowledge that statements about the genic origin of characters in complex multicellular animals are meaningful primarily when they refer to the

differences between different animals in the same environment as evidence that there are differences in the genomes.

The preceding series of examples of different kinds of modification of development, leading to different kinds of outcomes, is not at all intended to show that anything about the structure of the eyes, or of the joints, or of the visual system, or of the adrenal glands, must be "learned." It is a series of examples of rearing in abnormal environments which lead, through mechanisms of varying degrees of specificity and generality, occurring at different developmental stages, affecting growth processes with different degrees of directness, to developmental outcomes that represent interferences with the normal adaptive characteristics of the animal. All of these outcomes, however, throw light upon the manner in which those adaptive characteristics develop, and on the extent to which environmental influences may play a role in their development. The distinction between "learning" and other forms of "experience" is not sharp, although it is possible to see characteristic differences among different examples; the differences between environmental influences involving effects upon the activity of the nervous system and those not involving such effects are also not sharp, and many intermediates are possible. This is not to say that no classifications are possible, and that no distinctions can be made, among the various kinds of environmental influence. It is to say, however, that the distinction between "morphological ontogeny producing structure" and "trial-and-error behavior" producing learning (Lorenz, 1965), is not a realistic way of surveying the actual range of developmental processes that are involved in the ontogenetic origins of behavior, or of illuminating the varying processes that produce the phenotypic appearances of behavior.

To or From? The Perception of Development

Lorenz asserts that "some American psychologists" avoid the concepts of survival value and phylogenetic adaptation for no other reason than that they regard them as "finalistic." As Lorenz says, "they are not finalistic in the least. If a biologist says that the cat has crooked, pointed claws 'with which to catch mice,' he is not professing a belief in a mystical teleology, but succinctly stating that catching mice is the function whose selection pressure caused the evolution of that particular form of claws."

The concepts of adaptation and of natural selection are of course not teleological or preformationist, and Lorenz is quite wrong in asserting that Schneirla or I ever regarded them so. I believe, however, that a scientist who is interested in the analysis of development must have quite a different attitude toward some problems of causality and abnormality than that which is appropriately characteristic of a biologist who, like Lorenz, is primarily interested in the facts of adaptation and of evolutionary variation. If the observer's perception of a developmental process is wholly dominated by his pre-knowledge of the outcome (i.e., of the adaptive form of the fully developed behavior or structure), then it is very easy for an alteration of the developmental process, caused by a change in the environment, to seem like merely a "deviation" from the "normal" course of development. It would then seem quite natural for such an observer to say that any environmental treatment that led to a maladaptive outcome was of no interest to him, since it merely consisted in the production of a pathology, and not in the illumination of a normal course of development. To such an observer, the development of the normal genome in the normal environment to the normal outcome is merely the *background* for the production of what he is really interested in: the details of structure and behavior by which the organism is intricately adapted to the details of its environment.

The investigator who wishes to understand developmental processes analytically must, however, have quite a different attitude toward the "normal." For him, the normal and the abnormal environment are simply two ways of treating the developing organism, and it is precisely by considering how the development is changed by any particular variation in the environment that he arrives at some understanding of the mechanisms of development. It would be overstating the case (but not by very much) to say that for the student of development, the normal environment and the normal path of development are no more meaningful or significant or perceptually prominent than any other environment or outcome. The reason why it would be overstating the case is that the student of development is, of course, interested in understanding the ontogenetic origin of the actual characteristics of real species, and he cannot (as Lorenz quite rightly points out) apply experimental treatments entirely at random, regardless of whether they will throw light upon the normal developmental outcome. It remains true, however, that insight into normal developmental processes comes from comparing normal and abnormal treatments and outcomes so that they are illuminated by the differences between them. From this point of view it is not at all true to say that the more abnormal a developmental outcome, the less insight it gives into the normal outcome.

One way of putting this point is that the student of adaptation and evolutionary variation often regards development as proceeding, at any stage, *toward* the functional form, since it is the functional form that is adaptive and since natural selection has acted to select genomes which, in a natural environment, produce the functional form; while for the analytic student of ontogeny, development must be seen as proceeding, at any stage, *out of* the immediately preceding stage, and as being produced by processes going on at that time, since it is *his* aim to understand the processes that create developmental change. The student of adaptation and evolution may, therefore, be talking in entirely legitimate and meaningful terms about the problem in which he is interested, while the application of the same terms and concepts to the problems in which the student of ontogeny is interested may accurately be seen as the intrusion of teleology and preformationism. It is in this sense that Lorenz's use of the concepts of adaptation in the discussion of development may be seen as "finalistic" by writers who do not in the least lack understanding of biological concepts.

CONCLUSION: SOME PROBLEMS OF COMMUNICATION AMONG SCIENTISTS

It is clear that at least some of the difficulties in the discussions of the concept of "innateness" arise from the fact that while the various writers believe, and convey to their readers, that they are arguing about matters of fact or interpretation with respect to which one side or the other must be wrong, they are in fact talking about different problems. To some biologists interested primarily in the functions of behavior and in the nature of behavioral adaptations achieved through natural selection, many developmental effects of experience seem trivial and uninteresting, and do not appear to bear upon *their* central question, of the role of natural selection in the establishment of the specific details by which the behavior of specific species is adapted to the necessities of specific environments. To the student of development, however, experiential effects, no matter how diffuse and no matter how remote from the specific details of any particular sensory discrimination or motor act, must be seen as part of the network of causes for the development of any behavior pattern or behavioral capacity to which they are relevant. Further, the rest of the network seems, to such a biologist, definable only in terms of the outcome of experiments on individual development. It ought to be possible to agree on ways of formulating the concepts with which we work so that confusing meanings will be avoided, and so that mutual misunderstandings could be minimized.

One difficulty in the way of such harmony lies in the fact that intelligent people, like unintelligent ones, resent implications that they are illiterate or incompetent, and tend to defend themselves against them. In this essay, I have not succeeded in concealing my resentment at Lorenz's repeated implication that Schneirla and I are or were ignorant of biological concepts; and Lorenz's book repeatedly expresses his sense of outrage at being reminded, as if he did not already know them, of concepts of development which no biologist would deny. It is true, however, that in my earlier contribution to this discussion, I did not deal adequately with the conceptual problems posed by the adaptive character of species-specific behavior, and that what I have said in this essay represents what is, for me, a formulation of problems which I had not considered in detail previously. It is equally true that the formulation of the concept of "innateness" in his recent book is quite different from the concepts found in Lorenz's earlier writings, which always strongly implied developmental fixity as an essential criterion of innateness. There is, therefore, something unbecoming about my implication in this essay that I always knew what I now say, and that I am merely clarifying it for the benefit of people who could not understand it; just as there is something unbecoming in Lorenz's insistence that all criticisms of his point of view are based upon the ignorance of its critics, even while he changes some of his conceptions to meet the criticisms.

I do not think that this kind of problem can be eliminated in scientific communication, and I mention it now for no other reason than to try to make explicit something which is very often inherent in complex discussion. We do not lightly give up ideas which seem central to us, and when they are attacked, we tend to mobilize defenses against the attacks. This means restating the attacked ideas in such a form as to make them seem again convincing to an audience whose confidence in them might have been weakened by the criticism. But when we change the formulation of the ideas in such a situation, we may also be modifying the ideas themselves, in response to criticisms which really may have been leveled against weaknesses in the original formulations. The distinctions are not at all sharp or clear between restating an idea in a clearer form, modifying the idea so as to meet criticism, adding to the formulation things which were previously known but left out as unnecessary, and actually seeing new relationships and new concepts as a consequence of grappling with criticism. I think that no participant in active scholarly discussion can be absolutely sure that everything that he now says represents his long-standing knowledge, without the incorporation of any criticism, and without the inclusion of new ideas made possible only by coping with criticism. If this is a criticism of anybody's writing, it must also apply to some of my own writing in this very essay.

Lorenz believes that what students of behavior are primarily, or even solely, concerned with should be "to elucidate . . . the amazing facts of adaptedness." People interested in analytic studies of the causation and the development of adaptive behavior, however, also have other interests, which are equally legitimate, and for which the concepts derived solely from the study of function and adaptation may not be centrally useful. It is not necessary that all problems fit into the same conceptual framework. It is not required of any theory based on watching intact lower vertebrates that it explain the causes of war, the physiology of the nervous system, and *also* the mode of action of the genes; and it is not an affront to any theory to point out that there are some questions that it cannot answer because it has not asked them.

If it seems that this essay has been oriented primarily to a discussion of Lorenz's recent book, this is only because that book is the most recent attempt at a comprehensive discussion of the so-called "heredity-environment" issue in relation to animal behavior. My central aim has been, not to criticize any other point of view, but to present a positive statement of the relationships between the problems of adaptiveness and of development as seen from the point of view identified with, and best exemplified by, the late T. C. Schneirla.

ACKNOWLEDGMENTS

The preparation of this paper was aided by a Research Career Award from the National Institute of Mental Health, which is gratefully acknowledged.

I am indebted to J. Rosenblatt and D. Dinnerstein for their helpful criticism of this paper during its preparation.

Contribution No. 58 from the Institute of Animal Behavior, Rutgers University.

REFERENCES

Anastasi, A., and J. P. Foley, Jr. 1948. A proposed reorientation in the heredity-environment controversy. *Psychol. Rev.* 55: 239–249.

Bateson, P. P. G. 1966. The characteristics and context of imprinting. *Biol. Rev.* 41: 177–220.

Beach, F. A., and J. Jaynes. 1954. Effects of early experience upon the behavior of animals. *Psychol. Bull.* 51: 239–263.

Berlyne, D. E. 1960. *Conflict, arousal and curiosity.* New York: McGraw-Hill.

Bykov, K. M. 1957. *The cerebral cortex and the internal organs.* New York: Chemical Pub.

Carmichael, L. 1946. The onset and early development of behavior. In L. Carmichael, ed., *Manual of child psychology.* New York: Wiley. Pp. 43–166.

Denenberg, V., and A. E. Whimbey. 1963. Behavior of adult rats is modified by the experiences their mothers had as infants. *Science* 142: 1192–1193.

Drachman, D. B., and L. Sokoloff. 1966. The role of movement in embryonic joint development. *Develop. Biol.* 14: 401–420.

Eibl-Eibesfeldt, I. 1961. The interactions of unlearned behaviour patterns and learning in mammals. In J. F. Delafresnaye, ed., *Brain mechanisms and learning*. Oxford: Blackwell. Pp. 53–73.

Ford, E. B., and J. S. Huxley. 1927. Mendelian genes and rates of development in *Gammarus chevreuxi*. *Brit. J. Exp. Biol.* 5: 112–133.

Fuller, J. L., and W. R. Thompson. 1960. *Behavior genetics*. New York: Wiley.

Gos, E. 1935. Les reflexes conditionnels chez l'embyron d'oiseau. *Bull. Soc. Sci. Liege* 3: 194–199; 4: 246–250.

Gottlieb, G. 1968. Prenatal behavior of birds. *Quart. Rev. Biol.* 43: 148–174.

Haldane, J. B. S. 1946. The interaction of nature and nurture. *Ann. Eugen.* 13: 197–205.

Hamburger, V. 1963. Some aspects of the embryology of behavior. *Quart. Rev. Biol.* 38: 342–365.

Hamburger, V. 1970. Development of embryonic motility. In E. Tobach, L. R. Aronson, and E. Shaw, eds., *Biopsychology of development*. New York: Academic Press, in press.

Hamburger, V., and M. Balaban. 1963. Observations and experiments on spontaneous rhythmical behavior in the chick embryo. *Develop. Biol.* 7: 533–545.

Hamburger, V., M. Balaban, R. Oppenheim, and E. Wenger. 1965. Periodic motility of normal and spinal chick embryos between 8 and 17 days of incubation. *J. Exp. Zool.* 159: 1–14.

Hamburger, V., E. Wenger, and R. Oppenheim. 1966. Motility in the chick embryo in the absence of sensory input. *J. Exp. Zool.* 162: 133–160.

Harlow, H. F., and M. K. Harlow. 1965. The affectional systems. In A. M. Schrier, H. F. Harlow, and F. Stollnetz, eds., *Behavior of nonhuman primates*, vol. 2. New York: Academic Press. Pp. 287–334.

Harnly, M. H. 1941. Flight capacity in relation to phenotypic and genotypic variations in the wings of *Drosophila melanogaster*. *J. Exp. Zool.* 88: 263–273.

Hebb, D. O. 1953. Heredity and environment in animal behaviour. *Brit. J. Anim. Behav.* 1: 43–47.

Held, R., and A. Hein. 1963. Movement-produced stimulation in the development of visually guided behavior. *J. Comp. Physiol. Psychol.* 56: 872–876.

Hinde, R. A. 1962. Some aspects of the imprinting problem. *Symp. Zool. Soc. Lond.* 8: 129–138.

Hinde, R. A. 1966. *Animal behaviour: a synthesis of ethology and comparative psychology*. New York: McGraw-Hill.

Hinde, R. A. 1968. Dichotomies in the study of development. In J. M. Thoday and A. S. Parkes, eds., *Genetic and environmental influences on behaviour*. Eugenics Soc. Symp. no. 4. Edinburgh: Oliver and Boyd. Pp. 3–12.

Hochberg, J. 1963. Nativism and empiricism in perception. In L. Postman, ed., *Psychology in the making*. New York: A. A. Knopf. Pp. 255–330.

Hubel, D. H., and T. N. Wiesel. 1959. Receptive fields of single neurones in the cat's striate cortex. *J. Physiol.* 148: 574–591.

Hubel, D. H., and T. N. Wiesel. 1962. Receptive fields, binocular interaction and functional architecture in the cat's visual cortex. *J. Physiol.* 160: 106–154.

50   DANIEL S. LEHRMAN

Hubel, D. H., and T. N. Wiesel. 1963. Receptive fields of cells in striate cortex of very young, visually inexperienced kittens. *J. Neurophysiol.* 26: 994–1002.

Hunt, E. L. 1949. Establishment of conditioned responses in chick embryos. *J. Comp. Psychol.* 42: 107–117.

Jensen, D. D. 1961. Operationism and the question "Is this behaviour learned or innate?". *Behaviour* 17: 1–8.

Kimble, G. A. 1967. *Foundations of conditioning and learning.* New York: Appleton-Century-Crofts.

Koopman, K. F. 1950. Natural selection for reproductive isolation between *Drosophila pseudoobscura* and *Drosophila persimilis. Evolution* 4: 135–148.

Kuo, Z.-Y. 1932a. Ontogeny of embryonic behavior in Aves. I. The chronology and general nature of the behavior of the chick embryo. *J. Exp. Zool.* 61: 395–430.

Kuo, Z.-Y. 1932b. Ontogeny of embryonic behavior in Aves. II. The mechanical factors in the various stages leading to hatching. *J. Exp. Zool.* 62: 453–489.

Kuo, Z.-Y. 1932c. Ontogeny of embryonic behavior in Aves. III. The structure and environmental factors in embryonic behavior. *J. Comp. Psychol.* 13: 245–272.

Kuo, Z.-Y. 1932d. Ontogeny of embryonic behavior in Aves. IV. The influence of embryonic movements upon the behavior after hatching. *J. Comp. Psychol.* 14: 109–122.

Lehrman, D. S. 1953. A critique of Konrad Lorenz's theory of instinctive behavior. *Quart. Rev. Biol.* 28: 337–363.

Lehrman, D. S. 1957. Nurture, nature and ethology: review of W. H. Thorpe, *Learning and instinct in animals,* 1st ed. *Contemp. Psychol.* 4: 103–104.

Levine, S. 1962. The effects of infantile experience on adult behavior. In A. J. Bachrach, ed., *Experimental foundations of clinical psychology.* New York: Basic Books. Pp. 139–169.

Lorenz, K. Z. 1935. Der Kumpan in der Umwelt des Vogels. *J. Orn.* 83: 137–213, 289–413.

Lorenz, K. Z. 1937. Ueber die Bildung des Instinktbegriffes. *Naturwissenschaften* 25: 289–300, 307–318, 324–331.

Lorenz, K. Z. 1950. The comparative method in studying innate behaviour patterns. *Sympos. Soc. Exp. Biol.* 4: 221–268.

Lorenz, K. Z. 1961. Phylogenetische Anpassung und adaptive Modifikation des Verhaltens. *Z. Tierpsychol.* 18: 139–187.

Lorenz, K. Z. 1965. *Evolution and modification of behavior.* Chicago: Univ. Chicago Press.

Mayr, E., and T. Dobzhansky. 1945. Experiments on sexual isolation in *Drosophila.* IV. Modification of the degree of isolation between *Drosophila pseudoobscura* and *Drosophila persimilis* and of sexual preferences in *Drosophila prosaltans. Proc. Nat. Acad. Sci. U.S.* 31: 75–82.

Mayr, E. 1942. *Systematics and the origin of species.* New York: Columbia Univ. Press.

Miller, N. E. 1969. Psychosomatic effects of specific types of training. *Ann. N.Y. Acad. Sci.* 159: 1025–1040.

Miller, N. E., and L. di Cara. 1967. Instrumental learning of heart rate changes in curarized rats: shaping and specificity to discriminative stimulus. *J. Comp. Physiol. Psychol.* 63: 12–19.

Morgan, T. H., A. H. Sturtevant, H. J. Muller, and C. B. Bridges. 1923. *The mechanism of Mendelian heredity,* 2nd ed. New Haven: Yale Univ. Press.

Prechtl, H. F. R. 1965. Problems of behavioral studies in the newborn infant. *Adv. Stud. Behav.* 1: 75–98.

Razran, G. H. S. 1961. The observable unconscious and the inferable conscious in current Soviet psychophysiology: interoceptive conditioning, semantic conditioning and the orienting reflex. *Psychol. Rev.* 68: 81–147.

Riesen, A. H. 1960. Effects of stimulus deprivation on the development and atrophy of the visual sensory system. *Am. J. Orthopsychiat.* 30: 23–36.

Riesen, A. H. 1961. Stimulation as a requirement for growth and function in behavioral development. In D. W. Fiske and S. R. Maddi, eds., *Functions of varied experience.* Homewood, Ill.: Dorsey Press. Pp. 57–105.

Riesen, A. H., and L. Aarons. 1959. Visual movement and intensity discrimination in cats after early deprivation of pattern vision. *J. Comp. Physiol. Psychol.* 52: 142–149.

Riesen, A. H., M. I. Kurke, and J. C. Mellinger. 1953. Interocular transfer of habits learned monocularly in visually naive and visually experienced cats. *J. Comp. Physiol. Psychol.* 46: 166–172.

Rogers, K. T. 1957. Optokinetic testing of cyclopean and synophthalmic fish hatchlings. *Biol. Bull.* 112: 241–248.

Schneirla, T. C. 1956. The interrelationships of the "innate" and the "acquired" in instinctive behavior. In P.-P. Grassé, ed., *L'Instinct dans le comportement des animaux et de l'homme.* Paris: Masson. Pp. 387–452.

Schneirla, T. C. 1957. The concept of development in comparative psychology. In D. B. Harris, ed., *The concept of development.* Minneapolis: Univ. Minnesota Press. Pp. 78–108.

Schneirla, T. C. 1966. Behavioral development and comparative psychology. *Quart. Rev. Biol.* 41: 283–303.

Schutz, F. 1965. Sexuelle Prägung bei Anatiden. *Z. Tierpsychol.* 22: 50–103.

Sedláček, J. 1962. Temporary connections in chick embryos. *Physiol. Bohemoslov.* 11: 300–306.

Sedláček, J. 1964. Further findings on the conditions of formation of the temporary connection in chick embryos. *Physiol. Bohemoslov.* 13: 411–420.

Spelt, D. K. 1948. The conditioning of the human fetus *in utero. J. Exp. Psychol.* 38: 338–346.

Stockard, C. R. 1909. The development of artificially produced cyclopean fish —"the magnesium embryo." *J. Exp. Zool.* 6: 285–337.

Thorpe, W. H. 1963a. *Learning and instinct in animals,* 2nd ed. Cambridge, Mass.: Harvard Univ. Press.

Thorpe, W. H. 1936b. Ethology and the coding problem in germ cell and brain. *Z. Tierpsychol.* 20: 529–551.

Tinbergen, N. 1963. On aims and methods of ethology. *Z. Tierpsychol.* 20: 410–433.

Whitman, C. O. 1919. The behavior of pigeons. *Publ. Carnegie Inst. Wash.* 257 (3): 1–161.

Wiesel, T. N., and D. H. Hubel. 1963. Single-cell responses in striate cortex of kittens deprived of vision in one eye. *J. Neurophysiol.* 26: 1003–1017.

JAMES W. ATZ
Department of Ichthyology
The American Museum of Natural History
New York

The Application of the Idea of Homology to Behavior

"But the concept of homology, connoting a significant evolutionary relationship between comparable mechanisms among species, has not been validated as yet for behavior and its organization."

T. C. Schneirla (1957)

Ever since they were first used by zoologists, "homology" and "homologous" have served as verbal tools that have been confusing as well as clarifying. They were first applied to the structures of animals whose similarities and differences were believed to have resulted from special creation, but with the recognition of evolution such similarities became evidence for common ancestry, and "homology" and "homologous" acquired new, dynamic meanings (Boyden, 1943, 1947; Haas and Simpson, 1946). Not all present-day biologists have accepted these new extended meanings, but for the great majority the evolutionary definition given by Simpson (1961, p. 78) would serve as a succinct expression of their views: "Homology is resemblance due to inheritance from a common ancestry."

Simpson's definition implies another extension of the original meaning of "homology"; it applies to functions as well as structures. A few biologists, including Boyden and Haas, have opposed this further extension of meaning, but they seem to be in the minority. As if in response to Hubbs' (1944) call for a wider application of the concept of homology, the present tendency is to classify more and more kinds of things as homologous or not, for example, DNA, proteins and parts of proteins, nonprotein hormones, tissues that respond to the same hormone, physiological processes, and behavior.[1]

That the concept of homology can be meaningfully applied to behavior is believed by many of today's zoologists and behaviorists, among them Baerends (1958), Bock (1963), Emerson (1938), Etkin and Livingston (1947), Hubbs (1944), Lorenz (1935, 1950), Mayr (1958), Michener (1953), Remane (1961), Simpson (1958a, b), Tinbergen (1942, 1962), and Wickler (1961a). The principal analyses of behavioral homology have been presented by Baerends and Wickler and they agree that to be homologous, two behaviors exhibited by two phylogenetically related forms must have been present as a single behavior in their common ancestor (Baerends, 1958, p. 403; Wickler, 1961a, p. 313). All other definitions, explicit or implied, are also in substantial agreement. Although several criteria have been proposed to facilitate the recognition of homologous behavior, this operation has proved to be a difficult one. The purpose of this paper is to evaluate these criteria and, in so doing, inquire whether the process of homologizing behavior is indeed a legitimate one.

BIOLOGICAL RELATIONSHIPS INVOLVED WITH HOMOLOGY

Structure and Function

The structures and functions of living matter are, in a real sense, inseparable, and this led Simpson (1946, p. 323) to state that he could "see no objection to speaking of homologous functions, as an expression of opinion as to the origin of a dynamic complex including the functions along with the structures performing them." On the other hand, "functions, considered as abstractions and without consideration for the structures that perform these functions, should not be spoken of as homologous." [2] Whether function alone, unattached to any morphological entity, can be homologized is certainly another matter, but most functions are easily and habitually associated with some sort of structural entity—although it need not be a morphological one, for example, digestion with enzymes, secretion with hormones, and immunity with antibodies—and so even if one speaks only of the process (function) of digestion, a structure of some sort (enzymes, glands, gut) is implicit. This is frequently not true of behavior, however. Not only is the connection between behavior and a specific part of the nervous system usually extremely tenuous, because of lack of knowledge

---

[1] Some of these are now in the process of acquiring special, clear-cut meanings, just as the terms "gene homology" and "chromosome homology" did during the 1920s and 1930s.

[2] Despite these views, Simpson subsequently approved of the application of homology to behavior.

(see p. 61), but behavior is habitually treated as an abstraction without any reference to structure. This disjunction perhaps lies at the heart of the homology-behavior problem.

Ontogeny and Phylogeny

The evolution of an animal may be looked upon as a series of consecutive ontogenies, each of which consists of the development of an individual from egg to adult with the ultimate production of another developing egg, the next generation (Simpson, 1958b, p. 525). Natural selection may operate on any part of this process and, over the generations, produce changes in structure and function. At the same time that natural selection is producing changes in one part of an animal's ontogeny, it may be maintaining another part unchanged. Harland (1936, p. 105) called attention to one of the important consequences of this process of selection with his evidence that in two closely related species, the genetic basis of the same character or organ may become widely different. Harland then pointed out that although organs, such as the eye, are homologous throughout the vertebrates, the genes responsible for them must have changed many times during the course of evolution (see also Dobzhansky, 1959, p. 23). The constancy of the phenotype that species show despite much turnover in their gene pools (Mayr, 1963, p. 280) is another manifestation of the same phenomenon. It was Sir Gavin de Beer who first recognized the significance of this complex relationship between ontogeny and phylogeny to the concept of homology. Since (1) homologous characters need not be controlled by identical genes and characters controlled by identical genes are not necessarily homologous, and (2) homologous structures do not necessarily arise from similar embryonic cells, parts of the egg, or developmental mechanisms, it follows that even though homology of structure "implies affinity between organisms in phylogeny, it does not necessarily imply similarity of genetic factors or of ontogenetic processes in the production of homologous structures" (de Beer, 1958, p. 153). In practice this means that developmental or hereditary similarities may be taken as evidence for, but not proof of, homology, while differences in development or genome cannot be considered as evidence against it. One should always keep in mind that "the essential notion of homology . . . is the continuity of structures in phylogeny and not the resemblance between homologous structures, for structures undoubtedly homologous may be very different anatomically and histologically (e.g., the pineal eye in reptiles and the pineal gland in mammals)" (de Beer, 1938: 65; 1958: 146). Although de Beer considered homology only in relation to structure, a homologous function must exhibit phylogenetic continuity just as a homologous structure does.

DIFFICULTIES IN ACQUIRING AND HANDLING BEHAVIOR DATA

Even at the simplest operational level, the problems associated with recording and analyzing behavior are many times more serious than those involving structure. Exactly what part of the animal's behavioral repertory, including its many variations, should be recorded and measured, and how should this be done? Is repeated access to the same events feasible for possible reinterpretation or deeper analysis? How can the events be described, in reasonable yet pertinent detail, to persons who have never witnessed any like them? The student of structure faces all of these questions, too, but for him satisfactory answers are much more easily come by than for the student of behavior, because under the usual observational or experimental conditions, each behavioral act is essentially evanescent. Tinbergen's (1942, p. 48) statement that "movements of limbs, etc. can be measured just as well as the form of a structure" seems like whistling in the dark, and Cullen's (1959, p. 136) declaration that "behaviour is no more than four-dimensional morphology" is a gross oversimplification.

Perhaps the most difficult task of all is to extract from the continuum of the animal's total behavior, parts to measure, count, and compare. Although these abstractions vitally affect the entire course of investigation, the methods by which they are accomplished must be complex, for very few investigators have ever seriously attempted to explain how they have proceeded. Ethologists, e.g., Tinbergen (1942, p. 47; 1951, p. 189) and Lorenz (1950, pp. 238, 261), have generally equated the process with the activities of the comparative anatomist. Their homologizing has almost entirely been confined to the relatively stereotyped motor behavior they have called "fixed action patterns." Dilger (1964, p. 158) attempted to be objective by dividing the animal's behavior into parts that could not be analyzed into finer components. Dilger quoted Russell, Mead, and Hayes (1954, p. 163), and defined each of these "acts" as "a set of observable activities in different effectors (e.g., muscle tensions), *regularly observed in combination* (thus not analysable into separately occurring components . . .), hence recognisably different from other such acts observed in the same species." Indivisibility cannot be an all-sufficient criterion, however, and underlying all attempts to homologize behavior there exists a question of the significance of what is being compared.

CRITERIA FOR HOMOLOGY IN BEHAVIOR

Criteria Applicable to Structure as Well as Behavior

REMANE'S PRINICIPAL CRITERIA

On the basis of a detailed review of the literature and original work of his own, Remane (1952) formalized a set of criteria by which the homology of plant and animal structures could be determined. Baerends (1958), Wickler (1961a), and Albrecht (1966), attempted to apply these criteria to behavior.

According to Remane (1952, 1961), there are three principal criteria for determining homology:[3]

1. Criterion of position: Homology results from the same position in comparable systems of structures (that is, structures that occupy the same position within comparable anatomical units are likely to be homologous).

2. Criterion of special quality: Similar structures can be homologized, without regard to position, if they agree in several unusual characteristics. The greater the complexity and degree of correspondence, the more likely the homology.

3. Criterion of constancy or continuity: Even dissimilar and differently located structures can be considered homologous if intermediate, connecting forms can be shown to exist, so that Criterion 1 or 2 can be met by comparing the adjacent forms. The connecting forms may be ontogenetic stages or members of taxonomic groups.

Remane (1961, p. 450) believed that neither his first nor third criterion could be applied to behavior with any frequency, but Wickler (1961a, p. 313) disagreed with this view. Although Wickler was able to present a number of apparently excellent examples of species whose behavior bridges the gap between two other behaviors (Criterion 3), he did not cite any cases of behavioral homology that had been determined by means of Criterion 1. Baerends (1958, p. 408), on the other hand, provided five examples of behavior patterns he considered homologous because they occur at the same place in similar sequences of activity, but the four that concern fishes are not convincing. In fact, behavioral patterns of fishes seem to be characterized by the variability with which the different elements precede or follow one another, even though there is overall progression from one

---

[3] Translated from Remane (1961, p. 449), which is identical with Remane (1952, p. 63) except for added subheadings. The parenthetical expression is mine.

major activity toward another (e.g., see Aronson, 1949; Barlow, 1961, 1962b; Liley, 1966; Morris, 1958; Tavolga, 1954).[4]

Remane's Criterion 2 is the most generally applicable of all and, in fact, could be considered to include the other two as special cases. Unless behaviors are not to be homologized at all, there is no question of the applicability of Criterion 2, and this has been recognized by Baerends (1958), Tinbergen (1959, 1962), Wickler (1961a), and Albrecht (1966). According to Albrecht (1966), Criterion 1 is seldom applicable to behavior, but Criterion 3 often is, more often in fact than it is to anatomy.

REMANE'S SUPPLEMENTAL CRITERIA

In addition to his Principal Criteria, Remane (1952, p. 64; 1961, p. 449) proposed supplemental ones, which may be summarized as follows: the probability of the homology of even simple structures increases with the frequency of their distribution among related species and decreases with the frequency of their occurrence in unrelated ones. These criteria clearly bring out the tautology inherent in the evolutionary definition of homology, namely that evolutionary relationships are established by means of homology while homologies are explained on the basis of phylogeny. This has been decried by Boyden (1943, 1947) and Woodger (1945), defended by Haas and Simpson (1946), Simpson (1961), and Inglis (1966), and recognized as something to be kept in mind and avoided, when circular reasoning is directly involved, by Baerends (1958, p. 404), Hinde and Tinbergen (1958, p. 252), and Grimstone (1959). In practice, the student of behavior works with animals already classified, and it is natural that he should use the distribution of the behavior patterns he is studying as evidence for or against his proposed behavioral homologies, as Tinbergen (1959, p. 55; 1962, p. 2) and Wickler (1961a, p. 322) have advocated.

INHERITANCE IN HYBRIDS

The appearance of parental behavior patterns in interspecific hybrids sometimes provides evidence that is indicative of their homology. As Hinde (1956, p. 210) has pointed out, if a complex behavior, that is present in both parental species, is exhibited unchanged by the F_1, the behaviors of the two parents most likely depend on very similar hereditary mechanisms and are therefore likely to be homologous. If the parental species exhibit different behavior patterns, however, the F_1 may show an intermediate form of behavior, and Hinde presented good evidence that intermediate behavior was characteristic of hybrids between fairly closely related species of birds. On the basis of much more limited data, he also believed that "if the parent

---

[4] Wickler (1961a, p. 312) also questioned whether Remane's first criterion could legitimately be applied to the examples given by Baerends and he concluded that the homology of these patterns could be judged better under Criteria 2 and 3.

species differ very markedly in a particular aspect of behaviour, either the hybrid resembles one or other parent, or there is a total breakdown and the behaviour does not appear at all." Tinbergen (1962, p. 3), referring to Hinde's work, stated that intermediate behavior of interspecific hybrids is an argument for the homology of that behavior—a conclusion that does not follow from Hinde's views and is not supported by what is known about polygenic inheritance. Detailed quantitative studies on the inheritance of behavior in a piscine and in an avian cross (Clark, Aronson, and Gordon, 1954; Sharpe and Johnsgard, 1966) have shown that some parts of the behavior of the F_1 were intermediate and others resembled one parent or the other. Moreover, there is some evidence that the more closely the parents are related, the *greater* is the frequency with which F_1 hybrids resemble either parent, rather than being intermediate (Atz, 1962, p. 176).

ONTOGENY

Both Baerends (1958, p. 407) and Wickler (1961a, pp. 319, 329) believed that ontogeny can aid in the determination of behavioral homologies. Wickler (1961a, b), well aware that Haeckel's Biogenetic Law has been discredited for behavior as well as anatomy (Schneirla, 1957, p. 84), presented a documented list of examples in which ontogeny seems to recapitulate phylogeny and another in which it does not. Wickler emphasized the great care with which this criterion obviously must be applied.

Since de Beer (1938, 1958) has convincingly shown that the ontogeny of homologous structures need not be the same, it becomes more and more necessary to take into consideration the different periods of each animal's life history. We divide the continuous process of transition from egg to adult into stages that are abstractions with essentially arbitrary limits, although, in the development of structure, there are times (stages) at which divisions may more readily be made (fertilization, hatching, appearance of secondary sex characters, etc.). Meaningful separations are especially hard to make with behavior, however, because significant, interdependent developmental changes in behavior may occur continuously through most of the life cycle. The problem is most acute in the mammals with their exceedingly plastic behavioral organization including great ability to learn (Schneirla, 1957). In effect, this imposes another dimension of difficulty on the abstracting of meaningful parts from the continuum of the animal's total behavior (see p. 56).

Tinbergen (1951; p. 189) maintained that only the "innate basis" of behavior could be homologized, but Wickler (1961a, p. 313) believed that whether or not two behavior patterns are homologous is independent of the extent to which either of them is learned (acquired); in fact, he implied that two behavior patterns, one of them innate and the other learned (in the usual, uncritical sense of these terms) could theoretically be homologous. Although the capacity to react to the environment is genetically determined, whether the final outcome is a stereotyped behavior that makes its appearance in nearly definitive form or one that gradually develops in variable and obviously changing interactions with the environment (Schneirla, 1957), Wickler is more correct than Tinbergen in emphasizing the behavior patterns themselves and not their mode of development.

Ontogeny may provide indications of behavioral homologies, even though it cannot serve as a decisive determinant. Unfortunately, however, very few ethologists or other animal behaviorists have more than casually incorporated ontogenetic data into their comparative analyses.

Criteria Applicable Only to Behavior

RELATIONSHIP TO HOMOLOGOUS STRUCTURES

Peripheral structures. "Movements employing homologous structures in a similar way may be considered homologous" (Tinbergen, 1962, p. 3), but conversely, nonhomologous behavior patterns may in some cases be based on homologous structures (Lehrman, 1953, p. 351). The fact that homologous structures are involved in behaviors may be an indication that the behaviors are homologous, but it cannot be considered proof. As Simpson (1946, p. 324) has pointed out, in more general terms, "the functions of homologous structures may not be homologous." That organs can become adapted to entirely new functions is almost an evolutionary commonplace. For example, the anal fin of the male guppy and other fishes of the Family Poeciliidae is modified into a gonopodium with which to fertilize the female. Although anal fin and gonopodium are homologous, the motions of the latter in courtship and copulation can hardly be homologized with the relatively limited movements of the unmodified fin—the more so because the functions of some of the muscles have been changed (Rosen and Gordon, 1953, p. 39).

On the other hand, it is manifestly impossible for nonhomologous structures to have homologous functions. Concerning this point, a terminological problem arises—one that also may be illustrated by the gonopodium. The anal fin of male four-eyed fishes (Family Anablepidae) is modified to form a gonopodium, but in a distinctly different way from the poeciliid structure. Consequently, the anal fins of poeciliids and anablepids are homologous as fins but not as gonopodia. We might therefore assume that the movements of the two kinds of gonopodia cannot be homologized since no continuity can be expected in the functioning of structures that have arisen independently in the course of evolution, as these two presumably have (Rosen and Bailey, 1963, p. 24). It is likely that the poeciliids arose from some egg-laying cyprinodontoid fish (Rosen and Bailey, 1963, p. 24), and perhaps the Anablepidae arose from the same group. Many male cyprino-dontoids use their more or less unmodified anal fin during spawning to establish contact with the female and guide the milt, presumably in order to insure fertilization of the eggs as they are laid. In the evolution of the anal fin toward the gonopodium, change in behavior may well have *preceded* change in structure.[5] If this did occur, some of the common cyprinodontoid behavior probably evolved along with each of the two kinds of gonopodia as they gradually took form. This would seem to lead to a situation in which two nonhomologous structures (gonopodia) are associated with homologous behavior patterns. It has already been pointed out, however, that the two gonopodia are obviously homologous as fins, and so it can still be maintained that completely nonhomologous structures cannot have homologous functions.

Strict application of de Beer's essential notion of continuity (p. 55) makes possible reliable homologizing, even in problematic cases such as these. Nevertheless, the conceptual difficulties inherent in dealing with an immaterial entity like behavior are apparent. It is also interesting and instructive to note that the reproductive behavior associated with the gonopodium and that associated with the more or less unmodified anal fin cannot in practice be homologized, as was noted above, even though theoret-ical considerations indicate the possibility of doing so. This is a good ex-ample of how much easier it is to establish homologies between structures (the unquestionable homology between anal fins and gonopodia) even in the absence of any fossil record.

Since the behavior we are dealing with consists mostly of motor activities, the muscles involved in them might logically be considered indicative of their homology. Baerends (1958, p. 409) concluded, "it is most essential for homology that the patterns of muscle contractions should be largely identical. If these show important differences, which cannot be explained as secondary adaptations to specialised functions, homology is less likely, even when the topography and function of the elements are identical." Very few behavioral studies, however, have included analyses of muscle function or homology. Moreover, there is as yet no reason to believe that muscles exhibit any more fixed evolutionary relationship to associated be-havior than do other structures.

Nervous system.   Peripheral structures are not the only ones with which behavior is involved; the nervous system plays an essential role in every pattern of behavior. Here one soon passes into terra incognita. In fact, Bullock (1965, p. 451) not long ago declared that, "The gulf between

---

[5] See Mayr (1958, p. 355) for a discussion of the problem of whether structure or function comes first in evolution.

our present level of physiological understanding and the explanation of behavior as we see it in higher forms . . . is indeed the widest gap between disciplines in science." For example, although it is possible to establish homologous nuclear regions in the brains of animals as diverse as fishes and mammals (Schnitzlein, 1964), this provides no information regarding the homologies of the behavior of these animals. So little is known about the neural basis of memory, learning, perception, and motor patterns of behavior in all animals, except some of the simple invertebrates, that one could doubt the propriety of Sperry's (1958) basic assumption that structural correlates for all behavioral events will eventually be found. Nor can Pribram (1958, p. 142) be followed safely when he reasons that because behaviors are served by homologous neural structures they may be assumed to have homologous elements until proved otherwise. Even if a great deal more were known about what activities of the central nervous system could properly be associated with a given behavior, the relationship of these two correlates in evolutionary time would still not necessarily be understood. In the light of the fundamental uncertainties in historically relating ontogenetic and phylogenetic processes, which de Beer (1958) has so clearly pointed out (p. 55) how can one ever hope confidently to assume that any particular behavior has continuously been associated with a particular condition (whatever it may be) of the central nervous system? Neural-behavioral relationships may in the not too distant future prove helpful in establishing homologies in animals low on the psychological scale, but as far as higher forms are concerned, not enough is at present known to make even speculation worthwhile.

RELATIONSHIP TO MOTIVATION OR DRIVE

Baerends (1958, p. 406) and Tinbergen (1959, p. 29) claimed that "motivational factors" or motivation could be used as an indication of homology, and Wickler (1961a, pp. 312, 332) agreed that these could be one of the special qualities included under Remane's second criterion (p. 58). Baerends, however, recognized that the same behavior may occur under the influence of different motivations, and that under the influence of the same motivation, different behaviors may become very similar in appearance. He therefore concluded that two behaviors are not necessarily nonhomologous if they are associated with different motivations, nor can identity of motivation ever be more than a general indication of homology. On the other hand, Hinde and Tinbergen (1958, p. 265) maintained that the study of the function and causation of behavior enables the comparative behaviorist to distinguish between homology and convergence, and this despite the fact that Tinbergen (1959, p. 64) cited cases among the gulls in which the motivation underlying a certain posture may have often changed during the course of evolution. Wickler (1961a) has provided examples of behaviors that are associated with different motivations although supposed to be homologous.

The concept of motivation or drive is a complex one, and the term has frequently been used in vague and unspecified ways (Hinde, 1966, p. 140ff). For comparative purposes, some sort of identification of different drives becomes necessary. Investigators have therefore been forced to make gratuitous assumptions when comparing two behavior patterns and the drives believed to be associated with them. The kind of difficulties inherent in this procedure are well illustrated by the example chosen by Baerends himself (1958, p. 406): "For instance, several cichlid species have a similar defensive frontal display which appears during boundary fights under the combined influences of aggressive and fleeing motivations. We also find in the same species, when they are leading young, another more or less similar activity, 'calling young', with an identical motivation."

SPECIAL PROBLEMS IN HOMOLOGIZING BEHAVIOR

Lack of Paleontological Evidence

With a very few peculiar exceptions, there can be no fossil record of behavior. The importance of this has been recognized by many workers, including Albrecht (1966, p. 271), Baerends (1958, p. 404), Blest (1961, p. 103), Boyden (1947, p. 667), Hinde and Tinbergen (1958, p. 253), Mayr (1958, p. 343), Simpson (1958a, p. 10), Tinbergen (1951, p. 186; 1963, p. 427), and Wickler (1961a, p. 317). Most informative is Simpson's discussion of the consequences of the almost total lack of historical documentation for evolutionary studies on behavior. He has pointed out some of the more serious problems that confront comparative anatomists when they try to reconstruct lineages entirely on the basis of nonhistorical data. Without doubt, their methods are equally pertinent for the historical study of behavior, but the many pitfalls in them "are probably . . . even more serious for behavioral studies than they have proved to be for morphological studies" (Simpson, 1958a, p. 12).

Convergence

Convergence is the evolutionary development of similar characteristics by unrelated groups of animals in response to similar environmental needs— one might almost say, in spite of not having any pertinent hereditary factors in common. Many biologists have recognized that behavior is often convergent, including Blest (1961, p. 119), Cullen (1959, p. 134), Hinde and Tinbergen (1958, p. 253), Mayr (1958, p. 351), Remane (1961, p.

450), Simpson (1958b, p. 533; 1961, p. 76), Tinbergen (1959, p. 60), and Wickler (1961a, p. 323). According to Lorenz (1935, p. 382; 1937, p. 309, display behavior is very unlikely to show convergence, but Hinde (1956, p. 210), Marler (1957, p. 28), and Tinbergen (1959, p. 60) have shown that it does so in an appreciable number of instances. Behavior is subject to particularly strong selection (Simpson, 1964, p. 1536), and recent field work has indicated the all-pervasiveness of natural selection as a force bringing about behavioral adaptations of animals to their environment (Tinbergen, 1965). Strong selection pressures, especially those associated with rigorous environments, tend to result in convergences.

Another factor favoring convergence is a limited number of possible responses by the animal to natural selection. Pantin (1951, p. 144) nicely epitomized this: "The fact that man and *Paramoecium* both have acetylcholine does not necessarily argue for a common ancestry in the same way that the pentadactyl limb does for a man and for a mole." At the molecular, biochemical, cytological, and early embryonic levels, similarities are much more likely to be the result of convergence than they are in morphology where many, many more configurations have been realized (Etkin and Livingston, 1947; Grimstone, 1959; Mayr, 1965; Pantin, 1951; Simpson, 1964). The spectrum of possible ways of accomplishing things seems smaller for behavior than for structure, except in man and perhaps some of the other mammals. Undoubtedly, the difficulty in detecting, recording, and analyzing the details and variations of behavior contributes to this impression. Nevertheless, the behavioral repertory of the invertebrates and lower vertebrates is certainly exceeded in variety by their structures. As better techniques are developed, more and more behaviors once thought to be identical will be revealed as different, but there will always be behaviors whose similarity must be attributed to convergence.

The teleost fishes provide excellent examples of how problematic the homologizing of behavior patterns may be when complicated by convergence of indeterminable extent. One of the characteristic postures of the live-bearing members of the Family Poeciliidae is the S-curve in which the fish's body assumes a double lateral flexure so that the head and tail point in approximately opposite directions (Clark, Aronson, and Gordon, 1954; Liley, 1966; Rosen and Tucker, 1961). Males perform the S-curve in male-to-male displays and in courtship and copulation, and the way it is used and the extent to which it is used varies from species to species. The females of some species also exhibit the S-curve. In a good many members of the egg-laying Families Cyprinodontidae and Oryziatidae, the S-curve is also used in aggressive displays, courtship, and mating (Barlow, 1961; Newman, 1907; Ono and Uematsu, 1957; Tait, 1965). These data tend to indicate that the S-curve is homologous throughout the Suborder Cyprinodontoidei, to which these three families belong, but its irregular distribution within the groups and its differences in form and apparent function, both within the same species and among different ones, make the actual tracing of homology highly conjectural. Moreover, the S-curve is exhibited by other, completely unrelated fishes—e.g., *Badis badis* (Barlow, 1962a) and *Serranus subligarius* (Clark, 1959)—as well as by fishes in the egg (Hamburger, 1963, p. 352), and this indicates strong possibilities for independent development, that is, convergence or parallelism. Liley (1966, p. 99), who studied the behavior of four poeciliid species, recognized that "many of the similarities in courtship movements may be due to similarities in the causal situation in which they occur, rather than to their having a common origin." Among such likely-to-be-convergent behaviors he included sigmoid and other lateral displays. Myers (1965) called attention to another type of lateral display, which he called body-wag and found to be practiced by fishes from at least thirteen widely divergent teleost families. Students of fish behavior have also had trouble distinguishing between the various behavior patterns performed by a single species (see, e.g., Breder and Rosen, 1966, p. 449), and their observations provide illustrations of Schneirla's (1955, p. 194) dictum that "what seems to be the same motor act can occur in very different contexts for the animal." The fantastically diverse structures of the teleosts do not appear to be accompanied by a similar diversity of behavior. The resemblances that the behavior patterns of fishes frequently show are more likely to make the establishment of homologies difficult than easy.

Psychological Levels

Hubbs (1944, p. 299) observed that "homologies in phylum, class, order and family characteristics are commonly much more obvious and secure than are the homologies in generic, specific and subspecific characters." This is not true as far as behavioral homologies are concerned: a monkey's arm, whale's fin, bird's wing, and salamander's forelimb are all indubitably homologous, but homologizing their movements would be risky in the extreme.

Lehrman (1953, p. 351), following concepts developed by T. C. Schneirla, showed that the differences in the psychological levels of far-distantly related animals make valid homologizing impossible. For instance, the whole neural-behavioral organization of the bird is so unlike that of the mammal that any similarities in behavior between them must be attributed to convergence. Baerends (1958, p. 411) and Tinbergen (1963, p. 428) stated that to be reliable, homologous behaviors must be limited to closely allied forms. Lorenz (1939, p. 72; 1950, p. 241) knew of only a single attempt to extend behavioral homology beyond the taxon of the family, and Wickler (1961a) did not record any additional ones in his extensive review of the homology of behavior. Lorenz had concurred with Oskar Heinroth in his attribution of decisive phyletic importance to the way in which vertebrates from the reptile to the mammal scratch the head with a hind foot. The matter is evidently not a simple one, however, as the investigations of Wickler (1961b, p. 340) have shown: "Because scratching over the wing is in its coordination identical with the scratching of reptiles and mammals, and also because of its apparent awkwardness, Heinroth and with him Lorenz assumed scratching over the wing to be primitive. A thorough study of the ontogeny of scratching revealed the unexpected fact that, while there are many birds known in whose ontogeny scratching under precedes scratching over, no case of the opposite sequence has come to our knowledge. On the basis of present information the question which of the two motor patterns is the primitive one cannot be decided."

The most convincing examples of homologous behavior comprise peculiar patterns of motor activities that are found exclusively in a single group of related species, and in all the members of the group. Not to conclude that such behavior was also practiced by the common ancestor of the group would violate the principle of parsimony. Nevertheless, as one goes beyond such a patent, ideal situation and begins to deal with less closely related species and with behavior patterns that are not unique and whose distribution is less coherent, the certainty with which the behavior can be homologized rapidly diminishes.

Products of Animal Behavior

The concept of homology has been applied to certain products of animal behavior such as the nests of termites by Emerson (1938, 1952), the cases of caddisworms by Ross (1964), the nests of bees by Michener (1964), and the songs of birds by Goodwin (1952). These are undeniably extensions of the animal's behavior and often remarkably invariable in structure, but their use in homologizing presents all of the problems inherent in behavioral homologies. These structures may sometimes reasonably be inferred to have had continuity in phylogenetic time and so, within the limits of the inferences, they could rigorously qualify as homologous. To suppose, however, that the behavior that produced them is therefore homologous involves unwarranted assumptions, in accordance with de Beer's essential notion (p. 55). Schneirla (1952) called attention to this fundamental weakness when he declared that "behavior patterns cannot be represented validly in terms of their structural results."

BEHAVIOR AND SYSTEMATICS

As Simpson (1958b, p. 524) has pointed out, "similarity of behaviors tends, like structural similarity, to be proportional to phylogenetic affinity," and so the comparative study of behavior has provided the zoological systematist with an entire group of characters to help in his classification. The number of instances in which behavior has provided valuable clues to systematic relationships has continued to grow (Amadon, 1959; Cullen, 1959; Marler, 1957; Mayr, 1958; Petrunkevitch, 1926; Simpson, 1958b), but it should be made clear that the establishment of detailed homologies was seldom, if ever, necessary to accomplish this. Such instances show that Michener (1953, p. 113) was wrong when he asserted that without homologies, behavior becomes useless in helping to understand phylogenetic relationships.[6] Morphological systematic characters usually involve demonstrable homologies, in fact they typically depend on them. In contrast, functional, and especially behavioral, characters usually do not involve demonstrable homologies, but depend instead on resemblances that may be detailed and specific but nevertheless cannot be traced, except in a general way, to a common ancestor. Hence the supremacy of morphology in systematics.

BEHAVIOR, STRUCTURE, AND HOMOLOGY

There is a prophetic aspect to Whitman's (1899, p. 329) statement that "instincts, like corporeal structures, may be said to have a phylogeny," for not until students of animal behavior began to treat behavior like structure did studies on the evolution of specific behaviors come to the fore.[7] The

---

[6] A similar error was made by Russell (1952) who implied that the search for behavioral homologies is the most important aspect of the comparative study of behavior. This apparent belief in the widespread existence of homologous behavior subsequently led Russell (1964) to declare that success in extrapolating from the behavior of animals to that of man in pharmacological tests depends, theoretically at least, on establishing the homology of the behavior under study.

[7] Even more prophetic is Lorenz's misquotation of another statement of Whitman: "Instinkte und Organe müssen von gemeingsamen Gesichtspunkt phyletischer Amstammung erforscht werden" (Lorenz, 1939, p. 71). This is, declared Lorenz (1950, p. 238), "the sentence that marks the birth of comparative ethology," and he then translated back into English his own German translation of it: "Instincts and organs are to be studied from the common viewpoint of phyletic descent." What Whitman (1899, p. 328) had actually written was: "Instinct and structure are to be studied from the common standpoint of phyletic descent, and that not the less because we may seldom, if ever, be able to trace the whole development of an instinct." For the sake of historical accuracy, it ought to be pointed out that although Whitman may have provided some foundation blocks for the ethological

early ethologists abstracted their observations to such an extent that they were able to treat species-typical behavior patterns as if they were morphological entities, and a strong tinge of preformationism colored their views on the relationship between patterns of behavior and the nervous system (Lehrman, 1953). This explains the ease with which they applied the concept of homology to behavior. Through the years, their point of view has changed very little. Tinbergen (1951, p. 191), directly following Lorenz (1937, p. 310), declared that the fixed patterns of behavior played "the same part as 'organs' in comparative anatomy," and twelve years later he emphasized "the striking justification of treating behavior patterns as 'organs' " (Tinbergen, 1963, p. 428). More recently, Albrecht (1966, p. 298) has written about the structure of behavior.[8] The ethologists are still the foremost students and advocates of the concept of behavioral homologies.

The extent to which behavior can be homologized is directly correlated with the degree to which it can be conceived or abstracted in morphological terms. Nevertheless, no morphological correlates have ever been found, either in the nervous system or peripheral structures, by which the homology of behavior can be established. On the other hand, to deny that homologous behavior exists would seem to deny that behavior is a characteristic of animals that is subject to evolutionary change. What is the basis of this paradoxical situation and can it be resolved? Its resolution rests on the recognition that the idea of homology is essentially a morphological concept. It is operationally impossible to conceive of two homologous entities without some kind of structure being involved. Therefore, when one thinks of homologous behavior, one is perforce thinking of behavior in morphological terms (fixed action patterns and the like) or as closely associated with some structure (neural or otherwise). The difficulties in homologizing behavior that have arisen are almost all the result of the

---

edifice, he did not conceive of the "instincts" with which he dealt as equivalent to morphological entities. The quotation from him that I have used above in the text has also been misunderstood, and I have used it for that very reason. It forms part of Whitman's (1899, p. 329) discussion of the Biogenetic Law and reads in full: "Although instincts, like corporeal structures, may be said to have a phylogeny, their manifestation depends upon differentiated organs." It should also be mentioned, for the sake of historical perspective, that Tinbergen (1949, p. 47; 1951, p. 191) twice used the same quotation from Whitman that Lorenz did, but quoted it correctly although not in full.

[8] "Die Homologisierung beruht auf einem Vergleich von Strukturen verschiedener Organismen und bildet die Basis jede, natürlichen Systems. Strukturen des Verhaltens in diese Betrachtungsweise einzubeziehen, kann zu einer Verfeinerung des Systems wesentlich beinatürlichen tragen."

Homologizing depends on a comparison of the structures of different organisms and forms the basis of every natural system. Structures of behavior that are implicated in this method of study can contribute substantially to a refinement of the system.

lack of morphological correlates in behavior. Until the time that behavior, like more and more physiological functions, can critically be associated with structure, the application of the idea of homology to behavior is operationally unsound and fraught with danger, since the history of the study of animal behavior shows that to think of behavior *as* structure has led to the most pernicious kind of oversimplification.

SUMMARY

1. Many zoologists and animal behaviorists believe that the concept of homology can validly be applied to behavior, and they define homologous behaviors as those that can be traced back to the behavior of a common ancestor.

2. Behavior is much more difficult to treat comparatively than is structure because of its variability, continuity, extended ontogeny, and the evanescence of each behavioral act.

3. The view of de Beer that the essential notion of homology is continuity in phylogeny must hold for behavior as well as structure.

4. The morphological criterion of identity of position in comparable systems of structures can seldom be translated into behavioral terms, but the criterion of special characteristics, with or without connecting links in related species, can be applied in homologizing behavior.

5. Although the following criteria may be indicative for the establishment of behavioral homologies, they can be neither essential nor sufficient, in accordance with de Beer's dictum:

   a) Inheritance in F_1 hybrids of behavior patterns that are the same in both parental forms.
   b) Similarity of ontogeny.
   c) Close association with homologous structures.
   d) Association with identical patterns of muscular contractions.

6. So little is known about the neural correlates of behavior that they have not yet proved of help in establishing homologies.

7. Convergence in behavior is prevalent, probably because of intense selection pressures and limited possible responses by the animal.

8. The only reliable homologizing of behavior has been confined to closely related forms. The more distantly related the animals, the less likely that judgments regarding homologies between their behaviors are valid.

9. The essentially morphological concept of homology cannot at present be applied to behavior in any meaningful (nontrite) way because of its lack of structural correlates.

ACKNOWLEDGMENTS

This paper in part owes its existence to the custom of a curatorial coffee time, each day after lunch, at The American Museum. Here I aired my views, and the responses I received and the discussions that ensued encouraged me to proceed. Above all, I am in the debt of Dr. William N. Tavolga whose thoughts helped clarify my own and who also gave vital strength to my convictions.

REFERENCES

Albrecht, H. 1966. Zur Stammesgeschichte einiger Bewegungsweisen bei Fischen, untersucht am Verhalten von *Haplochromis* (Pisces, Cichlidae). *Z. Tierpsychol.* 23: 270–301.

Amadon, D. 1959. Behavior and classification: some reflections. *Vjschr. Naturf. Ges. Zürich* 104 (Festschrift Steiner): 73–78.

Aronson, L. R. 1949. An analysis of reproductive behavior in the mouthbreeding cichlid fish, *Tilapia macrocephala* (Bleeker). *Zoologica* (N.Y.) 34: 133–158.

Atz, J. W. 1962. Effects of hybridization on pigmentation in fishes of the genus *Xiphophorus. Zoologica* (N.Y.) 47: 153–181.

Baerends, G. P. 1958. Comparative methods and the concept of homology in the study of behaviour. *Arch. Néerl. Zool.* 13 (Suppl.): 401–417.

Barlow, G. W. 1961. Social behavior of the desert pupfish, Cyprinodon macularius, in the field and in the aquarium. *Am. Midland Naturalist* 65: 339–359.

Barlow, G. W. 1962a. Ethology of the Asian teleost, *Badis badis*. III. Aggressive behavior. *Z. Tierpsychol.* 19: 29–55.

Barlow, G. W. 1962b. Ethology of the Asian teleost, *Badis badis*. IV. Sexual behavior. *Copeia,* 1962, pp. 346–360.

Blest, A. D. 1961. The concept of "ritualisation." In W. H. Thorpe and O. L. Zangwill, eds., *Current problems in animal behaviour.* New York: Cambridge Univ. Press. Pp. 102–124.

Bock, W. J. 1963. Evolution and phylogeny in morphologically uniform groups. *Am. Naturalist* 97: 265–285.

Boyden, A. 1943. Homology and analogy: a century after the definitions of "homologue" and "analogue" of Richard Owen. *Quart. Rev. Biol.* 18: 228–241.

Boyden, A. 1947. Homology and analogy, a critical review of the meanings and implications of these concepts in biology. *Am. Midland Naturalist* 37: 648–669.

Breder, C. M., Jr., and D. E. Rosen. 1966. *Modes of reproduction in fishes.* Garden City: Natural History Press.

Bullock, T. H. 1965. Physiological bases of behavior. In J. A. Moore, ed., *Ideas in modern biology.* Garden City: Natural History Press. Pp. 449–482.

Clark, E. 1959. Functional hermaphroditism and self-fertilization in a serranid fish. *Science* 129: 215–216.

Clark, E., L. R. Aronson, and M. Gordon. 1954. Mating behavior patterns in two sympatric species of xiphophorin fishes: their inheritance and significance in sexual isolation. *Bull. Am. Mus. Nat. Hist.* 103: 135–225.

Cullen, J. M. 1959. Behaviour as a help in taxonomy. *Pub Syst. Assoc.* (3): 131–140.

de Beer, G. R. 1938. Embryology and evolution. In G. R. de Beer, ed., *Evolution, Essays on aspects of evolutionary biology.* London: Oxford Univ. Press. Pp. 57–78.

de Beer, G. R. 1958. *Embryos and ancestors,* 3rd ed. London: Oxford Univ. Press.

Dilger, W. C. 1964. The interaction between genetic and experiential influences in the development of species-typical behavior. *Am. Zool.* 4: 155–160.

Dobzhansky, T. 1959. Evolution of genes and genes in evolution. *Cold Spr. Harb. Symp. Quant. Biol.* 24: 15–30.

Emerson, A. E. 1938. Termite nests—a study of the phylogeny of behavior. *Ecol. Monogr.* 8: 247–284.

Emerson, A. E. 1952. The supraorganismic aspects of the society. *Colloques Int. Centre. Natl. Rech. Sci.* 34: 333–353.

Etkin, W., and L. G. Livingston. 1947. A probability interpretation of the concept of homology. *Am. Naturalist* 81: 468–473.

Goodwin, D. 1952. A comparative study of the voice and some aspects of behaviour in two old-world jays. *Behaviour* 4: 293–316.

Grimstone, A. V. 1959. Cytology, homology and phylogeny—a note on "organic design." *Am. Naturalist* 93: 273–282.

Haas, O., and G. G. Simpson. 1946. Analysis of some phylogenetic terms, with attempts at redefinition. *Proc. Am. Phil. Soc.* 90: 319–349.

Hamburger, V. 1963. Some aspects of the embryology of behavior. *Quart. Rev. Biol.* 38: 342–365.

Harland, S. C. 1936. The genetical conception of the species. *Biol. Rev.* 11: 83–112.

Hinde, R. A. 1956. The behaviour of certain cardueline F_1 inter-species hybrids. *Behaviour* 9: 202–213.

Hinde, R. A. 1966. *Animal behavior, a synthesis of ethology and comparative psychology.* New York: McGraw-Hill.

Hinde, R. A., and N. Tinbergen. 1958. The comparative study of species-specific behavior. In A. Roe and G. G. Simpson, eds., *Behavior and evolution.* New Haven: Yale Univ. Press. Pp. 251–268.

Hubbs, C. L. 1944. Concepts of homology and analogy. *Am. Naturalist* 78: 289–307.

Inglis, W. G. 1966. The observational basis of homology. *Syst. Zool.* 15: 219–228.

Lehrman, D. S. 1953. A critique of Konrad Lorenz's theory of instinctive behavior. *Quart. Rev. Biol.* 28: 337–363.

Liley, N. R. 1966. Ethological isolating mechanisms in four sympatric species of poeciliid fishes. *Behaviour,* Suppl. 13: 1–197.

Lorenz, K. Z. 1935. Der Kumpan in der Umwelt des Vogels. *J. Ornithol.* 83: 137–213, 289–413. [English translation in Schiller, C. H. 1957.]

Lorenz, K. Z. 1937. Über die Bildung des Instinktbegriffes. *Naturwissenschaften* 25: 289–300, 307–318, 324–331. [English translation in Schiller, C. H. 1957.]

Lorenz, K. Z. 1939. Vergleichende Verhaltensforschung. *Zool. Anz. Suppl.* 12: 69–102. [English translation in Schiller, C. H. 1957.]

Lorenz, K. Z. 1950. The comparative method in studying innate behaviour patterns. *Symp. Soc. Exp. Biol.* 4: 221–268.

Marler, P. 1957. Specific distinctiveness in the communication signals of birds. *Behaviour* 11: 13–39.

Mayr, E. 1958. Behavior and systematics. In A. Roe and G. G. Simpson, eds., *Behavior and evolution.* New Haven: Yale Univ. Press. Pp. 341–362.

Mayr, E., 1963. *Animal species and evolution.* Cambridge: Harvard Univ. Press.

Mayr, E. 1965. Classification and phylogeny. *Am. Zool.* 5: 165–174.

Michener, C. D. 1953. Life-history studies in insect systematics. *Syst. Zool.* 2: 112–118.

Michener, C. D. 1964. Evolution of the nests of bees. *Am. Zool.* 4: 227–239.

Morris, D. 1958. The reproductive behaviour of the ten-spined stickleback (*Pygosteus pungitius* L.). *Behaviour,* Suppl. 6: 1–154.

Myers, G. S. 1965. The body-wag, an innate behavioral characteristic of bony fishes. *Tropical Fish Hobbyist* 13 (9): 21–25.

Newman, H. H. 1907. Spawning behavior and sexual dimorphism in Fundulus heteroclitus and allied fish. *Biol. Bull.* (Woods Hole), 12: 314–348.

Ono, Y., and T. Uematsu. 1957. Mating ethogram in *Oryzias latipes. J. Fac. Sci. Hokkaido Univ.* IV, 13: 197–202.

Pantin, C. F. A. 1951. Organic design. *Advanc. Sci.* (London) 8: 138–150.

Petrunkevitch, A. 1926. The value of instinct as a taxonomic character in spiders. *Biol. Bull.* (Woods Hole), 50: 427–432.

Pribram, K. 1958. Comparative neurology and the evolution of behavior. In A. Roe and G. G. Simpson, eds., *Behavior and evolution.* New Haven: Yale Univ. Press. Pp. 140–164.

Remane, A. 1952. *Die Grundlagen des natürlichen Systems, der vergleichenden Anatomie und der Phylogenetik.* Leipzig: Akad. Verlags.

Remane, A. 1961. Gedanken zum Problem: Homologie und Analogie, Praeadaptation und Parallelität. *Zool. Anz.* 166: 447–470.

Rosen, D. E., and R. M. Bailey. 1963. The poeciliid fishes (Cyprinodontiformes), their structure, zoogeography, and systematics. *Bull. Am. Mus. Nat. Hist.* 126: 1–176.

Rosen, D. E., and M. Gordon. 1953. Functional anatomy and evolution of male genitalia in poeciliid fishes. *Zoologica* (N.Y.) 38: 1–47.

Rosen, D. E., and A. Tucker. 1961. Evolution of secondary sexual characters and sexual behavior patterns in a family of viviparous fishes (Cyprinodontiformes: Poeciliidae). *Copeia,* 1961, pp. 201–212.

Ross, H. H. 1964. Evolution of caddisworm cases and nets. *Am. Zool.* 4: 209–220.

Russell, R. W .1952. *The comparative study of behaviour.* Inaugural Lecture, Univ. Coll. London. London: H. K. Lewis.

Russell, R. W. 1964. Introduction. Extrapolation from animals to man. In H. Steinberg et al., eds., *Animal behaviour and drug action.* Boston: Little, Brown. Pp. 410–418.

Russell, W. M. S., A. P. Mead, and J. S. Hayes. 1954. A basis for the quantitative study of the structure of behaviour. *Behaviour* 6: 153–205.

Schiller, C. H. 1957. *Instinctive behavior, the development of a modern concept.* New York: International Universities Press.

Schneirla, T. C. 1952. [In discussion of Emerson, A. E. 1952, p. 352.]

Schneirla, T. C. 1955. [In discussion of paper by K. Z. Lorenz in B. Schaffner, ed., *Group Processes.* New York: Josiah Macy Foundation. P. 194.]

Schneirla, T. C. 1957. The concept of development in comparative psychology. In D. B. Harris, ed., *The concept of development.* Minneapolis: Univ. Minnesota Press. Pp. 78–108.

Schnitzlein, H. N. 1964. Correlation of habit and structure in the fish brain. *Am. Zool.* 4: 21–32.

Sharpe, R. S., and P. A. Johnsgard. 1966. Inheritance of behavioral characters in F$_2$ mallard × pintail (*Anas platyrhynchos* L. × *Anas acuta* L.) hybrids. *Behaviour* 27: 259–272.

Simpson, G. G. 1946. [Note in Haas, O., and G. G. Simpson. 1946, pp. 323–324.]

Simpson, G. G. 1958a. The study of evolution: methods and present status of theory. In A. Roe and G. G. Simpson, eds., *Behavior and evolution.* New Haven: Yale Univ. Press. Pp. 7–26.

Simpson, G. G. 1958b. Behavior and evolution. In A. Roe and G. G. Simpson, eds., *Behavior and evolution.* New Haven: Yale Univ. Press. Pp. 507–535.

Simpson, G. G. 1961. *Principles of animal taxonomy.* New York: Columbia Univ. Press.

Simpson, G. G. 1964. Organisms and molecules in evolution. *Science* 146: 1535–1538.

Sperry, R. W. 1958. Developmental basis of behavior. In A. Roe and G. G. Simpson, eds., *Behavior and evolution.* New Haven: Yale Univ. Press. Pp. 128–139.

Tait, C. C. 1965. Notes on the species *Nothobranchius brieni* Poll (Cyprinodontidae). *Puku* (3): 125–131.

Tavolga, W. N. 1954. Reproductive behavior in the gobiid fish, *Bathygobius soporator. Bull. Am. Mus. Nat. Hist.* 104: 427–459.

Tinbergen, N. 1942. An objectivistic study of the innate behaviour of animals. *Bibl. Biotheor., Leiden* 1: 37–98.

Tinbergen, N. 1951. *The study of instinct.* London: Oxford Univ. Press.

Tinbergen, N. 1959. Comparative studies of the behaviour of gulls (Laridae): a progress report. *Behaviour* 15: 1–70.

Tinbergen, N. 1962. The evolution of animal communication—a critical examination of methods. *Symp. Zool. Soc. Lond.* no. 8, pp. 1–6.

Tinbergen, N. 1963. On aims and methods of ethology. *Z. Tierpsychol.* 20: 410–433.

Tinbergen, N. 1965. Behavior and natural selection. In J. A. Moore, ed., *Ideas in modern biology.* Garden City: Natural History Press. Pp. 519–542.

Whitman, C. O. 1899. *Animal behavior.* Biological Lectures, Wood's Hole, Summer Session, 1897 and 1898. Pp. 285–338.

Wickler, W. 1961a. Ökologie und Stammesgeschichte von Verhaltensweisen. *Fortschr. Zool.* 13: 303–365.

Wickler, W. 1961b. Über die Stammesgeschichte und den taxonomischen Wert einiger Verhaltensweisen der Vögel. *Z. Tierpsychol.* 18: 320–342.

Woodger, J. H. 1945. On biological transformations. In W. E. Le Gros Clark and P. B. Medawar, ed., *Essays on growth and form, presented to D'Arcy Wentworth Thompson.* London: Oxford Univ. Press. Pp. 95–120.

LESTER R. ARONSON
Department of Animal Behavior
The American Museum of Natural History
New York

Functional Evolution of the Forebrain in Lower Vertebrates

". . . the evolution of mind-stuff, that is, the precursors and instruments of mentation, is an open book, though but little of it has been deciphered."

C. J. Herrick (1956)

The growing need felt by Darwin and his contemporaries to demonstrate morphological continuity among animals was a powerful impetus to the study of comparative anatomy. At the same time, the introduction of evolutionary theory changed comparative anatomy from a static subject to a dynamic one with far-reaching possibilities. For obvious reasons the greatest successes were achieved in the analyses of skeletal systems, but comparative studies of soft structures, although limited for the most part to living species, also yielded many promising results. Among these were comparative studies of the vertebrate brain. By the turn of the century, comparative vertebrate neurology was well established as a new and exciting branch of comparative anatomy. The names of Ludwig Edinger, Elliot Smith, and Ariëns Kappers in Europe, and C. J. Herrick, Johnston, Huber, Crosby, and Papez in the United States lead the roster of famous comparative neuroanatomists. The structural evolution of the forebrain became a favorite topic, since this research clearly showed how the very small forebrain of primitive vertebrates evolved into the great cerebrum of mammals and man, the major site of all higher mental processes.

Most of the neuroanatomists expressed interest in the functional properties of the structures that they described, but they tended to base their conclusions on only the most elementary facts. If, for example, an area of the brain received secondary or higher order fibers from the retina, it was called a visual area. If the area also had auditory connections, then it served to correlate vision and hearing. Physiological and behavioral studies of the brains of lower vertebrates, which during the early decade of this century were admittedly rather crude, were largely overlooked by the neuroanatomists. This seemed particularly true in studies concerning the forebrain.

Morphologists studying the evolution of the forebrain concentrated on those primitive forms that seemed to be closest to the main line of vertebrate evolution, such as the cyclostomes (lampreys), elasmobranchs (sharks and rays), dipnoans (lung fishes), chondrosteans (sturgeons), and holosteans (gars and bowfins). Considerably less attention was paid to the great many highly specialized teleosts (bony fishes). Similarly among the amphibians the more primitive brains of salientians (salamanders) were of much more interest than the more specialized brains of anurans (frogs and toads).

Since fibers from the olfactory bulbs pervade all (or almost all) of the forebrain in the primitive fishes and in amphibians, the obvious conclusion of the neuroanatomists was that the forebrain of lower vertebrates serves primarily for the reception, dissemination, and intensification of olfactory information. In teleosts, gustatory fibers from the hypothalamus appear to reach the forebrain, leading Herrick (1924) and Papez (1929) to conclude that in these fishes the forebrain also serves to integrate taste and smell. Then in reptiles, birds, and mammals there is a progressive increase in the representation of other sensory modalities. Thus we see the cerebrum of higher vertebrates as a multi-sensory integrative mechanism, which formed the basis for the development of the higher nervous processes of mammals. This simple picture of the origin and evolution of the vertebrate forebrain became very popular and is so well embedded in the literature that one still finds statements in current review papers and textbooks that the forebrain of fishes and amphibians is concerned solely with the function of olfaction. Actually, evidence has been accumulating for a half century or more that the primitive forebrain serves other functions. Although these are still not clearly defined, they are presently the subject of much experimentation and discussion.

The morphologist can readily select his material on the basis of taxonomic position even though the needed individuals may be large, rare, or difficult to maintain in captivity. For most studies only a few well-preserved specimens are required. Requirements for specimens used by physiologists and behaviorists are much more stringent: many specimens are usually needed; it must be possible to study them in controlled conditions; they must remain in good health in captivity; and they must respond readily in a variety of experimental situations. In contrast to the neuroanatomists who, as noted above, were able to study a variety of primitive fishes, the experimentalists, because of these limitations, have concentrated on the highly specialized teleosts, particularly the common goldfish. Among amphibians, anurans have been used most widely in functional studies while the neuroanatomical picture has been worked out in great detail for the simpler salientians, which are closer to the main line of vertebrate evolution. Unfortunately, this situation makes it necessary to extrapolate backwards from the more functionally specialized forebrain to the less functionally specialized. We also need to make the assumption that forebrain processes seen in the specialized species exist to some extent in the primitive and extinct species from which we build the evolutionary continuum. Because of these two qualifying considerations, few people have attempted to discuss the functional evolution of the forebrain. Meanwhile the structural evolution of the forebrain continues to receive attention, as for example, in the recent book on the subject edited by Hassler and Stephan (1967) and the evolution of the limbic system by Riss, Halpern, and Scalia (1968). The only significant discussion in recent years on the functional evolution of the forebrain is in the book by Karamian (1956), and in a subsequent article by the same author (Karamian, 1965).

PHYLETIC SURVEY

Hemichordates

Although the nervous system of most hemichordates is typically similar to that of invertebrates, many of the Enteropneusta (e.g., *Saccoglossus, Balanoglossus*—acorn worms) possess a nerve cord in the collar region that has strong resemblances to the vertebrate neural tube (Hyman, 1959; Bullock, 1965). These similarities are its dorsal position and derivation from the epidermis, its hollow construction, and the presence of anterior and posterior neuropores. This dorsal nerve cord (neurocord) contains a variable number of giant neurons whose arrangement compares favorably with the Mauthner and Müller cells of lower vertebrates as to position, decussation, and course of fibers (Bullock, 1965).

In a few species of the genus *Saccoglossus* the dorsal nerve cord of the collar extends into the proboscis. It may be significant in light of current ideas on forebrain function in vertebrates that damage to this cord affects the spontaneity of peristaltic burrowing activity. According to Bullock (1940) and Knight-Jones (1952), this control of peristaltic burrowing, an activating process, is a property of the whole nerve plexus of the proboscis with the neurocord acting essentially as a conduction pathway. The latter author also proposes that certain cells at the periphery of the base of the proboscis, and others in the groove of the ciliary organ, may be chemoceptive. Presumably, sensory information from these cells is transmitted to the neurocord via the nerve net of the proboscis. Bullock (1940) demonstrated that the proboscis has definite negative phototactic properties, but no special sensory apparatus was identified.

It cannot be emphasized too strongly that a relationship of the proboscis to the vertebrate forebrain has not been established. In fact Bullock (1940, 1945) argues on neurophysiological grounds that since the neurocord lacks the typical integrative properties of a central nervous system, it should not be designated as such. Hence, the functional similarities just described, although suggestive of a very primitive condition, may have only limited phylogenetic significance.

Tunicates

The nervous system of most adult tunicates is also typical of invertebrates and bears few if any vertebrate characteristics. However, the neotenic appendicularians (class Larvacea) and the tadpole larvae of other tunicate classes possess a dorsally situated hollow nerve tube, which most investigators agree is homologous to the neural tube of vertebrates (Bullock, 1965). In many species a cerebral ganglion with anterior, middle, and posterior bulges is located at the rostral end of the neural tube. In the appendicularian *Oikopleura longicauda* the anterior bulge was labeled "forebrain" by Martini (1909). Two pairs of sensory nerves extend anteriorly, one to specialized sensory receptors on the upper lips of the mouth cavity, and the other to the anterior sensory organ and possibly to sensory receptors on the lower lips and endostyle. Martini (1909), following the suggestion of older investigators, considered the anterior sensory organ to be olfactory and the sensory receptors on the upper and lower lips to be chemoreceptors, possibly gustatory. Lohmann (1933) cited an experiment by Fol (1872) in which a black pigment was introduced into the water in which an appendicularian was swimming. As soon as the water carrying the pigment reached the olfactory groove, the animal reversed the current ejecting the water from the mouth. Lohmann (1933) also described a frontal sensory organ innervated by the forebrain and located just dorsal to the brain. However, the function of the sensory cells are unknown.

Here again it is not at all clear that the anterior bulge of the appendicularian brain is homologous to the vertebrate forebrain or even to the prosencephalon (telencelphalon plus diencephalon), and the experimental evidence concerning the chemoceptive nature of the sensory receptors is limited. Nevertheless, it is of considerable interest that the anterior part of the brain of these primitive animals seems to integrate special sensory information coming from the snout, lips, and pharynx.

Amphioxus

While some investigators have stated that there is no homologue of the forebrain in amphioxus (Delsman, 1918; Papez, 1929), the prevailing view is that the forward and expanded end of the neural tube represents the prosencephalon. The anterior end of the prosencephalon is, therefore, equivalent to the forebrain (Ariëns Kappers, Huber, and Crosby, 1936). Since lateral ventricles are not present, the anterior end of the brain in amphioxus is homologous to the unevaginated portion of the vertebrate forebrain known as the telencephalon medium. Ahead of the forebrain there is a groove containing special sensory cells which Kölliker (1843) identified on morphological grounds as an olfactory organ. From these sensory cells of the olfactory groove an unpaired nerve runs caudally to the forebrain. Neither Nagel (1894) nor Parker (1908) were able to demonstrate special chemical sensitivity for this organ, since they observed that all parts of the anterior portion of the body responded equally by withdrawal movements to the various taste and odorous substances that they used. Although Parker considered his experiments inadequate to demonstrate olfaction in fishes, he accepted the earlier morphological arguments that this structure is truly the homologue of the olfactory organ of vertebrates.

A paired sensory nerve (n. apicus) innervating the skin of the rostrum also runs caudally to the forebrain. It is now identified as the nervus terminalis (van Wijhe, 1919; Ariëns Kappers, Huber, and Crosby, 1936). The pigment spot that lies at the rostral end of the brain was formerly thought to be a primitive eye, but Parker (1908) was unable to demonstrate light sensitivity in this region.

The meager information available on amphioxus suggests that the limited forebrain of this primitive chordate may serve to disseminate and integrate chemical and tactile information from the snout.

Ostracoderms

These fossil agnatha (cyclostomes) from Silurian and Devonian deposits are the most primitive vertebrates known. The two major groups of ostracoderms are the Osteostraci, which are related to present-day Petromyzontida (lampreys), and the Heterostraci, related to present-day Myxinoidea (hagfishes). From extensive examination of the fossil material, Stensiö (1963) observed that representative species of both groups had well-developed forebrains that apparently resembled the forebrains of living cyclostomes. From the telencephalic hemispheres a pair of olfactory nerves extended anteriorly to an unpaired nasal cavity. These observations tell us that as far back as we are presently able to trace vertebrate ancestry, there was always a forebrainlike structure related to olfactory input. Whether this primitive "forebrain" also had other functional properties such as those found in modern fishes (pp. 84–88) is not known.

Modern Cyclostomes

Lampreys and hagfishes have the most primitively organized brains of all living vertebrates (Johnston, 1912). For this reason they have attracted the attention of the neuroanatomists, and numerous detailed neuroanatomical descriptions are available (for review, see Ariëns Kappers, Huber, and Crosby, 1936). On the other hand, functional studies of brain mechanisms on these forms are very limited, and, as far as we are aware, no behavioral studies have been related to forebrain function.

The telencephalon consists of a pair of large olfactory bulbs separated by a fissure from paired cerebral hemispheres, which, in lampreys, are smaller than the bulbs. According to some investigators (e.g., Heier, 1948) all of the fundamental pallial and subpallial areas of the vertebrate forebrain can be distinguished in this relatively small cerebrum. Secondary olfactory fibers from the bulb project on all regions of the cerebrum and on the diencephalon (Papez, 1929; Nieuwenhuys, 1967). From the pallium and subpallium, tracts extend to the epithalamus (habenula), thalamus, hypothalamus, and midbrain. From the hypothalamus, a large ascending tractus pallii projects on pallial and subpallial areas. These are fundamental connections that are typical of all vertebrates. Johnston (1906) notes that tertiary gustatory fibers terminate in the hypothalamus and he suggests that taste impulses are relayed to the forebrain by the ascending tractus pallii.

From the neuroanatomical evidence, especially from the fact that olfactory fibers reach all areas of the cerebrum, we can assume that, to a large extent, the lamprey forebrain is concerned with the organization and dissemination of olfactory information, and perhaps the integration of olfaction and taste. Since the nuclear configurations and fiber connections are basically similar to those of teleosts (although much less elaborate and differentiated), it is anticipated that the functions of the teleost forebrain, especially those concerned with the facilitation of lower centers (p. 86), will be found, at least in elementary form, in the cyclostome forebrain.

Based purely on anatomical considerations, Riss, Halpern, and Scalia (1968) recognize a limbic system in lampreys that they consider homologous to the limbic system of mammals. Moreover, they propose that the olfactory system arose from the limbic system in primordial vertebrates rather than either of the more common views that the limbic system is separable from the olfactory system or evolved from it. As central to their argument, they have described the nervus terminalis in lampreys as projecting on major limbic nuclei of the forebrain and diencephalon rather than on the olfactory bulb, which is the termination of all olfactory nerves (see also van Wijhe, 1919). While the olfactory system is responsive to *external* chemical changes, Riss and his coworkers argue that the accessory olfactory organ and terminal nerve are responsive to *internal* chemical changes. Hence, the terminal nerve is a direct afferent to limbic structures.

Illumination of the eyes of *Lampetra* evoked electroencephalographic (EEG) responses from the tectum, tegmentum, medulla oblongata, and spinal cord, but not from the forebrain (Karamian et al., 1966). This experiment supports the anatomical view that with the possible exception of taste the various sensory systems do not project anteriorly to the forebrain at this phylogenetic level.

Kleerekoper and van Erkel (1960) describe the olfactory apparatus in the lamprey *Petromyzon* as large and complex, indicating an elaborate and sensitive olfactory apparatus. Since the olfactory apparatus is small and simply organized in the ammocoetes larva, Kleerekoper and van Erkel suggest that olfaction is minimal or absent in the larval stage. Nevertheless the description of the brain of ammocoetes by Mayer (1897) and by Tretjakoff (1909) indicate that the larval telencephalon does not differ materially from that of the adult. This finding lends further support to the hypothesis that the cyclostome forebrain has functions that are additional to the reception and integration of internal and external chemical stimulation.

Elasmobranchs

In sharks and rays a huge folded olfactory mucosa adheres closely to large olfactory bulbs that are usually subdivided into medial and lateral parts. Olfactory tracts from both parts of the olfactory bulbs run caudally to all regions of the forebrain except possibly the small somatic area. Fiber tracts from the forebrain extend to the epithalamus (habenula) and hypothalamus, and a large tractus pallii runs from the hypothalamus to the forebrain hemispheres. The picture one obtains from the anatomical descriptions is that of a huge olfactory system dominating a relatively small cerebrum, which receives additional impulses (possibly gustatory) from the hypothalamus.

The forebrains of several elasmobranch species were partially or completely removed by Loeb (1891, 1900), Bethe (1899), Polimanti (1911), and Rizzolo (1929). No changes in locomotion and equilibrium or obvious changes in behavior were observed by these investigators. Steiner (1888), on the other hand, reported that his forebrainless sharks *Scyliorhinus* no longer searched for food. He attributed this to deficits in olfaction since he obtained the same results when only the olfactory tracts were severed. At the same time, Steiner stated that vision was unaffected by these operations.

Using a square wave potential, Chauchard and Chauchard (1927) found that electrical stimulation of various parts of the forebrain of the rays *Myliobatus* and *Trygon* did not cause muscular movements, thus supporting the more general conclusion that in lower vertebrates the forebrain is not directly involved in motor activity. Similar results were obtained by Springer (1928) on the dogfishes *Mustelus* and *Squalus*. Polimanti (1912) reported no changes in motor activity after applying curare or cocaine to the forebrain of the cat shark, *Scyllium,* and the electric ray, *Torpedo.*

Electroencephalographic recordings taken from several points on the dorsal surface of the forebrain and from 2 to 5 mm below the surface were made by Gilbert, Hodgson, and Mathewson (1964) and by Agalides (1967) in restrained lemon sharks, *Negaprion,* bonnet sharks, *Sphyrna,* and nurse sharks, *Gingylymostoma.* When chemical solutions which normally induce approach behavior were introduced into the nasal cavity, changes in amplitude and frequency of electrical potential followed. With photic stimulation, Veselkin (1964) and Karamian et al. (1966) could not obtain evoked potentials from the forebrain indicating that the optic system does not project forward to this part of the brain.

The experimental data for elasmobranchs, although much more extensive than in previous groups, still adds relatively little to the overall picture. It is clear that the forebrain functions as an olfactory mechanism and that it is not directly involved in motor activity. There is also a possibility that the elasmobranch forebrain functions as an arousal mechanism, as in teleost fishes, but so far no experiments have been performed to test this hypothesis.

Teleosts

In comparison with the amount of research done on ancestral groups previously considered, work on the forebrain of teleosts is extensive. Yet it is still minuscule when compared with the vast research on cerebral function in mammals. Since neurological research on teleosts has been reviewed several times (Ten Cate, 1935; Aronson, 1963; Aronson and Kaplan, 1968; Segaar, 1965), only the highlights will be considered here. We have arranged these studies under three main categories, namely, those concerned with (1) olfaction, (2) learning, and (3) social behavior.

OLFACTION

In teleosts, as in all vertebrates, olfactory nerves from the nasal mucosa terminate in the paired olfactory bulbs. From there, fiber tracts run to the cerebral hemispheres and hypothalamus. Streick (1925) trained blinded minnows, *Phoxinus,* to respond to an odor (wads of cotton soaked with musk, cumerin, or skatol) by using food as a reward. After the forebrain had been ablated the minnows no longer reacted to the odorous cotton wads. These experiments, however, do not tell us whether the deficits in olfaction resulted from loss of the olfactory nerves and bulbs or whether the cerebral hemispheres of the forebrain were involved in ways other than the direct transmission of information to lower centers. The importance of the olfactory bulbs is emphasized by Grimm (1960), who stimulated these structures electrically and induced normal feeding reactions in goldfish.

In a series of electrophysiological studies, Hara, Ueda, and Gorbman (1965) and Ueda, Hara, and Gorbman (1967) obtained a characteristic pattern of high amplitude EEG responses from the olfactory bulbs of adult spawning salmon when the nasal sacs were infused with water from the spawning sites where the fish had been obtained. Water from other spawning sites did not elicit this response. These results may be considered an example of olfactory conditioning to the water in which the salmon were living at the time of capture. These investigators believe, however, that the characteristic EEG responses of the bulb reflect long term memory of the spawning site, and provide neurophysiological evidence for Hasler's (1966) theory of salmon migration. According to this theory, migrating salmon actually store information about the chemical characteristics of the home streams that was acquired several years before, when as fingerlings they descended these streams on their way to the sea. Since similar EEG responses were obtained after the olfactory tracts were sectioned and the bulbs were isolated from the rest of the brain, Gorbman (personal communication) believes that the memory of the home stream resides in the olfactory bulbs.

In further studies on the goldfish *Carassius,* Hara and Gorbman (1967) recorded electrical responses from the olfactory bulbs when the olfactory epithelium was stimulated with solutions of sodium chloride, other organic salts, or sucrose. Section of the olfactory tracts augmented these electrical responses. Electrical stimuli applied to the opposite bulbs or to the cerebral hemispheres altered the characteristics of the potentials, indicating that normal afferent stimuli to the bulbs are modulated by influences from the other bulb and from the hemispheres.

Johnston (1911, 1912), Herrick (1922), and Papez (1929) describe an ascending tract from the hypothalamus to the forebrain. Since this tract seems to connect secondary gustatory centers of the hypothalamus to the forebrain, they concluded that one function of the forebrain is to integrate taste and smell. The experiments of Wrede (1932) support this conclusion. She found that skin slime from minnows could not be discriminated from other taste substances (salt and acetic acid) by forebrainless minnows; this skin slime normally activates both the olfactory and gustatory systems in an intact minnow.

LEARNING

As outlined in the reviews by Segaar (1965) and Aronson and Kaplan (1968), some investigators (e.g., Bernstein, 1962; Ingle, 1965) found no changes in learning and conditioning after forebrain injury or removal when olfactory stimuli were not involved. A few studies, however, describe rather profound and lasting differences between intact and forebrain-ablated subjects in the acquisition and performance of learned responses. In all of the latter studies, the forebrainless fish could still acquire the conditioned response or retain the response if operated on after training. What usually changed was the latency or speed of response, or the consistency of the behavior. Hale (1956) found that grouped and isolated forebrainless sunfish, *Lepomis,* took much longer than their intact counterparts to swim through a gate to a second compartment where they were subsequently fed. Warren (1961) found that forebrain-damaged paradise fish, *Macropodus,* were significantly inferior to controls during successive reversals in a T-maze and in 6 out of 12 detour problems. Aronson and Herberman (1960) found that the latency to strike an introduced paddle in an operant situation was very much higher and more variable in forebrainless *Tilapia* than in intact fish or those with just the olfactory bulbs removed. Using an avoidance conditioning paradigm, Aronson and Kaplan (1963, 1965) and Kaplan and Aronson (1967, 1969) trained *Tilapia* to swim through a hole in a partition to avoid or escape electric shock. Onset of either a light or a 500 Hz sound was used as the conditioned stimulus. Intact fish, sham operates, and those with just the olfactory bulbs removed, learned to avoid the shock consistently. When the entire forebrain was ablated, the operates took much longer to learn the task and they never learned to avoid the shock as consistently as the control subjects. Similarly, when the operations were performed after the *Tilapia* had been trained, the responses were much slower and erratic. Oftentimes the subjects received the shock for several seconds before escaping through the hole. Qualitative observations revealed that the operated fish no longer maintained efficient intertrial habits, such as waiting near the hole in the partition for the onset of the conditioned stimulus. Also, the first reactions to the light or sound were not so immediate or so well oriented as before operation, and the swimming speed of some subjects was considerably slower.

Experiments involving avoidance learning were also performed on the goldfish *Carassius* by Hainsworth, Overmier, and Snowdon (1967) and Savage (1968). They, too, found major deficiencies in forebrain-deprived subjects.

Largely on the basis of experiments involving simple classical conditioning, Karamian (1956) concluded that in fishes the associations between conditioned stimulus (CS), unconditioned stimulus (US), and conditioned response (CR) are formed in the cerebellum, while in birds and mammals this function is transferred to the forebrain. Amphibians and reptiles represent intermediate stages in this evolutionary transfer of function. Learning in fishes, according to Karamian, is quite independent of the forebrain since he could condition forebrainless fish (carp) just as rapidly and effectively as intact subjects. Karamian is correct in stating that the associations involved in conditioning are not formed in the forebrain, but the experiments cited above show that the forebrain does influence the acquisition and performance of complex conditioning and learning processes. Recent experiments in this laboratory (Kaplan and Aronson, 1969) support Karamian's views concerning cerebellar function, since we were also unable to establish a stable conditioned avoidance response after removing the body of the cerebellum.

SOCIAL BEHAVIOR

Extirpation of the forebrain in various species of telcosts resulted in marked deficits in reproductive behavior, parental activities, aggressive behavior, and schooling. In several of these experiments control fish with only the olfactory bulbs removed did not show these defects, indicating clearly that the changes in behavior following deprivation of the cerebral hemispheres cannot be attributed to loss of olfaction. In the reproductive and parental studies, Noble (1936, 1939), using cichlid, cyprinodont, and anabantid species, and Segaar (1961, 1965) and Segaar and Nieuwenhuys (1963), studying sticklebacks, concluded that the cerebral hemispheres regulate specific aspects of reproductive behavior, e.g., spawning, parental and aggressive acts. According to Segaar and his coworkers the timing and level of performance of these acts could be altered by relatively small, localized lesions, suggesting to these investigators that the cerebral hemispheres of the forebrain contain discrete, integrative areas. Aronson (1948) and Kamrin and Aronson (1955), working with the West African mouthbreeder, *Tilapia,* and the platyfish *Xiphophorus,* respectively, noted that extirpation of the forebrain caused a decline in the frequency of most acts associated with reproduction and parental care, but the operation did not eliminate any of these behaviors. It was concluded, therefore, that the forebrain does not contribute directly to the organization of behavior. Rather, it acts as a nonspecific regulatory mechanism that facilitates, by neural (or neurochemical) excitation or inhibition, behavior organized in lower centers. This arousal hypothesis will be discussed further in the next section.

Extirpation of the forebrain also affects schooling and aggressive behavior. Kumakura (1928), Hosch (1936), and Berwein (1941) found temporary deficits in schooling in forebrainless goldfish and minnows. More permanent deficits after forebrain injury were reported by Noble (1936) and Wiebalck (1937) in cichlid and percoid species. Following partial or complete extirpation of the forebrain in several teleost species including the fighting fish *Betta,* the stimuli of social situations that typically produce fighting were less effective, so that fights occurred infrequently; but when fighting did occur, it was as vigorous as in intact subjects (Noble, 1939, Noble and Borne, 1941). Similar results were obtained by Hale (1956a) in the sunfish *Lepomis,* and by Schönherr (1955) and Segaar (1961, 1965) in the 3-spined stickleback, *Gasterosteus.*

In the wrasse *Crenilabrus,* Fiedler (1967) reported that fighting intensity remained the same or increased when various areas of the forebrain (anterior commissure, anterior pole, area dorsalis) were destroyed by means of electrocoagulation. In the sparid *Diplodus annularis,* on the other hand, lesions that destroyed the connections between the olfactory bulbs and diencephalon reduced the intensity of fighting but did not cause qualitative changes in this behavior. Fiedler (1968) stimulated the anterior pole of the forebrain of *Crenilabrus* electrically and elicited body quivers and jerky fin movements. With the electrodes in various areas of the forebrain, he obtained yawning, quivering, forward thrusts, and backward movements. With an electrode in the anterior commissure of a sunfish, *Lepomis,* Fiedler (1965) recorded single unit potentials when the subject caught a crab, was chewing it, and when the fish swallowed it. In the sunfish, electrical stimulation of the olfactosomatic area of the forebrain elicited nestbuilding (Demski, 1969).

THE AROUSAL HYPOTHESIS

From the studies referred to in this report and in the several review articles cited earlier, it is evident that damage to, or complete removal of, the forebrain usually results in a quantitative reduction in the frequency of specific behavioral patterns and sometimes changes in the timing of these events. Since it is evident that no behavioral sequences (including those generated by learning experiments) are completely eliminated by the operation, one must conclude that the behavior in question is organized (or the conditioned connections established) in neural centers below the forebrain. Various bits of evidence suggest that the midbrain as well as the cerebellum may be important areas for such organizational functions.

While the forebrain does not participate directly in the organization of behavior, it is a nonspecific regulator of lower brain mechanisms. In the several conditioning experiments reported from the American Museum (Aronson and Herbeman, 1960; Aronson and Kaplan, 1963, 1965; Kaplan and Aronson, 1967, 1969), the fish still made some conditioned approach responses, avoidance, escapes, or correct discriminations in the first tests immediately after complete removal of the forebrain. In the approach conditioning, there was usually further deterioration in subsequent tests. This can be explained by hypothesizing either the existence of reverberating circuits in the lower centers which run down when no longer primed by forebrain activity or the depletion of a biogenic substance which is normally secreted by the forebrain, and which impinges on lower centers. For example, Schildkraut and Kety (1967) recently suggested that norepinephrine may produce a nonspecific state of arousal. Kusunoki and Masai (1966) found substantial quantities of monoamine oxidase (a catecholamine inactivator) in several areas of the goldfish forebrain, indicating that these portions of the brain may produce norepinephrine. Celesia and Jasper (1966) have shown that activation of the cerebral cortex of mammals in the aroused or alert state may be mediated by acetylcholine. In removing the forebrain we may be removing part of a system involved in nonspecific neurohumoral arousal.

The concept of a facilitatory action of the forebrain is also supported by neurophysiological evidence. While a number of investigators have reported changes in the EEG patterns of the forebrain in several species of fish during photic stimulation, Gusel'nikov, Onufrieva, and Supin (1964) and Voronin and Gusel'nikov (1963) have produced compelling evidence that these may all be artifacts. Timkina (1965), on the other hand, found that EEG responses recorded from the midbrain following photic stimulation in goldfish and carp were modified when the forebrain was strychninized. Similarly Zagorul'ko (1965) reported that forebrain extirpation in carp caused an increase in amplitude and duration of photically evoked responses from the tectum, while both electrical and strychnic stimulation of the forebrain inhibited these evoked responses to photic stimulation. Zagorul'ko concludes that the forebrain exerts a regulatory influence on the activities of the midbrain visual system. Thus, while the lower centers may exert only a limited influence on forebrain activity, all the evidence indicates that the forebrain exerts a profound effect on various infracerebral neural mechanisms.

The arousal hypothesis receives further support from the experiments of Bert and Godet (1963) on the African lungfish, *Protopterus*. Following olfactory stimulation they recorded from the forebrain of this primitive species a 15 Hz rhythm which they believe has characteristics similar to the rhinencephalic arousal mechanism of mammals. This alerting response could not be obtained when the fish were aestivating (Godet, Bert, and Collomb, 1964).

Anatomically the major portion of the forebrain of lower vertebrates represents the ancestral homologues of those rhinencephalic structures of mammals—hippocampus, pyriform cortex, amygdala, and septum—that form the core of the limbic system of mammals. Some investigators are unwilling to extend this homology to teleosts since in these forms the forebrain develops ontogenetically by *eversion* of the lateral walls of the neural tube. In all other vertebrates the forebrain forms by a lateral evagination of the median ventricle and *inversion* of the dorsal portion of the side walls of the neural tube (Herrick, 1922; Nieuwenhuys, 1967). Nevertheless, the major fiber connections are similar to those of other lower vertebrates (Schnitzlein, 1964), leading to the conclusion that the basic functions of the teleost forebrain have not changed even though the structural characteristics have become topographically rearranged (Aronson and Kaplan, 1968). The finding of Aronson (1963) and Nieuwenhuys (1967) that secondary olfactory fibers do not reach the dorsal areas of the forebrain in many teleosts lends support to this interpretation.

The limbic system in mammals is also considered a nonspecific modulator of behavior patterns organized in other parts of the brain (Gloor, 1960) and a regulator of the awareness of experience (Douglas and Pribram, 1966). The inadequate responses made by the forebrainless fish in the conditioning experiments cited above clearly reflects deficits in a comparable nonspecific system. In the avoidance situation, for example, there was frequently a delay of several seconds between the presentation of the conditioned stimulus and the first overt response to it. When the subject finally responded, the swimming was much too slow and the orientation too poor for the fish to reach the escape hole in time to avoid shock. Whereas the intact fish usually waited beside or directly in front of the escape hole during the intertrial interval, the forebrainless subjects rarely assumed an efficient position in preparation for the onset of the conditioned stimulus (Kaplan and Aronson, 1967, 1969). Other examples of inefficient function were described in these papers.

In summary, the rather extensive anatomical and experimental data from teleosts provide substantial evidence that we can expect to find in the forebrain of lower vertebrates the anatomical and functional primordium of the limbic system of mammals.

Amphibians

SPONTANEITY

As early as 1824, Flourens observed the loss of spontaneous movements in frogs when the forebrain was extirpated. This observation was confirmed in subsequent years by numerous other investigators (reviewed by Aronson and Noble, 1945), and has also been reported in forebrainless salamanders (Reznitchenko, 1962). Loss of spontaneity was characterized by a failure to move or respond unless strong stimulation was applied. When a forebrainless frog was aroused, the resulting behavior was, in most cases, similar to that of an intact animal. On land, for example, forebrainless frogs tend to sit in one place for hours or even days, but if they are poked with a rod, they will leap in normal fashion. Sexually active male leopard frogs with the forebrain ablated will not swim to a receptive female even if close by, but if the male should accidentally touch the female, he will immediately turn, clasp her, and a normal spawning usually follows (Aronson and Noble, 1945). Forebrainless frogs merely snap at insects without ingestion although they seem perfectly capable of swallowing food. Similarly, ablation of the telencephalon of tadpoles of *Alytes* interferes with feeding behavior in some individuals (Rémy and Bounhiol, 1962).

A few researchers, notably Desmoulins (1825), Schrader (1887), and Loeser (1905) did not find this loss in spontaneity after removal of the forebrain. Later investigators have attributed these negative findings to faulty surgical procedures and failure to analyze the extent of the brain damage postoperatively.

Blankenagel (1931) noted that if the anterior third of both lobes of the forebrain is removed, or if just the pallial (roof) portions of both hemispheres are removed, spontaneity will not be affected. In further experiments he was able to localize the critical area for spontaneous activity in the caudal portion of the subpallium. The importance of this region for spontaneous action was confirmed by Diebschlag (1934). While most of the research on forebrain function in amphibians has been carried out on anurans (frogs and toads), this author extended the observations to the urodeles (salamanders) and found that here, too, complete removal of the forebrain resulted in a marked reduction in spontaneous actions.

Aronson and Noble (1945) conducted a more detailed study of spontaneity in the grass frog, *Rana pipiens,* particularly in relation to sexual behavior. In their experiments, control and operated males and test females were all treated with anterior pituitary implantations to insure high levels of sexual arousal. They found that after extensive or complete extirpation of the forebrain (as verified by histology), only a few males still swam to the receptive female during a two-hour test period. When the forebrain was ablated except for the preoptic area, 100% of the experimental males swam to the female. Clearly, the preoptic area is critical for this response. However, the situation is more complicated, for removal of just the preoptic area alone did not materially reduce the behavior. Therefore, other parts of the forebrain must be able to substitute for the missing preoptic area. Aronson and Noble interpret their results in terms of a regulatory or facilitatory function of the forebrain, with special emphasis on the hippocampus-septum-preoptic area-hypothalamus continuum. They relate this system in frogs to the telencephalic component of the Papez (1937) circuit of emotions, which was the forerunner of the current concept of the limbic system as a nonspecific arousal mechanism.

In recent publications by Schmidt (1966, 1968) on several species of frogs and toads, the sex or mating call and "mate orientation" were studied in experiments involving forebrain stimulation or ablation. Mate orientation is defined as orientation to the female before clasping, and is thus similar to the behavior that Aronson and Noble called "swimming to the female." Unfortunately, Schmidt did not observe the orientation movements directly in his experimental subjects; rather he assumed that these movements must have occurred if the male was found clasping a receptive female (or a substitute). When the ablations included the dorsal magnocellular nucleus of the preoptic area, mate orientation was eliminated. When the ablations included the region of the ventral magnocellular nucleus, the mating call was eliminated. Moreover, electrical stimulation of the preoptic area evoked the mating calls. Schmidt suggests that the preoptic area of the forebrain contains androgen-sensitive cells which activate critical centers located in posterior regions of the brain which control these behavioral processes (e.g., the trigemina-isthmic tegmentum for vocalizations). This interpretation is, therefore, in accord with the view stated above that the forebrain is primarily an activating mechanism.

A number of early investigators including Blankenagel (1931) and Diebschlag (1934) removed just the olfactory bulbs or the anterior third (or less) of both hemispheres. These operations, which eliminate olfaction, do not effect spontaneity. This means that the forebrain arousal mechanism can function independently of olfactory stimulation in amphibians as in teleosts.

LEARNING

Although anurans are notoriously difficult to train, several authors have claimed success and were able to compare intact and forebrainless individuals. Using various instrumental conditioning paradigms, with either light or sound as the conditioned stimulus, intact frogs and toads were successfully conditioned but forebrainless individuals failed to respond (Burnett, 1912; Blankenagel, 1931; Diebschlag, 1934). This, of course, was attributed to loss of spontaneity or malfunction of the arousal mechanism. A classical conditioning paradigm was therefore used with light or sound as the CS, and poking with a rod (Blankenagel, 1931) or electric shock (Diebschlag, 1934) as the US. The conditioned response (CR) in these experiments was hopping or jumping. This response was completely lost after forebrain extirpation. Bianki and Dushabaev (1964) employed a similar response which they called the conditioned motor defense reaction. They used electric shock as the US and intermittent light shining into the left eye as the CS. The other eye was removed. The CR was readily established, but was almost completely eliminated when the contralateral right hemisphere was ablated. In other subjects removal of the homolateral left hemisphere caused a partial reduction in the performance of the CR. Karamian (1956) reported that A. V. Baur had successfully conditioned forebrainless frogs to light or the sound of a bell although with greater difficulty than in normal subjects.

Several other investigators including Noll (1931), Chernetski (1964), and Karamian et al. (1966) reported that spinal reflexes are delayed when the forebrain is ablated. Although much of this evidence is highly variable, it would seem that normally the forebrain must facilitate or inhibit the reflex mechanisms of the spinal cord.

NEUROPHYSIOLOGICAL EXPERIMENTS

Using a flashing light as a stimulus, several Soviet investigators have described slow wave evoked responses of 50–250 msec from the surface of the forebrain of frogs (Zagorul'ko, 1957; Voronin and Gusel'nikov, 1959). Karamian et al. (1966) identified the critical area for these responses as the dorsal surface of the primordium hippocampus. Voronin and Gusel'nikov (1963) and Supin and Gusel'nikov (1964) obtained similar responses from the hippocampus by stimulating the sciatic nerve (somatosensory). Weaker responses in the forebrain following auditory clicks could not be localized. The evoked responses following optic and somatosensory stimulation presented almost simultaneously do not block one another and can be summated. This would indicate that associative properties of the hippocampus are minimal. Fast waves of lower amplitude were also recorded from the forebrain after photic stimulation. These are considered by the last two investigators to be artifacts resulting from the spread of evoked potentials from the optic tectum. Marked seasonal variations in the spontaneous electrical activity of the olfactory bulbs and cerebral hemispheres were reported by Segura and de Juan (1966). An alpha-like rhythm was strongest in spring and summer. At this time of the year, evoked potentials recorded from the hemispheres showed a high sensitivity to changes in the level of illumination.

As noted in the previous section, similar evoked potentials have not been obtained in teleosts, indicating that the afferent projection of the sensory systems to the forebrain is much better developed in amphibians. Supin and Gusel'nikov (1964) conclude from their experiments that afferent pathways to the forebrain for different forms of sensation are beginning to differentiate at this phylogenetic level. In this respect the neurophysiological evidence agrees with the older anatomical findings.

ANATOMICAL CONSIDERATIONS

The structure of the amphibian brain has been extensively studied by many investigators. C. J. Herrick (1948) spent many years concentrating on the salamander brain. This great effort was ably summarized in his book, *The Brain of the Tiger Salamander*. Here he not only reviewed a vast body of minute anatomical detail, but he also discussed many ideas concerning vertebrate brain function. Herrick, nevertheless, seemed singularly unaware of, or disinterested in, the considerable body of experimental and neurologic evidence then available, preferring, for the most part, to deduce his ideas on brain function mostly from the anatomical picture. Concerning the forebrain, he was impressed by (1) the relatively large olfactory bulbs from which secondary and tertiary olfactory fibers pervade all parts of the forebrain and (2) the poorly differentiated thalamic fasciculus (thalamo-frontal tract), which is the only afferent sensory projection from lower centers to the forebrain. He viewed this tract as the common primordium of all of the thalamo-cortical projection systems of mammals. While Herrick stressed the olfactory dominance of forebrain activities and the very limited influence of the other sensory modalities on forebrain function at this phylogenetic level, he also emphasized the nonspecific arousal properties of olfaction. He concluded that ". . . the olfactory sense, lacking any localizing function of its own, co-operates with other senses in . . . the activation and sensitizing of the nervous system as a whole and of certain appropriately attuned sensori-motor systems in particular, with resulting lowered threshold for excitation for all stimuli and differential reinforcement or inhibition of specific types of response. The olfactory cortex (and its predecessors in lower vertebrates) may, then, serve for nonspecific facilitation of other activities. . . . This facilitation may involve both general excitatory action and general inhibition" (p. 95). In mammals the "cooperation with other senses" occurs primarily in the forebrain where all the senses are well represented, but in amphibia the flow is mostly downward to the epithalamus and hypothalamus, so that the sensory integration that Herrick speaks of takes place in the diencephalon, midbrain, and lower centers.

Although Herrick rarely mentioned the functional experiments, his conclusions are nevertheless largely in agreement with these studies. As reviewed in this section and in the previous section on teleosts, a major function of the forebrain of lower vertebrates is nonspecific facilitation of lower centers. Herrick viewed this as essentially an olfaction-dominated process. Apparently, Herrick was unaware that this system can function effectively in teleosts and amphibians in the absence of all olfactory input. It appears to this reviewer that we are dealing here with a system that developed historically through the influence of olfactory stimulation, but at the teleost and amphibian levels the system had gained a considerable degree of independence from this sensory modality. On the other hand, we should recall again the unorthodox position taken by Riss, Halpern, and Scalia (1968) that the limbic system was the primary mechanism of the forebrain in ancestral vertebrates and that the olfactory system evolved from the limbic system.

Reptiles

SPONTANEITY

Ten Cate (1937), in reviewing the older literature (e.g., Fano, 1884; Sergi, 1905) clearly showed that forebrainless reptiles displayed few spontaneous movements, but those that did occur appeared normal (Bickel, 1901). More recently, Goldby (1937) and Diebschlag (1938) also demonstrated the lack of spontaneous movements in forebrainless lizards and snakes. With strong stimulation, however, locomotion did occur. As with frogs and toads, the posterior third of the subpallium is the critical area. Removal of one hemisphere or the pallial (roof) portions of both hemispheres did not affect spontaneity. Completely forebrainless subjects did not eat, but if the posterior third of the ventral portion of the forebrain was not disturbed, then feeding continued. Diebschlag explains the deficit as a reduction in response to environmental changes, and presents several other experiments which support this conclusion. Spigel and Ellis (1966) reported that lesions in the forebrain of the turtle *Chrysemys* that involved the general cortex and extended into the striatum produced decrements in climbing (escape) responses. They conclude that the reptilian forebrain normally participates in "emotionally integrative behavior patterns." Hertzler and Hayes (1967) tested box turtles, *Terrapene,* and red eared turtles, *Pseudemys,* for optokinetic and visual cliff responses. Extensive lesions in the cerebral cortex or tectum alone did not significantly alter these responses, but combined forebrain and tectal lesions produced notable decrements in behavior. The importance of the general cortex for visual function is emphasized. Goldby (1937) reported that forebrain stimulation or extirpation influenced the motor processes of the tectum.

Masai, Kusunoki, and Ishibashi (1966) found the enzyme monoamine oxidase (a catecholamine inhibitor) in those areas of the forebrain and diencephalon of turtles that are believed to be precursors of the limbic system of mammals. Decrements in the arousal mechanism (i.e., loss of spontaneity) in forebrainless reptiles is probably related to disruption of this system.

LEARNING

Hasratian and Alexanian (1933) studied the defensive reflex to mechanical stimulation of the carapace in turtles. After forebrain removal, these reflexes were retained or new ones were established. Diebschlag (1938) trained emerald lizards to discriminate between yellow and blue bowls. When the forebrain was completely removed, the lizards did not respond to either bowl, reflecting a loss in spontaneous movement. When one hemisphere or both hemispheres except for the posterior third of the subpallium were removed, then the discrimination was retained and re-training to the opposite color was successful (see pp. 89, 90).

ANATOMICAL CONSIDERATIONS

In reptiles, a general cortex is clearly differentiated from the medially situated hippocampus (archipallium) and from the laterally situated pyriform area (paleopallium). All three structures have developed a cortex-like arrangement (Ariëns Kappers, Huber and Crosby, 1936; Goldby and Gamble, 1957). The subpallium consists of the septum and a strio-amygdaloid complex which is variously subdivided.

The presence of a well-defined general cortex has stimulated considerable interest among the electrophysiologists whose experiments will be described in the next two sections. The presence of a general cortex also introduces the possibility in reptiles of thalamic relay nuclei that project sensory impulses of differing modalities on the cortex. Powell and Kruger (1960) and Kruger and Berkowitz (1960) have investigated this question using the methods of retrograde degeneration and the Bodian silver stain. They found that the nucleus rotundus and nucleus anterior medialis of the thalamus project upon the lateral paleostriatum via components of the lateral forebrain bundle. They did not find pathways from the thalamus to any parts of the pallium. On the other hand, Motorina (1965) has demonstrated, by using the Nauta method, that all parts of the pallium of *Varanus* receive an extensive number of fibers from the hypothalamus.

MOTOR RESPONSES

Most of the early investigators were able to demonstrate some motor activity after chemical or electrical stimulation of the forebrain (Johnston, 1916; Bagley and Langworthy, 1926; Tuge and Yazaki, 1934). Koppanyi and Pearcy (1925) were unable to obtain any reactions following electrical stimulation of the general cortex in turtles, but they did obtain positive responses by stimulating the striatum. Bremer, Dow, and Moruzzi (1939) demonstrated, by cocainizing the cortex, that any movements obtained from the pallial structures are due to the spread of current to subcortical areas. Using permanently implanted electrodes, Shapiro (1965) also obtained locomotory activity by stimulating the striatum of the alligator *Caiman,* but not after stimulating the dorsal pallium. Motor responses resembling integrated flight behavior were obtained by Kormann and Horel (1967) following electrical stimulation of the amygdalar region of the alligator. Goldby and Gamble (1957) believe that electrical stimulation of the reptilian forebrain produces only total responses which are not individuated. These are mediated by the medial forebrain bundle that runs to the tegmentum. They doubt that there are direct connections between the forebrain and motor nuclei of the spinal cord.

NEUROPHYSIOLOGICAL STUDIES

In recent years EEG studies of the cerebral cortex have provided new insights into forebrain function in reptiles. It became evident several years ago, especially from the studies of Voronin and Gusel'nikov (1959, 1963) and Gusel'nikov and Supin (1964) that stimuli of several modalities, particularly optic, auditory, and somatosensory, could evoke electrical potentials from the same points on the surface of the pallium. The most effective area for recording these evoked potentials is the hippocampus, particularly, according to Voronin and his coworkers, the dorsomedially situated fascia dentata. However, Kruger and Berkowitz (1960), Moore and Tschirgi (1962), and Mazurskaya, Davydova, and Smirnov (1966) obtained their best recordings from an area coextensive with the general cortex. This area lies dorsally between the hippocampal and pyriform regions. Karamian and Belekhova (1963) recorded evoked potentials from both the general and hippocampal cortices. It is apparent, therefore, that several sensory modalities project in superimposed fashion on these parts of the forebrain.

The most striking feature that the above studies have revealed is some manner of cortical association between the optic, auditory, and somatosensory systems. This has been demonstrated in several ways. When recording from the hippocampus of the lizard *Agama,* Voronin and Gusel'nikov (1963) showed that a second stimulus of like modality, which followed rapidly after the first, was blocked by the refractory period that normally follows the first evoked potential. A similar blocking action was found when a different sensory modality was used for the second stimulus, suggesting to these authors a true convergence of the three sensory modalities. In reptiles, therefore, the forebrain becomes the connecting associative link between various sensory systems, the nuclear zones of which are still at diencephalic or lower levels. By recording from single units, Gusel'nikov and Supin (1964) demonstrated true convergence of two sensory modalities on the same cell.

Belekhova (1963), Karamian and Belekhova (1963), and Karamian et al. (1966) describe a retino-thalamo-cortical system in the turtle *Emys* and the lizard *Varanus,* which they consider on the basis of their electrophysiological evidence to be a nonspecific projection system. They liken this mechanism to the secondary, multisynaptic thalamo-cortical system of mammals, which is also nonspecific, and they contrast it with the highly specific primary projection areas of mammals.

In the experiment by Mazurskaya et al. (1966) an interaction effect of different sensory systems on potentials evoked from the dorsal (general) cortex of the pond turtle, *Emys,* was not obtained. They found that responses to simultaneous combinations of photic and cutaneous stimuli were equal to the sum of the responses to each stimulus separately. They conclude on the basis of this experiment that the various afferent impulses forming the evoked potentials do not possess common pathways to the cortex and that they are bound for separate cells. The convergence reported in earlier studies occurs on cortical neurons having special dendritic characteristics. Contrary to the numerous studies cited above, Weisbach and Schwartzkopff (1967) have located an area that is electrically responsive to sound stimuli in the somatic striatum, just ventral to the general cortex.

Sollertinskaya (1967) established a defensive conditioned reflex in the lizard *Varanus,* using electric shock to the tail as the US and light or sound as the CS. Marked increases in frequency and amplitude of the background EEG rhythm occurred in the general and hippocampal cortex during the formation of the conditioned response.

Tauber, Rojas-Ramirez, and Peón (1968) recorded an EEG arousal reaction from the pyriform and hippocampal lobes of the forebrain of the lizard *Ctenosaura* in response to general environmental stimuli (presence of investigator). This consisted of an increase in voltage associated at times with an increase in frequency. Sleep, in contrast with wakefulness, was characterized by diminution in amplitude of electrical activity. Both REM and non-REM sleep could be distinguished by their special EEG characteristics. Belekhova and Zagorul'ko (1964) found an intensification of electrical activity in the anteromedial part of the cerebral cortex (hippocampus) of the tortoise *Emys* following rapid photic stimulation (50 flashes per second). This reaction was similar to the after-discharge that follows the potential evoked by a single photic stimulus.

OLFACTION

Since a pair of prominent olfactory nerves enter the forebrain, it seems obvious that the cerebral hemispheres must serve this function, and, according to the classical picture of the reptilian forebrain, it is largely an olfactory correlation center. Unfortunately, we have not found any functional studies in support of this fundamental and widespread assumption. In a few of the EEG studies, however, the olfactory nerve was stimulated electrically with rather interesting results. Whereas the optic, auditory, and cutaneous discharges concentrate in the same zone of the anterior dorsal cortex (hippocampus and general cortex), evoked potentials following electrical stimulation of the olfactory nerve were recorded from the posterior lateral portion of the hemisphere (Mazurskaya et al., 1966) and from the septum (Voronin and Gusel'nikov, 1963). Moreover, the so-called tertiary olfactory tracts to the hippocampus do not conduct olfactory information. Hence, there is only minimal olfactory representation in the hippocampus where the other sensory systems converge. As in teleosts (p. 82), the olfactory mechanism seems confined to a limited portion of the forebrain, mostly in the subpallium.

GENERAL CONSIDERATIONS

Since the only clearly defined sensory afferents to the forebrain of lower vertebrates are olfactory and possibly gustatory, the classical neuroanatomists viewed this part of the brain as a site primarily for the reception and dissemination of olfactory information (Papez, 1929) and for the correlation of olfaction and taste (Herrick, 1922). Starting with amphibians, there is evidence for the projection of other sensory afferents from the dorsal thalamus to the strio-amygdaloid area of the forebrain (Herrick, 1948). In reptiles these sensory projections are more extensive and they obviously are the evolutionary forerunners of the massive sensory integrative mechanisms characteristic of the mammalian cerebrum. There is considerable truth in this early conceptualization of the functional evolution of the forebrain, but it should be clear from the present review that it presents only a portion of the total picture.

As far back among the invertebrates as we can confidently trace the direct ancestors of vertebrates, there is evidence of a chemosensory input into the anterior end of the neural tube. The input in these forms is not yet defined as olfactory, nor is the front end of the neural tube differentiated into a forebrain. In *Amphioxus,* the beginnings of a forebrain are recognized, and there are suggestions of an olfactory nerve. In the ostracoderms, the earliest vertebrates known, paleontologists describe a pair of olfactory nerves entering the forebrain, and this fundamental relationship persists in all later vertebrate groups. There is little doubt that from its inception the forebrain was concerned with olfaction, although an alternative view (see pp. 80, 92) has recently been presented by Riss et al. (1968). The question still to be answered is whether the primordial forebrain had functions other than olfaction. If it did, when and how did these functions arise?

The second major function of the forebrain is that of an arousal mechanism that energizes the functional activities of lower centers. This is well demonstrated in teleosts, is more noticeable in amphibians, and seems most pronounced in reptiles, where the so-called loss in spontaneity of forebrainless subjects is a striking phenomenon. This function has not been investigated in species below teleosts, but we would expect to find it, perhaps to a lesser extent, even in the cyclostomes and elasmobranchs, as their nuclear masses and fiber connections are basically similar to the bony fishes (Ariëns Kappers et al., 1936).

We have noted elsewhere (Aronson, 1963) that olfaction consists of two physiological processes: one a specific process providing the basis for discrete discriminations, associations, and orientation; the other is a diffuse, nonspecific process activating other sensorimotor systems which in their simplest forms guide the animal toward or away from a stimulus. In ancestral vertebrates the center for this olfactory arousal mechanism was the forebrain which projected its impulses downward to the diencephalon, midbrain, and lower centers. This formed the basis of the primitive limbic system (Segaar, 1965; Aronson and Kaplan, 1968). The olfactory arousal function gradually enlarged, and somewhere in early vertebrate history became independent of olfaction. This independence is clearly established in bony fishes where the nonolfactory arousal mechanism of the forebrain seems to influence almost every kind of behavior that has been investigated.

Although the major flow of impulses from the forebrain of fishes is downward, there is some forward projection, but this comes mainly from the hypothalamus and presumably consists of visceral and gustatory information. The bulk of the anatomical and neurophysiological evidence suggests that little other sensory information reaches the forebrain even indirectly via the hypothalamus, but this point has not been definitely settled. In amphibians, fibers carrying sensory information from the dorsal thalamus to the striatum are clearly identified and neurophysiological experiments confirm this, particularly in respect to visual, auditory, and somatic sensory stimulation. In reptiles the dorsal thalamus is enlarged, and a well-defined thalamic projection system has been described, but again, this seems to go only to the enlarged striatum. The bulk of the electrophysiological evidence, however, indicates that the sensory impulses of light, sound, and touch (but not olfaction) reach the hippocampal and general cortices where a considerable level of integration occurs, including the true convergence of two or more sensory modalities on the same cell. This may still be a nonspecific mechanism comparable to the secondary, multisynaptic thalamo-cortical system of mammals. Specific primary projection areas, which are characteristic of the mammalian cortex, are evidently not present at the reptilian level. It seems unfortunate that so few experiments have been performed on the role of the forebrain in learning and other behavioral entities in reptiles, for such studies might provide insight into the functional significance of the original forward projections of the sensory systems.

A third forebrain mechanism, namely, a motor system, appears for the first time in reptiles. Evidence from older studies of a true motor system in the forebrain of fishes is now attributed to experimental artifacts (Healy, 1957). In reptiles this motor system is located in the striatum. It produces only total responses which are not individuated. The system projects on the tegmentum of the midbrain and not directly on the motor nuclei of the spinal cord. It most probably represents the precursor of the extrapyramidal system of mammals. True pyramidal (cortico-spinal) tracts do not appear at this phyletic level.

SUMMARY

The forebrain arose in the progenitors of vertebrates in relation to olfactory input. It apparently started as a diffuse system that activated sensorimotor processes organized in other portions of the nervous system. This olfactory activating mechanism persists in almost all vertebrates in relatively unmodified form, but early in vertebrate evolution it gave rise to a general and nonspecific arousal system which could function independently of olfaction. This new system, which is well-defined in teleosts, is elaborated in amphibians and reptiles and constitutes the forerunner of the so-called limbic arousal system of mammals. In fishes the forebrain receives olfactory, visceral, and gustatory input, but in amphibians and even more so in reptiles, projections of several other sensory modalities to the hippocampal and general cortices are evident. Here, true associative functions have been revealed by electrophysiological experimentation. These sensory projections are still thought to form a nonspecific system that is apparently the precursor of the secondary thalamo-cortical system of mammals. A motor system, which is most probably the precursor of the mammalian extra-pyramidal system, also makes its appearance for the first time in reptiles.

ACKNOWLEDGMENTS

The author wishes to thank Dr. Ethel Tobach and Harriett Kaplan for reading the manuscript and for many helpful suggestions. Several students of The American Museum's Undergraduate Research Training Program, supported by Special Projects in Science Education, National Science Foundation Grant GY-350, participated in the author's recent experiments that are referred to in this article.

REFERENCES

Agalides, E. 1967. The electrical activity of the forebrain of sharks recorded with deep chronic implanted electrodes. *Trans. N.Y. Acad. Sci.* 29 (4): 378–389.

Ariëns Kappers, C. U., G. C. Huber, and E. Crosby. 1936. *The comparative anatomy of the nervous system of vertebrates including man.* New York: Hafner.

Aronson, Lester R. 1948. Problems in the behavior and physiology of a species of African mouthbreeding fish. *Trans. N.Y. Acad. Sci.* 2, II: 33–42.

Aronson, Lester R. 1963. The central nervous system of sharks and bony fishes with special reference to sensory and integrative mechanisms. In P. W. Gilbert, ed., *Sharks and survival.* Boston: D. C. Heath. Chap. 6.

Aronson, Lester R., and R. Herberman. 1960. Persistence of a conditioned response in the cichlid fish, *Tilapia macrocephala,* after forebrain and cerebellar ablations. *Anat. Rec.* 138 (3): 332.

Aronson, Lester R., and H. Kaplan. 1963. Forebrain function in avoidance conditioning. *Am. Zool.* 3 (4): 483–484.

Aronson, Lester R., and H. Kaplan. 1965. Effect of forebrain ablation on the acquisition of a conditioned avoidance response in the teleost fish, *Tilapia h. macrocephala. Am. Zool.* 5 (4): 654.

Aronson, Lester R., and H. Kaplan. 1968. Function of the teleostean forebrain. In D. Ingle, ed., *The central nervous system and fish behavior* Chicago: Univ. Chicago Press. Pp. 107–125.

Aronson, Lester R., and G. K. Noble. 1945. The sexual behavior of Anura. 2. Neural mechanisms controlling mating behavior in the leopard frog, *Rana pipiens. Bull. Am. Mus. Nat. Hist.* 86: 87–139.

Bagley, C., and O. Langworthy. 1926. The forebrain and the midbrain of the alligator with experimental transections of the brainstem. A study of electrically excitable regions. *Arch. Neur. Psychiat.* 16: 154–166.

Belekhova, M. G. 1963. Electrical activity in cerebral hemispheres of *Varanus* evoked by diencephalic stimulation. *Fiziol. Zh. SSSR* 49 (11): 1318. (*Fed. Proc.* 24 (1): T159–T165.)

Belekhova, M. G., and T. M. Zagorul'ko. 1964. Correlations between background electrical activity, after-discharge and EEG activation response to photic stimulation in tortoise brain (*Emys lutaria*). *Zh. Vyssh. Nerv. Deyatel' nosti* 14 (6): 1079. (*Fed. Proc.* 24 (6): T1028–T1032.)

Bernstein, J. J. 1962. Role of the telencephalon in color vision of fish. *Exp. Neurol.* 6: 173–185.

Bert, J., and R. Godet. 1963. Réaction d'éveil télencéphalique d'un Dipneuste. *Compt. Rend. Soc. Biol.* 157 (10): 1787–1790.

Berwein, M. 1941. Beobachtungen und Versuche über das gesellige Leben von Elritzen. *Z. Vergleich. Physiol.* 28: 402–420.

Bethe, A. 1899. Die Lokomotion des Haifisches (*Scyllium*) und ihre Beziehung zu den einzelnen Gehirnteilen und zum Labyrinth. *Arch. Ges. Physiol.* 76: 470–493.

Bianki, B. L., and Z. R. Dushabaev. 1964. Morphophysiological structure of visual analysor of amphibians in relation to its paired function. *Vestnik Lenin. Univ.* (*Ser. Biol.*) 3 (1): 88. (*Fed. Proc.* 24 (2): T351–T356.)

Bickel, A. 1901. Beiträge zur Gehirnphysiologie der Schildkröte. *Arch. Physiol.* 1901: 52–80.

Blankenagel, F. 1931. Untersuchungen über die Grosshirnfunctionen von Rana temporaria. *Zool. Jahrb. Abt. Allg. Zool. u. Physiol. der Tiere* 49: 271–322.

Bremer, F., R. S. Dow, and G. Moruzzi. 1939. Physiological analyses of the general cortex in reptiles and birds. *J. Neurophysiol.* 2: 473–487.

Bullock, T. H. 1940. The functional organization of the nervous system of the Enteropneusta. *Biol. Bull.* 79 (1): 91–113.

Bullock, T. H. 1945. The anatomical organization of the nervous system of the Enteropneusta. *Quart. Microscop. Sci.* 86 (341): 55–111.

Bullock, T. H. 1965. Chaetognatha, Pogonophora, Hemichordata, and Tunicata. In T. H. Bullock and G. A. Horridge, eds., *Structure and function in the nervous systems of invertebrates.* San Francisco: W. H. Freeman and Company. Chap. 27.

Burnett, T. C. 1912. Some observations on decerebrate frogs with special reference to the formation of associations. *Am. J. Physiol.* 30: 80–87.

Celesia, G., and H. H. Jasper. 1966. Acetylcholine released from cerebral cortex in relation to state of activation. *Neurology* 16 (11): 1053–1064.

Chauchard, A., and B. Chauchard. 1927. Recherches sur les localisations cérébrales chez les poissons. *Compt. Rend. Acad. Sci. Pons* 184: 696–698.

Chernetski, K. E. 1964. Facilitation of a somatic reflex by a sound in *Rana clamitans:* Effects of sympathectomy and decerebration. *Z. Tierpsychol.* 21 (7): 813–821.

Delsman, H. C. 1918. Short history of the head of vertebrates. *Proc. Konin. Akad. v. Weten.* (Sci. Sec.) 20 (2): 1005–1020.

Demski, Leo. 1969. Behavioral effects of electrical stimulation of the brain of the sunfish, *Lepomis macrochirus. Anat. Rec.* 163 (2): 177.

Desmoulins, A. 1825. *Anatomie des systèmes nerveux des animaux à vertébres appliquée à la physiologie et à la zoologie.* Paris: Méquignon Marvis.

Diebschlag, E. 1934. Zur Kenntnis der Grosshirnfunktionen einiger Urodelen und Anuren. *Z. Vergleich. Physiol.* 21: 343–394.

Diebschlag, E. 1938. Beobachtungen und Versuche an intakten und grosshirnlosen Eidechsen und Ringelnattern. *Zool. Anz.* 124 (1/2): 30–40.

Douglas, R. J., and K. H. Pribram. 1966. Learning and Limbic Lesions. *Neuropsychologia* 4: 197–220.

Fano, G. 1884. Saggio sperimentale sul meccanismo dei movimenti voluntari nella testuggine palustre (*Emys europaea*). *Pubbl. Ist. Stud. Super.,* Florence.

Fiedler, Kurt. 1965. Versuche zur Neuroethologie von Lippfischen und Sonnenbarschen. *Zool. Anz.,* Suppl. 28: 569–580.

Fiedler, Kurt. 1967. Degenerationen und Verhaltenseffekte nach Elektrokoagulationen im Gehirn von Fischeu (*Diplodus, Crenilabrus*—Perciformes). *Zool. Anz.,* Suppl. 30: 351–366.

Fiedler, Kurt. 1968. Verhaltenswirksome Strukturen im Fischgehirn. *Zool. Anz.,* Suppl. 31: 602–616.

Flourens, J. P. M. 1824. *Recherches expérimentales sur les propriétés et les functions du système nerveux dans les animaux vertébrés.* Paris.

Fol, H. 1872. Études sur les Appendiculaires du Détroit de Messine. *Mem. Soc. Phys. et d'hist. nat. de Genève* 21 (2): 1–55.

Gilbert, P. W., E. S. Hodgson, and R. F. Mathewson. 1964. Electroencephalograms of sharks. *Science* 145 (3635): 949–951.

Godet, R., J. Bert, and H. Collomb. 1964. Apparition de la réaction d'éveil télencéphalique chez *Protopterus annectens* et cycle biologique. *Compt. Rend. Soc. Biol.* 158 (1–4): 146–149.

Goldby, F. 1937. An experimental investigation of the cerebral hemispheres of *Lacerta viridis. J. Anat.* 71: 332–355.

Goldby, F., and H. J. Gamble. 1957. The reptilian cerebral hemispheres. *Biol. Rev.* 32 (4): 383–420.

Gloor, P. 1960. Amygdala. In J. Field, H. Magoun, and V. Hall, eds., *Handbook of physiology.* Section 1, Neurophysiology, vol. 2, Washington, D. C.: Am. Physiol. Soc. Pp. 1395–1420.

Grimm, R. J. 1960. Feeding behavior and electrical stimulation of the brain of *Carassius auratus. Science* 131 (3394): 162–163.

Gusel'nikov, V. I., M. I. Onufrieva, and A. Ya. Supin. 1964. Projection of visual, olfactory and lateral line receptors in the fish brain. *Fiziol. Zh. SSSR* 50 (9): 1104. (*Fed. Proc.* 24 (5-2): T768.)

Gusel'nikov, V. I., and A. Ya. Supin. 1964. Somatosensory and olfactory representation in the forebrain of the lizard (*Agama caucasia*). *Fiziol. Zh. SSSR* 50 (2): 129. (*Fed. Proc.* 24 (3): T426–T430.)

Hainsworth, F. R., J. B. Overmier, and C. T. Snowdon. 1967. Specific and permanent deficits in instrumental avoidance responding following forebrain ablation in the goldfish. *J. Comp. Physiol. Psych.* 63 (1): 111–116.

Hale, E. B. 1956. Social facilitation and forebrain function in maze performance of green sunfish, *Lepomis cyanellus. Physiol. Zool.* 29 (2): 93–107.

Hale, E. B. 1956a. Effects of forebrain lesions on the aggressive behavior of green sunfish, *Lepomis cyanellus. Physiol. Zool.* 29 (2): 107–127.

Hara, T. J., and A. Gorbman. 1967. Electrophysiological studies of the olfactory system of the goldfish, *Carassius auratus* L. Part I. Modification of the electrical activity of the olfactory bulbs by other central nervous structures. *Comp. Biochem. Physiol.* 21: 185–200.

Hara, T. J., K. Ueda, and A. Gorbman. 1965. Electroencephalographic studies of homing salmon. *Science* 149 (3686): 884–885.

Hasler, A. D. 1966. *Underwater guideposts.* Madison: Univ. Wisconsin Press.

Hasratian, E., and A. Alexanian. 1933. Bedingtreftexlorische Tätegkeit bei Schildkröten mit entfernten Grosshirnhemisphären und Zwischenhirn. *Fiziol. Zh. SSSR* 16: 887–891.

Hassler, R., and H. Stephan. 1967. *Evolution of the forebrain.* New York: Plenum Press.

Hayes, W., and D. Hertzler. 1967. Role of the optic tectum and general cortex in reptilian vision. *Psychonomic Sci.* 9 (9): 521–522.

Healey, E. G. 1957. The nervous system. In M. E. Brown, ed., *The physiology of fishes,* vol. 2. New York: Academic Press. Chap. 1.

Heier, P. 1948. Fundamental principles in the structure of the brain; a study of the brain of *Petromyzon fluviatilis. Acta Anat.* 5 (Suppl. 8): 1–213.

Herrick, C. J. 1922. Functional factors in the morphology of the forebrain of fishes. In *Libro en honor de D. S. Ramón y Cajal,* vol. 1. Madrid: Jiménez y Molina. Pp. 143–204.

Herrick, C. J. 1924. *Neurological foundations of animal behavior.* New York: Henry Holt.

Herrick, C. J. 1948. *The brain of the tiger salamander Ambystoma tigrinum.* Chicago: Univ. Chicago Press.

Hertzler, D., anl W. Hayes. 1967. Cortical and tectal function in visually guided behavior of turtles. *J. Comp. Physiol. Psych.* 63 (3): 444–447.

Hosch, L. 1936. Untersuchungen über Grosshirnfunktionen der Elritze (*Phoxinus laevis*) und des Gruendlings (*Gobio fluviatilis*). *Zool. Jahrb.* 57 (3): 57–98.

Hyman, L. H. 1959. *The invertebrates,* vol. 5. *Smaller coelomate groups.* New York: McGraw-Hill.

Ingle, D. 1965. Behavioral effects of forebrain lesions in the goldfish. *Proc. Am. Psychol. Assoc.* 1: 143–144.

Johnston, J. B. 1906. *Nervous system of vertebrates.* Philadelphia: Blakiston.

Johnston, J. B. 1911. The telencephalon of ganoids and teleosts. *J. Comp. Neurol.* 21 (6): 489–591.

Johnston, J. B. 1912. On the teleostean forebrain. *Anat. Rec.* 6 (11): 423–438.

Johnston, J. B. 1912a. The telencephalon in cyclostomes. *J. Comp. Neurol.* 22 (4): 341–404.

Johnston, J. B. 1916. Evidence of a motor pallium in the forebrain of reptiles. *J. Comp. Neurol.* 26 (5): 475–479.

Kamrin, R. P., and L. R. Aronson. 1955. The effects of forebrain lesions on mating behavior in the male platyfish, *Xiphiphorus maculatus. Zoologica* (N.Y.) 39 (4): 133–140.

Kaplan, H., and L. R. Aronson. 1967. Effect of forebrain ablation on the performance of a conditioned avoidance response in the teleost fish, *Tilapia h. macrocephala. Anim. Behav.* 15 (4): 436–446.

Kaplan, H., and L. R. Aronson. 1969. Function of the forebrain and the cerebellum in the acquisition and performance of a conditioned avoidance response in the teleost fish, *Tilapia h. macrocephala. Bull. Amer. Mus. Nat. Hist.* 142 (2): 141–208.

Karamian, A. I. 1956. *Evolution of the function of the cerebellum and cerebral hemispheres.* Medgiz, Leningrad. Translated by National Science Foundation, Washington, D.C.

Karamian, A. I. 1965. Evolution of functions in the higher nervous divisions of the central nervous system and their regulating mechanisms. In W. Pringle, ed., *Essays on physiological evolution.* New York: Pergamon Press.

Karamian, A. I., and M. G. Belekhova. 1963. Functional evolution of nonspecific thalamocortical system. *Zh. Vyssh. Nerv. Deyatel'nosti* 13 (5): 904. (*Fed. Proc.* 23 (6): T1189–T1194.)

Karamian, A. I., N. P. Vesselkin, M. G. Belekhova, and T. M. Zagorul'ko. 1966. Electrophysiological characteristics of tectal and thalamocortical divisions of the visual system in lower vertebrates. *J. Comp. Neurol.* 127: 559–576.

Kleerokoper, H., and G. A. van Erkel. 1960. The olfactory apparatus of *Petromyzon marinus* L. *Can. J. Zool.* 38 (1): 209–223.

Knight-Jones, E. W. 1952. On the nervous system of *Saccoglossus cambrensis* (Enteropneusta). *Phil. Trans. Roy. Soc. (London)*, B, 236: 315–354.

Kölliker, A. 1843. Ueber das Geruchsorgan von Amphioxus. *Arch. f. Anat., Physiol. u. Wiss. Med.* 1843: 32–35.

Kôppanyi, T., and J. F. Pearcy. 1925. Comparative studies on the excitability of the forebrain. *Am. J. Physiol.* 71 (2): 339–343.

Kormann, L. A. and J. A. Horel. 1967. Electrical stimulation of the amygdaloid complex of the alligator (*Caiman sklerops*). *Psychonomic Bull.* 1 (2): 23.

Kruger, L., and E. Berkowitz. 1960. The main afferent connections of the reptilian telencephalon as determined by degeneration and electrophysiological methods. *J. Comp. Neurol.* 115 (2): 125–142.

Kumakura, S. 1928. Versuche an Goldfischen, denen beide Hemisphären des Grosshirns exstirpiert worden waren. *Nagoya J. Med. Sci.* 3: 19–24.

Kusunoki, T., and H. Masai. 1966. Chemoarchitectonics in the central nervous system of goldfish. *Arch. Hist. Jap.* 27: 363–371.

Loeb, J. 1891. Über den Anteil des Hörnerven an den nach Gehirnverletzung auftretenden zwangsbewegungen, Zwangslagen und assoziierten Stellungsänderungen der Bulbi und Extremitäten. *Arch. ges. Physiol.* 50: 66–83.

Loeb, J. 1900. *Comparative physiology of the brain and comparative psychology.* London: G. P. Putnam's Sons.

Loeser, W. 1905. A study of the functions of different parts of the frog's brain. *J. Comp. Neurol.* 15: 355–373.

Lohmann, Hans. 1933. Appendiculariae. In W. Kükenthal and T. Krumbach, eds., *Handbucher der Zoologie,* 5 (2): 15–202. Berlin: Walter de Gruyter.

Martini, E. 1909. Studien über die Konstanz histologischer Elemente. I. Oekopleura longicauda. *Z. Wiss. Zool.* 92: 563–626.

Masai, H., T. Kusunoki, and H. Ishibashi. 1966. The chemoarchitectonics in the forebrain of reptiles. *Experientia* 22 (11): 745–746.

Mayer, F. 1897. Das Centralnervensystem von Ammocoetes. *Anat. Anz.* 13 (24): 649–657.

Mazurskaya, P. Z., T. V. Davydova, and G. D. Smirnov. 1966. Functional organization of exteroceptive projections in the forebrain of the turtle. *Fiziol. Zh. SSSR* 52 (9): 1050. (*Neurosci. Trans.* 1967–68 (1): 109–117.)

Moore, G. P., and R. D. Tschirgi. 1962. Non-specific responses of reptilian cortex to sensory stimuli. *Exp. Neurol.* 5 (3): 196–209.

Motorina, M. 1965. Development of hypothalamic-cortical interrelationships in reptiles. *Zh. Evolyuts Biokhim. Fiziol.* 1 (3): 262–268 (Engl. summ.).

Nagel, W. A. 1894. Geruchs und Geschmackssinn und ihre Organe. *Bibliotheca Zool.* 7 (18): 1–207.

Nieuwenhuys, R. 1967. Comparative anatomy of olfactory centers and tracts. In Y. Zotterman, ed., *Progress in brain research,* vol. 23. New York: Elsevier. Pp. 1–64.

Noble, G. K. 1936. Function of the corpus striatum in the social behavior of fishes. *Anat. Rec.* 64: 34.

Noble, G. K. 1939. Neural basis of social behavior in vertebrates. *Collecting Net* 14: 121–124.

Noble, G. K., and R. Borne. 1941. The effect of forebrain lesions on the sexual and fighting behavior of *Betta splendens* and other fishes. *Anat. Rec.* 79 (3) Suppl.: 49.

Noll, A. 1931. Über den Einfuss des Grosshirns auf den Wischreflex des Frosches. *Arch. Ges. Physiol.* 229: 198–204.

Papez, J. W. 1929. *Comparative neurology.* New York: Thomas Crowell.

Papez, J. W. 1937. A proposed mechanism of emotion. *Arch. Neurol. Psychiat.* 38: 725–743.

Parker, G. H. 1908. The sensory reactions of Amphioxus. *Proc. Am. Acad. Arts Sci.* 43 (16): 415–455.

Polimanti, O. 1911. Contributi alla fisiologia del sistema nervoso centrale e del movimento dei pesci. I. Selacoidei. *Zool. Jahrb., Abt. Allg. Zool. u. Physiol. der Tiere* 30: 473–716.

Polimanti, O. 1912. Contributi alla fisiologia del sistema nervoso centrale e del movimento dei pesci. II. Batoidei. *Zool. Jahrb., Abt. Allg. Zool. u. Physiol. der Tiere* 32: 311–366.

Powell, T. P. S., and L. Kruger. 1960. The thalamic projection upon the telencephalon in *Lacerta vividis. J. Anat.* 94: 528–542.

Rémy, C., and J. J. Bounhiol. 1962. Quelques résultats de l'ablation du télencéphale chez le têtard d'*Alytes obstetricians* (Laur.). *J. Physiol.* (Paris) 54 (2): 404–405.

Reznitchenko, P. N. 1962. Analysis of the role of different sections of the brain of the axolotl in locomotor function. *Akad. Nauk. SSSR Trudy Inst. Morph.* 40: 78–122.

Riss, W., M. Halpern and F. Scalia. 1969. Anatomical aspects of the evolution of the limbic and olfactory systems and their potential significance for behavior. *Ann. N.Y. Acad. Sci.* 159 (3): 1096–1111.

Rizzolo, A. 1929. A study of equilibrium in the smooth dogfish (*Galeus canis* Mitchell) after removal of different parts of the brain. *Biol. Bull.* 57 (4): 245–249.

Savage, G. E. 1968. Temporal factors in avoidance learning in normal and forebrainless goldfish (*Carassius auratus*). *Nature* 218: 1168–1169.

Schildkraut, J. J., and S. Kety. 1967. Biogenic amines and emotion. *Science* 156: (3771): 21–30.

Schmidt, R. S. 1966. Central mechanisms of frog calling. *Behaviour* 26 (3–4): 251–285.

Schmidt, R. S. 1968. Preoptic activation of frog mating behavior. *Behaviour* 30 (2–3): 239–257.

Schnitzlein, H. N. 1964. Correlation of habit and structure in the fish brain. *Am. Zool.* 4 (1): 21–32.

Schönherr, Josef. 1955. Über die Abhängigkeit der Instinkthandlungen vom Vorderhirn und Zwischenhirn (Epiphyse) bei *Gasterosteus aculeatus* L. *Zool. Jahrb. abt. Physiol.* 65 (4): 357–386.

Schrader, M. E. G. 1887. Zur Physiologie des Froschgehirns. *Arch. Ges. Physiol.* 41: 75–90.

Segaar, J. 1961. Telencephalon and behavior in *Gasterosteus aculeatus*. *Behaviour* 28 (4): 256–287.

Segaar, J. 1965. Behavioral aspects of degeneration and regeneration in fish brain: a comparison with higher vertebrates. In M. Singer and J. P. Schadé, eds., *Progress in brain research,* vol. 14. New York: Elsevier. Pp. 143–231.

Segaar, J., and R. Nieuwenhuys. 1963. New ethophysiological experiments with male *Gasterosteus aculeatus. Anim. Behav.* 11 (2/3): 331–344.

Segura, E. T., and A. de Juan. 1966. Electroencephalographic studies in toads. *Electroencephalog. Clin. Neurophysiol.* 21 (4): 373–380.

Sergi, S. 1905. Il sistema nervoso centrale nei movimenti della testudo. *Arch. Farmacol. Sper.* 4: 179–192.

Shapiro, H. 1965. Motor functions and their anatomical bases in the forebrain and tectum of the alligator. *Diss. Abstr.* 25 (10): 5492.

Sollertinskaya, T. N. 1967. Influence of the hypothalamus on conditioned reflexes in the reptile, *Varanus griseus. Zh. Vyssh. Nerv. Deyatel'nosti* 17 (3): 473–482. (*Neurosci. Trans.* 1967 (3): 305–312.)

Spigel, I. M., and K. R. Ellis. 1966. Cerebral lesions and climbing suppression in the turtle. *Psychonomic. Sci.* 5 (5): 211–212.

Springer, Mary G. 1928. The nervous mechanism of respiration of the Selachii. *Arch. Neurol. Psychiat.* 19: 834–864.

Steiner, J. 1888. Die Fische. In *Die Functionen des Centralnervensystem und ihre Phylogenese.* Part 2. Braunschweig.

Stensiö, E. 1963. The brain and the cranial nerves in fossil, lower craniate vertebrates. Skrifter Utgett av Det Norske Videnskops—*Akademi i Oslo I. Mat. Naturv. Klasse Ny.* Serie #13.

Streick, F. 1924. Untersuchungen über den Gerruchs- und Geschmackssinn der Elritze (*Phoxinus laevis A.*). *Z. Vergleich. Physiol.* 2: 122–254.

Supin, A. Ya., and V. I. Gusel'nikov. 1964. Representation of visual, auditory and somatosensory systems in frog forebrain. *Fiziol. Zh. SSSR* 50 (4): 426. (*Fed. Proc.* 24 (2): T357–T362.)

Tauber, E. S., J. Rojas-Ramirez, and R. Hernández Peón. 1968. Electrophysiological and behavioral correlates of wakefulness and sleep in the lizard, *Ctenosaura pectinata. Electroenceph. Clin. Neurophysiol.* 24: 424–433.

Ten Cate, J. 1935. Physiologie des Zentralnervensystem der Fische. *Ergeb. Biol.* 11: 335–409.

Ten Cate, J. 1937. Physiologie des Zentralnervensystem der Reptilien. *Ergeb. Biol.* 14: 225–279.

Timkina, M. I. 1966. Relationship between different sensory systems in bony fish. *Zh. Vyssh. Nerv. Deyatel'nosti.* 15 (5): 927. (*Fed. Proc.* 25 (5-2): T750–T752.)

Tretjakoff, D. 1909. Das Nervensystem von Ammocoetes. II. Gehirn. *Arch. Mikro. Anat.* 74 (3): 636–779.

Tuge, H., and M. Yazaki. 1934. Experimental note on the presence of electrically excitable areas in the reptilian cerebral hemispheres: *Clemmys japonica. Sci. Rep. Tohoku Univ.* 9: 79–85.

Ueda, K., T. Hara, and A. Gorbman. 1967. Electroencephalographic studies on olfactory discrimination in adult spawning salmon. *Comp. Biochem. Physiol.* 21: 133–143.

Veselkin, N. P. 1964. Electrical responses in skate brain to photic stimulation. *Fiziol. Zh. SSSR* 50 (3): 268–271. (*Fed. Proc.* 24 (2): T368–T370.)

Voronin, L. G. and V. I. Gusel'nikov. 1959. Some comparative physiological data on the biological reactions of the brain. *Zh. Vyssh. Nervnoi Deyatel'-nosti* 9 (3): 398–408. (Office of Technical Service, Washington, D.C.)

Voronin, L. G., and V. I. Gusel'nikov. 1963. Phylogenesis of internal mechanisms of analysis and integration in the brain. *Zh. Vyssh. Nerv. Deyatel'-nosti* 13 (2): 193. (*Fed. Proc.* 23 (1): T105–T112.)

Warren, J. M. 1961. The effect of telencephalic injuries on learning by paradise fish, *Macropodus opercularis. J. Comp. Physiol. Psych.* 54: 130–132.

Weisbach, W., and J. Schwartzkopff. 1967. Nervöse Antworten auf Schallreiz im Grosshirn von Krokodilen. *Naturwissenschaften* 54 (24): 650.

Wiebalck, Ute. 1937. Untersuchungen zur Funktion des Vorderhirns bei Knochenfischen. *Zool. Anz.* 117: 325–329.

van Wijhe, J. W. 1919. On the nervous terminalis from man to amphioxus. *Pro. Konin. Akad. v. Weten* (Sci. Sec.) 21 (1,2): 172–182.

Zagorul'ko, T. M. 1957. On the localization of cerebral centers of the visual analyzer in the frog. *Fiziol. Zh. SSSR* 43 (12): 1165.

Zagorul'ko, T. M. 1965. Interaction between the forebrain and visual centers of the midbrain in teleosts and amphibians. *Zh. Evolyuts Biokhim. Fiziol.* 1 (5): 449–458 (Engl. summ.)

II

DEVELOPMENT OF BEHAVIOR

GILBERT GOTTLIEB

The Psychology Laboratory
North Carolina Dept. of Mental Health,
Division of Research
Raleigh, North Carolina

Conceptions of Prenatal Behavior

"It is not the gale but the set of the sail
that determines the direction of the ship."

Author unknown

In the classical usage of the term, all present-day theories of prenatal behavioral development can be characterized as epigenetic.[1] This term denotes the fact that patterns of activity and sensitivity are not immediately evident in the initial stages of embryonic development and that the various behavioral capabilities of the organism become manifest only during the course of development. However, major disagreement exists with regard to the fundamental character of the epigenesis of behavior. One viewpoint holds that behavioral epigenesis is predetermined by invariant organic factors of growth and differentiation (particularly neural maturation), and the other main viewpoint holds that the sequence and outcome of prenatal behavior is probabilistically determined by the critical operation of various endogenous and exogenous stimulative events. The purpose of the present chapter is to review the several theories encompassed by each of these viewpoints.[2]

Before presenting the main theories of prenatal behavior, it should be noted that thus far there have been relatively few manipulative experiments performed in this area and that the theories are in most instances strong

---

[1] For a scholarly historical review of the opposing concepts of epigenesis and preformation, the reader is referred to Needham's history of embryology (1959)—no comparable treatise is available for behavioral embryology. The empirical aspects of behavioral embryology are thoroughly reviewed by Carmichael (1963).

[2] Due to the abstract focus of the essay, as well as space limitations, literature citations are highly selective and I apologize to the many researchers whose empirical work is not cited in the text. Readers interested in an exhaustive review of the empirical literature on prenatal behavior are referred to Carmichael's (1963) paper.

inductive generalizations stemming from observation. So, while the observational basis and techniques for behavioral embryology are now well established, the relative lack of manipulative or deductively designed experiments suggests that the theoretical positions described below are highly tentative formulations.

PREDETERMINED EPIGENESIS OF BEHAVIOR

There are two main versions of the concept that the sequence of behavioral epigenesis is predetermined. The first view is that the motility of the embryo or fetus is *autogenous*. This term means that the behavior of the embryo or fetus derives solely from factors intrinsic to the embryo's neuromusculature. The second view is that the motility of the embryo is *reflexogenous,* which means that the behavior of the embryo (fetus) reflects the establishment of particular neural sensorimotor pathways. Both views have the assumption that the behavioral sequence is predetermined by factors of neural growth and differentiation (maturation), which have an essentially invariant schedule of appearance. In this view, nonsensory environmental factors merely support development or allow behavioral development to occur and sensory stimulation does not influence or determine the course of behavioral development in any significant way.

Autogenous Motility

There are two main sets of observations that pertain to the view that motility is an intrinsic derivative of the embryo's neuromusculature (i.e., that early embryonic motility is not activated or modified by sensory stimulation).

1. In 1885, Preyer published a monumental tome on behavioral embryology. Among his conclusions was the generalization that the early bodily movements of the embryo (particularly the chick embryo) begin before the embryo is capable of showing an overt response to exteroceptive sensory stimulation. This generalization suggested that the higher order synaptic connections of the sensorimotor reflex arcs must be incomplete during the early stages of the chick embryo's development and that the embryo's early movements resulted solely from endogenous stimulation or an autonomous discharge of motor neurons.

In 1926, Tracy studied the ontogeny of behavior in toadfish and observed that rhythmic bodily movements occurred long before hatching and that the fish did not respond to tactile stimulation until well after hatching. Tracy concluded that the rhythmic activity of the larvae is uninfluenced by exteroceptive stimulation and that the fundamental basis of motor activity is of endogenous origin. In Tracy's view (as well as in Preyer's), exteroceptive sensory stimulation becomes important and effective only in the later stages of development, and behavior is fundamentally based upon the early autonomy of the embryo's neuromotor system or, at most, upon the endogenous stimulation that affects the motor system.

In the Preyer-Tracy idea concerning the onset of motor movement in birds and teleosts, one can discern the (theoretical) ontogenetic forerunner of the Fixed Action Pattern of modern ethology. Ethologists hold that the postnatal motor movements of many species of animals are autonomously derived, in the sense that they are highly invariant or stereotyped in form, independent of sensory stimulation once initiated, spontaneous (may occasionally occur in the complete absence of exteroceptive sensory stimulation), and are not dependent upon learning or practice (Moltz, 1965).

2. In later years the Preyer-Tracy hypothesis of autogenous motility has been revived and extended by several experiments. In 1934, Windle and Orr reported that the multisynaptic reflex arcs between the dorsal root fibers and the internuncial neurons of the chick embryo are not completed until after six or seven days of embryonic development. In 1939, Visintini and Levi-Montalcini confirmed this fact. Thus, from the neuroanatomical standpoint, the multisynaptic exteroceptive sensory system is unable to influence the motor system of the chick during the early stage of embryonic motility, as indeed suggested by Preyer's behavioral observations and required by the Preyer-Tracy hypothesis of autogenous motility. However, these latter observations did not rule out the possibility that the early movements of the chick embryo resulted from (a) monosynaptic proprioceptive self-stimulation or (b) simply represented passive movements arising from mechanical or vibratory stimulation. For example, during the earliest stage of development the head of the chick and the duck embryo rests on the exposed heart muscle and appears to nod or bob in passive response to the beating of the heart; the embryos are moved by contractions of the amnion, etc.

In addition to their neuroanatomical work, Visintini and Levi-Montalcini noted that the chick embryo's early movements seemed to be periodic or rhythmic, very much like those Tracy had observed in toadfish larvae. This point concerning the temporal periodicity of movement has recently been viewed as especially significant evidence for the autogenous hypothesis.

In an interesting series of experiments, Hamburger and Balaban (Hamburger, 1963) studied the periodicity of the general bodily movements (or movements of parts of the body) of the chick embryo with a view toward ruling out exogenous and endogenous stimulation as participating factors, and establishing autonomous motor neuron discharge as the main causal factor. The main dependent variable of the Hamburger-Balaban studies is the duration of activity and inactivity phases within motility cycles of embryos of the same age. These phases have been found to be quite regular in a statistical sense (treating the data so that the great individual variability is minimized) and the general trend is toward longer periods of activity beginning about day 7. Though this increase in the duration of activity phases corresponds to the time when the multisynaptic sensorimotor junctions are established via the internuncial neurons, Hamburger and Balaban do not feel that exteroceptive sensory stimulation plays a role in the behavior of the embryo at this time (or even at 11 days). They obtained some support for this contention by lightly stroking the skin of the embryo's wing or leg and observing no change in the relative amount of activity or inactivity within the motility cycle as compared to unstimulated control embryos.

Other experiments were performed involving transections of the embryo's spinal cord at various levels including decapitation at 11 days. This latter operation had no immediate effects on the duration of the activity phase of the cycle but resulted in a significant increase in the length of the inactivity phases.[3] The results of these experiments are interpreted as further support for the view that the activity of the embryo and fetus is an autonomous function of its neuromusculature and that exteroceptive sensory stimulation plays no role in the prenatal development of behavior until some very late stage. In other words, according to the modern extension of the autogenous hypothesis, not only is the onset of activity autonomous but behavior remains an autonomous function of the motor system even after the embryo (fetus) is anatomically capable of responding to exteroceptive sensory stimulation. Thus, according to the latest version of the autogenous hypothesis, behavioral epigenesis is predetermined solely by factors intrinsic to the embryo's neuromusculature (probably the rhythmic discharge of motor or internuncial neurons).

For the sake of clarity, it may be useful to list the defining characteristics of autogenous behavior:

1. Autogenous behavior is automatic (does not require sensory stimulation for its initiation).

2. Autogenous behavior is persistent (does not require sensory stimulation for its maintenance).

---

[3] The results of a more recent study by Hamburger et al. (1965) are at variance with these specific findings. However, the newer findings do not absolutely force the authors to change their view that exteroceptive stimulation plays no role in embryonic motility cycles until Day 17 and this view still awaits parametric stimulative examination at the behavioral level. According to the autogenous hypothesis, exposing chick embryos to exteroceptive stimulation at any time prior to Day 17 should cause no change in the activity or inactivity phases of the motility cycle, and this aspect of the autogenous hypothesis remains to be definitively tested.

3. Autogenous behavior is unmodifiable (is impervious to sensory stimulation).

4. Autogenous behavior has its own rhythm or periodicity (which is not dependent upon and cannot be altered by sensory stimulation).

5. Autogenous behavior is spontaneous (defined by the above four characteristics—it is in no sense a function of sensory stimulation).

The above five points are relatively unambiguous and are clearly susceptible to further experimental examination when stated in these terms. To avoid a possible misunderstanding, it should be emphasized that the propositions of the autogenous hypothesis refer only to generalized and rather abstracted motility cycles (temporally defined phases of activity and inactivity) and not to the intensity of activity, to the form or pattern of the behavior, or to the relationship or succession between movements of one part of the body and another. Motility cycles are conceived as the "underlying" or fundamental property of behavioral function within which or upon which the particulars of behavior occur and are dependent. In this sense, for example, though the Fixed Action Patterns studied by ethologists may be tentatively considered autogenous in their derivation and share certain defining characteristics of motility cycles, the ethological emphasis is on the characteristic form or pattern of movement of specific parts of the animal, not only on the periodicity and spontaneity of these movements. From the standpoint of the analysis of prenatal motility cycles, the particular characteristics of the activity are irrelevant—nodding, bending, turning-to, turning-away, jerking, flipping, swallowing, paddling, bill-clapping, etc., are considered to be of equal significance in the measurement of the temporal duration of an activity phase. Thus, motility cycles have relevance only to the temporal or rhythmic aspect of behavior. Consequently, an understanding of the specific actions of the embryo may require a broader conceptual basis than an understanding of the temporal activities of the embryo, though there is a clear connection between the two problems.

It remains to be added that the most recent evidence does confirm the first two defining characteristics of autogenous behavior as listed above: that embryonic motility can be (a) initiated and (b) maintained for a fairly long period in the absence of sensory input. These points were demonstrated in an experiment (Hamburger, Wenger, and Oppenheim, 1966) in which chick embryos were deafferented at the leg level very early in development, and the embryos showed more or less normal changes in the cyclic activity of their legs for the first two-thirds of embryonic development. While this experiment does demonstrate that cyclic activity can be instigated and maintained in the absence of sensory stimulation, the question is still open whether the motility of the *intact* embryo is in fact impervious to the influence of sensory stimulation (points 3, 4, and 5 above). In the light of other evidence it would seem that the temporal aspect of embryonic motility can be either augmented or inhibited by sensory stimulation. For example, in contrast to the findings of Hamburger and Balaban in using tactile stimulation, Bursian (1964) found that intense light (or possibly heat) stimulation altered the duration of movement periods in chick embryos beginning as early as the fourth day of incubation. In the younger embryos (4–7 days) the radiant stimulus increased the duration of periods of movement, while at 9 days the total duration of movement over a 5-minute period was reduced as a function of stimulation. (Though Bursian used a heat-insulating filter consisting of a 20 mm thickness of copper sulphate between two glass plates, in the absence of thermal measurements it is difficult to be certain that the filter was completely effective in reducing the heat from the 500-watt photographic lamp used as the light stimulus.)

Less questionable evidence for the possible inhibitory influence of sensory stimulation comes from an experiment (Oppenheim, 1966) in which the amnion was removed from chick embryos in order to determine if amniotic contractions affect the duration of motility cycles. When the data from the amnion removal study are analyzed in the usual way, i.e., using necessarily arbitrary temporal limits to define phases of activity and inactivity (Hamburger et al., 1965), no change is observed in the temporal aspects of embryonic activity before and after amnion removal (Oppenheim, 1966). However, when the arbitrary definitions of activity-inactivity phases are ignored and the overall activity of the embryos is examined, some of the embryos show a statistically reliable *increase* in the duration of general activity after they are removed from the influence of amniotic contractions (Oppenheim, 1967, personal communication).[4] Since these results seem to conflict with those of Kuo (1939) who found an increase in rate of local leg movement but a decrease in *rate* of general embryonic movement after amnion removal, determination of the "actual" effects of amniotic removal must await an analysis that combines the behavioral measures used by both Oppenheim and Kuo. For

---

[4] Dr. Oppenheim kindly performed this analysis at my request following publication of his paper in 1966. The re-analysis showed that (after amnion removal) activity increased in the Day 8 and 11 embryos, but not in the Day 9 and 10 embryos. Consequently, the possibility that the intact amnion exerts an inhibitory influence on duration of general activity is subject to these qualifications. A recent study (1968) by C. H. Narayanan and R. Oppenheim (*J. Exp. Zool.*, 168: 395–402) lends further substance to the suggestions that (1) the arbitrary criterion for designating activity-inactivity phases can significantly distort the picture of the actual amount of embryonic activity and (2) stimulation can have inhibitory as well as excitatory effects on ongoing activity. In the face of the present evidence, it becomes necessary to ask whether the de-afferented embryos (Hamburger, Wenger, and Oppenheim, 1966) would have shown the same activity level as the control group if the arbitrary criterion of activity-inactivity had not been employed.

the present purpose, however, it is significant to note that neither the results of Kuo, nor part of Oppenheim's results, support those features of the autogenous hypothesis that assert the noninfluence of sensory stimulation on embryonic motility. On the basis of all the present evidence it seems safe to tentatively conclude that while embryonic motility is in some basic or primary sense self-generated, it may also be susceptible to the inhibitory and/or excitatory influence of sensory stimulation or mechanical factors.

Reflexogenous Behavior

According to the reflexogenous view, prenatal behavior reflects the establishment of particular pathways between the sensory and motor components of the nervous system.

As pointed out above, in terms of underlying assumptions there may not be any essential disagreement between the autogenous hypothesis and the reflexogenous hypothesis. Both hypotheses assume that behavioral epigenesis is predetermined by an inevitable sequence of neural growth and differentiation. And, while the modern form of the autogenous hypothesis relates to the temporal periodicity of any kind of embryonic activity, the reflexogenous hypothesis pertains to the type and spatial extent of the observed action. Adherents of the reflexogenic approach have used exteroceptive stimulation (usually touch) mainly as a diagnostic tool and they have not been unaware that the embryo (fetus) moves in the apparent absence of exteroceptive stimulation. Another difference is that the workers studying reflexogeny have used species in which motility does not occur until after some of the neural sensorimotor reflex arcs are established (these arcs being established very early in prenatal development in these species) and in which, therefore, it can be said only that "spontaneous" behavior occurs (i.e., the source of stimulation, if any, being unidentified).

In the reflexogenous study of behavior development, there have been two alternative views. One view is that behavior begins as a total response or a mass movement to stimulation out of which local movements later develop, and the other view is that the embryo (fetus) first exhibits local responses (movements) to stimulation and these are compounded during ontogeny such that total integrated behavior is the outcome of prenatal development.

TOTAL PATTERN

In 1929, George Ellett Coghill, after twenty-five years of studying the development of aquatic and terrestrial locomotion and feeding in the larval salamander *Ambystoma,* concluded: "Behaviour develops from the beginning through the progressive expansion of perfectly integrated total pattern and the individuation within it of partial patterns which acquire various degrees of discreteness" (Coghill, 1929, p. 38).

Coghill was a neuroanatomist and the strategy of his experimental work was (a) to observe the response of the larval salamander to stimulation at given stages of development and (b) to correlate functional neuroanatomy with the behavior observed at these stages of development. On this point he was able to conclude: "The behavior pattern develops in a regular order of sequence of movements which is consistent with the order of development of the nervous system and its parts" (ibid., p. 36). It follows that, if behavior is solely a function of the nervous system, then an explanation of the development of the nervous system will explain behavior. This represents a subtle point of difference in the strategic approach of psychologists and anatomists and has undoubtedly caused some serious and unexplicated misunderstandings, particularly in the nature-nurture controversy. Psychologists have tended to work from the outside in and the anatomists from the inside out. Behavior is the primary datum for most psychologists, while the nervous system is the primary datum for the anatomist. Attesting to the validity of this strategic distinction, Coghill (ibid., p. 76) states: "The burden of this lecture is to place an old problem on a new basis and to present a working hypothesis for its solution. How do conduction paths of the central nervous system come to be where they are, or how do they acquire their definitive function?" This is of course the key question of developmental neuroanatomy and, like most of the other central issues touched upon in this chapter, it remains unanswered even today. On this major question, Coghill (ibid., pp. 76–78) seems to have accepted the complementary hypotheses of Bok, Ariëns Kappers, and Child on preneural and neural growth. These hypotheses will be fully described in the subsequent section on the probabilistic epigenesis of behavior. Coghill's overall guiding assumption regarding the relationship of the nervous system and behavior is summarized by his statement that the total behavior pattern is or becomes ". . . organized through a regular order of sequence of definite phases in the growth of the nervous system" (ibid., p. 16).

Coghill's theory of prenatal behavior was based on behavioral observations and neuroanatomical analyses which indicated that during the larval stage, when the larval salamander first becomes responsive to exteroceptive stimulation, the organism's entire functional neuromuscular apparatus responds in an undifferentiated way (total pattern) to the stimulation. It is only later on in development that discrete or independent movements of parts of the body occur (partial pattern) in response to stimulation. According to Coghill, during development the partial patterns become emancipated from the initial total pattern response; the increasingly differentiated (independent) responses reflecting structural differentiation in the nervous system. Since the simplicity of Coghill's formulation may be deceptive, it might be useful to offer several concrete examples of what the term total pattern is intended to convey.

Salamander larvae first becomes responsive to tactile stimulation in the vicinity of the snout region, where gentle tactile stimulation is usually followed by the head turning away from (though sometimes toward) the side stimulated. According to Coghill's neuroanatomical analysis, the first reflex activity occurs in the neck region because it is the first region in which the motor neurons have established functional connection with muscles. As subsequent motor connections are established caudally, the response to stimulation also spreads caudally. Thus, for Coghill, the spatial extent of development of the reflex mechanism determines the spatial extent of the embryo's response to stimulation. Consequently, one meaning of Coghill's term *total pattern* is that at any given time the entire functional neural reflex mechanism responds to stimulation and that this principle is operative even in the later stages of development. The generality of Coghill's principle for all vertebrate development has been ably reviewed by Hooker (1952), who feels that his observations and those of Humphrey (1964) on human fetuses support the neuroanatomical aspect of Coghill's formulation at the primate level.

A second meaning of total pattern is derived strictly from overt observation and is best exemplified by the fact that when the gills, forelimbs, and hindlimbs first appear in the larval salamander (in that order), these parts of the body are at first moved only passively by the sinusoidal swimming-like movements of the embryo's trunk. Later on in development these parts of the body become capable of active movement independent of trunk movement and Coghill referred to this later development as the individuation of partial patterns from the whole or total pattern. So, for Coghill, local movements become emancipated from total or mass movement of the embryo and the capability for local movements represents a later refinement of neural as well as behavioral development. Thus, according to Coghill, behavioral epigenesis follows a predetermined unidirectional course, beginning with neurally integrated total movement and proceeding to the emancipation of local movements from (or within) the total pattern. Exteroceptive stimulation was used by Coghill (and many of the neuroanatomists who followed his paradigm) only as a diagnostic device which revealed the functional components of the neural sensorimotor apparatus. Exteroceptive, interoceptive, or proprioceptive stimulation was not considered in any way to directly influence the course of behavioral development, either by way of facilitating development or, more strongly, channeling behavior along certain lines rather than other ones. As a neuroanatomist,

Coghill was interested in the workings of the nervous system and, in the neuroanatomical tradition, behavior is at best an epiphenomenon of the nervous system. Consequently, the methodological stress is on standardizing a particular stimulative event and observing behavioral changes as a function of neural maturation, a strategy that is appropriate to the assumption that neural maturation (and therefore behavior) is predetermined. As is well known, holding any viewpoint at all will necessarily (logically) lead one to favor certain research strategies rather than others. In the case of Coghill's strategy and his findings, among the salient highly specific unanswered questions are the following: (1) Is the intensity of stimulation positively correlated to the spatial extent of the movement? (2) Is the intensity of stimulation also related, for example, to the direction of head-turning, weak stimulation resulting in turning toward the stimulative source (ipsilateral flexion of the head) and stronger stimulation resulting in turning away (contralateral flexion of the head)?

The above two questions are by no means especially critical to Coghill's theory, but they do lead to consideration of an important meta-theoretical point. If simple intensity (or other) differences in stimulation in the same sense modality are correlated with differences in pattern or kind of response, then it is incomplete (insufficient) to assert that the observed behavior is a function of the nervous system. It would be in the same sense incomplete to assert that the observed behavior is a function of stimulation. It is obvious that the observed behavior is a joint function of the kind or type of stimulation employed and the developmental state of the nervous system at the time of stimulation. A change in either variable (stimulation or nervous system) is or may be correlated with a change in behavior. Therefore, to assert the primacy of either variable to the exclusion of the other variable represents an incomplete specification of the observed phenomenon. It may be this very elementary point of incomplete specification that causes needless conceptual difficulties in the study of prenatal behavior. For many neuroanatomists the stimulative variable is a convenient but otherwise insignificant tool which makes the embryo behave so something can be inferred about its nervous system; for many psychologists the nervous system is a convenient but otherwise unimportant tool to demonstrate the differential effects of stimulation in determining behavior.

PARTIAL PATTERN (Independent or Local Movements)

It will be recognized that the terms *total pattern* and *partial pattern,* though appropriate and necessary for the initial stage of inquiry, are relatively imprecise phenomenological descriptions of the embryo's movements. They are not quantitative terms, nor have they been reduced to distinct graphic patterns of squiggly lines which can be generalized across species.

In other words, it is difficult to generalize the meaning of these terms to the prenatal behavior of other species in those instances when the patterns of movement are different from *Ambystoma*. Nonetheless, it was Coghill's belief that the total pattern-partial pattern principle was a universal phenomenon of vertebrate development: ". . . There is nothing in our knowledge of the development of behavior to indicate that the principle does not prevail universally in vertebrates, including man" (Coghill, 1929, p. 89).

In the late 1920s, two of Coghill's graduate students attempted to replicate his *Ambystoma* findings with albino rat fetuses. One of the students (Angulo y Gonzalez, 1932) found that the prenatal behavior of rat fetuses supported Coghill's inductive generalization from salamander larvae, while the other student (Swenson, 1929) gained the impression that local or "simple" movements occurred first in response to stimulation and that these local movements were the building blocks for the more totally integrated movements which he observed later in development.

The additive viewpoint suggested by Swenson was subsequently greatly expanded and developed by Windle (e.g., 1944) and his coworkers based on work utilizing a variety of species. Windle also developed the view that total movements were experimental artifacts produced by anoxia in mammalian and other embryos, and that the normal (nonanoxic) sequence of prenatal movement was the compounding or additive expansion of local reflexes during the course of ontogeny. Windle (1944) also raised questions about the nature of the stimulation required to elicit certain movements, but this was by way of a methodological critique where variable methods of stimulation represented experimental artifacts, not significant points for behavioral inquiry. It seems that no one asked why behavioral development should inevitably follow either the total pattern or the additive course. If function (behavior) is determined by structure and structure is predetermined, there is of course nothing to ask. It is only in the case where one conceives of the possibility of a bidirectional relationship between structure and function, so that function can determine structure as well as the reverse, that it becomes an investigative question to ask why behavioral development takes the course(s) it does. (This point will be expanded in the subsequent section on probabilistic epigenesis.)

The 1930s was a particularly prolific period for the study of behavioral embryology and it is not possible in the space allotted to adequately describe all the studies bearing on Coghill's formulation. It is fair to say, however, that the most extensive study of mammalian prenatal behavior during that era (Carmichael's 1934 study of the fetal guinea pig) failed to exclusively or clearly support either Coghill's view or the additive view of Swenson and Windle: "The present study does not give unqualified support to any of the more general theories of the development of behavior, such as those summarized by the words "individuation" or "integration," but suggests rather that the formulation of such generalizations is at present premature" (Carmichael, 1934, p. 466).

Systemogenous Behavior

During the late 1930s and 1940s it became evident that if it was to be at all possible to assert general principles of vertebrate behavioral development, then, in order to subsume species differences, these principles would have to be couched in more abstract terms than the relatively concrete notions of mass movement and local reflex. Such a theory has been proposed by Anokhin (1964) in relation to developmental neuroanatomy, using behavior and ecology as convenient predictors of the selective acceleration of certain neurological processes according to the adaptive requirements of the species under consideration. In essence Anokhin's main inductive generalization is similar to Carmichael's (1963) principle of *anticipatory morphological maturation*. Anokhin's principle holds that given a knowledge of the behavior that is requisite for the survival of the newborn in its usual ecological setting, then one can predict that the neuroanatomical substrate of the behavior is assured of being completed before birth or hatching. Phrased in this abstract way, the anticipatory growth principle has the generality which is required to incorporate many important species differences in behavioral capability at birth and the principle also has instigated a number of corroborative neuroanatomical investigations on a comparative basis (Anokhin, 1964).

Anokhin's concept of systemogenesis is meant as an alternative to the rule of organogenesis which, he believes, implies that there is a proportional maturation of each organ during prenatal development. Systemogenesis means that ". . . there are . . . processes of selective and accelerated growth of substrates which in future will combine to create a fully developed and arborized functional system with positive adaptive effect for the newborn" (Anokhin, 1964, p. 66). For Anokhin the sequence of behavioral epigenesis is predetermined by the systemogenetic principle and prenatal systemogenesis itself is predetermined. "It is true that the systemogenetic type of the maturation and the growth is the most marked for those functional systems of the organism which must be mature exactly at the moment of birth. They are evidently inborn, the preparation for their consolidation is preformed, and in fact, in the process of the ontogenesis, they correspond demonstrably to the ecological factors of that species of animal . . ." (ibid., p. 85).

PROBABILISTIC EPIGENESIS OF BEHAVIOR

The term *probabilistic epigenesis* is chosen to designate the view that the behavioral development of individuals within a given species does not follow an invariant or inevitable course, and, more specifically, that the sequence or outcome of individual behavioral development is probable (with respect to norms) rather than certain.

A degree of uncertainty in behavioral development is demanded by the overwhelming importance ascribed to stimulative factors at almost all stages of prenatal development. In this view, stimulative factors are regarded, in a weak sense, as merely facilitating the development of behavior along certain lines or, in a much stronger sense, as channeling or forcing the development of behavior along certain lines. Stimulative events are broadly construed and include six main factors: (1) presensory or nonsensory mechanical agitation; (2) interoceptive stimulation; (3) proprioceptive stimulation; (4) exteroceptive stimulation; (5) neurochemical (e.g., hormonal) stimulation; and (6) the musculoskeletal effects of use or exercise.

A second way in which probabilistic epigenesis differs from predetermined epigenesis is that certain theorists implicitly assume that probabilistic epigenesis necessitates a bidirectional structure-function hypothesis. The conventional version of the structure-function hypothesis is unidirectional in the sense that structure is supposed to determine function in an essentially nonreciprocal relationship. The unidirectionality of the structure-function relationship is one of the main assumptions of predetermined epigenesis. The bidirectional version of the structure-function relationship is a logical consequence of the view that the course and outcome of behavioral epigenesis is probabilistic: it entails the assumption of reciprocal effects in the relationship between structure and function whereby function (exposure to stimulation and/or movement or musculoskeletal activity) can significantly modify the development of the peripheral and central structures that are involved in these events. (Evidence on this point is presented later.)

The third main difference between the assumptions of predetermined and probabilistic epigenesis concerns the discontinuous or continuous nature of behavioral development. Whereas adherents of the predetermination hypothesis emphasize the possibility of sharp or abrupt transitions in the development of behavior due to growth spurts which, e.g., suddenly cause the embryo to become sensitive to certain types of stimulation, or abruptly cause the embryo's motility to become coordinated, holders of the probabilistic viewpoint emphasize the gradual and continuous development of behavior in line with the hypothesis that stimulation or motor patterning of some sort is almost always involved in one way or another in the activity of the embryo or fetus. Along this same vein, adherents of the probabilistic viewpoint assume that the "preadapted" quality of early postnatal behavior can be traced to (and is in some sense dependent upon) activities that occurred or began prior to birth or hatching, and that the "adaptive" features of postnatal behavior are not attributable to maturational changes alone.

In brief, then, the main assumptions of probabilistic epigenesis of behavior are: (a) the critical importance ascribed to stimulative factors in facilitating or, perhaps, shaping behavioral development; (b) the bidirectional or reciprocal structure-function relationship; (c) the gradual and continuous nature of the development of behavior, especially the idea that the early postnatal behavior capabilities of the neonate stem from activities that occurred or began prior to birth or hatching.

Though a number of independent hypotheses relevant to the view of probabilistic epigenesis have been offered by individual investigators, only the more comprehensive theoretical accounts (those of E. B. Holt and Z.-Y. Kuo) are considered here. After a review of the theories of Holt and Kuo, the final section of the chapter summarizes the very few (and very recent) prenatal experiments regarding the bidirectional nature of the structure-function relationship, as this latter point would seem to represent a particularly important change in conceptualizing the developmental relationships of anatomy, physiology, and behavior.

E. B. Holt's Theory

For Holt (1931), like Coghill, the problem of behavioral development is tied to the problem of neural sensorimotor development, particularly to an explanation of the processes by which pathways are established between certain sensory organs and certain muscles or groups of muscles. From Holt's point of view, the establishment of such major pathways is highly regular, but by no means predetermined, and depends on the operation of three principles of stimulation during the earliest (truly formative) stage of ontogeny. These three principles are Child's (1924, reprinted 1964) notion of axial gradients, Bok's (1915) idea of stimulogenous fibrillation, and Ariëns Kappers' (1936) conception of neurobiotaxis.

Child's contribution to embryology was his axial gradient theory, the five main assumptions of which are: (1) that all protoplasm contains energy; (2) that this energy is released only by stimulation or excitation (e.g., heat, light, electricity, chemical substances); (3) that all protoplasm requires stimulation (excitation) in order to grow; (4) that stimulation determines the apparently selective course of growth; and (5) that within experimentally determinable limits stimulation determines the extent of growth of any particular part of the embryo.

During the initial stages of embryonic development the head-tail or apico-basal relationship is the primary structural axis of growth. It is only later on in development that radial, bilateral, and dorso-ventral axes make their appearance. According to Child, the apical region or head of the embryo arises from the region of the greatest activity or highest metabolic or oxidative rate, and at this stage the apical region is therefore defined as the high end of the polar gradient. According to Child's theory, alteration of the stimulative conditions will change the height or slope of the gradient and the embryo should therefore deviate from the norm in appearance. In line with this prediction, the main way that Child sought empirical support for his theory was to modify the axial gradient of growth by introducing or enhancing stimulation at certain points of the axes and producing disproportionate increases of growth in these areas. In mild cases, the embryo remains recognizable, but in the more extreme cases the embryo is made to look almost unrecognizable by being subjected to atypical stimulation or by being subjected to typical stimulation for atypically long periods. In Child's terms, therefore, the observed stability of organismic growth patterns in members of the same species is attributable to the highly stable stimulative parameters to which embryos of the same species are subject during the course of prenatal development.

While Holt utilized Child's axial gradient principle to account for the general features of organismic growth, he turned to the ideas of Bok and Ariëns Kappers for a developmental account of the establishment of neural pathways between sensory and motor nerves. The "laws" of neuronal growth postulated by Bok and Ariëns Kappers are conceptually related to Child's theory in the sense that electrical potentials are identified as the guiding forces of dendritic and axonal growth. From the developmental point of view, the fundamental building block of a neural network is the establishment of a synapse (contact) between the axon of one nerve cell and the dendrite or cell body of another nerve cell. Bok's notion of stimulogenous fibrillation refers to axonal growth. It suggests that a growing bundle of nerve fibers activates less developed neuroblasts, such that the axon of the newly developing nerve cell grows in the same direction taken by the electrical current that radiates from the growing bundle. Ariëns Kappers' conception of neurobiotaxis refers to dendritic growth and asserts that the developing dendrites of the nerve cell grows *toward* nerve bundles from which they receive the greatest amount of stimulation. Not only do the dendrites "reach out," so to speak, in the direction of the activating source, but the nerve cell itself is said to migrate in the direction of greatest neighboring stimulation.

Holt, reasoning from the idea that nerve impulses travel in only one direction, specifically, from the axon of one neuron to the dendrites of another and thence (via the axon of the same nerve) to muscle, made particular use of Ariëns Kappers' suggestion concerning the growth of dendrites. The directionality of function at the synapse, combined with the neurobiotaxic principle, Holt felt to be the definitive law of association between stimuli and responses. Holt said (1931, p. 27): "In my opinion this growth of dendrites under the stimulus of nerve impulses is the sole basis of learning," by which he meant (erroneously, I believe) Pavlovian conditioning.

For Holt, behavioral development, both in its prenatal and postnatal phases, represents the establishment, concatenation, and elaboration of reflexes. In this view, which can be contested, Holt followed the assertion of Pavlov, namely, that conditioned reflexes ". . . are being acquired all during the life of the individual, and are the education and the development of the individual." Though there are a number of objections to the view that behavioral development mainly consists of the acquisition of conditioned reflexes, the one that requires mention is Holt's neglect of the motivational component (deprivation or noxious condition) which is a prerequisite for the establishment of the Pavlovian conditioned reflex. Pavlov (1927) himself, in his own writings, appears not even to have mentioned, much less emphasized, the motivational component, e.g., the need to starve his dogs in the conditioned salivation experiments, and apparently Holt overlooked this crucial step in the establishment of a conditioned reflex. (For avoidance conditioning, a painful or noxious stimulus must be used, which involves fear, anxiety, or the escape from pain as the motivational component.) The basic requirement of a so-called drive state for conditioning to occur is the main reason for doubting Holt's statement that the conditioned reflex is the recurrent and ubiquitous building block of prenatal development. This was apparently an unwitting oversight on Holt's part because he faithfully gives William James' (1890, Vol. II, pp. 584–5) "drainage hypothesis" priority as the original statement of the conditioned reflex, which it was not. On the other hand, James' drainage hypothesis probably was or is the accurate behavioral analogue of the neurobiotaxic law, which is what Holt apparently intended. Neither the neurobiotaxic law nor the drainage hypothesis require the inference of a motive, need, or drive. James simply asserted that afferent (sensory) impulses have a tendency to spill over into simultaneously active efferent or motor pathways and are thereby discharged through muscular contractions. Ariëns Kappers simply asserted that dendrites grow toward an active neuron or nerve bundle provided that the neuron from which the dendrite grows and the neuron toward which it grows are in excitation simultaneously or in close succession. Thus, at different levels of discourse, contiguous *stimulation* (not motivation) is the explanatory key. Though motivation could possibly enhance or disrupt this process, there is no evidence to suggest that it is necessary to invoke motivation to account for prenatal behavior as it occurs under typical conditions of development or stimulation. (It is a paradox that Holt's book is titled *Animal Drive and the Learning Process,* whereas in current terms Holt would probably be known as the originator of a psychology without motivation.)

Be that as it may, a brief summary of the main tenets of Holt's novel theory of prenatal behavior development follows. Holt's basic assumption was that the Hobbes-Locke *tabula rasa* ("blank slate") doctrine rested on firm embryological as well as psychological ground. First, the initial movements of the fetus are random and, second, during early ontogenesis the sensory system is not yet connected to the motor system. Thus Holt's primary (and probably too liberal) assumption was "that the possibility is open for any sense-organ to acquire functional connection with any muscle" (Holt, 1931, p. 33). During ontogeny, as the organism stimulates itself or is stimulated by environmental forces, the process of neurobiotaxis establishes certain connections which become reflex arcs. Fetal and neonatal movements become less random throughout development because the initial spread of afferent impulses becomes narrowed down as a result of junctions being formed between sensory and motor pathways on the basis of neurobiotaxis and lowered synaptic resistance. For Holt, all learning is based on random movement, so (for Holt) random movements never entirely cease throughout the life of the individual.

Learning or Holt's version of Pavlovian conditioning is based on the existence of neural reflex circles. As an example, Holt analyzes the prenatal establishment of the infantile grasping reflex as follows:

> Consider a random impulse reaching the flexor muscle of a finger. In the foetal position the fingers are often closed over the palm of the hand, and the least random flexion of a finger will cause it to press on the palm. Then (what is not random) afferent impulses ("tactile") from the two surfaces in contact (palm and finger) will be sent back to the central nervous system, where by the principle already cited they will find an outlet in the motor paths that were just now excited, that is, those of the flexor muscle of the finger in question. When this has happened a few times (as it is bound to happen) the reflex circle will be established; and then a pressure stimulus on either palm or finger will cause the finger to flex and so to close down on the object that caused the pressure. Such is the origin of the "grasping reflex," which is so useful through all the latter life. This reflex is regularly established before birth. With Pavlov's law and the external anatomy of a baby in mind the reader can discover for himself the genesis of many of the earliest reflexes,—lip closure, "extensor-thrust" when the bottom of the foot is touched . . . and many others. (Ibid., pp. 38–39.)

In the above description, we have a good example of what Holt erroneously considered a Pavlovian conditioned reflex. On the other hand, Holt's description of the self-stimulative basis of the grasping reflex is an ingenious and provocative one, whether the particular explanation is valid or not.

Though the above account of Holt's view is sketchy, it does give the flavor of his thought concerning the prenatal development of behavior. By way of final summary, the five essential ingredients of the behavioral aspects of Holt's theory appear to be as follows: (1) During the earliest stages of development, the ongoing motor activity of the embryo (fetus) is susceptible to becoming attached to various sorts of interoceptive, proprioceptive, and exteroceptive stimulation. (2) This attachment is based on temporal or spatial contiguity of stimulation and motor activity. (3) "Every organism *at first* responds positively towards (i.e. so as to get more of) *any* stimulus" (ibid., p. 94). (4) "If, however, the stimulation is, or becomes, very intense, the transmitted excitation spreads . . . in the central nervous system, overflowing into diffuse and random motor channels, and the young organism is thrown into a state of more or less general and inco-ordinated activity . . ." (p. 94), which, though random, eventually carries the organism out of range of the stimulus. (5) As development proceeds, behavior (the functional relationship between stimuli and responses) becomes more and more canalized (Holt's term) and less and less random as a function of previously established contiguity relations.

From the standpoint of the relationships between prenatal and postnatal behavior, Schneirla (1965) has made significant innovations and detailed extensions of the five major premises of Holt's theory of prenatal behavior. Both the theories of Holt and Schneirla rest importantly on the postulate of a stimulus intensity dynamism in the development of behavior (points (3) and (4) above) and this idea requires definitive investigation particularly in view of Coghill's (1929) and Humphrey's (1964) impression that the embryo's initial response to stimulation is an avoidance or "protective" reflex (i.e., turning-away). As stated earlier in this chapter with reference to Coghill's theory, variations in stimulus intensity in provoking turning-to and turning-away have not yet been explored at the prenatal level.

At the morphological and neurological levels, Holt's theory rests on the postulates of Child, Bok, and Ariëns Kappers. The principles of morphological and neuronal growth are not yet clarified, and it appears that Ariëns Kappers' hypothesis still remains to be definitively tested as it applies to ontogeny [see, e.g., the discussion following Levi-Montalcini's paper in Purpura and Schadé (1964)]. On the other hand, there is some evidence for and against Bok's hypothesis (e.g., Hamburger, 1946).

Regarding Holt's emphasis on prenatal conditioning, the work of Gos

(1933), Hunt (1948), and Sedláček (1962) suggests that some kind of conditioning is possible using *noxious* stimulation in highly developed chick embryos, but apparently the effects of these procedures are transient and disappear soon after hatching. By using essentially the same noxious procedures at the mammalian level, Sedláček et al. (1964) have presented evidence of prenatal conditioning in highly developed guinea pig fetuses and Spelt (1948) has offered suggestive evidence of prenatal conditioning of human fetuses *in utero*.

Z.-Y. Kuo's Theory

Though initially identified in the 1920s with the most radical form of Watsonian (and Holtian ?) behaviorism, Kuo altered his theoretical position in the 1930s after intensive study of the development of behavior in the chick embryo. Kuo was apparently so profoundly impressed with the intricacies of prenatal behavior as observed in the chick embryo, that in the 1930s he steadfastly disavowed all existing theories, whether they were Pavlovian, Coghillian, Swensonian, or Preyerian in origin. Since the 1940s, however, Kuo has been engaged in formulating a system of his own and the fruits of this work have recently been brought together in book form (Kuo, 1967).

According to Kuo, prenatal (as well as postnatal) behavior is a function of three simultaneously interacting factors: (1) the embryo's developmental history, (2) the present (or immediate) environmental context, and (3) the physiological condition of the embryo. The importance of these three factors in prenatal behavioral development emerged from Kuo's purely observational (nonmanipulative) research on the chick embryo. Therefore, the precise interactional relations between the three factors is not yet known (i.e., it is not possible to deduce a set of specific predictions for any given stage of development). Nonetheless, Kuo's ideas would appear to have a certain heuristic value and Kuo himself does not make any other than programmatic claims for the importance of his system.

In Kuo's usage, the term *developmental history* is intended to cover the various prior internal and external stimulative events that impinge on or determine the embryo's behavior at any given stage and, in particular, to include selective sensory and perceptual responses of the embryo to intercurrent stimuli. From Kuo's point of view, it is particularly important that the various organs and muscles of the embryo have a developmental history, as well as the central nervous system of the embryo. Kuo's notion of environmental context stresses the situational or "field" determinants of the organism's immediate behavior. Such a notion is in contrast to the problematical conception of "*a* stimulus" in certain versions of stimulus-response psychology or "*the* releaser" of classical ethology. Finally, Kuo's inclusion of the physiological condition of the embryo as a determinant of behavior points to the metabolic, biochemical, and other (typically unobserved) interior events that play a role in the embryo's overt behavior.

Kuo claims that such notions as local reflex or total pattern are misleading abstractions and he has proposed the conception of "behavioral gradient" to call attention to the fact that the embryo is more or less active all the time. More specifically, the gradient notion is a dynamic reference to relative differences in both extent and intensity of activity between different parts of the body during various stages of prenatal development. These differences and changes in activity are ascribed to morphological changes (growth and decline) and responses to sensory and other kinds of stimulation during the course of prenatal development. The concept of behavioral gradient, which at first may seem quite vague, is very well exemplified (though not used) in Kuo's (1939) study of the various factors influencing simultaneous changes in massive and local movements of the chick embryo.

The other major concept of Kuo's proposal is the more familiar idea of "behavioral potentials," which is meant to imply that the range of possibilities of behavioral development always exceeds the range of behavior that is actualized during the course of individual development. The developmental history of the individual is the major (though not the sole) limiting factor on the range of behavioral potentials, and Kuo, like Holt, favors the view that successive stages of development are correlated with a diminished plasticity or narrowing of the range of potentialities. Along this line, Kuo has developed the novel idea that the behavioral epigenicist should strive to produce conditions that favor the development of behavioral mutants or behavioral neophenotypes by way of gaining knowledge of the outside limits of the range of behavioral potentials. Kuo's suggestion is another way of calling attention to the probable importance of a standard prenatal environment as a requirement for the manifestation of species-typical behavior, a point previously expounded and documented by Child in reference to the development of species-typical morphology.

As can be inferred from the above accounts of Holt's and Kuo's ideas, theories pertaining to the view that the course of behavioral epigenesis is probabilistic rather than predetermined are even more lacking in manipulative experimental support than the theories that favor the predetermined view of behavioral epigenesis. Thus far, both theoretical camps have been forced to rely mainly upon strictly observational accounts of the prenatal development of behavior and, if one may be permitted some hopeful optimism, perhaps the current decade will see the execution of deductively designed, manipulative experiments that will subject the tenets of the various theoretical proposals outlined above to systematic examination.

BIDIRECTIONALITY OF STRUCTURE-FUNCTION RELATIONSHIP

In recent years the results of several manipulative experiments have suggested a reciprocal relationship between structure and function, i.e., each influencing the development of the other, whereas the traditional assumption has been that structure influences function in a unidirectional manner. This important change in conceptualizing the structure-function relationship is demanded (theoretically) by the probabilistic view of epigenesis which places a strong emphasis on stimulative factors in fostering or, perhaps, channeling prenatal physio-anatomical growth and differentiation. This emphasis on the significance of the operation of stimulative factors during prenatal development should not be misconstrued as a modern version of the Hobbes-Locke *tabula rasa* ("blank slate") theory of development. In keeping with the presently known facts of organic development, the effects of stimulation presumably operate within a limited (not infinite) range of organic growth possibilities, a range that is subject to experimental determination in any given instance.

Including studies of intact embryos or fetuses only, there would appear to be five prenatal experiments which pertain to the bidirectional structure-function relationship.

The avian embryo is very active during prenatal development and it may be asked whether movement (function) of the skeletal musculature plays a role in fostering adequate development of the limb and neck joints, among other things. In 1962 Drachman and Coulombre, building on the previous work of a number of other investigators, paralyzed otherwise undisturbed chick embryos for single periods of 24 to 48 hours during the second third of the incubation period (7 to 15 days) by infusing curare into the chorioallantoic vessels of the egg. Though there were relationships between the age of the embryo, dosage level, and duration of the dose, in general the findings upon hatching revealed deformities of the toe joints, ankle joints and knee joints. In the more severe cases, wing joints and neck joints were also deformed. The authors report no other teratological deformities, nor did they find spinal or other neural changes. In line with the results of this experiment, as well as others like it (Drachman and Sokoloff, 1966), it may be concluded that articular cavity formation and fine sculpturing of the cartilaginous surfaces require the mechanical action usually provided by movements of the embryo's own skeletal muscles.

While the above experiment indicates the effect of function on skeletal structure, the following two experiments implicate changes in brain and receptor structures as a consequence of visually stimulating the highly developed chick and duck embryos. These experiments are also important because they show an acceleration of normal developmental changes via the presence of stimulation rather than the production of abnormalities as a result of the absence of stimulation.

As an incidental finding, Peters, Vonderahe, and Powers (1958) noted the premature appearance of electrical responses of the eye and optic lobe of the chick embryo after continued exposure to a flickering light on the 17th or 18th days of incubation (2–3 days before hatching). In view of these findings, Paulson (1965) performed a similar experiment with duck embryos, which included systematic exposure or nonexposure to a flickering light on each of several days prior to hatching. Paulson found small but apparently consistent enhancement of electrical responses at the retina, telencephalon, and optic lobe in the embryos that were prematurely exposed to optical stimulation, thereby supporting and extending the incidental findings of Peters et al. (1958) with chick embryos. The positive contribution of relevant sensory stimulation at an appropriate developmental stage is further attested to by the postnatal findings of Hubel and Wiesel (1964) with kittens. In these experiments it was found that visual deprivation for two to three months after birth resulted in marked cellular atrophy (30–40%) in the lateral geniculate body, while similar deprivation experiments with adult cats did not produce these particular changes. (The literature concerning the effects of postnatal stimulative deprivation has been summarized by Riesen, 1961. An intensive study of the effects of postnatal stimulative enrichment is reviewed by Bennett et al., 1964.)

Another prenatal experiment that should be mentioned in the present context pertains to the role of hormonal factors in stimulating (in this case, channeling) growth and differentiation of genital structures, as well as affecting later sexual behavior. Goy, Bridgson, and Young (1964), building on the results of a previous experiment, found that administration of testosterone propionate during particular periods near mid-gestation in guinea pigs resulted in masculinization of the genital structures of the genotypic female fetuses. These prenatal treatments were also associated with a heightened incidence of male-like mounting behavior and reduced estrous behavior at adulthood in the female fetuses thus affected. Along this same line, an earlier and less extensive prenatal hormone experiment by Willier Gallagher, and Koch (1937) produced organic indications of sex reversal in male chick embryos.

The incomplete account offered above suggests the effects of function or stimulation upon structural development during the prenatal period—the many everyday examples of the effects of structure upon function at all stages of development are probably more numerous and better known.

Perhaps in the near future it will be possible to answer some of the specific questions raised by the bidirectional operation of the structure-function relationship during prenatal development. As an example of the very preliminary and rudimentary kind of interdisciplinary experiments that can be undertaken in the light of our changing conception of the relationship between structure and function, Appel, Paulson, and Gottlieb (1965, unpublished observation) found that optical or acoustical stimulation accelerates the appearance of protein synthesis at the neural junctions in the brain stem, optic lobe, etc., of highly developed duck embryos. A similar experiment might be performed to examine Langworthy's hypothesis (1933) that the rate and degree of neural myelinization is enhanced by exposure to relevant sensory stimulation during the prenatal period.

While experiments such as the ones reviewed above support the idea that function or stimulation is an important factor in fostering or enhancing typical structural and physiological changes, other prenatal experiments are required to determine the role of stimulation (if any) in channeling subsequent physiological responses along certain lines rather than other ones. From the postnatal literature, for example, Lindsley et al. (1964) have shown that monkeys that have been deprived of light for long periods show a reversal of the usual electrophysiological response to optical stimulation: instead of showing a blocking of cortical alpha-like waves to light stimulation, these monkeys showed alpha-blocking to the absence of optical stimulation, whereas light tended to produce synchronous waves. Along the same vein, Heron and Anchel (1964) found that adult rats exposed to five-per-second synchronous acoustic clicks and optical flashes since infancy showed a predominant brain wave frequency of five-per-second instead of the typical seven-per-second frequency.

In both of the above experiments the sensory environment was unusually uniform and persistent as compared to the usual postnatal rearing environment of these animals. Relative uniformity in the range and persistence of stimulation are ever present features of the *prenatal* sensory environment, so it is not inconceivable that some of the rhythmic physiological regularities observed postnatally in neonates of a given species are partially attributable to the persistent rhythmic sensory and mechanical occurrences to which the embryo has been exposed during prenatal development. This is a particularly important point in relation to establishing the general validity of a probabilistic conception of epigenesis.

SUMMARY AND CONCLUSIONS

To those with an historical perspective, it comes as no surprise that modern theories of prenatal behavior readily fall into only two categories representing the ancient dualism of nature and nurture. A hopeful sign for those of us who find this dichotomy inadequate, either as a single term or taken in "interaction," is the relatively operational (i.e., testable) assertions that can be derived from an analysis of theories representing each point of view. Though the area of prenatal behavior abounds in a fund of excellent descriptive-observational research (including correlative neuro-anatomy) involving a great variety of vertebrate species, it is far from suffering the luxury of a surplus of manipulative or deductively designed experiments. On this score, theories holding the assumption that behavioral epigenesis is predetermined would seem at present to have a greater claim to experimental support than those theories that assume the probabilistic character of behavioral epigenesis.

The single most important issue dividing the two theoretical camps is the role of endogenous and exogenous stimulation in behavioral, neuro-anatomical, and musculoskeletal development. At the level of functional anatomy, there are a few recent prenatal experiments that suggest that relevant sensory and musculoskeletal stimulation may be essential to the initiation and maintenance of normal (typically observed) maturational changes. These and other experiments with neonates would seem to indicate a particularly significant shift in our conception of the structure-function relationship from a unidirectional one (structure → function) to a bidirectional one (structure ←→ function). This shift is consonant with a probabilistic conception of epigenesis and raises an important question for future resolution at the prenatal level, namely, the degree to which stimulation or activity merely fosters or enhances development and the degree to which (or areas or stages in which) stimulation or activity channels future development.

ADDENDUM

After submission of the manuscript I have had the opportunity to further explore the topic in connection with reviews of the most recent and comprehensive evidence from developmental neurobiology as well as behavior. These works are listed here for the benefit of interested readers: *Quarterly Review of Biology,* 1968, 43:148–174; *Development of Species Identification in Birds: An Inquiry into the Prenatal Determinants of Perception,* 1970, University of Chicago Press; Ontogenesis of sensory function in birds and mammals (chapter in *Biopsychology of Development,* E. Tobach et al., eds., Academic Press: New York, 1970). Viktor Hamburger's most recent thoughts and experiments on autogenous motility are also summarized in a chapter in the last named book.

ACKNOWLEDGMENTS

This paper was written in connection with research activities supported by U. S. Public Health Service grant HD-00878 from the National Institute of Child Health and Human Development. I am most grateful to Leonard Carmichael, Tryphena Humphrey, Zing-Yang Kuo, Ronald W. Oppenheim, George W. Paulson, Marvin L. Simner, and John G. Vandenbergh for their helpful comments on an earlier draft of the paper.

REFERENCES

Angulo y Gonzalez, A. W. 1932. The prenatal development of behavior in the albino rat. *J. Comp. Neurol.* 55: 395–442.

Anokhin, P. K. 1964. Systemogenesis as a general regulator of brain development. In W. A. Himwich and H. E. Himwich, eds., *The developing brain.* New York: American Elsevier. Pp. 54–86.

Ariëns Kappers, C. U. 1936. *The comparative anatomy of the nervous system of vertebrates including man,* vol. 1. New York: Macmillan. (Reprinted in 1960 by Hafner, New York.)

Bennett, E. L., M. C. Diamond, D. Krech, and M. R. Rosenzweig. 1964. Chemical and anatomical plasticity of brain. *Science* 146: 610–619.

Bok, S. T. 1915. Stimulogenous fibrillation as the cause of the structure of the nervous system. *Psychiat. en Neurol. Bl. Amsterdam* 19: 393–408.

Bursian, A. V. 1964. The influence of light on the spontaneous movements of chick embryos. *Bull. Exp. Biol. Med.* 7: 7–11 (translation).

Carmichael, L. 1934. An experimental study in the prenatal guinea-pig of the origin and development of reflexes and patterns of behavior in relation to the stimulation of specific receptor areas during the period of active fetal life. *Genet. Psychol. Monogr.* 16: 338–491.

Carmichael, L. 1963. The onset and early development of behavior. In L. Carmichael, ed., *Manual of child psychology.* New York: John Wiley. Pp. 60–185.

Child, C. M. 1924. *Psysiological foundations of behavior.* New York: Holt, Rinehart & Winston. (Reprinted in 1964 by Hafner, New York.)

Coghill, G. E. 1929. *Anatomy and the problem of behaviour.* Cambridge, England: Cambridge Univ. Press. (Reprinted 1964 by Hafner, New York.)

Drachman, D. B., and A. J. Coulombre. 1962. Experimental clubfoot and arthrogryposis multiplex congenita. *Lancet,* 523–526.

Drachman, D. B., and L. Sokoloff. 1966. The role of movement in embryonic joint development. *Develop. Biol.* 14: 401–420.

Gos, M. 1933. Les reflexes conditionnels chez l'embryon d'oiseau. *Bull. Soc. Sci. Liége* 4 (7) : 194–199; 246–250.

Goy, R. W., W. E. Bridgson, and W. C. Young. 1964. Period of maximal susceptibility of the prenatal female guinea pig to masculinizing actions of testosterone propionate. *J. Comp. Physiol. Psychol.* 57: 166–174.

Hamburger, V. 1946. Isolation of the brachial segments of the spinal cord of the chick embryo by means of tantalum foil blocks. *J. Exp. Zool.* 103: 113–142.

Hamburger, V. 1963. Some aspects of the embryology of behavior. *Quart. Rev. Biol.* 38: 342–365.

Hamburger, V., M. Balaban, R. Oppenheim, and E. Wenger. 1965. Periodic motility of normal and spinal chick embryos between 8 and 17 days of incubation. *J. Exp. Zool.* 159: 1–14.

Hamburger, V., E. Wenger, and R. Oppenheim. 1966. Motility in the chick embryo in the absence of sensory input. *J. Exp. Zool.* 162: 133–160.

Heron, W., and H. Anchel. 1964. Synchronous sensory bombardment of young rats: effects on the electroencephalogram. *Science* 145: 946–947.

Holt, E. B. 1931. *Animal drive and the learning process.* New York: Holt.

Hooker, D. 1952. *The prenatal origin of behavior.* Lawrence: Univ. Kansas Press.

Hubel, D. H., and T. N. Wiesel. 1964. Receptive field studies in the visual system of newborn and monocularly deprived kittens. *Proc. XVII Int. Cong. Psychol.,* pp. 304–306.

Humphrey, Tryphena. 1964. Some correlations between the appearance of human fetal reflexes and the development of the nervous system. In D. P. Purpura and J. P. Schadé, eds., *Growth and maturation of the brain.* Amsterdam: Elsevier. Pp. 93–133.

Hunt, E. L. 1949. Establishment of conditioned responses in chick embryos. *J. Comp. Psychol.* 42: 107–117.

James, W. 1890. *Principles of psychology,* vol. II. New York: Macmillan.

Kuo, Z.-Y. 1939. Studies in the physiology of the embryonic nervous system: II. Experimental evidence on the controversy over the reflex theory in development. *J. Comp. Neurol.* 70: 437–459.

Kuo, Z.-Y. 1967. *The dynamics of behavior development.* New York: Random House.

Langworthy, O. R. 1933. Development of behavior patterns and myelinization of the nervous system in the human fetus and infant. *Contr. Embryol., Carnegie Inst. Wash.* 24 (139) : 1–57.

Levi-Montalcini, Rita. 1964. Events in the developing nervous system. In D. P. Purpura and J. P. Schadé, eds., *Growth and maturation of the brain.* Amsterdam: Elsevier. Pp. 1–26.

Lindsley, D. B., R. H. Wendt, D. F. Lindsley, S. S. Fox, J. Howell, and W. R. Adey. 1964. Diurnal activity, behavior and EEG responses in visually deprived monkeys. *Ann. N. Y. Acad. Sci.* 117: 564–587.

Moltz, H. 1965. Contemporary instinct theory and the fixed action pattern. *Psychol. Rev.* 72: 27–47.

Needham, J. 1959. *A history of embryology.* New York: Abelard-Schuman.

Oppenheim, R. W. 1966. Amniotic contraction and embryonic motility in the chick embryo. *Science* 152: 528–529.

Paulson, G. W. 1965. Maturation of evoked responses in the duckling. *Exp. Neurol.* 11: 324–333.

Pavlov, I. P. 1927. *Conditioned reflexes.* London: Oxford Univ. Press.

Peters, J. J., A. R. Vonderahe, and T. H. Powers. 1958. Electrical studies of functional development of the eye and optic lobes in the chick embryo. *J. Exp. Zool.* 139: 459–468.

Preyer, W. 1885. *Specielle Physiologie des Embryo.* Leipzig: Grieben.

Riesen, A. H. 1961. Stimulation as a requirement for growth and function in behavioral development. In D. W. Fiske and S. R. Maddi, eds., *Functions of varied experience.* Homewood, Ill.: Dorsey Press. Pp. 57–80.

Sedláček, J. 1962. Temporary connections in chick embryos. *Physiol. Bohemoslov.* 11: 300–306.

Sedláček, J., Hlaváčková, V., and Švehlová, M. 1964. New findings on the formation of the temporary connection in the prenatal and perinatal period in the guinea-pig. *Physiol. Bohemoslov.* 13: 268–273.

Schneirla, T. C. 1965. Aspects of stimulation and organization in approach/withdrawal processes underlying vertebrate behavioral development. In D. S. Lehrman, R. A. Hinde, and E. Shaw, eds., *Advances in the study of behavior,* vol. 1. New York: Academic Press. Pp. 1–74.

Spelt, D. K. 1948. The conditioning of the human fetus *in utero. J. Exp. Psychol.* 38: 338–346.

Swenson, E. A. 1929. The active simple movements of the albino rat fetus: the order of their appearance, their qualities, and their significance. *Anat. Rec.* 42: 40 (abstract).

Tracy, H. C. 1926. The development of motility and behavior reactions in the toadfish (*Opsanus tau*). *J. Comp. Neurol.* 40: 253–369.

Visintini, F., and Rita Levi-Montalcini. 1939. Relazione tra differenziazone strutturale e funzionale dei centri e delle vie nervose nell'embrione di pollo. *Arch. Suisses Neurol. Psychiat.* 43: 1–45.

Willier, B. H., T. F. Gallagher, and F. C. Koch. 1937. The modification of sex development in the chick embryo by male and female sex hormones. *Physiol. Zool.* 10: 101–122.

Windle, W. F. 1944. Genesis of somatic motor function in mammalian embryos: a synthesizing article. *Physiol. Zool.* 17: 247–260.

Windle, W. F., and D. W. Orr. 1934. The development of behavior in chick embryos: spinal cord structure correlated with early somatic motility. *J. Comp. Neurol.* 60: 287–307.

JACK P. HAILMAN

Department of Zoology
University of Wisconsin
Madison, Wisconsin

Comments on the Coding of Releasing Stimuli

"Thus configurational (Gestalt) organization of at least visual
sign stimuli seems to be the rule."

N. Tinbergen (1951, p. 78)

"Animal experiments and the human clinical data alike
indicate that the perception of simple diagrams as distinctive wholes
is not immediately given but slowly acquired through learning."

D. O. Hebb (1949, p. 35)

American comparative psychology nearly died with the death of the *Journal of Animal Behavior.* That the science survived its infancy was largely due to a handful of dedicated investigators, of whom T. C. Schneirla was certainly a leader. Dr. Schneirla was one of the few American psychologists to take serious interest in nonmammalian animals; the study of natural behavior under field conditions had otherwise been left largely to ornithologists, entomologists, and their kin. Happily, as a prominent exception to the rule, Dr. Schneirla could not only bring to America the studies of European colleagues, but also evaluate their explanatory concepts. It is a privilege to continue on the trail he has blazed.

I shall discuss some recent studies of how sensory systems encode environmental stimuli that elicit or direct stereotyped, species-common behavior patterns. Marler (1961) reviewed thoroughly the relevant earlier studies, which rely largely on inference from behavioral measures; some recent studies provide more direct physiological evidence. I shall review selected examples of recent studies, including studies of ontogenetic changes, and discuss the implications of these findings for several theories of animal perception.

CODING OF PERCEPTUAL PREFERENCES

When an animal responds differentially to various stimuli, I say that the animal is showing a *perceptual preference* or *choice,* without implying any connotations to these terms. The neurophysiological processing that leads to perceptual preferences I refer to as *perceptual coding.* The following studies, which concern preferences of sounds, colors and visual forms, have been chosen as particularly instructive examples of perceptual coding.

Examples of Coding

THE FLEEING FROG'S COLOR PREFERENCE

Frogs (*Rana* spp.) leap into the water when frightened. Pearse (1910) and others noted that the escape response is partly directed by the water's

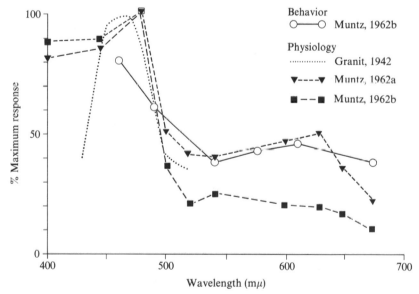

FIGURE 1

Comparison of behavior and physiology of the frog's color choice. The behavioral measure is the percentage of the total number of choices that could be made for each medium-band Wratten filter used (cf. Fig. 4). The physiological measures in Muntz's curves (two units) are the percentage of the spike rate given to the most effective narrow-band interference filter used. The physiological measure in Granit's curve is based on reciprocal threshold values, the maximum having been raised to 100%. The behavioral measures were taken from *Rana temporaria,* Muntz's physiological measures from *R. pipiens,* and Granit's physiological measures from *R. temporaria.* (Redrawn or calculated from references indicated in figure.)

blue color. Muntz (1962b) found that the behavioral preference for blue is stable over a great range of intensities and spectral bandwidth. The activity of single units recorded in the frog's diencephalon is contingent upon the same wavelength parameters of light (Figure 1) as the behavior of the whole animal (Muntz, 1962a). Evidence indicates that recordings were made from terminal arborizations of the optic tract, so that coding of the blue preference is complete by the second synapse (third cell) in the afferent visual pathway. Granit (1942) recorded a similar blue response from the frog's retinal ganglion cell bodies, the cells that give rise to the axons that constitute the optic tract.

Muntz (1963a, 1963b) suggests that the primary coding lies in interactions between receptors themselves. Since a blue light becomes less effective in releasing and orienting escape when a green light is added to it (Muntz, 1962b), green wavelengths are inhibitory and the color preference cannot be due to a single type of receptor. Since the red rods have photolabile pigments absorbing strongly in the green, these may furnish the inhibitory input. The remaining receptors (cones and green rods) may act synergistically to furnish the excitatory input.

THE PECKING GULL CHICK'S COLOR PREFERENCE

Chicks of the laughing gull (*Larus atricilla*) peck at their parent's red beak, a form of begging that elicits parental regurgitation of partially digested food upon which the chicks feed. Chicks peck most frequently at red and blue wavelengths, less at green and yellow. Like the frog's color preference, the gull chick's is largely independent of wavelength purity (Hailman, 1964, 1967). Also, as with the frog, adding energy of middle wavelengths to a preferred stimulus decreases the response, indicating that at least two receptor-types are involved in the coding.

A tentative coding mechanism, based on the filtering characteristics of two classes of colored oil droplets imbedded in the cones, was hypothesized in quantitative form (Hailman, 1964). More recent evidence suggests that a third type of cone may contribute to the coding, in a manner analogous to the proposed coding of the frog's color preference, except that only one receptor is excitatory, and two provide inhibitory input synergistically.

The coding scheme is based on the assumption that the physiological response of a single cone receptor is proportional to the product of (1) the *I*ntensity (I) of incoming light, (2) the percent transmission (O) of the colored *O*ildroplet filter through which the light passes to reach the cone's outer segment, and (3) the absorption spectrum (P) of the photolabile *P*igment contained in the outer segment. Since most sensory systems respond approximately to the logarithm of stimulus intensity, the *C*one's response (C) becomes:

$$C = k \log IOP \tag{1}$$

where k is a constant of proportionality.

Three colors of oildroplets (red, yellow-orange, and greenish-yellow) appear in the gull chick's retina (Hailman, unpublished data). Several considerations suggest that the coding scheme for the color *Response* (R) might be:

$$R = C_r - C_y - C_g \tag{2}$$

where the subscripts denote the color of the oildroplet contained in the cone. Substitution of eq. (1) into eq. (2) for each cone type, and simplification by canceling based on the assumptions that (a) the intensity of light reaching each oildroplet is the same, and (b) the same photopic pigment is contained in each cone, yields:

$$R = k' \log \frac{O_r}{IO_yO_gP} \tag{3}$$

where k' is a combined constant of proportionality. The coding scheme cannot yet be tested critically, as the values of O_g and P are not fully determined.

A preliminary test has been made. Since all values of eq. (3) except I are constant for a given wavelength, the response rate should be proportional to $-\log I$. That is, as the stimulus becomes dimmer by log units, the pecking rate should rise arithmetically. A preliminary test of this prediction, using a single chick tested at two wavelengths each over four log units of intensity, yielded the predicted function (Hailman, unpublished data).

If an achromatic stimulus-object is placed before a colored background, the chick pecks at the object differentially according to the background color. This color preference for the background is nearly the spectral mirror-image of the preference for the colored stimulus-object. The parameters of the object-background preferences strongly resemble the center-periphery characteristics of retinal units recorded from other vertebrates (Hailman, 1966).

THE FROG'S INSECT DETECTOR

Frogs strike at insects with their tongues; they obviously detect and track the insects largely by vision. Although no explicit behavioral experiments have characterized the form-movement stimulus that elicits striking, it seems likely that a small, convex, dark moving shape is the principal relevant element in the stimulus.

The elegant electrophysiological work of a team at M.I.T. (e.g.,

Maturana et al., 1960; Lettvin et al., 1961) is now so well known that only a sketch of their results is necessary here. The frog's retina contains four functionally distinct ganglion cell types (plus an intensity detector) that respond to patterned vision, one unit in particular responding only to a dark, convex border moving across the visual field. Probably all four types contribute some information about the identification of prey; none has been completely characterized.

Because the receptive fields of these units differ in area, the functional units can be tentatively identified histologically by the size of ganglion cell dendritic arbors in the retina. Their connections with bipolar cell types then offer a tentative scheme for retinal interactions (Lettvin et al., 1961).

Going centrally from the receptor toward the central nervous system, fourth-order neurons of the optic tectum are arranged in four vertical layers, cells from each layer being functionally similar to a particular ganglion cell type. Fifth- and higher-order neurons are of two types: (a) "newness" detectors, that respond when an insect-like object enters anywhere in the visual field, and (b) "sameness" units, that respond so long as such an object is still within the field even if it no longer moves for a short period; when it moves again, the unit is more active.

THE PECKING GULL CHICK'S FORM-MOVEMENT PREFERENCE
COMPARED WITH VISUAL UNITS OF OTHER VERTEBRATES

Behavioral experiments (e.g., Hailman, 1967) have shown that the gull chick pecks with greatest frequency to stimuli having these characteristics: (1) elongate shape; (2) stimulus-object darker than background; (3) vertical orientation of the long axis; (4) specific width (about 9 mm) of the short axis; (5) movement; (6) horizontal movement in preference to vertical; (7) movement across long axis in preference to movement along it; (8) specific speed of movement; and (9) independence of frequency of flickering light. These characteristics are summarized diagrammatically in Figure 2A. Response to color was discussed above, and there are many other parameters still to be tested.

The form-movement preference of the gull chick could be coded by visual units similar to those found in the striate cortex of cats (Hubel, 1959, 1963b; Hubel and Wiesel, 1959, 1963a). These units (Figure 2B) have receptive fields whose elongate centers respond either to the onset of light (some units) or the cessation (other units), and whose peripheries respond oppositely to the center. Within a single columnar area of the cortex, all cells are aligned with their long axes in the same direction. Similar but more complex units have been described in pigeons (Matuana and Frenk, 1963; Maturana, 1964).

Most of the known parameters of the gull chick's form-movement

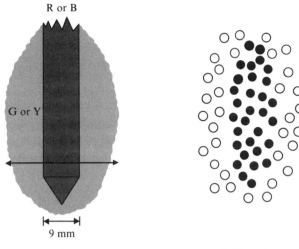

R or B

G or Y

9 mm

A. Behavior   B. Physiology

FIGURE 2

Comparison of the optimal visual stimuli that elicit a behavioral response in the gull chick and a physiological response in a cortical neuron of the cat. A. The optimal stimulus for pecking is a red or blue vertical rod or stripe against a green or yellow background. If either the background or the rod is achromatic, then the chromatic component interacts with the achromatic component in a simple way: if the chromatic component is of short wavelength (e.g., blue), then the best achromatic component is black, whereas if the chromatic component is of long wavelength (e.g., yellow, orange or red), then the best achromatic component is white. If both rod and background are achromatic, then the best combination is a black rod on a white background. (For explanation of other characteristics of the optimal stimulus, see text.)
B. A diagramatic representation of the "map" of a receptive field of a cortical unit with vertically aligned axis. The center responds to the cessation of light, the periphery to the onset. A stripe moving horizontally across the long axis would be an optimal stimulus. (From data of Hubel and Wiesel, 1959.)

preference would be coded by cat-like visual units with vertically aligned elongate axes. This suggestion predicts that a stimulus rod projecting vertically down from above with its tip at eye level and a similar rod projecting up from below with its tip also at eye level should elicit equal pecking rates. Although rods placed in both vertical positions do elicit higher pecking rates than horizontally placed rods, the vertical rod projecting down from above elicits a higher peck rate than the one projecting up from below (Schmerler and Hailman, 1965). Since the retinal image is inverted, this result might be explained by supposing that the gull chick's cat-like units receive more input from the lower half of the retina than from the upper half, but the entire speculation must be tested electrophysiologically.

In a number of other experiments, Hubel and Wiesel (e.g., 1961, 1965; Hubel, 1963a) have shown that receptive fields of precortical visual units (ganglion cells and lateral geniculate cells) are symmetrical, target-shaped fields with ON-centers and OFF-peripheries, or vice versa. More centrally located than the units with elongate receptive fields are the "complex" and "hypercomplex" units; one of these, for instance, responds to a particular shape regardless of where in the receptive field the shape is placed. These units suggest a possible coding mechanism for complicated perceptions involving "object constancies" and related phenomena.

THE BULLFROG'S STIMULUS FOR CALLING

Male bullfrogs (*Rana catesbeiana*) answer one another's calls in spring congregations sometimes termed "breeding choruses." The effects of the calling upon the female are unknown, but the coding of the call in the auditory nerve is identical in the sexes. The auditory (dorsal) portion of the eighth cranial nerve contains two classes of first-order neurons that

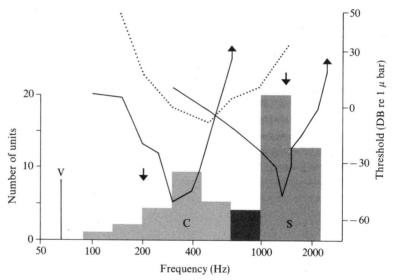

FIGURE 3
Relationship of behavior and physiology in bull frog auditory stimuli. Two kinds of neurons are found in the frog's ear: simple units, for which a sample "tuning curve" of threshold response is shown above the letter S in the diagram, and complex units. Complex units have a tuning curve of excitation similar to that of the simple unit (sample curve above the letter C), but also have a tuning curve of inhibition that has a much higher average threshold (broken curve). The tuning curves refer to the right ordinate, with the threshold sound pressure in decibels (db) relative to 1.0 microbar. V indicates the frequency of vibration to which complex units also respond. The two histograms (reference left ordinate) plot the peak frequencies of many simple (S) and complex (C) units, with some of both kinds of units being found at frequencies just below 1000 Hz. The two arrows indicate the tone frequencies combined to elicit calling from normal frogs in behavioral experiments. (Physiological data redrawn from various figures in Frishkopf and Goldstein, 1963.)

respond to the onset of sound (Frishkopf and Goldstein, 1963). "Simple" units have narrow "tuning curves" of spectral sensitivity centered about 1500 Hz maximum, while "complex" units have broader tuning curves maximal at about 400 Hz (Figure 3). The frog's call has considerable energy at both these frequencies, but little energy in the intermediate spectrum. Firing of the "complex" units is, in fact, inhibited by energy of this intermediate band (Figure 3). Therefore, only stimuli having relatively great energy at about 1500 and 400 Hz and relatively little in the intermediate band will evoke simultaneous activity of both simple and complex units.

Anatomical study (Geisler et al., 1964) revealed that simple units originate at the basilar papilla, while complex units originate at the structurally more complex amphibian papilla. Since the auditory neurons are first-order neurons without collaterals, the inhibition of complex units is non-neural and presumably a mechanical process in the amphibian papilla.

Correlative behavioral studies (Capranica, 1965) show that calling may be elicited from a male frog by essentially the same stimuli that activate simultaneously both complex and simple units. The minimum effective behavioral stimulus (Figure 3) is merely two tones, one at about 1400 Hz and the other at about 200 Hz. If a sufficiently intense tone of a middle frequency is added to the stimulus, the evoked calling ceases. In both behavioral and physiological studies, only frequency coding has been studied, temporal factors of the frog's trilling call having been held constant.

Discussion of Coding

The few examples give only a hint about coding complexities; yet, some generalities already emerge. For instance, (a) coding may take place in serial steps as neural information is sent centripetally to the brain. In the visual systems of the cat and frog, units become more complex, with regard to the external stimuli effective in exciting them, the further centrally from the receptors they are placed. Perhaps more surprisingly, (b) a given degree of complexity may be found in different loci in the sensory systems of different animal species, and (c) a given degree of complexity may occur earlier in the sensory system of a phylogenetically "lower" animal than in a higher one. Thus, the frog's ganglion cells of the eye (third order) require a more complex stimulus input than do cells in the lateral geniculate (fourth order) of the cat. Is this difference due to general evolutionary trends, or is it due rather to the different functions which the visual systems have been selected to serve? Cats depend far less on vision in their everyday lives than they do upon other senses; one suspects that their visual systems serve mainly in tracking and pouncing upon prey (see, e.g., Leyhausen, 1960).

As a working concept, we may now preliminarily define the coding mechanism for releasing stimuli as *the entire sensory system up to and including the level at which single units respond to a stimulus that is isomorphic with the releasing stimulus for behavior of the whole animal.* The working definition purposely excludes internal "motivating" factors and efferently imposed modifying factors in the sense organs, with full recognition that a complete control mechanism of a behavioral pattern must make allowance for such factors.

It should be pointed out that the definition offered above is predicated upon an assumption of how the nervous system works in controlling stereotyped, environmentally resistant behavior. This assumption is that there does exist a single unit that responds to a stimulus isomorphic with the releasing stimulus for the behavior of the whole animal. Based on studies of the coding of learnable sensory discriminations, R. P. Erickson (e.g., 1963; Erickson, Doetsch, and Marshall, 1965) denies that such single units are to be found. He argues that the neural message for sensory quality is made of the amount of neural response across many units, and that it would be inefficient for the nervous system to have single units that respond to each possible stimulus (e.g., all the chemicals in gustation) among which discriminations are made. The question cannot be settled until further information is available. I suggest that the coding for stereotyped, species-common stimulus-choices discussed in this contribution could be organized differently from the coding that underlies learnable discriminations; or that, at least, a continuum of coding schemes exists of which these two are end-points.

From the examples, the frog's color preference can be said to be coded by the visual system before the optic tract terminal arborizations of the midbrain (third-order neurons); the gull chick's color coding mechanism might likewise be entirely within the eye (as soon as second order); the frog's "bug" coding mechanism has many steps (perhaps up to sixth- or higher-order neurons); and the gull chick's coding mechanism for form-movement is undetermined, as is the frog's call-recognizing mechanism.

ONTOGENY OF PERCEPTUAL PREFERENCES

If an adult animal shows appropriate perceptual preferences, it is of interest to know how these preferences developed in the individual; that is, primarily whether perceptions are fixed before birth or whether they subsequently change, and if they change, whether the changes are due to usual kinds of learning or to other experiential factors. Furthermore, it is of ultimate importance to trace the perceptual changes to any changes in the underlying coding mechanism.

Examples of Ontogeny

THE FROG'S COLOR PREFERENCE

Muntz (1963a) tested the responsiveness of tadpoles at different ages. Figure 4 shows the responsiveness of the tadpoles at three different ages, and how the choice changes from an initial preference for green to the adult preference for blue (p. 140). The selective advantage of the early green-response is unknown, although it might have something to do with orientation to vegetation. Histologically, cones develop first. The frog's cones apparently contain a visual pigment, the absorption spectrum of which resembles the spectral response of the swimming tadpole, suggesting that the preference is mediated by a single visual pigment. Subsequently, two kinds of rods develop in the frog's eye; Muntz believes one kind to be excitatory (the green rods), acting synergistically with the cones, and the other inhibitory (red rods). Some other evidence for this view comes from similar experiments on urodeles (Muntz, 1963b). *Triturus,* which has green rods, shows a blue-preference similar to that of the frog, while *Salamandra,* which lacks green rods, shows a green-avoidance response which is the mirror image of the tadpole's green-preference.

THE CAT'S FORM ANALYZERS

The newborn kitten possesses all the general kinds of visual units found in the adult cat, except that these units are somewhat sluggish and less well oriented. Early deprivation of patterned light causes histological atrophy

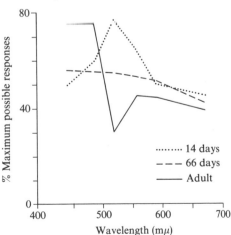

FIGURE 4
Ontogeny of the phototactic response in *Rana temporaria.* Ages are approximate; compare adult's responses with the curve in Figure 1. (Redrawn from Muntz, 1963a.)

in the lateral geniculate, which in turn disrupts visual information from reaching the visual cortex; late visual deprivation (after three months of age) has no effect. Therefore, some ontogenetic process takes place in the first three months which "sharpens" visual units (i.e., makes them more responsive and better orientated) and prevents their degeneration. Visual experience with patterned light seems to be necessary, since translucent ocular covers allowing diffuse illumination still prevent proper maintenance and development (Hubel and Wiesel, 1963b; Wiesel and Hubel, 1963a, 1963b).

THE GULL CHICK'S PECKING PREFERENCE

Laughing gull chicks tested at hatching and at a week or more of age show no consistent differences in color choices (p. 140) but show dramatic changes in form-movement preferences (p. 142) (Hailman, 1967). The form preference of older chicks can no longer be described in the simple terms used to characterize the cat's or the pigeon's visual units. Rather, a qualitatively more precise *Gestalt* develops, so that in order to elicit pecking, models must more closely resemble the actual parent's head (Hail-

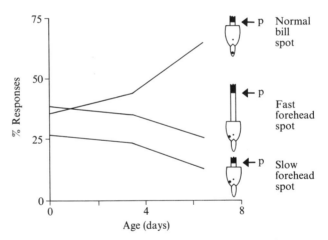

FIGURE 5

Ontogeny of a configural perception in chicks of the herring gull. As was found by Tinbergen and Perdeck (1950) newly hatched chicks prefer a "bill spot" to a "forehead spot," but only because the forehead spot moves slower, and through a smaller arc. Wild chicks of about a week old strongly prefer the bill spot, regardless of the movement parameters of the forehead spot. The small triangular arrows labeled "p" indicate the pivot point for movement of the models. (Adapted from Hailman, 1967.)

man, 1962, 1967). That such changes are due to a form of conditioning is indicated by that fact that chicks come to prefer a model of a species other than their own parent when fed in response to pecking at this model (Hailman, 1967) and can also learn to respond preferentially to an originally nonpreferred stimulus (Schmerler and Hailman, 1965).

In a related species, the herring gull (*Larus argentatus*), the parent's head is white, and its yellow beak has a subterminal red spot at which the chicks peck. Tinbergen and Perdeck (1950) found that models with the red spot on the forehead received fewer pecks than normal ones; they attributed this result to the disruption of spatial relations among the stimulus elements. Following a suggestion by Schneirla (1956), Hailman (1967) repeated the experiment, adding another model with a "forehead" spot in which the spot's movement was equated to that of the bill spot. This third model received just as many pecks as did the normal bill-spot model, showing that the degree of movement, not configuration, was important. The preference of newly hatched herring gull chicks is thus essentially identical to that of the newly hatched laughing gull chick, and could be coded by "simple" visual units. When older herring gull chicks are tested, however, they prefer the normal bill-spot model to both models with forehead spots, regardless of velocity or arc of movement (Figure 5); that is, the chick does develop a Gestalt-like preference. Thus the herring gull chick's perceptual ontogeny also resembles that of the laughing gull chick.

Discussion of Ontogeny

These few examples suggest that some perceptual preferences may be "built in" and function completely at birth in the same way as they do later (gull chick's color preference). Others may develop because of maturation of sensory apparatus (frog's color response). Still others require experience for development. The experience may be general visual familiarity with forms during a critical period (as apparently in the kitten), or specific conditioning to an environmental object (as in the gull chick's form preference).

NOTES ON SOME CONCEPTS ABOUT ANIMAL PERCEPTION

The studies cited suggested some new ways of looking at perception in animals. Some old concepts about animal perception may no longer be very useful; four such examples are discussed.

The Innate Releasing Mechanism

Although now usually conceived of as a brain center that responds to a specific configural stimulus without the aid of previous learning, the innate releasing mechanism (IRM) was not meant originally to be so specific (Schleidt, 1962). It is clear from the examples (p. 146) that releasing mechanisms are not necessarily innate. Schleidt (1961) toppled one of the commonly cited examples of an innately recognized configuration; and the gull chick experiments do away with another. Nor are coding mechanisms liable to be found only in a discrete center of the central nervous system; rather, they comprise segments of the afferent sensory pathway, the information being processed in steps along the way. Finally, it is clear that some degree of configuration is inherent in all visual stimuli other than pure color or brightness. The question is no longer whether or not a perception is configural, but how closely the configuration matches the environmentally appropriate object and how the configuration is coded.

Hebb's Learning of Patterned Stimuli

D. O. Hebb's (1949) magnificent neuropsychological theory assumes that even simple pattern perception in animals is learned—the antithesis of the innate releasing mechanism. Hebb's evidence (e.g., pp. 31–37) comes primarily from studies of animals who were deprived of early visual experience and were later found to have poor form vision. However, Hubel and Wiesel's studies on the cat's visual system now show that the situation is more complex than originally thought since learning (in the usual sense of the word) may not be involved; but, rather, merely experience with patterned visual input. Furthermore, the cat's nervous system does not begin as a *tabula rasa,* but has initial organization of form which may further sharpen through some kind of experience in normal development but degenerates irreversibly during early deprivation.

Intensity-coded Approach-Withdrawal

Recognizing the difficulties in assuming an extreme (i.e., all innate or all learned) position about perceptual development, Schneirla (1959, 1965) proposed an intermediate position: simple perceptions are operative at birth or hatching and later develop through experiential interactions into more specific and complex perceptions. This is essentially my position, too. More specifically, Schneirla suggested that many early responses are basically approach or withdrawal in relation to the stimulus source, and that, initially, low intensities of stimulation elicit approach, higher intensities elicit withdrawal.

In this basic form, the theory is certainly true. In order to position itself at a given distance from a stimulus source, an organism may move along a gradient of intensity. The stimulus intensity of light or sound increases by the square as the distance to a point source decreases. Therefore, in order to select a certain distance from a stimulus source, an animal may select a certain intensity that correlates with the proper distance. In order to reach this intensity, the animal must approach the source (i.e., move along a gradient of increasing intensity) when the intensity is weak, and must withdraw from the source when the intensity is strong. It thus follows from the physics of stimulus propagation that low intensities elicit approach and high intensities withdrawal.

In a more general context, the theory runs into certain difficulties of which three may be distinguished. (1) The first difficulty is in defining the stimulus, if attention is restricted to only the physical stimulus. Schneirla (1965, p. 3) gives the stimulus eliciting approach as "low or decreasing" or "low-intensity energy change." Clearly a stimulus could be of high intensity and decreasing simultaneously; or high, with a low-intensity change, etc. Some more consideration must be given to exact definition of the stimulus parameters.

Perhaps even more difficult is the possibility (2) that circular formulation may creep into the definition of the stimulus. Thus (ibid., p. 5) stimulus intensity is defined as the "*effective stimulus input* affecting neural input," or as (ibid., p. 3) "variations adequate for Approach-system arousal." If the stimulus is thus defined in terms of the response given to it, any stimulus that elicits approach could be said to be of low stimulus intensity, and there is no logical possibility of finding a behavioral system that does not fit the theory.

The last point (3) that requires more consideration in order to make the theory fully useful concerns the definition of the approach response and the extent of the concept's applicability. Must the entire animal approach the stimulus, or can pecking by gull chicks (ibid., p. 12 ff) and other orientated behavior logically be included? (If the peck is "approach," what is "withdrawal"?) It is thought that the theory applies to "all animals" (ibid., p. 5). However, there should be more consideration of Lehrman's (1953, pp. 347–48) criticisms of applying behavioral theories to all levels of animal organization, since he rightly asserts that ignoring species differences "hinders rather than helps analysis of the behavior patterns" (ibid., p. 348). In fact, Schneirla himself (1961) emphasized this point when he wrote: "The task of the developmental psychologist, therefore, is to find formulae appropriate to the ontogenetic patterns of the respective phyletic levels." He also asserted that "(if) the same terms are used for adjustments on very different levels (then the result will be) that important differences are obscured."

Classical Conditioning

Classical, or Pavlovian, conditioning was one of the earliest learning paradigms to be formulated (Pavlov, 1926). The 1949 Round Table on behavioral nomenclature decided upon the following as a definition of classical conditioning (Thorpe, 1956, p. 72): "the process of acquisition by an animal of the capacity to respond to a given stimulus with the reflex reaction proper to another stimulus (the reinforcement) when the two stimuli are applied concurrently for a number of times." Usual examples are cross-modal stimuli, such as Pavlov's dog salivating upon hearing a bell, instead of seeing or smelling food.

Taken in this way, classical conditioning has been demonstrated in few cases to be critical in the normal development of species-common behavior. But in cases of "perceptual sharpening," such as that of the gull chick, a conditioning-like process is certainly involved. The interesting thing to note is that the unconditional stimulus (parent's beak) is the same as the conditional stimulus; they overlap in time and space because they are physically identical. Is it not possible that the capacity to form conditional responses has evolved in animals as a mechanism of such perceptual sharpening, and that the usual laboratory cross-modal conditioning is merely a biproduct of this ability?

SUMMARY

Improved behavioral testing procedures, the recording of single-unit activity with microelectrodes, and the precise control over the physical stimulus are tools that have led to new results in the study of animal perception. Coding schemes for the stimuli that direct or elicit the meaningful species-common behavior patterns of food-begging, escape, and prey-capture have been studies using color, form-movement, and auditory stimuli. The "coding mechanism" may be the whole sensory pathway, with the processing of information occurring at all intermediate stations. A given level of stimulus complexity does not occur at the same locus in homologous sensory systems of different animals.

The ontogenetic development of perception is variable. A coding system may be formed by birth, and undergo only minor improvements with general sensory experience, but atrophy considerably in the absence of such experience (cat's form-vision). Or, a coding system may change during ontogeny because of the differential growth rate of different receptor types (frog's color preference). Some simple perceptions are adequate at hatching, but improve immensely by conditioning, or perceptual sharpening (gull chick's form preference).

The results from coding and ontogenetic studies cast doubt on the adequacy of certain concepts of animal perception, namely, the innate releasing mechanism, Hebb's learning of patterned stimuli, intensity-mediated approach-withdrawal theory, and classical conditioning.

ADDENDUM

Since the time that the final manuscript of this contribution was submitted, further evidence concerning nearly all of the examples cited has appeared in the literature. Since full discussion of these recent studies would require a complete reworking of the manuscript and a considerable lengthening of the bibliography, only highlights of the studies will be summarized here without bibliographic notation.

The frog's color preference has been further studied by Muntz and reported in several papers, and by R. Chapman, who provided unequivocal evidence that it is a true color preference and not just a spectral sensitivity. Unpublished data by Jaeger and Hailman on more than a dozen species of many different families suggests that the blue-preference is virtually universal among anuran amphibians, although some tropical species may show different spectral preference curves. Microspectrophotometric data by Liebman and Entine on the individual receptors of ranid frogs provide solid data for Muntz's receptor-coding hypothesis. Muntz's hypothesis is also strengthened by its ability to predict changes in the response curve due to the intensity of the stimuli and the state of light or dark adaptation of the frog.

Microspectrophotometry of the cones and oildroplets in the laughing gull retina by Liebman and Hailman have identified one cone pigment and three color-classes of oildroplets. The data allow evaluation of the coding equations given in the text above, and the equations are found to be only approximately predictive, and thus in need of revision. Hailman has also shown that the bill of the parent gull reflects only red wavelengths, and has no blue reflection corresponding to the secondary blue peak in the chick's preference curve. This finding reinforces the suggestion that the blue peak arises due to receptor-neural limitations of coding in the visual system.

An extraordinary amount of research has been published on visual receptive fields, mainly in mammals. Some of the papers on cats report greater complexities of receptive field organization at the retinal level than found by Hubel and Wiesel. Most of the recent findings have no critical bearing on the points raised in this contribution.

In general, none of the general points in the discussion portion of the contribution are affected by more recent results. However, recent results do show that the study of stimulus-coding is being actively pursued, and may soon demand another review of our concepts. More attention is being given to making studies of sensory physiology relevant to the animal's behavior (e.g., the work of Griffin on bat sonar, and of Roeder on auditory mechanisms of escape responses in moths), and one hopes that such trends will continue.

Another important paper is that by O.-J. Grüsser and Ursula Grüsser-Cornehls (1968. Neurophysiologische Grundlagen visueller angeborener Auslösemechanismen beim Frosch. *Z. Vergleich. Physiol.* 59: 1–24), in which they compare stimuli evoking fleeing and prey-catching reactions in ranid frogs with the stimuli that drive the classes of visual neurons. In particular, this paper shows that every effective visual stimulus that evokes a particular kind of reaction behaviorally also drives several of the classes of neurons. This result indicates that the perception of prey and predator is not coded by a single class of neurons, but presumably by the relative patterns of activities in the four classes.

ACKNOWLEDGMENTS

I am very grateful to Dr. Colin G. Beer for careful and critical comments on the final manuscript; Dr. Daniel S. Lehrman criticized an earlier version. Unpublished work cited in the text was completed during tenure as a USPHS Postdoctoral Research Fellow.

REFERENCES

Capranica, R. R. 1965. *The evoked vocal response of the bullfrog—a study of communization by sound.* Cambridge: M.I.T. Press.
Erickson, R. P. 1963. Sensory neural patterns and gustation. In Y. Zotterman, ed., *Olfaction and taste.* New York: Pergamon Press. Pp. 205–213.
Erickson R. P., G. S. Doetsch, and D. A. Marshall. 1965. The gustatory neural response function. *J. Gen. Physiol.* 49: 247–263.,
Frishkopf, L. S., and M. H. Goldstein, Jr. 1963. Responses to acoustic stimuli

from single units in the eighth nerve of the bullfrog. *J. Acous. Soc. Am.* 35: 1219–1228.

Geisler, C., W. A. Van Bergeijk, and L. S. Frishkopf. 1964. The inner ear of the bullfrog. *J. Morph.* 114: 43–58.

Granit, R. 1942. Colour receptors of the frog's retina. *Acta Physiol.* 3: 137–151.

Hailman, J. P. 1962. Pecking of Laughing Gull chicks to models of the parental head. *Auk* 79: 89–98.

Hailman, J. P. 1964. Coding of the colour preference of the gull chick. *Nature* 204: 710.

Hailman, J. P. 1966. Mirror-image color-preferences for stimulus-object and background in the gull chick. *Experientia* 22: 257.

Hailman, J. P. 1967. The ontogeny of an instinct: the pecking response in chicks of the laughing gull (*Larus atricilla L.*) and related species. *Behaviour,* Suppl. 15.

Hebb, D. O. 1949. *Organization of behavior.* New York: John Wiley.

Hubel, D. H. 1959. Single unit activity in striate cortex of unrestrained cats. *J. Physiol.* 147: 226–238.

Hubel, D. H. 1963a. Integrative processes in central visual pathways of the cat. *J. Opt. Soc. Am.* 53: 58–66.

Hubel, D. H. 1963b. The visual cortex of the brain. *Sci. Am.* 209 (Nov. 1963): 54–62.

Hubel, D. H., and T. N. Wiesel. 1959. Receptive fields of single neurons in the cat's striate cortex. *J. Physiol.* 148: 574–591.

Hubel, D. H., and T. N. Wiesel. 1961. Integrative action in the cat's lateral geniculate body. *J. Physiol.* 155: 385–398.

Hubel, D. H., and T. N. Wiesel. 1963a. Shape and arrangement of columns in cat's striate cortex. *J. Physiol.* 165: 559–568.

Hubel, D. H., and T. N. Wiesel. 1963b. Receptive fields of cells in striate cortex of very young, visually inexperienced kittens. *J. Neurophysiol.* 26: 944–1002.

Hubel, D. H., and T. N. Wiesel. 1965. Receptive fields and functional architecture in two nonstriate visual areas (18 and 19) of the cat. *J. Neurophysiol.* 28: 229–289.

Lehrman, D. S. 1953. A critique of Konrad Lorenz's theory of instinctive behavior. *Quart. Rev. Biol.* 28: 337–363.

Lettvin, J. Y., H. R. Maturana, W. H. Pitts, and W. S. McCulloch. 1961. Two remarks on the visual system of the frog. In W. A. Rosenblith, ed., *Sensory communication.* Cambridge and New York: M.I.T. Press and John Wiley. Pp. 757–776.

Leyhausen, P. 1960. Verhaltenstudien on Katzen. *Z. Tierpsychol.,* Suppl. 2.

Marler, P. 1961. The filtering of external stimuli during instinctive behavior. In W. H. Thorpe and O. L. Zangwell eds., *Current problems in animal behaviour.* Cambridge, England: Cambridge Univ. Press. Pp. 150–166.

Maturana, H. R. 1964. Functional organization of the pigeon retina. In *Information processing in the nervous system,* vol. III, pp. 170–178. Proc. XXII Int. Physiol. Cong. Leiden, 1962.

Maturana, H. R., and S. Frenk. 1963. Directional movement and horizontal edge detectors in the pigeon retina. *Science* 142: 977–978.

Maturana, H. R., J. Y. Lettvin, W. S. McCulloch, and W. H. Pitts. 1960. Anatomy and physiology of vision in the frog. *J. Gen. Physiol.* 43: 127–177.

Muntz, W. R. A. 1962a. Microelectrode recordings from the diencephalon of the frog (*Rana pipiens*), and a blue-sensitive system. *J. Neurophysiol.* 25: 699–711.

Muntz, W. R. A. 1962b. Effectiveness of different colors of light in releasing the positive phototactic behavior of frogs, and a possible function of the retinal projection to the diencephalon. *J. Neurophysiol.* 25: 712–720.

Muntz, W. R. A. 1963a. The development of phototaxis in the frog. (*Rana temporaria*). *J. Exp. Biol.* 40: 371–379.

Muntz, W. R. A. 1963b. Phototaxis and green rods in urodeles. *Nature* 199: 620.

Pavlov, I. V. 1926. *Conditioned reflexes.* Cambridge, England: Oxford. Univ. Press.

Pearse, A. S. 1910. The reactions of amphibians to light. *Proc. Am. Acad. Arts. Sci.* 45: 159–209.

Schleidt, W. 1961. Reaktionen von Truthühnern auf fliegende Raubvögel und Versuche zur Analyse ihrer AAM's. *Z. Tierpsychol.* 18: 530–560.

Schleidt, W. 1962. Die historische Entwicklung der Begriffe "Angeborenes auslösendes Schema" und "Angeborener Auslösemechanismus" in der Ethologie. *Z. Tierpsychol.* 19: 697–722.

Schmerler, S., and J. P. Hailman. 1965. Discrimination of orientation in the Laughing Gull, *Larus atricilla* L. *Am. Zool.* 5: 655.

Schneirla, T. C. 1956. Interrelationships of the "innate" and the "acquired" in instinctive behavior. In P.-P. Grassé, ed., *L'Instinct dans le comportement des animaux et de l'homme.* Paris: Masson. Pp. 387–452.

Schneirla, T. C. 1959. An evolutionary and developmental theory of biphasic processes underlying approach and withdrawal. In M. R. Jones, ed., *Nebraska symposium on motivation.* Lincoln, Nebraska: Univ. Nebraska Press.

Schneirla, T. C. 1961. Instinctive behavior, maturation—experience and development. In B. Kaplan and S. Wapner, eds., *Perspectives in psychological theory.* New York: International Universities Press.

Schneirla, T. C. 1965. Aspects of stimulation and organization in approach withdrawal processes underlying vertebrate behavioral development. In D. S. Lehrman, R. A. Hinde and E. Shaw, eds., *Advances in the study of behavior,* vol. I. New York: Academic Press. Pp. 1–74.

Thorpe, W. H. 1956. *Learning and instinct in animals.* Cambridge, Mass.: Harvard Univ. Press.

Tinbergen, N. 1951. *The study of instinct.* Oxford, England: Clarendon Press.

Tinbergen, N., and A. C. Perdeck. 1950. On the stimulus situation releasing the begging response in the newly hatched herring gull chick. (*Larus argentatus argentatus* Pont.) *Behaviour* 3: 1–39.

Wiesel, T. N., and D. H. Hubel. 1963a. Single-cell responses in striate cortex of kittens deprived of vision in one eye. *J. Neurophysiol.* 26: 1003–1017.

Wiesel, T. N., and D. H. Hubel. 1963b. Effects of visual deprivation on morphology and physiology of cells in the cat's lateral geniculate body. *J. Neurophysiol.* 26: 978–993.

ERIC A. SALZEN

Department of Psychology
University of Aberdeen
Aberdeen, Scotland

Imprinting and Environmental Learning

"My very chains and I grew friends,
So much a long communion tends
To make us what we are:—even I
Regain'd my freedom with a sigh."

Byron, "THE PRISONER OF CHILLON"

IMPRINTING AND A NEURONAL MODEL HYPOTHESIS

The term *imprinting* was coined by Lorenz (1937), in translation from the German "Prägung" (1935), to describe a process in the development of social attachments in precocial birds. By social attachments I mean the fixation of the class of objects to which social responses, both present and future, are directed and confined. Imprinting is commonly misconceived as the eliciting of the following response or as the acquisition of a response, e.g., Bindra (1959, p. 64) and Moltz (1963). In a frequently cited review, Moltz (1960) used the term imprinting ". . . to denote a particular experimental operation and *not* a process or a mechanism." Even more frequently quoted are the papers by Hess (e.g., 1959, 1964) whose views on imprinting are different from those of many other workers whom he has claimed (1959) are really studying associative learning and not imprinting. The reviews by Fabricius (1962), Hinde (1961, 1962), Salzen (1962), Sluckin (1965), Sluckin and Salzen (1961) and Thorpe (1956) follow the original concept of imprinting somewhat less rigidly than those by Hess but generally regard it as a process in which innate social responses that can be elicited by and directed to a wide variety of objects come to be elicited by and directed to only the class or classes of objects experienced in a limited neonatal period. Thus imprinting is a process of goal or object acquisition and *not* response acquisition. In view of the misunderstandings that may arise from selective reading of imprinting literature it might help to recall that in 1963 Hinde was able to say that most workers regard imprinting as not involving just supraindividual learning, that the sensitive period is by no means clearly marked, that the irreversibility of imprinting had been overestimated, and that it is not clear to what extent imprinting as studied in the context of the following response of the newly hatched bird affects later behavior. The same views had been expressed earlier by Thorpe (1956).

There remains one important distinguishing feature of imprinting; namely the apparent absence of reinforcement. This aspect has received little or no attention in the literature and most writers simply assert that it plays no part in the process (e.g., Hinde, 1962). It is true, however, that conventional physiological reinforcers are not employed in imprinting experiments and Hinde, Thorpe, and Vince (1956) and Hinde (1961) have emphasised that parental activities such as brooding and feeding are not necessary for the establishment of following behavior to objects. It was this apparent absence of reinforcement that led me to regard imprinting as a process of perceptual learning (Sluckin and Salzen, 1961). Sluckin (1965) has recently suggested the term *exposure learning* for this process. As a result of further studies of avoidance and approach behavior shown by chicks in various manipulations of imprinting situations, I suggested (1962) that this process could be understood in terms of Sokolov's (1960) neuronal model hypothesis of the orienting reflex (Figure 1). This hypothesis postulates a model-forming system in which patterns of stimulation set up corresponding neuronal assemblies. Subsequently the input from a new stimulus pattern will not coincide with the model or cell assembly and the modeling system then excites an amplifying system which instigates the orienting reflex. Repetition of the new afferent pattern sets up a corresponding new neuronal model in the modeling system which at the same time develops an inhibitory effect on the amplifying system by blocking its collateral afferent supply. Although Sokolov regarded the modeling system as representing the mammalian cortex and the amplifying system as being the reticular formation of the brainstem, the hypothesis of the interaction of a modeling and an activating system is independent of such specification of neural structures.

The orienting reflex can be regarded as part of a complex of adjustive or orienting behavior that the animal shows in relation to its environment and this behavior includes the approach-withdrawal mechanisms to which Schneirla (1959, 1965) has drawn attention. Schneirla has attempted to explain the development of such adjustive behavior in terms of the reinforcement of approach responses through their association with normal low-tonus and parasympathetic internal states. Similarly, withdrawal re-

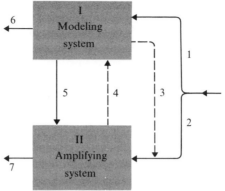

FIGURE 1

Schema for the orienting reflex proposed by Sokolov (1960). I. Modeling system; II. Amplifying system. 1. specific pathway from sense organ to cortical modeling system; 2. collateral afferent path to reticular formation amplifying system; 3. negative feedback from modeling system to synaptic connections of collateral afferents with reticular formation; 4. ascending activating influences from amplifier to modeling system; 5. pathway from modeling system to amplifier signaling discordance between afferent stimulation and neuronal models; 6. pathway to specific responses caused by concordance between afferent stimulation and neuronal models; 7. pathway to vegetative and somatic responses resulting from activation of the amplifier.

sponses come to be reinforced by their association with high-tonus and sympathetic processes. Low intensity stimuli evoke approach processes and high intensity stimuli evoke withdrawal processes. The associations between responses and the somaesthetic states or processes are established during embryonic maturation interactions of the kind postulated by Kuo (1932) on the basis of his studies of the chick embryo and of the kind more recently employed by Lehrman (1953) as a means of accounting for apparently "innate" behavior. The analysis of behavior in terms of developing interactions between changing structural, physiological and environmental features is a most appealing one to the biologist and has been ably attempted by Schneirla (1956, 1957) and by Lehrman (1956). In such an analysis the terms *innate* and *learned* have little use and the term *reinforcement* means the association by contiguity of patterns of afferent feedback of proprioception, interoception, and exteroception. It is interesting that in the analysis of imprinting there has arisen some confusion over the use of these same three terms. An analysis of imprinting on the basis of the neuronal model hypothesis, however, largely avoids the use of these terms and instead deals with the adjustive or orienting behavior involved and its relation to prior and present patterns of exteroceptive stimulation. This hypothesis can be brought closer to the maturational type of explanation of behavior by extending the model-forming concept to include the proprioceptive and interoceptive systems. In what follows, therefore, I shall try to extend and develop the neuronal model hypothesis to explain the manner in which the neonatal chick adjusts to its environment. It will be suggested that the chick learns the details of its environment, both physical and social, through exposure to stimulation of its distance receptors. It will also be suggested that the formation of a social attachment in the special sense of imprinting may differ from such environmental learning in requiring a "reinforcement" of adjustive responses to distant stimuli by the achievement of adequate patterns of contact stimulation.

NEURONAL MODELS AND APPROACH-WITHDRAWAL BEHAVIOR IN THE NEONATE

Contact Receptors and Neuronal Models

The neuronal model hypothesis states that repeated stimulus patterns serve to fashion neuronal models of the environment. The chick, therefore, may be hatched with models for contact and warmth stimulation already set up during maturation in the egg. Hatching drastically changes this input with the result that the model-forming system excites an activating system (Sokolov's amplifying system) and the chick shows struggling movements and gives distress calls. Collias (1952) has suggested that the first distress calls of the chick are due to cooling of the moist down and the loss of contact with the egg shell and that during the first hour after hatching these calls can be reduced in frequency by providing contact, warmth, or clucking noises. As far as providing contact and warmth are concerned, it would seem that this represents a restoration of the familiar patterns of stimulation which match the modeling system and remove its excitatory action on the activation system. With regard to the effect of clucking it should be noted that Collias also states that the tapping noises heard in an egg about to hatch (made by the bill striking the shell as the chick begins to chip its way out of the egg) may approximate in rate and frequency to the clucking of a hen. Thus a neuronal model of this auditory input could develop before hatching. This is the kind of self-stimulatory process given such importance in the development of behavior by Schneirla (1956, 1957) who has made the suggestion that processes of mechanical and auditory self-stimulation may provide one basis for the attraction of naive isolated neonates to their species mates.

The neuronal model hypothesis seems to apply to behavior prior to hatching; for example, Collias (1952) writes that if an egg near hatching is alternately cooled and warmed, distress and pleasure notes can be heard in company with the changes and occasionally a pipped egg will respond to clucking with a cessation of distress-calling. Moving an egg may also result in a flurry of pleasure notes. I shall suggest later that pleasure calling is not the antithesis of distress but rather that it occurs when there is a mild mismatch between stimulus input and the neuronal model system with a resulting low level of action by the activating system. Hence pleasure- or contentment-calling tends to occur just as the familiar input is being reestablished and the noncoincidence between input and model is being reduced to a minimum. The system, in this case, provides a self-restoring or homeostatic mechanism for the maintenance of stable conditions of warmth and contact. However, if new patterns of stimulation endure and persist, new neuronal models will be fashioned. Thus the newly hatched chick ultimately settles down (i.e., stops struggling and distress-calling) in a different spatiotemporal pattern of contact and surface temperature from that experienced in the egg.

Distance Receptors and Neuronal Models

The neuronal modeling system becomes particularly important when we consider the distance receptors. The amount of information that is handled by auditory and especially by visual receptors is much greater than that dealt with by the body contact receptors (see Sherrington, 1947, p. 333). The problems involved in producing neuronal models representing the animal's visual world and even its auditory world through embryogenetic processes may be so great that such processes may not have evolved. The patterns of stimulation involved are highly variable, for any one object can present a large number of such patterns according to the position of viewing, the character of the lighting, etc. Further, the nature and details of the visual and auditory environment of any particular species will also vary widely in space and time. It would seem more economical to provide the individual organism with a system that would rapidly fashion models representing the particular environment experienced. The genetic problem is thus reduced to that of providing, through maturational processes, sufficient organisation of neuronal models to ensure orientation to appropriate aspects of the environment for further differentiation of the models to occur. Thus the auditory neuronal models of the chick seem to be sufficiently unstructured as to require at most slow repetitive sounds of low pitch and short pulse form (Collias and Joos, 1953), although Gottlieb (1963, 1965) has claimed that ducklings and chicks respond preferentially to the parental call even if given prior experience of other calls. The chick's visual neuronal model system may be poorly organized for although the eyes and optic lobes are functional from the 18th day of incubation (Peters, Vonderahe, and Powers, 1958) little other than movements of light and shadow is likely to be seen inside the egg. This may explain why the newly hatched chick will approach a source of flashing light with pleasure calls and will stay by it (James, 1959; Smith, 1960). It could be that at hatching, the visual system is constructed simply to instigate orientation to and fixation of any source of stimulation that meets a few simple requirements of the visual receptor mechanisms. Sackett (1963) has outlined the possibilities of such peripheral filtering but the same principles have been previously considered by Marler (1961). Microelectrode records from ganglion cells of the pigeon retina (Maturana and Frenk, 1963) have demonstrated the existence of retinal mechanisms responding to features such as on-off, directional movement, and oriented contour or brightness contrast. These are the kinds of stimulation that elicit approach and following in chicks and are provided by all moving objects and by the classes of stimuli used by Smith (1962) in his studies of the approach response in chicks. Moltz and Stettner (1961) were able to produce better following of an object by 2- and 4-day chicks reared in diffuse light than by ones reared in social isolation but with normal visual experience. It is as though the lack of a structured visual environment maintained the neonatal unstructured visual system and its initial response properties. An alternative suggestion is that the distance receptor neuronal model systems are sufficiently unstructured at hatching so that almost any focal source of stimulation will give only a mild degree of mismatch and so give mild activation in the form of orientation, fixation, and approach responses. Thus Bateson (1964a) has reported that positive responses with no avoidance are shown by chicks in an imprinting situation during the first day after hatching. When first exposed after one day the chick gives avoidance responses which decrease before changing to approach and following behavior. He attributes this initial avoidance to the fact that after one day the chicks have a familiar environment and that the unfamiliarity of the object elicits this avoidance. This conclusion is identical with that of Salzen's study (1962) of the approach and fear behavior of chicks in imprinting situations. It would support the notion of an unstructured visual neuronal model system responding to focal stimulation with orienting and approach responses and becoming more structured and specific with experience, so that changes in the experienced pattern of stimulation produce more extensive and intense orienting behavior away from the focus of any new stimulation and towards the familiar. This is an alternative to the suggestion of Schneirla (1965) that there is a biphasic response system with withdrawal responses to strong stimulation and approach to weak or optimal stimulation.

The Development of Approach-Withdrawal Behavior

Schneirla (1965) has developed his theory of a biphasic response mechanism for approach-withdrawal behavior in an excellent review of such behavior in neonatal vertebrates. He concludes that "the stimuli adequate for eliciting A-processes are those existing neural input effects that are quantitatively low, regular, and limited in their ranges of magnitude. Stimuli adequate for eliciting W-processes are those existing neural input effects that are quantitatively high, irregular, and of variable extensive ranges" (p. 45). Most studies of the development of withdrawal behavior deal with neonates that have had some distance sensory experience before testing. What is striking when tests are made immediately after hatching and before the neonate has had much experience of distance receptor stimulation is the surprising lack of negative or withdrawal type responses to intense stimulation. Schaller and Emlen (1962) made extensive studies on precocial birds and found that avoidance responses are absent at first but appear by about 10 hours. They report that domestic chicks ignored the stimulus object during the first 5 or 6 hours. The conclusion was that it is the property of strangeness in an object that is essential for eliciting avoidance behavior and that intensity of response was not appreciably affected by the color, size, or form of the test object. The same authors had previously (1961) reported for an altricial species that all objects were accepted and given the gaping response (positive) rather than the crouching response (negative) until several days after pattern vision was possible. They suggested that after this time it was unfamiliar objects that elicited the negative response. Fishman and Tallarico (1961b) were able to elicit flinching responses in only 1/3 to 1/2 of their 3-hour old chicks and in a previous study (1961a) they failed to find differences in blinking of prematurely hatched chicks in response to an advancing, a retreating, or no test object.

I have made some preliminary studies of the responses of chicks hatched in separate dark boxes when exposed to a 1-foot-square board that was advanced very rapidly right up to the chick, moving through a distance of 3 feet in approximately 1/4 second. Even without controlling for air-

TABLE 1.
Responses of newly-hatched chicks to a large approaching object.

Behavior	Experiment 1	Experiment 2	
		Before striking	After striking
No response	18/30	10/33	4/33
Flinching	15/30	21/33	15/33
Distress-calling, Withdrawal	6/30	7/33	5/33
Pleasure-calling, Approach	11/30	16/33	26/33

The table gives the proportions of chicks that showed on at least one occasion each of the behavior items listed. In Expt. 1, a 1-foot-square board was advanced rapidly up to each chick 24 times immediately after removal from individual hatching boxes. In Expt. 2, the chick was struck with the board on the 20th trial.

pressure effects, the results indicate surprisingly few withdrawal responses. Table 1 gives the results of two experiments. In Experiment 1 the stimulus was repeated rapidly 24 times. In Experiment 2 the chick was struck sharply with the board on the 20th trial and the responses after this treatment are given separately in the table. It can be seen that although flinching responses were common, many chicks showed no response at all and many gave pleasure calls and even approach responses. Usually flinching ceased within a few trials. Only a minority showed distress calls and turning away, while flight was seen in only two chicks. Striking the chicks seemed to increase or elicit pleasure-calling and approach responses. It is possible that these chicks, having been taken from their hatching boxes for testing, showed no withdrawal responses to the approaching object because they were crouching or showing the freezing response to the new general environment. Even so, one might expect that the presentation of a focus of intense stimulus change would precipitate flight in such birds. Instead, on repetition, this stimulus induced, if anything, pleasure-calling and approach responses.

In another two experiments, chicks were hatched in separate boxes and, when 1–12 hours old, were placed singly in isolation rearing cages. As each chick was placed in its cage, it was tested for its reaction to the experimenter's hand thrust into the cage and up to the chick. The same test was made at different subsequent times. The results are given in Table 2. They confirm that the initial response to a large rapidly approaching

TABLE 2.
Responses of isolate chicks to the experimenter's hand thrust into their cages.

Time of test	0 min	5 min	15 min	1 hr	24 hr	7 days	Experiment
No response	9/9	—	—	2/16	—	—	
Flinching	—	9/9	—	—	2/22	—	Expt. 1
Flight	—	—	—	14/16	20/22	22/22	
No response	19/22	—	11/22	—	4/22	—	
Flinching	3/22	—	3/22	—	6/22	—	Expt. 2
Flight	—	—	8/22	—	12/22	—	

The table gives the times of the tests from the moment that the chicks were placed in their cages. Each test consisted of a single insertion of the hand into each cage. The proportions of birds that showed each of the listed responses is indicated for the appropriate test time.

object is, at most, a slight flinching movement. What is of interest, however, is that after 15 minutes of experience in an illumined environment, distinct signs of withdrawal behavior were seen in response to this same stimulus. This behavior becomes more distinct and intense and is shown by more chicks the longer the period before testing. It is as though the chick rapidly forms neuronal models of the environment and that the test stimulus then represents a disturbance of this environment and results in activation in the form of flinching, turning away, and finally flight.

NEURONAL MODELS AND IMPRINTING

Modeling the Environment

The neuronal model hypothesis can handle the data of imprinting experiments. It postulates that repeated patterns of distance receptor stimulation will fashion appropriate models of the environment. There are two aspects to this environment: the physical, inanimate or nonsocial aspect may be distinguished from the animate or social aspect. Jacobs and Smith (1960) have suggested that the amount of stimulus movement and the number of activated sense modalities are important elements in the gradient of complexity between inanimate and animate stimulus objects. They cite Krech and Crutchfield (1948) who have written: ". . . Person objects differ from other objects in an individual's field, because they have, among other characteristics, the properties of mobility, capriciousness, unpredictability. . . ." This complexity may be due also in part to the difference between apparent movement of the environment resulting from the animal's own movements and the real movement of the stimulus object relative to other parts of the environment and independently of the animal's own movements. Hein and Held (1962) and Held and Hein (1963) have shown very clearly how important this factor can be, for they report that kittens reared with experience of environmental motion resulting solely from being passively carried through the environment were subsequently deficient in visually guided paw placement, visual cliff behavior, and blink responses. There is also the possibility, argued by Gibson (1959), that animate objects can be sensorily differentiated as moving and deforming surfaces that provide a physical stimulus pattern distinct from that provided by the static and stable surfaces characteristic of inanimate objects. It is possible, therefore, that the neonatal chick develops neuronal models of both the static physical environment and the independently-moving social environment. Their behavior in imprinting experiments would involve an interaction of changes in both static and moving aspects of the test situation. After analyzing the behavior of socially experienced and inexperienced chicks in an imprinting situation (Salzen, 1962), I concluded that their behavior towards static, moving, and social objects depended on the difference between the test situation and their previous environment. Subsequent studies by Bateson (1964a, b) have resulted in a similar conclusion, namely that the behavior of birds in the imprinting situation can best be understood in terms of the relation between the testing situation and the rearing environments. Briefly, unfamiliarity with either static or moving aspects of the environment provoke distress and searching behavior. This behavior becomes differentiated according to the nature of the stimulus concerned. If the whole environment is strange, the chick shows freezing; with continued exposure, he shows searching, which is characterized by erect, alert-walking and distress-calling. Attempts to leave an enclosed situation occur in the form of jumping up at the walls. The initial freezing behavior has been described by Andrew (1956) as occurring in *Emberiza* spp. in response to unlocated fear stimuli. Until a focus has been located, the chick cannot show a directional response. When such a source is located, flight can occur. Strange objects thus evoke flight and in an imprinting situation cause the chick to distress-call and to retreat to and stay with a familiar object (Sluckin and Salzen, 1961). Candland, Nagy, and Conklyn (1963) have also shown that chicks reared with other chicks and/or with manipulable objects will go to the familiar chick or object when disturbed by a fear stimulus in the form of an unfamiliar moving object. The hypothesis of the development of neuronal models through repeated experience will fit the conclusions of many imprinting studies; conclusions such as: repeated exposure is necessary to induce approach and following (Jaynes, 1958a; Salzen and Sluckin, 1959); readiness to follow increases with experience (Sluckin and Salzen, 1961); selectiveness or discrimination increases with experience (Jaynes, 1958b), which may be massed or distributed with equal effect (Sluckin and Salzen, 1961). The hypothesis of activation, through change of stimulus input, resulting in orienting movements that restore the original stimulus input fits the fact that the imprinted bird searches for the imprinting object (Weidmann, 1958) and follows it. Further it is likely that the arousal or activating system is shared by all the neuronal models for the various sensory input patterns from the environment since Collias (1952) has reported that distress-calling in chicks that are cold can be inhibited for a time by hen-like clucking noises. In other words, the activation due to a mismatch of temperature receptor input can be inhibited by matching an auditory model. I have also observed that a similar temporary inhibition can be produced by providing contact stimulation with a cold object. Conversely, a variety of environmental changes seem to effect the same increase in activation, for there are reports that following of the imprinting object can be restored or intensified by changing the stimulus input to the modeling system, e.g., by introducing a strange object into the familiar test alley (Sluckin and Salzen, 1961), or by testing in an unfamiliar run (Moltz and Rosenblum, 1958b), or by giving electric shocks in association with the test run (Moltz, Rosenblum, and Halikas, 1959). A model-forming system of this kind should continue to operate throughout the life of the individual, but withdrawal from strange stimuli and searching for the familiar environment would make such a system resistant to change once models have been established. Where flight does not succeed in removing the bird from strange stimulus patterns, the neuronal models should adapt and new attachments should develop. Such changes have been reported by Jaynes (1957), Sluckin and Salzen (1961), Baron and Kish (1960), and Waller and Waller (1963). The model-forming hypothesis is also consistent with Hebb's (1946) hypothesis concerning fear of the strange and unfamiliar, and with the use of Hebb's hypothesis by Bindra (1959) and Moltz and Stettner (1961) to account for the development of fear and for the decline of following tendencies in imprinting studies.

Social Attachment

The neuronal model hypothesis does not necessarily account for social attachments of the kind originally implied in the imprinting concept of Lorenz. There is some evidence that lasting social attachments may be influenced by, or even depend upon, the consequences of approach and following responses as mediated by the distance receptors. Sluckin, Taylor, and Taylor (1966) have performed an experiment with chicks that was modeled on the studies by Harlow (1959, 1960, 1961) of monkeys reared with wire and cloth-covered mother surrogates. In the chick study, isolate birds were reared each with two boxes into which they could, and did, crawl. One box had a smooth lining and one a rough lining. If any of these chicks was separated from its boxes for a short time, it would tend to return to the smooth-lined box rather than to the rough-lined one on removal of the separating screen. Further, when chicks were reared each with only one type of box available, there was a greater tendency for those with a smooth-lined one to go to their box on removal of the separating screen. This would suggest that the chick is hatched with a particular contact preference which is not easily changed by experience and that strong attachment to an object will only occur if the object meets this contact preference. The implication is that the development of a social attachment to an object, i.e., imprinting, requires that the object provide an adequate contact stimulus pattern. If this is the case, then the chick experiment of Sluckin, Taylor, and Taylor provides information on the role of contact stimulation in the formation of *visual* attachments rather than on "tactile imprinting" as proposed by Taylor and Sluckin (1964), and the term *imprinting* should be reserved for the development of preferences for patterns of distance receptor stimulation and involves vision and perhaps hearing and smell. A corresponding interpretation of Harlow's studies would be that the monkey has an innate preference for a particular pattern of contact stimulation and so spends time with the cloth-covered mother rather than with the wire one when both are present. Further, this preference is not changed by rearing with a wire mother alone since Harlow reports that such monkeys show no affection for, and obtain no emotional comfort from wire mothers. Monkeys reared with both types of mother surrogate come to respond to them and to discriminate between them *at a distance*. If imprinting may be said to occur in these monkeys, then it is to the *sight* of the cloth-covered mother surrogate.

Imprinting and Reinforcement

It has been persistently stated that no reinforcement is involved in imprinting in birds (Hinde, 1961, 1962; Fabricius, 1962). Thorpe (1956) has considered this point and warned that few studies have controlled for secondary reinforcement. Most of the laboratory studies of imprinting use the rather special situation described by Moltz (1960). This involves first eliciting the approach and following responses to one object, isolating the bird for as little as 24 hours and often in the dark, too (e.g., Hess, 1959), and finally making a test of approach and following responses to the same object either alone or with a second, previously unseen, object in a simultaneous choice situation. One cannot regard such testing as indicating social attachment in the way that Harlow has tested such attachment in monkeys.

Although it is plain that in most experiments there is no food or temperature reinforcement, we are rarely able to be sure from the published accounts whether the chicks could contact the object during their first exposure. There is no doubt that striking the chick with the object increases the intensity of approach and following responses and produces bursts of pleasure-calling. Some studies of imprinting without following have made it impossible for the chick to contact the object (Baer and Gray, 1960; Collins, 1965; Moltz, Rosenblum, and Stettner, 1960; Thompson and Dubanoski, 1964). It should be noted that Thompson and Dubanoski found the imprinting effect of exposure without following to be very weak. Smith (1962) and Smith and Bird (1963) have been unable to demonstrate any significant effects of allowing chicks to crawl under or contact the stimulus source in their studies of the approach response. In Collins' study (1965), the chicks failed to show any effect of contacting the object on subsequent following performances. What may be happening in so many of these studies, however, is that the initial approach and following behavior is elicited, and in the absence of true social fixation by the attainment of adequate contact and because of a usually short (rarely more than five days) isolation period before retesting, these responses occur again. The neuronal model hypothesis would explain the discrimination shown in such tests as a process of environmental learning through distance receptor action.

It could be, then, that contact reinforcement is necessary for fixation of a social attachment, i.e., for imprinting, but not for persistent performance of approach and following by the "unattached" chick to a perceptually learned familiar stimulus pattern. It is possible that such behavior might wane in the absence of adequate contact reinforcement. An early study by Pattie (1936) showed that the tendency for chicks to approach other chicks declines in isolated birds but increases in socially reared ones. Most imprinting experiments never test the chicks long enough for this waning to occur. There is the study by Moltz and Rosenblum (1958a) in which Peking ducklings were exposed daily to a moving object and showed a marked waning in following in the last half of the 15 day test series. Hinde, Thorpe, and Vince (1956) report that coots and moorhens maintained their following over a period of 35 days. Abercrombie and James (1961) report persistent approach responses to a flashing light in chicks over a 12 day period. Some waning occurred when they were tested three times a day to a stationary object previously associated with the flashing light. In all these experiments it is not clear to what extent the birds achieved contact with the object. Marr (1964) has found that varying stimulation provided by rubbing, rocking, or by a flashing light is necessary for dogs to become attached to a visual pattern. On the other hand, Stanley (1965) and Stanley and Elliot (1962) have shown that contact with a passive human being can serve as a reinforcer for the approach and social behavior of puppies. Stanley, Morris, and Trattner (1965) have even suggested that the role of possible reinforcers in mammalian and avian social attachments should be reexamined. Thus the possibility that contact is necessary for the object fixation involved in developing social attachments remains and the question as to whether the approach and following responses ultimately wane in the absence of this kind of reinforcement also remains unanswered. It is interesting to recall that Collias (1952) has reported that chicks will not readily approach one another until they have experienced several minutes of mutual contact. More recently he has repeated this claim (Collias, 1962) and has said that chicks isolated for the first ten days and then reared socially appeared to keep somewhat apart from the flock when observed at 2 months.

It is possible, then, that the reinforcement of the approach response is the achievement of body contact. The notion of consummatory stimulation as reinforcement is not new (e.g., Sheffield and Roby, 1950) and has been discussed by Bindra (1959), Marler (1957), and Watson (1961). Through such reinforcement the sight of the reinforcing object may become secondarily reinforcing (cf. Campbell and Pickleman, 1961; Peterson, 1960). In a previous paper (Sluckin and Salzen, 1961) I took the view that imprinting was a nonreinforced process and used the term *perceptual learning* and subsequently (1962) I put forward the neuronal model hypothesis to account for such learning and for the chick's behavior in imprinting. It is still possible that such nonreinforced perceptual learning could occur and it may be that the following response studies already quoted are in fact investigating this phenomenon. I feel that until the role of contact reinforcement in the fixation of the social object or companion has been tested satisfactorily and experimentally, there remains the possibility that it is involved and that imprinting does not differ from other learning phenomena in the need for some kind of reinforcement. It has already been pointed out that imprinting is no longer regarded as involving a strict sensitive period, as being instantaneous, as being supra-individual learning, nor as being irreversible, and that its effect on subsequent behavior may be indirect. If it can also be said that it does involve some kind of reinforcement, then there surely can be no further reason for regarding it as a distinct concept. After thirty years of life it would seem that the concept of imprinting may no longer be viable and we may have to consider giving it a decent burial.

THE NEURONAL MODEL HYPOTHESIS AND ENVIRONMENTAL LEARNING

If true social attachment requires the reinforcement of contact stimulation, then the neuronal model hypothesis can be seen to play a more general role in the establishment of the norm of the individual's world as sensed through its distance receptors. Deviations from the norm produce activation in the form of orienting or adjustive movements. This is the approach-withdrawal behavior dealt with by Schneirla (1965). The neuronal model hypothesis, however, deals with this as a single process rather than a biphasic one (Table 3). At low levels of activation the orienting behavior takes the form of approach and at high levels withdrawal. Sympatheticoadrenal activity is postulated as accompanying this activation and behavior, in approach as well as withdrawal. A mild focal change in the environment produces mild adjustive movements that serve to recenter and restore the familiar focus and forms the basis of later exploratory behavior. In the chick this behavior includes looking, turning to, approach, pecking and pushing against, and may be accompanied by so-called pleasure calls or by calls transitional between pleasure and distress calls. A drastic change in the environment produces intense activation with large scale orienting movements that appear as withdrawal from the strange aspects but could equally be interpreted as attempts to find and approach the familiar environment. In the chick such behavior includes

TABLE 3.
Responses of chicks to changes in the familiar environment.

Sympatheticoadrenal and somatic orienting behavior			
Level of activation	Calls	Localized stimulus	Nonlocalized stimulus
High	Shriek	Flight	Freeze
⬇	⬇	⬇	⬇
	Distress	Withdraw ⟶ Avoid	Search
	⬇	⬇	
Low	Pleasure	Approach ⟶ Explore	Arrest ⟶ Alert

The response of the chick depends on the degree and localization of the disturbance. Activation decreases with persistence of the environmental change and the chicks behavior changes in the manner indicated in the table.

turning away, moving away, fleeing, avoiding, searching, escape jumping, and standing with head down in a corner. This behavior is accompanied by distress-calling, squealing or shrieking according to intensity of activation. It is interesting that Andrew (1964) has taken a similar view in the case of the chick's vocalizations. He has given evidence that the pleasure calls, which he terms "twitters," and the distress calls, which he terms "peeps," are simply variants of a single vocal system which includes the "alarm trill" and the "shriek." He suggests that they are elicited by increasing degrees of "stimulus contrast," by which he means the degree to which the focal stimulus stands out against its background. "Stimulus contrast" includes stimuli that intrinsically contrast with background stimulation, any persisting change in the surroundings to which the chick is accustomed, food signals, cold, absence of imprinting object, first contacts with an imprinting object, and pain.

The similarity between Andrew's concept of "stimulus contrast" and the neuronal model hypothesis is so evident as not to require iteration. Interestingly, Andrew suggests that vocalizations and approach or avoidance responses are evoked independently by stimulus contrast and that there is no invariable relationship between the type of call and approach or avoidance. Thus, Andrew reports that the well fed chick feeds silently and that twitters tend to occur only with intense alerting responses such as given to signals preceding food and water. It is common in imprinting experiments for twitters or pleasure calls to be given just before approaching begins, yet at other times chicks may give peeps or distress calls while actively following. I agree entirely with Andrew's treatment of approach and avoidance responses and vocalizations. To complete the parallels with the neuronal model hypothesis it should be noted that Andrew also considers the "orienting reflex," the "defense reaction," and autonomic responses as similarly produced by stimulus contrast. If noncoincidence of stimulus and established neuronal models be substituted for "stimulus contrast," the views presented in this paper will fit Andrew's hypothesis of vocalization in the chick (Andrew, 1964). Both hypotheses treat behavior, vocal, postural, and locomotory, as occurring in a continuum with "approach" patterns at the lower end of the scale. Both give importance to the degree of change in environmental stimulus patterns. While Andrew suggests that hunger gives the chick a set to perceive food cues, I would suggest that there is a brainstem neuronal model system for interoceptive patterns and that hunger is a change in the patterns of interoceptive action giving a model mismatch and resulting in activation. The fixation of approach behavior to food cues would then be the result of learning processes in which consummatory stimulation of food in the mouth would be the reinforcing agent just as contact reinforcement would be the agent in the fixation of approach behavior to social objects. In this way the modeling system concept may be used for homeostatic behavior and accounts for the activation of the organism in response to both internal and external environmental changes. The concept contributes to conditioning and learning processes that take place in appetitive and orienting behavior by suggesting that it is the restoration of neuronal model input from interoceptors and contact exteroceptors that is the reinforcing process at least in the neonate. When models are matched, the animal is quiet and parasympathetic actions related to normal body metabolic processes prevail. When models are mismatched, activation occurs with sympatheticoadrenal action. So-called pleasure responses are really final orienting actions and it is significant that they occur just before or at the moment of obtaining consummatory stimulation (Table 3). The same system operates with respect to neuronal models based on distance receptor input. The process of establishing these models of the distant environment takes place in the neonate and is the process of environmental learning. In this way the individual acquires a familiar environment, to which it orients and shows appetitive behavior, and which becomes a secondary reinforcer for such behavior.

REFERENCES

Abercrombie, B., and H. James. 1961. The stability of the domestic chick's response to visual flicker. *Anim. Behav.* 9: 205–212.
Andrew, R. J. 1956. Fear responses in *Emberiza* Spp. *Brit. J. Anim. Behav.* 4: 125–132.
Andrew, R. J. 1964. Vocalization in chicks, and the concept of "stimulus contrast." *Anim. Behav.* 12: 64–76.

Baer, D. M., and P. H. Gray. 1960. Imprinting to a different species without overt following. *Percept. Mot. Skills* 10: 171–174.

Baron, A., and G. B. Kish. 1960. Early social isolation as a determinant of aggregative behavior in the domestic chicken. *J. Comp. Physiol. Psychol.* 53: 459–463.

Bateson, P. P. G. 1964a. Changes in chicks' responses to novel moving objects over the sensitive period for imprinting. *Anim. Behav.* 12: 479–489.

Bateson, P. P. G. 1964b. Effect of similarity between rearing and testing conditions on chicks' following and avoidance responses. *J. Comp. Physiol. Psychol.* 57: 100–103.

Bindra, D. 1959. *Motivation: A systematic reinterpretation.* New York: Ronald Press.

Campbell, B. A., and J. R. Pickleman. 1961. The imprinting object as a reinforcing stimulus. *J. Comp. Physiol. Psychol.* 54: 592–596.

Candland, D. K., Z. M. Nagy, and D. H. Conklyn. 1963. Emotional behavior in the domestic chicken (White Leghorn) as a function of age and developmental environment. *J. Comp. Physiol. Psychol.* 56: 1069–1073.

Collias, N. E. 1952. The development of social behavior in birds. *Auk* 69: 127–159.

Collias, N. E. 1962. Social development in birds and mammals. In E. L. Bliss, ed., *Roots of behavior.* New York: Harper & Row. Pp. 264–273.

Collias, N. E., and M. Joos. 1953. The spectrographic analysis of sound signals of the domestic fowl. *Behaviour* 5: 175–187.

Collins, T. B. 1965. Strength of the following response in the chick in relation to degree of "parent" contact. *J. Comp. Physiol. Psychol.* 60: 192–195.

Fabricius, E. 1962. Some aspects of imprinting in birds. *Symp. Zool. Soc. Lond.* 8: 139–148.

Fishman, R., and R. B. Tallarico, 1961a. Studies of visual depth perception: I. Blinking as an indicator response in prematurely hatched chicks. *Percept. Mot. Skills* 12: 247–250.

Fishman, R., and R. B. Tallarico, 1961b. Studies of visual depth perception: II. Avoidance reaction as an indicator response in chicks. *Percept. Mot. Skills* 12: 251–257.

Gibson, J. J. 1959. Perception as a function of stimulation. In S. Koch, ed., *Psychology: a study of a science,* study 1, vol. 1. New York: McGraw-Hill. Pp. 456–501.

Gottlieb, G. 1963. The facilitatory effect of the parental exodus call on the following response of ducklings: one test of the self stimulation hypothesis. *Am. Zool.* 3: 518.

Gottlieb, G. 1965. Imprinting in relation to parental and species identification by avian neonates. *J. Comp. Physiol. Psychol.* 59: 345–356.

Harlow, H. F. 1959. Love in infant monkeys. *Sci. Am.* 200 (June, 1959): 68–74.

Harlow, H. F. 1960. Affectional behavior in the infant monkey. In M. A. B. Brazier, ed., *The central nervous system and behavior.* New York: Josiah Macy Jr. Foundation. Pp. 307–357.

Harlow, H. F. 1961. The development of affectional patterns in infant monkeys.

In B. M. Foss, ed., *Determinants of infant behaviour,* vol. 1. London: Methuen. Pp. 75–88.

Hebb, D. O. 1946. On the nature of fear. *Psychol. Rev.* 53: 259–276.

Hein, A., and R. Held, 1962. A neuronal model for labile sensorimotor co-ordinations. In E. E. Bernard and M. R. Kare, eds., *Biological prototypes and synthetic systems,* vol. 1. New York: Plenum Press. Pp. 71–74.

Held, R., and A. Hein, 1963. Movement-produced stimulation in the development of visually guided behavior. *J. Comp. Physiol. Psychol.* 56: 872–876.

Hess, E. H. 1959. Imprinting. *Science* 130: 133–141.

Hess, E. H. 1964. Imprinting in birds. *Science* 146: 1128–1139.

Hinde, R. A. 1961. The establishment of the parent-offspring relation in birds, with some mammalian analogies. In W. H. Thorpe and O. L. Zangwill, eds., *Current problems in animal behaviour.* Cambridge. England: Cambridge Univ. Press. Pp. 175–193.

Hinde, R. A. 1962. Some aspects of the imprinting problem. *Symp. Zool. Soc. Lond.* 8: 129–138.

Hinde, R. A. 1963. The nature of imprinting. In B. M. Foss, ed., *Determinants of infant behaviour,* vol. 2. London: Methuen. Pp. 227–233.

Hinde, R. A., W. H. Thorpe, and M. A. Vince. 1956. The following response of young coots and moorhens. *Behaviour* 9: 214–242.

Jacobs, H. L., and F. L. Smith. 1960. The classification of social stimuli: "social" and "non-social" distraction in the albino rat. *Anim. Behav.* 8: 134–140.

James, H. 1959. Flicker: an unconditioned stimulus for imprinting. *Can. J. Psychol.* 13: 59–67.

Jaynes, J. 1957. Imprinting: the interaction of learned and innate behavior: II. The critical period. *J. Comp. Physiol. Psychol.* 50: 6–10.

Jaynes, J. 1958a. Imprinting: the interaction of learned and innate behavior: III. Practice effects on performance, retention and fear. *J. Comp. Physiol. Psychol.* 51: 234–237.

Jaynes, J. 1958b. Imprinting: the interaction of learned and innate behavior: IV. Generalization and emergent discrimination. *J. Comp. Physiol. Psychol.* 51: 238–242.

Klopfer, P. H., and J. P. Hailman. 1964. Basic parameters of following and imprinting in precocial birds. *Z. Tierpsychol.* 21: 755–762.

Kuo, Z.-Y. 1932. Ontogeny of embryonic behavior in aves: IV. The influence of embryonic movements upon the behavior after hatching. *J. Comp. Psychol.* 14: 109–122.

Lehrman, D. S. 1953. A critique of Konrad Lorenz's theory of instinctive behavior. *Quart. Rev. Biol.* 28: 337–363.

Lehrman, D. S. 1956. On the organization of maternal behavior and the problem of instinct. In P.-P. Grassé, ed., *L'Instinct dans le comportement des animaux et de l'homme.* Paris: Masson. Pp. 475–520.

Lorenz, K. Z. 1935. Der Kumpan in der Umwelt des Vögels. *J. f. Orn* 83: 137–213, 289–413.

Lorenz, K. Z. 1937. The companion in the bird's world. *Auk* 54: 245–273.

Marler, P. 1957. Studies of fighting in chaffinches (4) appetitive and consummatory behaviour. *Anim. Behav.* 5: 29–37.

Marler, P. 1961. The filtering of external stimuli during instinctive behaviour. In W. H. Thorpe, and O. L. Zangwill, eds., *Current problems in animal behaviour.* Cambridge, England: Cambridge Univ. Press. Pp. 150–166.

Marr, J. N. 1964. Varying stimulation and imprinting in dogs. *J. Genet. Psychol.* 104: 351–364.

Maturana, H. R., and S. Frenk. 1963. Directional movement and horizontal edge detectors in the pigeon retina. *Science* 142: 977–999.

Moltz, H. 1960. Imprinting: empirical basis and theoretical significance. *Psychol. Bull.* 57: 291–314.

Moltz, H. 1963. Imprinting: an epigenetic approach. *Psychol. Rev.* 70: 123–138.

Moltz, H., and L. A. Rosenblum. 1958a. Imprinting and associative learning: the stability of the following response in Peking ducks (*Anas platyrhynchous*). *J. Comp. Physiol. Psychol.* 51: 580–583.

Moltz, H., and L. A. Rosenblum. 1958b. The relation between habituation and the stability of the following response. *J. Comp. Physiol. Psychol.* 51: 658–661.

Moltz, H., and L. J. Stettner. 1961. The influence of patterned-light deprivation on the critical period for imprinting. *J. Comp. Physiol. Psychol.* 54: 279–283.

Moltz, H., L. Rosenblum, and N. Halikas. 1959. Imprinting and level of anxiety. *J. Comp. Physiol. Psychol.* 52: 240–244.

Moltz, H., L. Rosenblum, and L. J. Stettner. 1960. Some parameters of imprinting effectiveness. *J. Comp. Physiol. Psychol.* 53: 297–301.

Pattie, F. A. 1936. The gregarious behavior of normal chicks and chicks hatched in isolation. *J. Comp. Psychol.* 21: 161–178.

Peters, J. J., A. R. Vonderahe, and T. H. Powers. 1958. Electrical studies of functional development of the eye and optic lobes in the chick embryo. *J. Exp. Zool.* 139: 459–468.

Peterson, N. 1960. Control of behavior by presentation of an imprinted stimulus. *Science* 132: 1395–1396.

Sackett, G. P. 1963. A neural mechanism underlying unlearned, critical period, and developmental aspects of visually controlled behavior. *Psychol. Rev.* 70: 40–50.

Salzen, E. A. 1962. Imprinting and fear. *Symp. Zool. Soc. Lond.* 8: 199–218.

Salzen, E. A., and W. Sluckin. 1959. The incidence of the following response and the duration of responsiveness in domestic fowl. *Anim. Behav.* 7: 172–179.

Schaller, G. B., and J. T. Emlen. 1961. The development of visual discrimination patterns in the crouching reactions of nestling grackles. *Auk* 78: 125–137.

Schaller, G. B., and J. T. Emlen. 1962. The ontogeny of avoidance behaviour in some precocial birds. *Anim. Behav.* 10: 370–381.

Schneirla, T. C. 1956. Interrelationships of the "innate" and the "acquired" in instinctive behavior. In P.-P. Grassé, ed., *L'Instinct dans le comporte-ment des animaux et de l'homme*. Paris: Masson. Pp. 387–452.

Schneirla, T. C. 1957. The concept of development in comparative psychology. In D. B. Harris, ed., *The concept of development*. Minneapolis: Univ. Minnesota Press. Pp. 78–108.

Schneirla, T. C. 1959. An evolutionary and developmental theory of biphasic processes underlying approach and withdrawal. In M. R. Jones, ed., *Nebraska symposium on motivation*, vol. 7. Lincoln, Nebraska: Univ. Nebraska Press. Pp. 1–42.

Schneirla, T. C. 1965. Aspects of stimulation and organization in approach/ withdrawal processes underlying vertebrate behavioral development. In D. S. Lehrman, R. A. Hinde, and E. Shaw, eds., *Advances in the study of behavior*, vol. 1. New York: Academic Press. Pp. 1–74.

Sheffield, F. D., and T. B. Roby. 1950. Reward value of a non-nutritive sweet taste. *J. Comp. Physiol. Psychol.* 43: 471–481.

Sherrington, C. S. 1947. *The integrative action of the nervous system*, 2nd ed. New Haven: Yale Univ. Press.

Sluckin, W. 1965. *Imprinting and early learning*. London: Methuen.

Sluckin, W., and E. A. Salzen. 1961. Imprinting and perceptual learning. *Quart. J. Exp. Psychol.* 13: 65–77.

Sluckin, W., K. F. Taylor, and A. Taylor. 1966. Approach of domestic chicks to stationary objects of different texture. *Percept. Mot. Skills.* 22: 699–702.

Smith, F. V. 1960. Towards definition of the stimulus situation for the approach response in the domestic chick. *Anim. Behav.* 8: 197–200.

Smith, F. V. 1962. Perceptual aspects of imprinting. *Symp. Zool. Soc. Lond.* 8: 171–191.

Smith, F. V., and M. W. Bird. 1963. Relative attraction for the domestic chick of combinations of stimuli in different sensory modalities. *Anim. Behav.* 11: 300–305.

Sokolov, E. M. 1960. Neuronal models and the orienting reflex. In M. A. B. Brazier, ed., *The central nervous system and behavior*. New York: Josiah Macy Jr. Foundation. Pp. 187–276.

Stanley, W. C. 1965. The passive person as a reinforcer in isolated beagle puppies. *Psychonomic Sci.* 2: 21–22.

Stanley, W. C., and O. Elliot. 1962. Differential human handling as reinforcing events and as treatments influencing later social behavior in Basenji puppies. *Psychol. Rep.* 10: 775–788.

Stanley, W. C., D. D. Morris, and A. Trattner. 1965. Conditioning with a passive person reinforcer and extinction in Shetland sheep dog puppies. *Psychonomic Sci.* 2: 19–20.

Taylor, K. F., and W. Sluckin. 1964. An experiment in tactile imprinting. (Abstract) *Bull. Brit. Psychol. Soc.* 17 (54): 10A.

Thompson, W. R., and R. A. Dubanoski. 1964. Imprinting and the "law of effort." *Anim. Behav.* 12: 213–218.

Thorpe, W. H. 1956. *Learning and instinct in animals*. London: Methuen.

Waller, P. F., and M. B. Waller. 1963. Some relationships between early experience and later social behavior in ducklings. *Behaviour* 20: 343–363.

Watson, A. J. 1961. The place of reinforcement in the explanation of behaviour. In W. H. Thorpe, and O. L. Zangwill, eds., *Current problems in animal behaviour.* Cambridge, England: Cambridge University Press. Pp. 273–301.

Weidmann, U. 1958. Verhaltensstudien an der Stockente (*Anas platyrhynchos* L.) II. Versuche zur Auslösung und Prägung der Nachfolge- und Anschlussreaktion. *Z. Tierpsychol.* 15: 277–300.

III

BEHAVIORAL PROCESS

ZING-YANG KUO
  The Need for Coordinated Efforts in Developmental Studies

D. M. VOWLES
  Neuroethology, Evolution, and Grammar

R. A. HINDE and JOAN G. STEVENSON
  Goals and Response Control

ETHEL TOBACH
  Some Guidelines to the Study of the Evolution and Development of Emotion

ELLIOT S. VALENSTEIN
  Pavolian Typology: Comparative Comments on the Development of a Scientific Theme

WILLIAM N. TAVOLGA
  Levels of Interaction in Animal Communication

HELMUT E. ADLER
  Ontogeny and Phylogeny of Orientation

KNUT LARSSON
  Mating Behavior of the Male Rat

ZING-YANG KUO
Hong Kong

The Need for Coordinated Efforts
in Developmental Studies

> *"If you tell me what you have been, I'll be able to foretell what you will be."*
>
> From a Chinese fortune teller

In 1930, after many months of failure to devise a technique by means of which I could observe the development of the avian embryo throughout incubation, I became despondent and took off one morning to visit one of the scenic, imposing Buddhist temples in Hangchow, China. I was mistaken for a wealthy tourist from Shanghai by the head of the temple, a learned Chinese Buddhist scholar, who came out to greet me personally. In good humor, I asked if he could forecast my future as he was well known as a fortune teller, and the above quotation was his instant answer.

After many years of developmental studies, it frequently occurs to me that this Buddhist monk might have been a better behavioral scientist than I, if he had been reared in an "enriched environment" instead of living in "total social isolation" from science and scientists (Harlow, Dodsworth, and Harlow, 1965).

The importance of studying the ontogeny of behavior is now commonly recognized by most students of behavior. Among ethologists, zoologists, learning psychologists, research workers on the effects of early experiences on later life, not to say psychiatrists, child psychologists, and many social scientists, no one would repudiate the influence of developmental history. Nevertheless, owing to the lack of an adequate theoretical perspective and coordination among various investigators, the current work on behavioral ontogeny has been disjunctive and fragmentary. In this chapter we shall critically examine the present approaches to the problems of the ontogeny of behavior and their theoretical bases. We will briefly outline a new theoretical setting which would require a concerted program of research by several branches of science to facilitate the solution of many difficult problems of the ontogenesis of behavior. These problems may remain unsolved for many decades to come if the current concepts of behavior and methods of approach are continued.

CURRENT EXPERIMENTAL APPROACHES

In recent years there has been a great deal of exchange between the European ethologists and American animal or comparative psychologists of opinions, information, and techniques of investigation. As a result, there has been mutual influence between the two schools of behavioral scientists. Nevertheless, the basic differences between the two groups are still distinguishable in their fundamental attitudes, theoretical assumptions, and methods of investigation. In our discussion of the ontogeny of behavior we shall divide the current approaches into two groups: the ethological approach and the American approach.

The Ethological Approach

THEORETICAL ASSUMPTIONS OF ETHOLOGY

While the current ethologists may differ among themselves in certain details, the dichotomy between innate and learned or acquired responses still remains a most outstanding principle of ethology.

Although in very recent years a large number of ethologists (e.g., Thorpe, 1965, and Tinbergen, 1965) have admitted that there are certain degrees of complexity in behavior and in environment, the fact remains that the fundamental assumptions of ethology are based on the theoretical conceptions of (1) uniformity of behavior and (2) uniformity of environment. As no ethologist is prepared to abandon the view of species-specificity of behavior and its survival value for the species as a result of natural selection, it would be a contradiction in terms were he to stress the extreme variability of behavior and environment.

EXPERIMENTAL APPROACHES

As is well-known, the ethologists, besides stressing naturalistic observations, have devised rather unique experimental methods to prove their a priori theoretical schemata, or what Schneirla has termed *innate schemata* (Schneirla, 1965). Two of the best known experimental devices are Lorenz's "hawk figure" model (Lorenz, 1939) and Tinbergen's and Perdeck's models for testing the pecking responses of Herring Gull hatchlings (Tinbergen and Perdeck, 1950). Numerous experiments of similar nature have been carried out by other workers with variations of procedure and the techniques of presentation of the stimulating models as well as variations of the forms or structures of the models. As I analyze the varying results of the experiments with models done by the original investigators (Lorenz, and Tinbergen and Perdeck, op. cit.) and by the later workers, I come to the conclusion that there are, at least, four groups of factors determining the responses of the hatchlings to the herring gull model and to the "hawk figure" model: (a) sensory factors, (b) developmental antecedents or previous experiences, (c) the environmental complex, and (d) the position of the hatchlings at the moment of tests.

a) Schneirla (1965) has rightly pointed out that in studies of this type, the visual modality, that is, the impact of the quantitative differences of the visual input to the retina must play an essential role. However, we must not overlook the effects of other sensory modalities, such as the change in air pressure as a result of the movement of the models; possible auditory stimulations, sounds produced by the movements of models, especially, when they are moved by hands; the variations in the distance between the beak and the herring gull model when moved by hand; and even the possible differences in odor among the different colors of paints in the herring gull model. Unless the effects of different sensory modalities are ruled out one by one by rigid control, we cannot say for sure that visual stimulation alone is responsible for the chick's response.

b) The hatchling does not "first" begin to peck after hatching. In chicks, ducklings, and pigeons, the beak has been active for at least two-thirds of the duration of incubation. The beak reacts to the heart beat, the amniotic contractions, the amniotic fluid, the yolk sac, and the embryo's feet, legs, and wings. In the later stages, it reacts to the egg membranes, the shell, and after part of the shell is torn, it comes into contact with external air. In the cases where the eggs have been incubated by the parents, I have made numerous observations that the parent's bill often gently touches the beak of the emerging chick many hours before hatching. The heart may not teach the embryo to peck, as Lorenz once joked, but certainly the beak has a long history of experience with the external world inside and outside the egg shell, including experience with the beaks of the brooding parents before hatching. As to the postnatal reaction to violent or intensive stimulation, I must mention the fact that one or two days before hatching a strong rocking of the incubator, a violent noise, or a very strong light suddenly turned on the face of the resting embryo, seldom fails to produce violent wriggling movements involving the whole organism, accompanied by eye movements and strong vocalizations or clapping of the beak.

Such prenatal historic antecedents must be taken into consideration whenever tests involving visual, auditory, or tactile stimulation are given to neonates.

c) The environmental complex of the room in which the hatchlings are tested is also influential. The brightness of the room, the constancy of light, the background of the area where the chick is placed for the test, and the breathing, body odor, and air movement or sound produced by the "unconscious" movements of the experimenter must also be taken into account.

d) As the experiments with models test the chick's reactions to visual objects, the chick's position during each test is also of considerable significance. Whether the chick is standing or sitting, whether its neck is bent (a carry-over from prenatal life) or straight, whether its head drops downward or lifts up, and the orientation and fixation of its eyes before the moving model comes within the range of its vision: all these are significant variables contributing to the response of the chick to the moving models.

Unfortunately, in their enthusiasm to present experimental evidence for innate behavior, resulting "from the experience of the species" and its innate releasing mechanisms (IRM), the ethologists inadvertently over-simplified their experimental procedures and ignored most of the variables of the environment and behavior. The result is a disastrously distorted account of the origin of behavior at the very outset of postnatal ontogeny.

When certain variables of behavior in response to the variables of the environment (e.g., the nest building of the canary) have not escaped the observation of the ethologists, the ethologists, instead of critically re-examining their basic tenets, pursue the same wavering and oscillating attitude adopted by a large number of American animal psychologists toward the problem of innate vs. learned behavior (Kuo, 1967, Chapter 5). That is to say, they conclude that every aspect of the nest-building activities in response to the environment and nest materials "can be reasonably described by the word 'innate' on the other hand, processes of conditioning and learned responses of various kinds are clearly also important" (Thorpe, 1965, p. 487).

Experimenting with models is merely one of the many methods of the ethological approach. However, our main concern here is not with their methods of study, observational or experimental, but rather with the fundamental assumptions upon which their methods of investigation are based. Such assumptions will hardly lead us to a genuine knowledge of processes and understanding of the causes of the ontogeny of behavior. Despite the fact that many ethologists are prone to express their views and facts of observations in terms of "ontogeny of behavior" (Thorpe, 1965), I am inclined to agree with Aronson (1961) that they pay only lip service to the concept of ontogeny.

American Approaches

As I have pointed out elsewhere (Kuo, 1967, Chapter 1) the behavioristic movement pioneered by Watson should eventually lead to interest in the study of the ontogeny of behavior. Although through unfortunate historic accidents, the term *behaviorism* is nowadays used almost as a synonym for operant reflexology, or the psychology of learning, I strongly believe that American workers in the development of behavior are more qualified to be called behaviorists than the Hullians or Skinnerians. At any rate, I am gratified to have witnessed, in the last forty years or so, a wide-spread and continuous tendency among many American workers to devote themselves to the investigations of behavior from the ontogenetic point of view. Nonetheless, there have been great shifts of emphasis, as I have previously pointed out, and most of the work is more or less fragmentary, or isolated. Therefore, a comprehensive picture of the developmental history with its causal factors experimentally worked out has not been achieved for any animal studied.

THEORETICAL ASSUMPTIONS

With the exception of a few investigators such as T. C. Schneirla (1965), D. S. Lehrman (1953), G. Gottlieb (1968), most, if not all, American workers on behavior development, animal and human, are predeterminists or believers in the dichotomy between innate and learned behavior, either by open admission or by implication. The influence of the concepts of conventional psychology of learning on the study of the ontogeny of behavior is also noticeable (e.g., the work by the University of Wisconsin group on the effect of social isolation on monkeys, Harlow, Dodsworth, and Harlow, 1965; and that of the University of California group at Berkeley on the effects of environmental differences on the structural and chemical development of the brain in the rat, Bennett, Diamond, Krech and Rosenzweig, 1964). The influence of traditional psychology on a number of current studies in behavior development is of scientific value and commendable. At the same time, this influence has become a handicap to many able investigators who have the facilities to work out their research programs in a deeper and more intensive manner than my crude experiments on the effects of early environment on later life. These experiments were carried out more than thirty years ago in a scientifically backward country and under conditions of constant social and political upheaval.

EXPERIMENTAL APPROACHES

In the 1930s, there was a great boom in prenatal studies as well as in studies of the new born human infant. Since World War II the interest in prenatal studies has greatly receded although none of the developmental problems in prenatal behavior has been experimentally solved. Our method of study has not reached the experimental level, and our observation has not gone beyond the molar or gross body movement level. We have not yet succeeded in devising a technique that would enable us to make continuous observations of the activities of the mammalian fetus *in utero*. This is probably due to two reasons: (1) the difficulty of devising such a reliable technique, and (2) the recent interest in the development of social behavior has shifted the attention of the behaviorists from prenatal to postnatal life with special emphasis on the effect of early experiences on later life and the interrelationship between individuals. The literature on postnatal development is voluminous and has been sufficiently reviewed by many authors. Our aim in this chapter is to make a very brief, but critical assessment of the current contributions on prenatal behavior to our scientific knowledge, pointing out their merits and shortcomings. We hope this assessment would lead to the realization of the necessity of theoretical reorientation and a concerted program of research in behavior development, prenatal and postnatal.

In my appraisal of the American work on behavior development, the reader may find that some of my critical remarks are severe or faultfinding. I hope, though, that he may adopt a forgiving attitude, if he notes that the same criticism is applied to my own work for the last forty years on both prenatal and postnatal behavior.

Prenatal studies.   In amphibian embryos we have had a rather comprehensive history of the development together with a useful chronology of the anatomical development of the nervous system. However, the functional development of the nervous system has yet to be worked out, as Coghill's interpretation of the correlation between morphological growth and behavioral changes is dubious (Kuo, 1939a, b, and 1967).

In the avian embryo we seem to have developed a rather satisfactory technique of observation by means of which we have obtained a comprehensive story of development during incubation. Nevertheless, the work has not gone beyond the observational level and the level of gross movements. Little has been done by controlled experiments to determine the basic causes for the changes in body movements from stage to stage. Still less have we studied the interrelationships between structural and functional differentiation and growth.

In the mammalian fetus we are still in search of a reliable technique for direct and continuous observation of the activities of the fetus *in utero*. It may be some time before we can by experimental control raise the same questions about causal factors of the mammalian fetus' behavior as we have regarding the avian embryo.

Postnatal studies.   When we come to the investigations on postnatal ontogeny, we find the following common characteristics:

1. Although there is no difficulty in employing a technique for continuous postnatal observation, we are still in urgent need of a comprehensive developmental history of a species from birth to adulthood, if not until death.

2. Most of the published work on early experiences was well planned, carefully carried out with adequate control and sufficient number of experimental animals, making the data of results statistically acceptable (except those cases dealing with nonhuman primates in which available specimens were apparently scarce).

3. Practically all the current research on behavior development is confined to a single problem, or even to certain phase or phases of a single problem such as sex behavior, nest building, brooding, "imprinting," the effect of social isolation, or restricted or enriched environments, etc. This is probably due partly to the terms of research grants, partly to personal interest of the investigators, and partly to inadequate financial support and laboratory facilities. However, it is also due either to the lack of a more forward looking theoretical perspective, or to the influence of the concepts of conventional psychology such as innate behavior, motivation, and learning theories. The results of such investigations are interesting and of scientific value. Nonetheless, they are fragmentary studies; they cannot be pieced together to form a coherent and comprehensive picture of the developmental history of an individual, or a species.

4. With some exceptions, such as studies of the relationship of experience and hormones in sex behavior, aggression, and fighting, the effect of early handling on the growth and functional changes of the adrenal glands, and the effects of different environmental contexts on the chemical and anatomical changes of the brain, current developmental studies have not gone beyond the level of gross body movement, or what is commonly known as molar level.

5. In a large number of publications on the effect of early experience on later life, the experimental procedures of both the experimental and the control groups are adequately described and the end results reported. Yet, the developmental processes involving behavioral changes in both groups during the period between the experimental treatment and the final tests have not been carefully observed and recorded. Thus a very large and essential part of the developmental history of the animal under investigation is missing. This is especially unfortunate, as we have no comprehensive ontogenetic history of postnatal life in any species of birds or mammals from the neonate to adulthood that can be used as a basis for comparison. This criticism applies also to the valuable studies of the

Berkeley group on the effects of different environmental settings on the chemical and structural changes of the brain. As a matter of fact, the Berkeley studies reveal an urgent call for an overall knowledge of chemical and morphological changes in response to external stimulations from the earliest beginning in prenatal life to adulthood. (For a more detailed discussion on this point see Kuo, 1967.)

6. With extremely rare exceptions such as the recent work by Bennett et al. (1964), and Gottlieb and Kuo (1965), interdepartmental cooperation on the same problem or the same phase of a problem with the same animal is practically nonexistent.

7. Many recent reports on developmental investigations place special emphasis on the effects of early environmental conditions, but none of them has given us any detailed and quantitative description and analysis of the physical or social environment in question. They refer to environment with such vague and generic statements as restricted vs. enriched environment, partial or total social isolation, deprivation of certain sensory stimuli, etc.

8. In the descriptions of the development of abnormal behavior patterns in consequence of living in an abnormal environment, we have seldom seen detailed accounts of the progressive changes in behavior from day to day during the period of the experiments.

THE EPIGENETIC APPROACH

The assessment of the work on the ontogeny of behavior for the last thirty or forty years as very briefly stated in the two preceding sections has led me to the conclusion that a new orientation or a new theoretical perspective in the study of behavior in general and in ontogenesis of behavior in particular, and a new experimental approach to the problems of behavioral ontogeny is needed, so as to arrive at an objective and un-biased solution to the problems of the functional processes and causes of behavioral development. This new theoretical perspective which I propose has been discussed at length in *The Dynamics of Behavior Development: An Epigenetic View* (Kuo, 1967). Here, the most essential points may be stated to serve as a theoretical background for the proposal for a concerted program for the investigation of behavioral ontogeny.

The New Theoretical Perspective

1. Behavior is a very complex process of the functional interaction between the organism and its environment.

2. There is great variability of behavior. No animal reacts to the same stimulus twice in the same way.

3. The environment of the animal both within and outside the body is constantly changing.

4. What has been commonly known as behavior (overt, or gross movement visible to the naked eye—the molar level) is merely an integral part of the total response of the animal to the environment, involving every part of the whole organism.

5. In this total response—*behavior*—of the organism, there are differentiations of intensity and extensity among the different parts of body. These are called *behavior gradients*.

6. Behavior gradients are organized into patterns in space and in series.

7. Ontogenesis of behavior is a process of modification, transformation, or reorganization of the existing patterns of behavior gradients in response to the impact of new environmental stimulation; and in consequence a new spatial and/or serial pattern of behavior gradients is formed, permanently or temporarily ("learning?") which oftentimes adds to the inventory of the existing patterns of behavior gradients previously accumulated during the animal's developmental history.

8. At the beginning of life the individual possesses an enormous range of behavioral potentials limited only by species differences in anatomical structures and functional capabilities. As ontogenesis proceeds, new behavior patterns are actualized out of the potential patterns and added to the existing *repertoire;* at the same time ontogenetic processes set a limit to the actualization of certain other patterns, either as result of the gradual loss of plasticity of the structures and function of the organs of the organism, or because of the incompatibility of such potential patterns with the actualized patterns.

9. Thus, in every stage of ontogenesis, every response is determined not only by the stimuli or stimulating objects, but also by the total environmental context, the status of anatomical structures and their functional capacities, the physiological (biochemical and biophysical) conditions, and the developmental history up to that stage.

10. In view of the fact that the animal, especially the social animal, is born with a tremendously wide range of behavioral potentials (the prenatal environment is relatively restricted and circumscribed), one major task of the student of behavior is to explore new potential patterns, unknown and unobserved before, by providing an appropriate environment at an appropriate stage for the actualization of such novel patterns, which I have named *behavioral epigenotypes*.

Practical Implications of the Epigenetic View

When the concept of behavior is so construed and so broadened, the study of behavior becomes a *synthetic science* in the same sense that space science is a joint undertaking of many branches of science and technology. There are multiple phases of behavior instead of the single-phase, the overt movement, as it has been traditionally understood. From the metabolic changes in the tissues to external muscular contractions, every event that takes place under and outside the skin is an integral part of the pattern of the behavior gradient. The question of intervening variables (in the physiological sense) or the question of motivation will cease to exist; for they are parts of the pattern of behavior gradients. As such, not only can they not be ignored, as the operant behaviorists do, but they must be systematically investigated as part of the behavior research program. We are inclined to think that under the concept of behavior gradients, physiology will become meaningless unless it is merged with and becomes part and parcel of the science of behavior.

In recent years we have begun to accumulate considerable evidence, most of which has been referred to by Gottlieb (see his chapter in this volume; also Kuo, 1932 and 1967), showing the interdependence between structural differentiation and functional activities during development. This inevitably leads to the conclusion that morphogenesis and functional development of different organs must be investigated in close conjunction with the study of behavioral ontogenesis. In other words, in developmental studies, structural and functional changes, and changes in overt movements are three inseparable phases of one single event, the modification or transformation of the pattern of behavior gradients.

Since the physical environment is of paramount importance to behavioral development, we must not content ourselves with such vague expressions as "restricted" vs. "enriched" environments, or the deprivation of certain sensory stimulation. Instead, we must take quantitative measures of stimulative effects of every sensory modality, and make qualitative analyses of the interactions of the component parts of the environmental context or complex. This requires the work of a trained physicist.

In the case of social animals, detailed daily records of behavior changes of the experimental animal in its reaction to other animals, including the experimenter, must be kept and analyzed quantitatively. Whenever possible, the responses of other animals to the experimental animal must also be observed in detail. The changes in the pattern of gradients of such responses must be detected and analyzed from day to day. The processes leading to the appearance of novel patterns (behavior epigenotypes) or of abnormal behavior patterns must receive the same analysis. Most publications on the effect of early environment on later social behavior seldom go beyond the description of the general condition of the early environment and the end-products after a certain period, leaving the much more important aspects, such as the developmental process leading to structural and functional changes, and the detailed interactions of the experimental animals and other animals in the environmental context, either unobserved or unreported. Such studies as those on effects of partial or total social isolation in monkeys, or on the maturation and critical period of socialization of dogs (Scott, 1962) are rather crude. My own studies on animal fighting (Kuo, 1960, I–VII) and creation of novel behavior patterns (Kuo, 1930, 1938, and 1967, Chapter 3) also belong to this category of crude experimental studies.

All the above discussions have come to bear on the view I have maintained for a good many years: that the key to the thorough knowledge of the ontogeny of behavior in breadth and in depth is a concerted research program participated in by psychologists, endocrinologists, neurophysiologists and sensory physiologists, experimental embryologists, zoologists, ecologists, and even cultural anthropologists and psychiatrists.

THE NEED FOR A RESEARCH CENTER IN DEVELOPMENTAL STUDIES

Since behavior has multiple phases besides the overt aspect and since all the phases of behavior interact with one another in a dynamic manner, each modifying the other, the ontogenetic process of behavior is an extremely complex and complicated event, so much so that comparative psychology alone or any other branch of biology alone cannot undertake any experimental study on development on a comprehensive scale. The development of the organism is a dynamic or functional process; structural and functional changes and changes in overt movements as well as changes in the environmental context modify one another. Some of these modifications are temporary, others have permanent effects on the organism. Developmental history is the summation, in a given stage, of the effects of such intermodifications throughout the preceding stages. Any isolated or independent study of development—structural, chemical and physical, or behavioral (in the conventional sense)—will not reach the root of the problem of ontogenesis.

For this reason, I have recommended that a research center be established in which specialists of the various scientific disciplines listed above could work out a common project on development and concentrate their efforts on it, attacking the same problems from different angles.

In such a research center, experimental facilities would be much more adequate than existing independent laboratories, working almost in the same field but with neither cooperation in their endeavour nor coordination towards a common goal. With the establishment of the research center, many difficult problems of experimental procedure and instrumentation may be solved speedily through group discussion by workers of different scientific disciplines, with the assistance of special technicians. Solutions of many such problems may take many years or even decades, if left to individual workers in independent laboratories.

The success in space science in the last few years has encouraged me to put forward this suggestion of establishing a research center for developmental studies. Through such common efforts many technical difficulties can be more easily overcome and many complicated research projects can be more readily carried out, than through individual efforts.

The recent widespread interest in animal behavior, the common recognition of the importance of the ontogeny of behavior and scattered and somewhat fragmentary investigations of the various aspects of the effects of early experiences have made it all the more essential that such a research center be established, so as to consolidate the developmental studies from various angles with a broader and deeper perspective.

In the Western world, the United States is at present the only country that has both financial and personnel resources for such a research center. As a matter of fact, the initial expenses for such a center should be far less than the total sum used for the establishment of the seven centers for primatological studies under the auspices of the Department of Health, Education and Welfare. And, there is no doubt that every primatologist will share my view that an objective and comprehensive knowledge of the ontogeny of behavior is of the same importance as our knowledge of the life of the primates.

REFERENCES

Aronson, L. R. 1961. Evolution of behavior. In Frank W. Blair, ed., *Vertebrate speciation*, Austin. Univ. Texas Press. Pp. 91–95.
Bennett, E. L., M. C. Diamond, D. Krech, and M. R. Rosenzweig. 1964. Chemical and anatomical plasticity of brain. *Science* 146: 610–619.
Cohill, G. E. 1929. *Anatomy and the problem of behavior*. Cambridge, England: Cambridge Univ. Press.
Gottlieb, G. 1968. Prenatal behavior of birds. *Quart. Rev. Biol.* 43: 148–174.
Gottlieb, G., and Z.-Y. Kuo. 1965. Development of behavior in the duck embryo. *J. Comp. Physiol. Psychol.* 50: 183–188.

Harlow, H. F., R. O. Dodsworth, and M. K. Harlow. 1965. Total social isolation in monkeys. *Proc. Nat. Acad. Sci. U.S.* 54: 90–97.

Kuo, Z.-Y. 1930. The genesis of the cat's responses toward the rat. *J. Comp. Psychol.* 15: 1–35.

Kuo, Z.-Y. 1932. Ontogeny of embryonic behavior in aves: V. The reflex concept in the light of embryonic behavior in birds. *Psychol. Rev.* 39: 499–515.

Kuo, Z.-Y. 1938. Further study on the behavior of the cat towards the rat. *J. Comp. Psychol.* 25: 1–8.

Kuo, Z.-Y. 1939a. Total patterns or local reflexes? *Psychol. Rev.* 46: 93–122.

Kuo, Z.-Y. 1939b. Studies in the physiology of the embryonic nervous system: II. Experimental evidence on the controversy over the reflex theory in development. *J. Comp. Neurol.* 70: 437–459.

Kuo, Z.-Y. 1960. Studies on the basic factors in animal fighting. I, II, III and IV, *J. Genet. Psych.* 96: 201–239. V, VI and VII, *J. Genet. Psych.* 97: 181–225.

Kuo, Z.-Y. 1967. *The dynamics of behavior development: An epigenetic view.* New York: Random House.

Lehrman, D. S. 1953. A critique of Konrad Lorenz's theory of instinctive behavior. *Quart. Rev. Biol.* 28: 337–363.

Lorenz, K. 1939. Vergleichende Verhaltungsforschung. *Zool. Anz.*, Suppl. 12: 69–102.

Schneirla, T. C. 1965. Aspects of stimulation and organization in approach withdrawal processes underlying vertebrate behavioral development. In D. S. Lehrman, R. A. Hinde, and E. Shaw, eds., *Advances in the study of behavior.* New York: Academic Press. Pp. 1–74.

Scott, J. P. 1962. Critical periods in behavioral development. *Science* 138: 949–958.

Thorpe, W. H. 1965. The ontogeny of behavior. In J. A. Moore, ed., *Ideas in modern biology.* New York: The Natural History Press. Pp. 485–518.

Tinbergen, N. 1965. Behavior and natural selection. In J. A. Moore, ed., *Ideas in modern biology.* New York: The Natural History Press. Pp. 521–539.

Tinbergen, N., and A. C. Perdeck. 1950. On the stimulus situation releasing the begging response in the newly-hatched herring gull chick (*Larus argentatus*). *Behavior* 3: 1–38.

Tobach, E., and T. C. Schneirla. 1968. The biopsychology of social behavior in animals. In R. E. Cooke, ed., *The biologic basis of pediatric practice.* New York: McGraw-Hill.

D. M. VOWLES

Department of Psychology
University of Edinburgh
Edinburgh, Scotland

Neuroethology, Evolution, and Grammar

"Not only speech, but all skilled acts seem to involve the same problems of serial ordering, even down to the temporal coordination of muscular contractions in such a movement as reaching and grasping. Analysis of the nervous mechanisms underlying order in the more primitve acts may contribute ultimately to the solution even of the physiology of logic."

Lashley (1951)

The relationship between a scientist and the intellectual climate of his society suggests that scientific fertility is not based only on merit. If his ideas are to breed, it is not enough that they be rooted in established order and technically testable. Apparently they must also be fashionable. If they are not, he may lack an active audience, fail to attract research students, gather no group around him, affect few minds beyond those of close colleagues, and become a fairly isolated figure. It was the misfortune of T. C. Schneirla that he was ahead of fashion by nearly two decades, and it is, perhaps, ironical that the fertility of his approach was widely appreciated only after the impact on American psychology of the school of European ethology in the 50s—a school whose work he strongly criticized.

Current opinion recognizes that the biological approach to behavior concerns four problems: how behavior evolved, how it develops in the life of the individual, what the immediate causal relationships of stimuli to responses are, and what the underlying physiological mechanism may be. These four problems are implicit in Schneirla's work from the late 20s onward. In his classical textbook on animal psychology with Maier (Maier and Schneirla, 1935), for example, he stressed the importance of relating behavior to neural organization. In considering neural mechanisms he showed that since different types of organization exist at different phylogenetic levels, it is dangerous to use similar conceptual explanations of behavior for species of different evolutionary status. Not the least cause of our present confusion about planarian learning (Thorpe and Davenport, 1964) is the attempt to press their behavior into the mold of conditioning. Although one can train flatworms by the techniques of conditioning, this does not mean the neural mechanism for learning is the same as that in rats. Schneirla indeed stressed the need for a genuine comparative psychology and urged the importance of studying different species rather than comparing one white rat with another or with a human. The differences between species are of interest not only in relationship to their neural mechanisms, but also because they throw light on the adaptiveness of behavior and the possible course of evolution—as shown in Schneirla's comparative studies of army ants.

When attempting the analysis of behavior in terms of its underlying physiological mechanism, it is advantageous if not essential to have a conceptual framework that is adequate both for behavior and for physiology. Following Schneirla's lead, one may suggest that a different framework may be needed for different phylogenetic levels. Schneirla asserted that the common theme underlying animal behavior at many levels is a tendency to turn toward weak stimulation and away from strong stimulation (Schneirla, 1959). He suggested that the complexities of behavior develop or evolve from this primitive stimulus-response system. This suggestion cannot yet have much immediate relevance to physiological psychologists since they usually work with mature individuals of highly evolved species. They are faced with the natural, present organization of complex behavior, and must seek their conceptual framework elsewhere.

CONTROL THEORY AND BEHAVIOR

Complex control devices show many analogies with both physiological and behavioral systems. The rapidly growing body of basic control theory provided a mathematical background that could be used to devise models of behavior mechanisms that can be tested both qualitatively and quantitatively. These possibilities were eagerly grasped by biologists hungry for solid concepts and techniques. The serious problems of translating such models into physiological terms have been discussed in detail in another paper (Vowles, 1964), but in some cases the approach may prove fruitful in both behavioral and physiological analyses (see for example McCann and MacGinitie, 1965, and Vowles, 1966).

Cybernetic analysis has been applied to phenomena ranging from the perception of movement in insects (Reichardt, 1962) to the regulation of food and water intake in doves (McFarland, 1965). It is significant, however, that in those cases where cybernetic analysis has been successfully used, the input and output variables possess a particular property. They can always be measured and assigned values on what is essentially a continuous numerical scale (e.g., the rate of movement of a visual stimulus and the rate of turn of a responding insect). If, in studies of behavior, we are dealing with variables that cannot validly be assigned a metric of this sort, models based on theories requiring metric variables may not be applicable.

ETHOLOGY, PHYSIOLOGY, AND FINITE STATE SYSTEMS

Quantitative measures are of course used in many studies of behavior. For example, these may take the form of counting the number of times a given action is performed, or the total time that bouts of action occupy. Such measures clearly give some information about the "tendency" of an animal to show a particular form of behavior, but one may doubt whether they represent the "intensity" or "magnitude" of behavior. It is not strictly correct to say that when a dove coos twelve times it is cooing more "strongly" than when it gives only six calls. Similarly when a stickleback proceeds further through its possible sequence of courtship actions than on another occasion, this has been interpreted as showing a higher level of sex "drive" but this *is* only an interpretation, and the behavior does not necessarily mean that the fish is courting more "strongly."

In fact, when dealing with complex behavior patterns, we are concerned with a finite number of possible actions which can be combined in a variety of spatial and temporal patterns. We might say that the animal behaves as if it can exist in one of a possible number of finite states, between which it moves when it shows a given sequence of behavior.

Some recent advances in mathematics have been in the field of finite state theory (Kemeny, Snell, and Thompson, 1956) and are concerned particularly with problems that do *not* involve those features such as infinite sets, limiting processes, continuity, etc. which involve common metrical systems and characteristically lead to the use of integral calculus. Possibly by trying to quantify behavior as if it were essentially a mathematically continuous quantity we may have been misled. Although behavior can be expressed quantitatively this does not mean that one must employ mathematical systems which involve relationships between metrical items.

This point might not be very significant if we had an adequate conceptual framework within which to discuss the mechanisms of behavior. Unfortunately this does not seem to be the case at either the physiological or the theoretical level. It is significant that several workers in the field of brain stimulation and behavior have recently expressed dissatisfaction with the unsuccessful attempts to fit their results into the classical electrophysiological mold (von Holst and St. Paul, 1959 and 1963; Delgado, 1964; Brown and Hunsperger, 1963). Von Holst particularly stressed the need to develop a terminology adequate for the level of complexity of the mechanism that is being studied.

The most influential physiological approach to the neural basis of behavior has been that of Sherrington. Much electrophysiological work has been devoted to unraveling the electrophysiological mechanisms underlying reflex systems, and has been particularly concerned with single unit analysis. Although it may be assumed that ultimately all mechanisms may be analyzed at this level, at present the techniques and ideas necessary for dealing with large neuronal networks seem inadequate. Even large modern computers are taxed to analyze the firing patterns of the seven cells in the crustacean cardiac ganglion. The Sherringtonian approach deals with the problem of complex patterns of reflexes in terms of individual connections between neurons involved in the simpler reflex arcs—but the mind boggles at identifying such connections in the brain. It is small wonder that Sherrington became lyrical when describing cortical function. Changes in the level of responsiveness of a reflex are mainly short term effects, which can be adequately explained in terms of delay circuits and direct postsynaptic events. But in instinctive behavior, the changes of responsiveness may be much more prolonged—from minutes to years. Clearly some other factors play a role.

Theories of instinctive behavior are also in a confused state. Some time ago a fairly simple proposal was put forward by ethologists in which motivation was regarded as some sort of energizer (Hinde, 1960; Lorenz, 1950); this proposal follows in a long line of such theories from Freud and MacDougall onwards. The sequential organization of behavior was the consequence of the sequential occurrence of stimuli, either in the external or internal environment (Tinbergen, 1951). This explanation fitted a modified reflex pattern rather well. But in the ethological literature since the 1950's, it is clear that the rapid accumulation of intractable new data led to the progressive abandonment of such concepts, and the development of a largely operational vocabulary such as "tendency" or "specific action potential." Although caution is well merited it has two consequences, firstly, in descriptive behavioral studies the data collected may be relatively trivial since it is concerned merely with obtaining information without reference to its significance (except perhaps in the field of genetics, or evolution), and secondly it abandons any attempt to suggest to physiologists what sort of phenomena should be looked for when studying the brain. This is unfortunate, since the development of techniques for stimulating, or recording from, small areas of the brain via chronically implanted electrodes in the conscious, free-moving animal has revolutionized our methods in this field. The neural substrate of behavior can now be studied directly. If such studies are to be fruitful, one needs not only adequate analyses of behavior but also adequate theoretical frameworks. It is the weakness of ethology that at present it provides no such framework, and its strength that it provides the most sophisticated methods yet for describing and analyzing behavior.

Most early work in classical comparative psychology was developed under the influence of learning theorists. Quantitative methods were developed for scoring an animal's performance in learning tasks. Three classical types of measure were developed—rate of pressing a lever, numbers of choices in a jumping stand or choice box, and error scores in maze running. As Hinde (1956) has pointed out these measures effectively apply to the consequences of the behavior (e.g., whether a bar is pressed, or the animal moves to a particular place) rather than to the behavior itself. In an automated Skinner box the rat might achieve the same scores by pushing a bar with its forepaws as by pirouetting on its hind legs and thumping with its tail. Although the behavior is clearly different, the counter does not discriminate. The ethologist, however, attempts to measure the behavior itself.

Ethological analysis proceeds by identifying "units" of behavior in terms of frequently occurring movements and postures. Each unit is composed of a combination of simple, effector actions. For example, "charging" in the ring dove consists of running with the body held in a low posture, the neck retracted, the bill held horizontally, the body puffed up (by inflation of internal air sacs), the wings held slightly away from the body, the feathers slightly erect (particularly on the rump), and the tail feathers slightly spread. The unit is thus characterized by a state vector (low posture, short neck, puffed up, etc.). A pattern of behavior is composed of a temporal sequence of such units, the transition from one to another involving some but not necessarily all of the effector components.

The technique for quantitative sequential analysis was pioneered by Baerends, Brouwer, and Waterbolk (1955) who introduced a flow diagram indicating the transition frequencies between the behavioral units. This type of representation has been widely used, particularly for insects; an example for the fruitfly is shown in Figure 1. Such descriptions are very similar to those used by electrical engineers or mathematicians for finite state devices (Gill, 1963; Kemeny and Snell, 1960). The mathematics for dealing with such systems is already available, and if the organism can be regarded as a machine that can exist in only a limited number of states, the use of such techniques would be appropriate. In this context the states of a finite state system would correspond to the behavior units.

Each unit (when it contains a variety of components) could be regarded as a vector; transitions then occurring between vectors. A flow diagram normally includes a statement both of the probability of any particular transition and the conditions (such as the machine input) that affect these probabilities. Although the stimulus situation clearly affects the observed transitions in behavior, the flow diagrams so far provided by ethologists unfortunately do not contain this information.

Another factor that may affect a transition probability is the recent past history of the device. When the previous states influence the next transition, this is a Markov-type system—of the first, second or nth order depending on the number (n) of previous states that influence the current transition. Influences of this sort are of great significance in finite state devices, but most ethological diagrams represent data pooled from a number of separate observations, which precludes an analysis of the Markov type. Nelson (1965) applied a Markov analysis to the reproductive behavior of fishes and concluded that only a first-order Markov system was involved. However, his method of time sampling, necessitated by the fishes' behavior, led him to ignore sequences of behavior units separated by more than a short time interval, and this may have forced his data artificially into a first-order mold.

In most analyses using finite state systems a time sampling method is normally adopted. This consists of measuring performance not continuously, as in many behavior experiments, but for finite periods at discreet time intervals. It is necessary for the sampling parameters to suit the performance, but this may not be easy to achieve in animal behavior, and sampling may have to be continuous.

One reason why behavior students have not used the notation of the Markov system for describing their results is that to make such an analysis worthwhile a great deal of data must be collected and transcribed. In a preliminary attempt by D. Harwood and the author to analyze reproductive behavior in the ring dove, approximately 30 hours of observation yielded records requiring nearly 500 hours manual analysis merely to obtain relevant figures, which then proved insufficient. Clearly, anyone wishing to embark on this voyage of discovery must be able to send his data fairly directly for computer analysis with a minimum amount of transcription.

The advantage gained by such labor is the development of a fairly precise and complete description of the behavior pattern studied. This description is quantitative and enables differences in the pattern produced, for example, by brain stimulation or lesions, to be compared with the normal system. However, a description of this sort says nothing about the nature of the mechanism that actually produces the behavior. One way of

gaining insight into a possible mechanism is to prepare a computer program that will simulate the system performance. The initial step in a
simulation study is to work out in detail the logical structure of the
hypothetical mechanism that one postulates to explain the behavior. It
will often be possible to write a number of different programs that simulate the mechanism equally well. The advantages of this procedure are
(1) that one is forced to develop a very clear statement of the nature
of the mechanism one is proposing, (2) that it will have implications about
what new data should be gathered and in what way the data should be
analyzed, and (3) it will enable new predictions to be made both about
the behavior itself and the way in which the underlying physiological
mechanism may function. For example, interference with the nervous

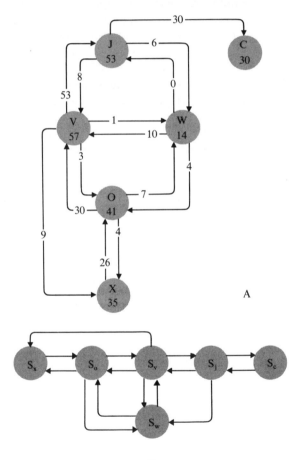

system by ablation or stimulation might produce behavioral changes more readily assimilable by one model than another.

GRAMMATICAL MODELS

The success of simulation studies will depend partly upon the complexity of the behavior being studied. If there are a large number of effector actions, which can be combined in a large number of different spatial and temporal patterns, the flow diagram representing the possible states and their transitions will become very large. If the model, or the

---

FIGURE 1

Diagram A is a representation of the sequences of behavior shown by male *Drosophila obscura* in successful courtship. X = Noncourtship behavior, O = orientation, V = Vibration of wings, J = jumping, C = copulation, W = wing posture. The arrows together with their numbers indicate the frequency with which a particular transition from one activity to another was observed. The numbers in each circle indicate the frequency with which that behavior item was observed. (After Brown, 1965).

Diagram B is an attempt to represent the behavior by a more formal finite state diagram. It appears from Brown's work that wing posture (W) is shown when the female actively repels courtship, and this may also cause the male to break off. However, there is apparently no stimulus of any specific sort provided by the female to release the normal links in the courtship chain, and the transitions presumably depend on the internal state of the male. The rules for these sequences are shown below.

S_x, S_o, S_v, S_r, S_c, S_w are states of the animal.

R = active repulsion by ♀, C = critical magnitude of any S which must not be exceeded if courtship is to proceed. The values of C are to be selected to give correct transition probabilities.

If $\overline{R}$ (no repulsion) and S_x, S_o, S_v, S_j, $S_c > C$,

then $S_x \rightarrow S_o$, $S_o \rightarrow S_v$, $S_v \rightarrow S_j$, $S_j \rightarrow S_c$.

If $\overline{R}$ (no repulsion) and S_x, S_o, S_v, S_j, $S_c \leq C$,

then $S_o \rightarrow S_x$, $S_v \rightarrow S_o$, $S_j \rightarrow S_v$, $S_c \rightarrow S_j$.

If R (repulsion) and S_o, S_v, S_j, $S_c > C$,

then $S_o \rightarrow S_w$, $S_v \rightarrow S_w$.

If R (repulsion) and S_o, S_v, S_j, $S_c \leq C$,

then $S_o \rightarrow S_x$, $S_v \rightarrow S_x$, $S_j \rightarrow S_x$.

If $S_w > C$,

then $S_w \rightarrow S_v$ or S_o.

If $S_w < C$,

then $S_w \rightarrow S_x$.

program, designed to simulate performance is closely isomorphic with the flow chart representing that system, it will presumably contain rules that say if in State A go to State B under conditions x, and to State C under conditions y. Each possible transition will need an equivalent rule. The more numerous the states and possible transitions, the larger the number of rules required. This will tax the capacity of the machine both to store the necessary information and rules and to search through the possible rules in each case. A further complication is the necessity for storing information about past states if the system is higher than first order: the greater the number of possible past states that can effect a transition, the more the storage capacity of the system will be strained.

It can therefore be argued that although rather simple behavior patterns may easily be simulated by finite state devices, it is unlikely that complex behavior patterns are organized in this way. The nervous system could possibly function as a finite state device, for animals low on the phylogenetic scale, or for those that show narrowly limited and stereotyped behavior patterns, such as *Drosophila*. In higher vertebrates, however, the number of possible units of behavior are more numerous, and they can be combined in many more different patterns. It seems unlikely that they can be economically mimicked by a finite state system. This was recognized some time ago, by Lashley (1951) and by Miller, Galanter, and Pribram (1960), who were particularly concerned with the problem of serial order in behavior. They suggested that when behavior involves a complex serial pattern, the animal has an overall "plan" of the total strategy, and that this plan is organized in the nervous system in a hierarchical manner—different levels of the hierarchy controlling different groups and subgroups of the basic units of behavior.

The problem of the serial complexity of behavior and the nature of its underlying mechanism reaches its climax in the analysis of human language. Work by modern American linguists (Roberts, 1964; Chomsky and Miller, 1963; Chomsky, 1963, 1964, and 1965; Postal, 1964) has been concerned with the construction of grammars (a set of formal rules) that will generate grammatical and meaningful sentences. A detailed account of their work cannot be given here, however, some of the points that have emerged from their work may be relevant to the study of neurobehavioral mechanism, and a naive outline of the position will be given.

The basic parts of spoken language are the phonemes and morphemes—the actual noises that make up words. In terms of animal behavior, words would be equivalent to "behavior units" and the phonemes and morphemes to effector actions. Sequences of words (terminal items) are sometimes referred to as terminal strings by the grammarians. For the moment we will ignore the problem of the selection of the specific words that are to be used in a terminal string, and discuss only the rules for arranging them in an appropriate sequence. This might be done using a finite state system of the Markov type which would choose the probability of occurrence of a word on the basis of preceding words. This could entail consideration of only one or two or three of the preceding words, and produce a statistical approximation to English. A typical fourth-order passage (one in which each word was dependent on the previous three) is:

> . . . the middle hole might often go out thoughtfully making his bed badly needs eating . . .

Clearly, a rule of this sort alone does not produce grammatical sentences.

An improvement could perhaps be made by writing into the rules factors that weight the probability of one word being chosen not only in terms of the actual words in its context but also in terms of the parts of speech involved. The disavanatge of such a Markov system lies in the enormous number of possible transitions from word to word that can occur. Every possible transition requires a rule to govern that transition. If the vocabulary of the language is very large this will need an even larger number of rules to allow for the possible transitions from word to word. The larger the context that can effect any transition, the greater the storage capacity needed by the machine and the greater the number of rules that it must contain.

When the rules governing the transition probabilities are stated formally in mathematical terms, they constitute a set of statements that can be called the grammar of the language to which they refer. In the case so far discussed, the rules make a finite state grammar, since they refer to the transition probabilities between a finite number of words. The use of finite state grammars for generating grammatical language has been found to be very limited indeed. If the rules are simple, meaningless jumbles of words are emitted; while the number of rules required to produce language in general, rather than a few specific sentences, is so enormous that not even the largest modern computers can provide sufficient capacity to store and operate them.

A number of different types of grammar have been suggested that economize on the methods for storing information and reducing the number of rules. Many of these may be reduced (Postal, 1964) to a common type called constituent or phrase structure grammars (PSG's). These grammars are essentially rules for rewriting symbols. The equations that express the rules are stated so that the right hand side of the equation is equal to or can replace the left, i.e., the symbols on the left can be written as the symbols on the right. Similar concepts are used in finite state grammars, but in PSG's the rewriting may involve not words but grammatical components. For example, if one starts with a symbol #S# to represent a whole sentence, #S# could be written as a noun phrase plus a verb phrase, which would have an equation of the form

$$\#S\# \longrightarrow NP + VP$$

Further rewritings of this rule can be proposed to transform the noun phrase into an indefinite article plus a noun. Finally, the parts of speech would be rewritten as the words that have been given (see above) for that sentence. The sequence of rules to produce one particular sentence are shown on the right side of Figure 2.

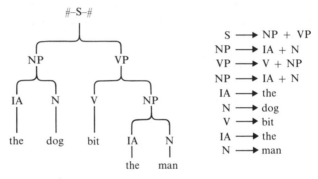

FIGURE 2

On the right, the rules generating one sentence; and on the left, the equivalent P marker for that sentence.

These rules can be represented diagrammatically by a tree. Such a tree is called a phrase marker (P marker) for that particular sentence, and every sentence has a single unique P marker. The appropriate tree or P marker is shown in Figure 2 next to the rules that it portrays.

The function of a phrase structure grammar is to provide a set of rules that generate a particular type of sentence irrespective of the actual words involved. The problem of selecting the lexicon (the words used in the sentence) will be discussed later. It is clear, however, that general rules of this sort will be economical, since the number of known types of sentence is limited. If the grammar is successful, it will produce language in which the transition probabilities between words are the same as those given by the finite state description of the language. In this case, however, it is the inevitable consequence of the "structural" properties of the rules as shown graphically by the P marker. It is normal convention to have a rule stating that the sentence must be read from left to right, and the word sequence is thus seen to depend upon the rules for rewriting in the higher levels of the tree.

Grammar and Animal Behavior

There is clearly a similarity between the structural diagrams of phrase structure grammars and the hierarchical systems of ethologists (Tinbergen, 1951). Both theories postulate a structural hierarchy, largely based on principles of economy, and on common causal factors that must generate the behavior observed. However, the phrase structure model makes various assumptions about the relationships between particular words, for example, those that make up a noun phrase, and these relationships are represented by the fact that they are grouped under a common node. One may perhaps assume that the number of nodes that must be traversed to pass from one word to another is reflected in the magnitude of the transition probability between words.

Using these assumptions it is possible to take the results of behavioral research that has yielded a picture of the transition probabilities between behavioral units, and to write a set of grammatical rules that would generate the observed patterns of behavior. Marshall has shown, in a preliminary study, that this can be done. Fabricius and Jansson (1963) have analyzed the reproductive behavior of the male pigeon in this way, and produced tables showing the frequencies with which various sequences of behavior occur. The most common courtship sequence is bowing—dis-

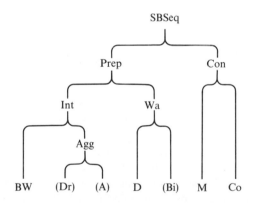

FIGURE 3

The P marker representing the rules suggested by Marshall (1966) for generating the most common sequences of behavior in the courtship of the male pigeon. There are two basic assumptions that allow predictions about the transition probabilities between the different items of courtship: (1) that the fewer the nodes to be traversed on the P marker when passing from one terminal item to another, the greater the probability of transition between those items, and (2) that, in the progression of items from left to right, each item has a higher probability of occurrence than the one following.

placement preening—mounting—copulation, but billing may be interposed between preening and mounting, and driving and attacking may follow bowing, and precede billing. Marshall gives a set of grammatical rules that generate these behavior sequences.[1] The rules are shown below; the P marker is shown in Figure 3.

KEY	Theoretical constructs		Actual behavior
SBSeq:	Sexual behavior sequence	Bw:	Bowing
Prep:	Preparatory behavior	Dr:	Driving
Con:	Consumatory behavior	A:	Attacking
Int:	Introduce	D:	Displacement preening
Wa:	Warm up	Bi:	Billing
Agg:	Aggressive behavior	M:	Mounting
		Co:	Copulation

Rules

1. SBSeq $\longrightarrow$ Prep ⌒ Con
2. Prep $\longrightarrow$ Int ⌒ Wa
3. Int $\longrightarrow$ Bw ⌒ (Agg)   ( )   optional choice
4. Agg $\longrightarrow$ $\begin{cases} (Dr) \frown A \\ Dr \frown A \end{cases}$ ⌒   concatenation
                                          $\{\}$ either or
5. Wa $\longrightarrow$ D ⌒ (Bi)
6. Con $\longrightarrow$ M ⌒ Co

Rules 2, 3, 4, and 5 are later expressed in a recursive form. For example, Rule 2 becomes

$$2a \quad Prep \longrightarrow Int \frown Wa$$
$$Prep \longrightarrow Prep \frown (Prep)$$

The reasons for these changes are partly because of the high frequency with which the early items in the chain (from bowing to billing) occur, compared with the frequency of mounting and copulation, and partly because individual items or pairs of items tend to be repeated in succession (e.g., bowing is often followed by more bowing).

There are numerous cases in which single items of behavior are successively repeated (this is one important difference between behavior and language; in language, the same word is seldom repeated in succession). In fact the number of times a behavioral action is repeated without any

---

[1] The author is indebted to Dr. John Marshall of the M.R.C. Speech & Communication Research Unit, Edinburgh, Scotland, for this analysis and permission to quote it.

other action intervening may be significant. In the ring dove, for example (personal observation), when the cock coos while bowing, the coo is normally repeated about 11 times, but when perch cooing the mean number of calls is 35, and when nest calling it is 84. In such cases further rules would be required which would operate recursively until the appropriate number of calls had been given. There are two obvious ways this might be done: the first would be part of the nervous mechanism generating the behavior, and the rule would simply say count the number of calls (or the total bout length) and stop when a particular number is reached; the second would make use of the external environment and state that the behavior should continue until a particular stimulus is perceived. It is well known that stimuli for ending behavior are important, a fact that makes the second rule probable. A rule of the first sort, however, might well be useful in imposing a safety factor that would cut off the behavior after a reasonable time if the appropriate external stimuli were not available.

Physiological Implications

It may be asked whether such structural explanations of behavior are of any value, either heuristic or explanatory. One important consequence of this type of analysis is that it provides a logical description of the rules governing behavior that are not implicit in, for example, the finite state description. In this sense, an advance has been made comparable perhaps to the transition from the general gas laws ($PV = RT$) to the kinetic theory of gases. The first provides merely a formal, general description of the observed behavior of gases, but allows no new type of prediction; while the second makes possible novel predictions that can be tested by observing "directly" the behavior of the molecules in a gas as well as its overall behavior. What sort of new prediction could one make about the physiological substrates of behavior based on a phrase structure grammar?

If one assumes an isomorphic relation between the P marker and the parts of the brain controlling the behavior pattern, then predictions can be made about the effect of localized stimulation or ablation of parts of the brain. One prediction could be based on the level in the equivalent of the P marker at which stimulation was applied. It would be assumed that stimulation would elicit (or interfere with) only those items below the point of stimulation. Thus, stimulation would produce single items of behavior if applied to the nodes near the terminal strings; if applied higher, it would produce a tendency to show particular groups of terminal items of behavior, although their specific appearance would depend upon the availability of the appropriate situation.

The predictions made on the basis of the phrase structure grammar agree well with the results of brain stimulation (for summary and review see Delgado, 1964). In unanaesthetized animals, stimulation of small regions of the brain may produce a variety of behavior such as simple motor acts (e.g., limb movements), coordinated behavior (e.g., feeding), complex and well coordinated sequences of behavior (e.g., hunting movements and attack), or general changes in the readiness to perform whole behavior patterns: the nature of the response depending on electrode position.

More complex predictions could be made, depending upon the derivation of an actual algorithm for computer simulation of the particular PSG involved. The flow diagram of such an algorithm would indicate the logical decisions that have to be made, and the temporal sequences by which they are interrelated. Thus, not only are predictions made about the spatial separation of different functional components of the system, but also about the temporal patterns of their activity, and the "logic" of the way in which these components must process the information they handle. These predictions could be investigated by using programmed patterns of stimulation with multiple, chronically implanted electrodes, and by recording the patterns of electrical activity in the brain.

One problem that arises in the computer simulation of a PSG is how to apply the rules that generate the terminal strings. First, the rules themselves must be stored, and second, a set of instructions is needed to specify the sequence in which the rules are applied. The storage of the rules themselves is not difficult, if the storage capacity of the machine is adequate. To solve the problem of sequential instructions, successful algorithms for generating speech have used a technique of "push-down storage." This works essentially by applying all the rules at one level in the P marker, and when these substitutions have been completed the outcomes are stored and the computer then proceeds to further substitutions at the next lower level (i.e., it is "pushed down" a level). When stored information is not subject to significant decay or distortion (as in most computers), this technique is successful. However, in the nervous system, transitory and dynamic states of storage are involved, which may well be subject to rapid decay. In animal behavior, such rapid decay may be disadvantageous. For example, suppose the mechanism for generating courtship in the cock pigeon had been activated along the lines suggested in the previous section. The performance of each item may well depend upon the perception of appropriate stimuli resulting from the hen's responses, and the courtship bout may be spread over 20–30 minutes. Thus, although the tendency to show copulation was established at an early stage it may not be appropriate until some considerable time later. If the stored information maintaining the state of readiness to copulate has decayed, this would clearly be embarrassing both for the cock, the hen, and the species. The mechanisms of information storage may well be related to those maintaining a readiness to respond.

A central problem in animal behavior is the mechanism that controls the continually varying responsiveness of the individual. Hinde (1959a) has provided a useful classification of the types of variation in responsiveness that animals exhibit. Variation may result from an increase or a decrease in responsiveness, and may be either specific to particular stimuli or specific to particular responses. Variations may be brief (a few seconds only) or of medium duration (about 2–20 minutes) or long term (months or years). It may be suggested that the physiological mechanisms involved in the first or short term variations might perhaps be correlated with postsynaptic facilitation or inhibition (typical of delay or reverberatory circuits in reflex system) and the last or long term with structural or chemical changes such as those underlying maturation and senility, or permanent memories, or with seasonal fluctuations in hormone levels. The changes of medium duration are more problematical, but seem to have at least three possible explanations—reverberatory electrical activity, D.C. shifts across blocks of tissue (see Brazier, 1963; Albert, 1965), and hormonal changes. These three factors are not necessarily independent, and may of course interact with each other or be causally interrelated. If D.C. shifts or levels of neurosecretion are responsible for relatively short term information storage (or changes in responsiveness), one may question the specificity of their action. The former presumably achieves its specificity through the localization of the tissue that it affects, and it is difficult to envisage this having a very selective influence on cells within that area. Similarly, hormonal secretions, if flowing in the blood stream, would have a potentially unspecific effect on neural mechanisms, unless the mechanisms themselves have a selective sensitivity to them. Although this may well be the case for overall sexual mechanisms, what is needed on the present theory is a different hormone for each behavioral item. Alternatively, specificity could be achieved by the localized secretion of small concentrations of hormones in restricted areas of the brain. The arrangement would then suffer the same restrictions as those for D.C. potentials.

One might speculate that the neural mechanisms underlying a PSG have some components that compute the various transformations according to the rules of the grammar. Some parts would perform at a relatively high speed with little delay in passing on information, while other parts—particularly those concerned with the actual generation of behavior items—may have to maintain their potential level of readiness as a mosaic of localized shifts of D.C. levels or hormone concentrations. Such speculations could be tested by direct stimulation or recording or analysis of small parts of the brain. Indeed, since (as suggested above) the information stored in this way must be relatively simple (i.e., not involve complex spatial and temporal firing patterns in the local neuron networks involved), one might anticipate a high degree of success in such attempts. Perhaps the brain is not really as complicated as we believe—this statement is an article of faith that is often tested by the finding that stimulation of the brain with large metal rods using gross electrical currents often produces complex coordinated and well adjusted patterns of behavior. Of course, the brain may be cunningly designed to allow for just such gross insults.

The PSG's so far discussed have been concerned with the rules for arranging any given set of terminal items (words) in a grammatical sequence. However, there are no suggestions as to why those words were chosen rather than others. The grammars could equally well generate "the bat hit the ball" as "the dog ate the rat," although the meanings of the two sentences are clearly different. The equivalent problem in instinctive behavior is to decide why, for example, an animal sometimes shows feeding behavior and at other times engages in reproductive activities. This decision is clearly related to its primary biological needs and is perhaps simpler for the student of animal behavior to make than for the linguist. Presumably a given P marker could be switched in when a particular need state arises, or a particular concentration of hormones occurs. The particular terminal items would then be the inevitable consequences of the activation of that mechanism. The rules governing such decisions would be partly under genetic control, this having been established in the course of evolution. In the life of the individual the environment would presumably play a role in determining the manifestations of the rules and perhaps also the rules themselves.

GRAMMAR AND EVOLUTION

Linguists have recently argued (see Postal, 1964) that for generating human speech, phrase structure grammars prove inadequate in a number of ways. For example, in moving from the active voice (the stone hit the man) to the passive (the man was hit by the stone), one needs two separate P markers although the meanings of the two statements are very similar. It is felt intuitively that a grammar should incorporate rules allowing for such transformations—e.g., by permitting the grammar to have more than a single P marker for a single expression. Similarly, the statement "John is easy to please" carries the semantic implications that "To please John is easy" and "They please John: it is easy." A grammar should also control the generation of such sentences by a common set of rules. Chomsky has therefore proposed the more powerful "transformational grammar" to generate language. This allows more than one P marker for each set of rules, and permits the substitution of one of these P markers for another (e.g., to go from the active to the passive statement). One advantage of this system is that it introduces much greater flexibility in the possible arrangement of terminal strings, and makes it possible to add extra terminal items to produce similar semantic results. Thus a common goal may be achieved via several different routes.

The implications of transformational systems for behavioral theory are too complex to discuss here in detail. However, it is clear that as one moves higher in the phylogenetic scale it is precisely the flexibility of behavior that increases, and the ability to show insightful responses to functional relationships in superficially distinct situations. It may tentatively be suggested that underlying this ability is the development of a type of grammar that has evolved in the higher mammals. Perhaps this type of grammar is a rudimentary form of the grammar underlying human language.

The arguments presented in the last few paragraphs have sketched the evolution in human thought of mathematical models for describing and generating complex behavior. It is attractive to see a parallel between the intellectual evolution of these ideas and actual phylogenetic changes. Although all behavior could be described in terms of a finite state system, an isomorphic neural mechanism rapidly becomes uneconomical and functionally cumbersome as the complexity of behavior increases. However, with simple behavioral systems the finite state model may be the most efficient. For example, if one considers the courtship of *Drosophila* as exemplified in Figure 1, this can (with some simplification) be expressed by the finite state diagram shown in Figure 4. The author has, however, been unable to write an appropriate phrase structure grammar to generate the behavior sequences found. The P marker shown on the right of Figure 1 would generate the main sequences of a successful courtship, but could not deal with the sequences found by Brown for unsuccessful courtships, nor with the frequent reverse transitions between states (the items of behavior). This suggests that for very simple systems involving only a few actions the finite state systems may be the more economical. It would also involve less storage of information, since each response would be very closely tied to the immediate stimulus situation.

It is possible that for primitive organisms the nervous system does behave as a finite state device. This may also be the case even in highly evolved invertebrates such as insects. Perhaps the Insecta represent the highest degree of complexity to which finite state devices can be carried within the size limitations imposed upon the nervous system by other factors. However, even when the brain can increase in size and complexity, as in the vertebrates, it may be that for expansion along the lines

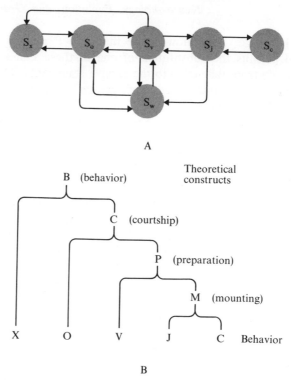

A

B (behavior)

Theoretical
constructs

C (courtship)

P (preparation)

M (mounting)

X   O   V   J   C   Behavior

B

FIGURE 4
Diagram A is the finite state diagram taken from Figure 1 to represent the courtship
of the male *Drosophila obscura;* diagram B is a P marker that would represent
rules generating the main courtship sequence. The finite state diagram has not proved,
however, to devise a simple grammar that will generate more than this simple sequence.

of finite state devices the returns are not commensurate with the cost; so
that in this group a more flexible and economical system of phrase
structure grammars has evolved. This would also make the behavior more
independent of the stimulus situation. Finally, perhaps only in the primates,
with the increasing flexibility in performance and the development of insight, a swing has been made to the rudiments of transformational grammars which may underlie human language. This proposed evolutionary
sequence has the advantage of implying a continuity between the behavior
mechanisms of human and lower primates, and makes more comprehensi-
ble the appearance of speech and language in our own species. It might
prove fruitful to examine the communication systems of other primates
from the grammatical point of view.

Theories about evolution are often supported by the study of the development of behavior: in some sense, ontogeny does recapitulate phylogeny. It is therefore interesting to see that in the development of language
in children, McNeill (1966) has found that the early speech can best be

simulated by a phrase structure system, while later language requires the more powerful transformational grammars.

It is uncommon nowadays for an author to have the luxury of unbridled speculation in print: but when the rider is lost his horse may take him safely home on a loose rein. It was suggested earlier that in studying the integrative action of the nervous system we are at present lost, because we are trying to apply the neo-Sherringtonian concepts of reflexology to levels of behavior and neural organization that are not reflex in nature. T. C. Schneirla pointed out the dangers of extending one type of concept or explanation to cover too wide a range of phenomena over too wide a range of species. At the same time he stressed the need to relate concepts of behavior both to the phylogenetic level of the organism and to the anatomical and physiological mechanisms involved. Schneirla provided not a theoretical map for the psychologist to follow but a conceptual compass and a sextant for charting the wilderness. If the analysis given here provides a rough map of use to future explorers, it is because those reliable navigational aids were available.

ACKNOWLEDGMENTS

The research reported here has been supported in part by the U. S. Air Force Office of Scientific Research through the European Office of Aerospace Research under Grant EOAR 65–14. I also received support from the Science Research Council. I am grateful to many friends and colleagues for helpful discussion of the ideas in the paper, particularly to D. Fender, D. Heenan, R. C. Oldfield, M. Clewes, and J. C. Marshall.

REFERENCES

Albert, D. J. 1966. The effects of polarising currents on the consolidation of learning. *Neuropsychologia* 4: 65–78.

Baerends, G. P., R. Brouwer, and T. J. Waterbolk. 1955. Ethological studies on *Lebistes reticulatus*. *Behaviour* 8: 249–335.

Brazier, M. A. B., ed. 1963. *Brain function,* vol. 1. Berkeley and Los Angeles: Univ. California Press.

Brown, R. G. B. 1964 Courtship behaviour in the *Drosophila obscura* group. I. *Behaviour* 23: 61–106.

Brown, J. L., and R. W. Hunsperger. 1963. Neuroethology and the motivation of agonistic behaviour. *Anim. Behav.* 11:439–449.

Chomsky, N. 1963. Formal properties of grammars. In R. D. Luce, R. Bush, and E. Galanter, eds., *Handbook of mathematical psychology*. II. New York: Wiley. Pp. 323–418.

Chomsky, N. 1964. *Current issues in linguistic theory*. The Hague: Mouton.

Chomsky, N. 1965. *Aspects of the theory of syntax*. Cambridge, Mass.: M.I.T. Press.

Chomsky, N., and G. A. Miller, 1963. Introduction to the formal analysis of natural languages. In R. D. Luce, R. Bush, and E. Galanter, eds., *Handbook of mathematical psychology*. II. New York: Wiley. Pp. 269–322.

Delgado, J. M. R. 1964. Free behavior and brain stimulation. *Int. Rev. Neurobiol.* 6: 349–438.

Fabricius, E., and A.-M. Jansson. 1963. Laboratory observation on the reproductive behaviour of the pigeon. *Anim. Behav.* 11: 534–547.

Gill, A. 1963. *An introduction to the theory of finite state machines*. New York: McGraw-Hill.

Hinde, R. A. 1959a. Unitary Drives. *Anim. Behav.* 7: 130–142.

Hinde, R. A. 1959b. Some recent trends in ethology. In S. Koch, ed., *Psychology: the study of a science*, vol. II. New York: McGraw-Hill. Pp. 561–611.

Hinde, R. A. 1960. Energy models of motivation. *Symp. Soc. Exp. Biol.* XIV: 199–214.

Holst, E. von, and U. von St. Paul. 1959. Vom Wirkungsgefüge der Triebe. *Naturwissenschaften* 18: 409–422. (*Translated as* Holst, E. von, and St. Paul, U. von. 1963. On the functional organisation of drives. *Anim. Behav.* 11: 1–20.)

Kemeny, J. G., and J. L. Snell. 1960. *Finite Markov chains*. New York: Van Nostrand.

Kemeny, J. G., J. L. Snell, and G. L. Thompson. 1956. *Introduction to finite mathematics*. Englewood Cliffs, N.J.: Prentice-Hall.

Lashley, K. S. 1951. The problem of serial order in behavior. In L. A. Jeffress, ed., *Cerebral mechanisms in behavior*. New York: Wiley.

Lorenz, K. 1950. The comparative method in studying innate behaviour patterns. *Symp. Soc. Exp. Biol.* IV: 221–268.

Maier, N. R. F., and T. C. Schneirla. 1935. *Principles of animal psychology*. New York: McGraw-Hill.

McCann, G. D., and G. F. MacGinitie. 1965. Optomotor response studies of insect vision. *Proc. Roy. Soc.* (London), B, 163: 369–401.

McFarland, D. J. 1965. Control theory applied to the control of drinking behaviour in the barbary dove. *Anim. Behav.* 13: 478–493.

McNeill, D. 1966. Development psycholinguistics. In F. Smith, and G. A. Miller, eds., *The genesis of language: in children and animals*. Cambridge, Mass.: M.I.T. Press (in press).

Miller, G. A., E. Galanter, and K. H. Pribram. 1960. *Plans and the structure of behavior*. New York: Henry Holt.

Miller, G. A., and N. Chomsky. 1963. Finitary models of language users. In R. D. Luce, R. Bush, and E. Galanter, eds., *Handbook of mathematical psychology*. II. New York: Wiley. Pp. 419–492.

Nelson, K. 1965. The temporal patterning of courtship behaviour in the glandulocaudine fishes. *Behaviour* 24: 90–147.

Postal, P. M. 1964. *Constituent structure: a study of contemporary models in syntactic description.* The Hague: Mouton and Co.

Reichardt, W. 1962. Nervous integration in the facet eye. *Biophys. J.* 2: 121–143.

Roberts, P. 1964. *English syntax.* New York: Harcourt, Brace and World.

Schneirla, T. C. 1959. An evolutionary and developmental theory of biphasic processes underlying approach and withdrawal. In M. R. Jones, ed., *Nebraska symposium on motivation.* VII. Lincoln, Nebraska: Univ. Nebraska Press. Pp. 1–41.

Thorpe, W. H., and D. Davenport, ed. 1964. Learning and associated phenomena in invertebrates. *Anim. Behav.* Suppl. No. 1.

Tinbergen, N. 1951. *The study of instinct.* London: Oxford Univ. Press.

Vowles, D. M. 1964. Models and the insect brain. In R. F. Reiss, ed., *Neural theory and modeling.* Stanford, California: Stanford Univ. Press.

Vowles, D. M. 1966. The receptive fields of cells in the retina of the housefly (*Musca domestica*). *Proc. Roy. Soc. Lond.* B 164: 552–576.

R. A. HINDE and JOAN G. STEVENSON
Subdepartment of Animal Behaviour
Madingley
Cambridge, England

Goals and Response Control

The theory of the evolution of species by natural selection raised two groups of questions for nineteenth century zoologists: those concerned with the origin of species and genera, and those concerned with the broad sweep of evolution, including the relations between the major animal phyla. Among students of behavior, the first group received scant attention until Heinroth (1911), Whitman (1919), Lorenz (1941), Tinbergen (e.g., 1959), and others showed that the species-characteristic "fixed action patterns" (Tinbergen, 1942) could be used as taxonomic characters, and that comparative study of the differences in such characters could provide information on the evolution and behavior of the species themselves. The second group of questions demanded immediate attention because they were associated with the most pressing issue that arose with the first impact of Darwinism: are the differences in behavior between species, and more especially between animals and man, of kind or of degree? Since this problem required comparison of the behavior of diverse species, more knowledge about the development and immediate causation of behavior in particular species was necessary: the search for this knowledge has occupied a major effort in psychological research ever since. It required refinement of techniques, increased precision of observation, and standardization of method. In pursuit of these, most psychologists inevitably forgot the aim of cross-phyletic comparison, which had provided their initial spur, and "comparative psychology" became a name for the study of animal behavior even though comparisons were in fact rarely made.

Since deep insight into the mechanisms underlying particular examples of behavior is of limited interest unless one can generalize further, many workers fell for the temptation of making broad generalizations from the study of one species. Such generalizations, of course, demand prior comparative study: wherever this is available it shows that breadth of generalization is inversely related to precision, and may obscure important differences. In a period when broad generalizations in the absence of comparative study were fashionable among psychologists, Schneirla has occupied a unique place in part because he gave repeated warnings against their dangers. In particular, he warned that difficulties may arise if terms such as "learning," "perception," or "instinct" are "used in sweeping fashion" across species to "encourage thinking conclusively of homogeneous agencies presumably understood in various cases, rather than processes of very different nature still requiring explanation" (1949, p. 248).

Uncritical use of terms such as learning can in fact lead to attempts to reduce organisms to but a few levels of complexity, with neglect of qualitative differences. Thus in seeking generalizations which could be made across species, learning theorists studied a variety merely as a means of cross-validation and did not look for differences between species. One of the most efficient methods of finding general principles of behavior, uncontaminated by species differences, has involved the use of a "free operant" situation, which consists of controlling and measuring a freely repeatable response in a restricted environment. This permits clear control of a learned response and has revealed similar conditioning and extinction functions in a wide variety of species. However, to find how such "pure behavior" (Skinner, 1959, p. 86) is related to other behavior, one must enlarge the situation and increase the number of responses observed. Both extensions within the operant field [1] and further experiments in more varied contexts are necessary for a full picture of the differences between, as well as the similarities in, "learning" in all its diversity.

Schneirla emphasizes the importance of this diversity as follows:

> Our theory of animal motivation is likely to be influenced strongly by the views we hold as to whether learning can differ qualitatively at different phyletic levels. Even the sparse existing evidence indicates that differences of a fundamental nature exist, not to be understood in terms of mere "complexity." Basic differences in the nature of learning seem involved in the fact that ants do not reverse a learned maze readily, although rats do, as also in the fact that the two animals initially acquire the habit very differently. Whereas ants learn a maze by stages, gradually mastering the local choice-points and only in a final stage integrating the segments, rats start local learning and integration together and can soon anticipate distant parts of a maze. (Schneirla, 1959, p. 5.)

The case of maze learning is of interest in that it had formed a focus

---

[1] Some relevant trends involve studies of: the topography and patterning of responses (reviewed in Blough and Millward, 1965), nonrigid reinforcement schedules that vary according to the subject's behavior (e.g., Blough, 1966), and reinforcers that are relevant to a particular species (e.g., Stevenson, 1967).

of discussion between those who, like Hull, proposed that it could be adequately described in terms of the formation of stimulus-response connections, either overt or covert, and the "cognitive" theorists who laid emphasis on cognitive processes without regarding these as basically similar to stimulus-response sequences. Although the differences between these approaches are not as great as might appear (MacCorquodale and Meehl, 1954), the controversy between them has thrown into relief the question of the extent to which it is proper to describe behavior as "purposive" or "goal-directed." One danger of the term "purpose" is that, like "learning" and "instinct," it may be applied indiscriminately across levels. Schneirla defines "purpose" as "acting persistently and appropriately with reference to anticipated results" (1949, p. 257). He suggests that such action is necessarily a product of ontogenetic development, and is less frequent in lower animals. "The kind of teleological doctrine which asserts that 'purpose' is present wherever adaptive behavior exists either has theological implications which are hardly relevant to scientific theory, or rest upon a gratuitous a priori endowment of all living animals with higher-level psychological functions" (1949, p. 257). From his argument, one would not choose rats as subjects about which to build a theory of purposive behavior.

"Purpose," like other terms discussed by Schneirla, has the danger that it may carry surplus meaning. Indeed Tolman, with whom the term is particularly associated, stated, "I wish now, once and for all, to put myself on record as feeling a distaste for most of the terms and neologisms I have introduced. I especially dislike the terms 'purpose' and 'cognition' . . . . Actually, I have used these terms . . . in a purely neutral and objective sense" (1932, pp. xi, xii). This did not protect him from being misunderstood as a subjectivist, or from having to defend his position on many subsequent occasions. Some of the difficulty centers on the teleological implications of "purpose." These were formerly anathema to purists, but since it has been recognized that even machines can be regarded as teleological mechanisms (e.g., Rosenblueth, Wiener, and Bigelow, 1943), teleology is no longer disreputable. With the swing of the scientific pendulum, there has been a growing feeling against over-simple explanations of the behavior of even lower organisms and a tendency once again to use concepts of "goal-direction" or "purpose" in describing it. It is instructive to consider two examples of this.

For Thorpe, "purpose" means "a striving after a future goal retained as some kind of image or idea" (1963, p. 3): since machines do not have "ideas," they can therefore act at most "as if" behaving purposefully. The question for the ethologist in his view is, "How much, if any, of the animal's behaviour is purposive and what is the relation of this behaviour to the rest?" (1963, p. 3). Since we cannot study "ideas" directly, Thorpe argues that we are entitled as biologists to use the term "purpose" for a mechanism that behaves "as if" it were purposive, reserving judgment on more fundamental issues. In this paper, therefore, we shall treat "goal-directed" and "purposive" as synonymous.

Both Thorpe's "striving after a future goal retained as some kind of image or idea" and Schneirla's "acting . . . with reference to anticipated results" imply an internal correlate of the goal situation. We could take these definitions to mean that goal-directed responses are governed by discrepancy between the present situation and a goal situation: the striving ceases when the two become the same (see also Craik, 1943). At first sight this suggests a simple criterion for goal-directed behavior: whereas behavior that is merely elicited by a stimulus increases in frequency or intensity the nearer the stimulus approaches a certain value, behavior directed toward a goal might be expected to decrease as the incident stimulus situation approximates that of the goal. In practice, things are not so simple. Rats often run fastest as a goal is approached (Miller, 1959), and human subjects may work hardest as a task nears completion (Welford, 1962). In such cases response strength bears no constant relation to the discrepancy, and we thus have no simple criterion for goal-directed behavior. Furthermore, this attempt to elaborate on Thorpe's definition falls into "the common error of entangling a definition with contemporary hypotheses of how the phenomenon is brought about" (Pantin, 1964, p. 2).

However, Thorpe does give additional, related criteria by which a set of actions may be judged as purposive: that there should be a large but unlimited series of alternatives for action, and that there should be some ability to choose the more expedient or simpler of two or more courses of action. It must be noted, however, that criteria for the existence of alternative courses of action are not defined by Thorpe: does it mean that the effectors for alternative courses are available, or that the central nervous mechanisms for alternative courses are available, or that the animal has the ability to develop new courses? And in what sense could a more or less expedient course be said to exist if the animal never used it?

Thorpe's criteria were of course intended only as rough guides for categorizing behavior with certain characteristics. They serve primarily to eliminate certain activities, which seem to show striving after a goal, from being called purposive. For instance, Lorenz and Tinbergen (1939) have described the movement used by various ground-nesting birds in rolling an egg back into the nest. The bird reaches out its bill, places it over the egg, and then draws it in toward the nest. The tendency for the egg to roll sideways is counteracted by lateral movements of the beak. Tinbergen (1953) emphasizes that the process is clumsy and inefficient, yet no ground-nesting bird has been known to retrieve an egg by using its apparently more appropriate legs or wings. Since the bird does not select between alternative courses of action, egg-rolling cannot be labeled as purposive in Thorpe's sense.

However, Thorpe does regard much species-characteristic behavior as satisfying his criteria for goal-directed behavior. He cites Tinbergen's field description of nest-building by the long-tailed tit as evidence that nest-building is not merely a stimulus-response chain, but depends on the birds having "some 'conception' of what the completed nest should look like, and some sort of 'conception' that the addition of a piece of moss and lichen here and there will be a step towards the 'ideal' pattern, and that other pieces there and there would detract from it" (Thorpe, 1963, p. 42). Similarly, he cites observations of nest-repair by wasps and tube-building by caddis fly larvae, and concludes: "It seems, then, that in much of appetitive behavior the animal's own activities, provided they are proceeding in a coordinated manner and in the *right direction,* must be self-rewarding and self-stimulating" (1963, p. 47).

Whether Thorpe's interpretation of nest-building is correct remains to be seen, but three types of observation raise some doubt.

1. Field (e.g., Howard, 1935) and laboratory (e.g., Hinde, 1958) observations of many species show that when motivation is low, nest-building often consists of incomplete sequences of behavior that are repeated even though not contributing in any way to the construction of a nest. Thus if building in the right manner has a special reinforcing value, this comes in only when motivation is above a certain level.

2. The building of even the most complex nests can be accounted for in much simpler terms than might appear at first sight. For instance, much of the shaping of the nest of the village weaver bird (*Ploceus* [*Sitagra*] *cucullatus*) is a consequence of the orientation of the building movements toward the edge of the growing structure (Collias and Collias, 1962; Crook, 1960, 1964).

3. Such experiments on nest-repair as have been carried out indicate that the repair behavior is essentially a stereotyped repetition of earlier phases of building. In weaver birds it is successful only as long as the orientation of the nest is not changed: if it is, the bird constructs a new initial ring and thus a new nest, rather than repairing the old one (Crook, 1964). Although the orientation of particular nest-building movements may be influenced by gaps or excrescences in the existing structure, this need not involve comparison between the present structure and an ideal pattern, and there is no clear evidence that they meet Thorpe's criteria of purposive behavior.

Further evidence concerning the "self-rewarding and self-stimulating" nature of behavior comes from asking whether a behavior pattern will still be maintained even when the end of the normal sequence is not allowed to occur. For example, Whalen (1961) has shown that naive male rats' intromission behavior was maintained, even though ejaculation was never allowed; and intromission without ejaculation continued to be an effective reward for maze running. Although this maintenance of units of a sequence produces behavior in the "right direction," whether the animal, as well as the experimenter, has a concept of the "right direction" remains to be shown. In this case, indeed, there is no need to regard the behavior leading to intromission as in any way a response guided towards a future goal of ejaculation, though it may be useful to describe the behavior leading to intromission as directed toward that short-term goal. Indeed, not only is it possible for individual responses to be goal-directed (by Thorpe's definition of striving after a future goal) when the whole sequence may not be, as in this case, but also for the whole sequence to be goal-directed when the individual responses are not, as when we try random solutions to a puzzle.

As we have seen, both Thorpe's and Schneirla's definitions imply an image or internal correlate of the goal: the usefulness of this is not always clear-cut. For example, in experiments in which rats learn to run a maze and to make a correct turn into the goal box, a stimulus-response theory such as Hull's would not postulate an internal model, and a cognitive theory, such as Tolman's, would. More relevant maze experiments sought to determine whether an animal that has learned one spatial or one behavioral route to a goal will transfer to the most direct alternative as soon as either the path or the response is blocked. Positive results were taken to support Tolman's theory of purposive behavior, namely that the animal had formed an "expectancy" of the goal and a "cognitive map" for reaching it. Following a discussion of conditions that gave evidence for "purposive" behavior in this sense, Osgood (1953, p. 460) pointed out that supporting results "probably become more typical as one goes up the evolutionary scale."

In a theory such as that of Miller, Galanter, and Pribram (1960) an internal model, and its discrepancy with the present situation, assumes prime importance. On their view, the fundamental unit in the organization of behavior is a negative feedback loop. An example of such a loop at an elemental level is provided by the control of graded muscular contractions. These depend on the sensory impulses from the muscle spindles whose frequency varies with the length of the muscle relative to that of the muscle spindles. As this changes in the course of a contraction and influences the further output of the motor neurons, so the further extent of the contraction is controlled. Since the length of the muscle spindles can be controlled via the "gamma-efferents," the point of balance can be altered.

Such a feedback loop can be described as a sequence of operations involving first the initiation of a test for congruity between present input and a required value. If incongruity occurs, a mechanism operates to reduce it: this is then followed by a further test. If congruity is found, no further action is taken. In the case of graded muscular contractions, the test consists of a comparison between the length of muscle spindles and muscle, and the operation of contraction is performed through the agency of nerve impulses. Miller, Galanter, and Pribram, however, generalize this further. At another level of analysis, the repeated sequence of test for congruity and operation may depend upon the transference not necessarily of nerve impulses but of *information:* since the transmission of information over a channel depends on correlation between input and output, we are concerned here not with particular physical structures but with correlations between events. Miller, Galanter, and Pribram go further than this to discuss, at a still higher level of analysis, the transference of *control.* In this case the test serves to ensure that the operational phase is appropriate to the task in hand: only when the test gives the appropriate outcome is control transferred to the operational phase. This more abstract level of application, though of primary concern to the authors, need not detain us here, but it must be noted that Miller, Galanter, and Pribram emphasize that the feedback involved is in any case quite different from "reinforcement." It does not reduce a drive or strengthen anything, but serves merely for comparison or testing; and it is not necessarily a stimulus, but may be "information" or "control."

Miller, Galanter, and Pribram suggest that complex types of behavior depend on a hierarchical system of such feedback units, the operational phase consisting of loops each with its own test and operational phases. Behavior is thus conceived as governed by feedback loops at all levels. In that such mechanisms are concerned primarily with the way in which responses come to an end, the main concern of Miller, Galanter, and Pribram is with "purposive" behavior, but apart from a discussion of the techniques involved in search, they do not really face up to the mechanisms responsible for selecting between responses.

Furthermore, they imply that mechanisms of this type are universal. While they acknowledge that much of the behavior of lower animals may depend on the provision by the environment of a succession of stimuli that release the successive stages in the sequence of behavior, they suggest that such cases fall within their system as follows: They suppose that any behavior must depend on a "Plan," which is a "hierarchical process in the organism that can control the order in which a sequence of operations is to be performed" (1960, p. 16). In the case of a sequence of responses integrated by successive environmental stimuli, they suppose that the Plan is partly environmental, dismissing as "something of a philosophical question" whether "we wish to believe in plans that exist somewhere outside of nervous systems" (1960, p. 38), though they are not consistent on this point (1960, cf. p. 38 with pp. 77–79). But by implying that feedback is universal, they fail to provide adequate criteria for the applicability of their system. As we shall see later, the concept of feedback cannot be applied to all behavior at all levels of analysis, and where it can, the similarity may be so superficial that use of the common term actually misrepresents the various phenomena considered by minimizing the existence of wide areas of difference among them (cf. Schneirla, 1949, p. 247).

Although generality is appealing, loss of discrimination between important differences is a high price to pay. We feel that any label, whether it be "feedback," "goal-direction," or "purpose," is useful here only as long as it serves to discriminate categories of behavior. Yet if such a label is given a precise definition, it has the danger of either implying a unitary mechanism to cover superficially a wide range of complexity, or being tied to one experimental design with neglect of other situations that would be relevant according to common usage of the term. In order to obtain both precision and flexibility, we shall simply begin with data, and consider in particular cases how individual responses are controlled. Since a basic property of "goal-directed" behavior is that it ceases when the goal is reached, we shall center attention on the ways in which responses come to an end. In each case it is necessary to consider a response at a particular level of analysis, and to assess whether control of the response is exerted by consequences at that level: clearly a response may be independent of its own consequences but still subject to precalibrated intrasystemic control. Responses differing widely in complexity will be discussed in an attempt not to produce a definitive classificatory system that would only allow us to attach a debatable label of "goal" or "no-goal," but rather to show the diverse ways in which cessation of a response is determined by its consequences. In order to do this, we have grouped examples into five categories which are not mutually exclusive or exhaustive, but which we believe to be useful for the present discussion.

I. *A response has no effect on itself or on the initial stimulus.*

Often the course of a response is independent of the consequences of the response, at least at the level of analysis in question. In some cases there is no apparent external stimulus, and any effect of the response itself can be ruled out. For example, Konishi (1963) found that the form of the crows of a bantam cock was disturbed little if at all by deafening. As Konishi points out, deafening does not rule out nonauditory feedback, but Nottebohm (personal communication) showed that crowing is still unchanged when deafening is followed by section of the recurrent branch of the vagus, which carries afferent fibers from the syrinx.

In other cases there is an external stimulus that elicits the response, but it cannot affect the course of the response either because it is removed before the response is made or because the response is too rapid to be controlled through exteroceptors. Thus a male firefly (*Photinus*) may orient correctly to a brief flash of light even though it does not start to turn until the flash is over (Mast, 1912); the consequences of the turn (the change in the relative direction of the flash) thus cannot influence the extent of the response. Similarly the extension of the legs in the strike of the praying mantis (Mantidae) occurs so rapidly that no information about the direction of the strike from the prey could be used to monitor its course: the movement must thus be determined from the start (Mittelstaedt, 1962).

In other cases the system may be intraorganismic, the output being assessed in terms of the discharge in motor nerves on the contractions of individual muscles. An example is provided by swallowing in dogs. This is a complicated activity involving a coordinated pattern of contraction in over twenty different muscles. Nevertheless the temporal pattern, duration, and amplitude of the contractions is independent not only of the means used to elicit swallowing, but also, at least within wide limits, of proprioceptive feedback. Strychnine, moderate asphyxia, excision, or procainization of participating muscles, and a wide variety of treatments likely to upset any feedback, produce only very minor changes in the temporal organization of swallowing (Doty and Bosma, 1956). Again, a simple nonpatterned input to the mesothoracic ganglion of the cicada *Graptopsaltria nigrofuscata* elicits a sequence of alternate responses in the two main muscles of the sound-producing mechanisms which lie on either side of the body (Hagiwara and Watanabe, 1956). Other examples of systems whose output is not influenced by feedback are provided by the lobster's cardiac ganglion, which give rise to a regular sequence of patterned bursts in the efferent nerve which continues even when the ganglion is isolated from outside influences (Maynard, 1955; Hagiwara, 1961), and the electric organs of Gymnotid fish which discharge at a constant rate uninfluenced by external feedback (Lissmann, 1958).

Of course, in none of the examples given is the possibility of feedback within the responding system completely ruled out: the behavior may reach completion either because the motor discharge has been appropriately predetermined, or because some intrinsic controlling system has been preset. Yet in all these cases the patterning and duration of each individual response is independent of the consequences of responding or of feedback mediated by the output of the system: in this sense, the behavior is not "goal-directed."

II. *The probability of repetition of a response is affected,*
   though the initial, external stimulus is not.

In the preceding category we considered cases in which the form or duration of an individual response was independent of its consequences. Here
we shall consider cases in which an individual response influences the
repetition of that response, even though the initial, external stimulus is not
affected.

If a captive chaffinch (*Fringilla coelebs*) is exposed to an owl, the
resulting "mobbing" response at first increases in intensity and then slowly
wanes. These changes occur although the owl is present continuously, and
there is no change in the external situation to which the chaffinch is exposed. Analysis shows that exposure to the stimulus is accompanied by
interacting effects which tend to produce increases or decreases in response
strength, and have time courses of decay ranging from seconds to months:
the response is thus a resultant of multiple processes (Hinde, 1960).

Comparable phenomena have been studied in a wide variety of responses
at all phyletic levels. In many cases, the changes in response strength are
more or less specific to the particular stimulus to which the animal is
exposed. Thus the equilibratory response of the frog *Rana esculenta* when
the substrate is tilted is mediated by both the eye and labyrinth. Responsiveness wanes when the animal is stimulated successively through one
modality, but it still responds through the other (Butz-Kuenzer, 1957). If
female mice are given ten presentations of a 1-day-old baby in close
succession, the number of subjects showing licking and nest-building
responses decreases with the successive presentations. If a 10-day-old baby
is given immediately afterwards, the proportion of females showing these
responses increases. Since naive subjects respond less strongly to a 10-day-
old than to a 1-day-old baby, most of the waning during the first ten
presentations must be more or less specific to the stimulus of a 1-day-old
baby (Noirot, 1965). As another example, if human newborns are exposed
to an olfactory stimulus, a breathing response which is elicited by the
stimulus gradually declines. If the initial stimulus consists of a mixture of
two odorants, and after the response has waned only one of the components is presented, the response increases. Thus the waning cannot be due
to simple sensory adaptation or fatigue, but is specific to the mixture of
stimuli (Engen and Lipsitt, 1965).

In some cases the rate of waning varies with the adequacy of the eliciting
stimulus. Thus the strike response of the mantis, referred to above, can
be elicited by paper dummies placed behind glass, but wanes in a few
minutes. With a live fly behind glass, however, the waning is very slow,

with the mantis continuing to strike for several hours (Rilling, Mittel-staedt, and Roeder, 1959). Similar data are available for the owl-mobbing response of chaffinches.

In most cases there is no direct means of knowing whether the changes in response strength depend on the occurrence of the whole response under study or merely on activity in only part of the responding system. Such evidence as there is indicates no simple answer. With toads (*Bufo bufo*), a light touch on the back with a hand elicits a wiping movement. Sudden stimulation with a bristle does not elicit wiping, but does lead to a reduction in responsiveness to subsequent stimulation of the same spot by the hand. The reduction thus seems to be at least partly independent of the response (Kuczka, 1956). Yet if stimuli which usually initiate head-turning are given to a human baby while he is sleepy or engaged in some other ac-tivity, so that the baby does not respond, there is no resultant decrement when he does (Prechtl, 1958).

Of special interest are certain responses that facilitate the perception of a stimulus and are elicited by a wide variety of stimuli with the common property that they are in some way novel to the animal (Berlyne, 1960). We refer here to the "orientation reflex," exploration, and so on. Sokolov (1960) has shown that after such a response to a novel stimulus has waned, it returns if any parameter of the stimulus changes, even if the change involves a reduction in physical intensity. Response to novelty im-plies storage of information about the familiar, and Sokolov pictures this as the building up of a "neuronal model" of the familiar by experience. He thus regards the elicitation of the orienting response as due to a dis-crepancy between the perceptual consequences of the stimulus and the "neuronal model." Whether this pseudophysiological term will direct future research into a fruitful path remains to be seen (see Horn, 1967).

Although the examples cited so far are all species-characteristic re-sponses, the extinction of learned responses also belongs in this formal category. For example, in the case of an operant response, which has been maintained by reinforcement, withdrawal of reinforcement will produce a slow decrease in responding, although temporary increases are often noted as well. The rate of decrease in responding is related to past conditions, such as previous training history (e.g., number of reinforcements, schedule of reinforcement, quality and amount of reinforcement) and amount of deprivation. Although extinction is usually studied in quite different con-texts, we are citing it here simply as an example of responding coming to an end without affecting the initial stimulus.

These examples differ widely in complexity, and the change in the probability of response with continued or repeated presentation of the stimulus is often not as simple as appears at first sight. We have already seen how exposure of a chaffinch to an owl produces both incremental and decremental effects on further responding, though the latter pre-dominate. In other cases as well, it seems likely that multiple incremental and decremental processes interact to produce the change in response strength. Furthermore, the rate of waning may depend on past conditions: we have already mentioned that the rate of extinction of an operant response varies with the training conditions. As a rather different example, the sucking response of human newborns to a piece of tubing does not wane over a long series of presentations of the tube, but does wane when blocks of tube-presentations alternate with blocks of ordinary nipple-presentations (Lipsitt and Kaye, 1965). The authors suggest that the initial higher-than-normal rate of responding to the tube when it was presented after the nipple involved a "perseverative" effect; and that the following decrease to a lower-than-normal rate involved a negative "con-trast" effect between a pleasurable (nipple) and less pleasurable (tube) stimulus. Thus incremental and decremental processes of diverse time courses interact, and the resulting change in response strength depends on past and present experimental conditions. In all cases the probability of a response changes as a consequence of repeated exposure to a stimulus, and the responding ceases although no change in the external situation is produced. The cessation of responding can thus hardly be ascribed to the reaching of a goal.

It must be noted that we are concerned here with the extent to which the reaching of a goal is responsible for bringing a response or succession of responses to an end. By some criteria, however, goals have an added role—they may act as reinforcers, making the repetition of the response on subsequent occasions more probable. Thus if, following a period of extinction, an operant response does in fact produce an appropriate stimulus (e.g., food), the operant will be more, rather than less, likely to be repeated. However, in other cases, such as the mantis's strike, infant's sucking, or chaffinch's owl-mobbing, repetition is influenced primarily by the strength of the eliciting stimulus (or the consequent response), rather than by changes in the external situation which might follow the response. The case for describing the operant response as goal-directed thus rests on its role in producing a new stimulus for a consummatory response and the resultant increase in probability of repetition of the operant: this is considered later.

III. *The response affects the initial stimulus, which in turn affects the course of the response.*

We shall now turn to some cases in which the response does affect the eliciting stimulus. The first concerns animals that adjust their spatial relationship to environmental stimuli by responses whose intensity or frequency is influenced by the strength of the incident stimulation. Thus woodlice (*Porcellio scaber*) move more slowly as the humidity increases, and they become aggregated in the damper regions ("orthokinesis"). The rate of movement of the animal is related to the conditions, and the animals are most likely to be found in regions in which their locomotion is slowest. A similar effect may be produced if the rate of turning (changing direction), as distinct from the rate of movement, is influenced by the external conditions ("klinokinesis"), though the precise nature of the effect may be influenced by a number of parameters including the rapidity of adaptation to the stimulus conditions (Fraenkel and Gunn, 1940).

Cases in which the incident stimulation influences the intensity or frequency of a response which produces a change in that stimulation are not limited to orientation. Thus hermit crabs (*Pagarus* and *Calcinus*) show a special type of appetitive behavior which leads to their righting shells and entering them. The amount of appetitive behavior shown (as assessed by the amount of motor activity and responsiveness to shells) depends on the characteristics of the shell (weight and internal configuration) it has already entered. The greater this "stimulus value" of the shell the crab already has, the smaller the probability of further appetitive behavior (Reese, 1963).

In these examples the change in the initial stimulus affects merely the intensity of the response. In others, however, the stimulus influences the direction or nature of the response. For example, a fish maintains its orientation about its longitudinal axis in part by maintaining equal illumination in its two eyes. If one eye is removed, it first rolls toward the intact side (von Holst, 1935). Similarly the crustacean *Armadillidium* moves toward a light by equating the light intensity on its two sides (Henke, 1930). In other cases insects orient not directly toward or away from the light, but at a constant angle to it. For instance, an ant or bee moving between nest and food source maintains direction in part by keeping a constant angle to the sun (e.g., Schneirla, 1933). Furthermore the angle that the insect maintains with a stimulus source may change, according to whether the animal is moving outwards or homewards (e.g., Jander, 1957), with time of day (e.g., Birukow, 1954), or with previous stimulation (von Frisch, 1954). It was thought earlier that the insect must possess a separate orienting mechanism for each possible angle to the stimulus source. A much simpler hypothesis, which accounts for all such cases, is that the tendency of the animal to turn at any moment, and the direction in which it will do so, is governed by the discrepancy between the current direction of the stimulus source and an optimal value. The turn reduces the discrepancy, and thus the likelihood of further turning. The different orientations that the animal takes up involve changes in the direction of the goal (e.g., Mittelstaedt, 1961). Similar principles apply to cases of orientation to complex configurational stimuli. Thus the homing of a diggerwasp (*Bembix rostrata*) involves response to discrepancy between a correlate of the local characteristics of the nest-site, learned previously, and those actually encountered (van Iersel and van den Assem, 1964).

We could extend this category to include cases in which discrepancy between the present stimulus situation and a distant stimulus situation influences the direction and intensity of the response. For instance, the ability of rats to take the shortest detour in a maze could depend on response to discrepancy between the present situation and the goal situation, and consequent correct orientation. Similarly, the ability of many birds to orient correctly from an unknown release point toward home within a few seconds of release implies that they must be responding to differences in at least two parameters between the release point and home. Although there are strong indications that the sun is an important stimulus source for orientation, the precise characters responded to are still not known with certainty (Matthews, 1955; see also Pennycuick, 1960; Schmidt-Koenig, 1965). Finally, response to discrepancy could involve selection from a number of possible responses (as opposed to intensities or directions of one response) one which produces greatest reduction in the discrepancy, and thus some of the most complex types of problem-solving behavior. Clearly it is here that many types of behavior that would generally be labeled "goal-directed" are to be found.

Nevertheless, this does not mean that all behavior that can best be described in terms of response to a discrepancy is profitably included under the single label "goal-directed." For instance, an orientation response (in Sokolov's sense, see category II) may be made to stimuli that differ from a "neuronal model," but the removal of the discrepancy comes about through a change in internal organization and not, as in the cases cited in this section, through a change in the external situation. It is hardly useful to subscribe both under the same rubric.

IV. *The response, while not affecting the initial stimulus, produces a second stimulus, which decreases the response directly.*

In the previous category the response was affected by a change it produced in the initial situation. In the next two categories, the response comes to an end by producing a new stimulus, not present at the initiation of the response. In the present case, since the initial stimulus is not affected, and since no incompatible response is elicited by the new stimulus, the effect of the new stimulus on the response can be considered to be direct. We shall consider later cases in which the new stimulus brings the first response to an end by initiating a second response.

When a young rhesus monkey is exposed to a fear-inducing situation, it shows a variety of fear responses, including running to its mother or mother-substitute and clutching her. Its fear-responses then wane more rapidly than they would if the mother were absent, although the fear-inducing stimuli are still present (e.g., Harlow and Zimmerman, 1959). In studies of eating and drinking responses, stimuli from the mouth and throat region and the stomach have been shown to bring about a reduction in the response in spite of the continued presence of both the environmental stimuli (food or water) and the tissue deficiencies which preceded it. Indeed, the feeding behavior of rats ceases after consumption of non-nutritive saccharin (Miller, 1957). As another example, a chaffinch influenced by gonadal hormones sings on his territory, and his song attracts females: once he is paired, he sings much less although his hormonal condition is (presumably) little changed (Marler, 1956).

On a longer time scale, the nest-building of canaries results in the formation of a nest-cup and thus in a change in the stimulus situation to which the female is exposed when she sits in the nest. On a short time scale, stimuli from the nest-cup produce a decrease in nest-building behavior even though the nest-material is still present and the hormonal conditions are still appropriate for it (Hinde, 1958).

In all these cases the animal encounters a new "consummatory" stimulus as a consequence of its response, and this new stimulus produces a reduction in the response although the eliciting conditions are still present. There is a procedural difficulty in the way of establishing that the effect is a direct one—the new stimulus could act by eliciting a new response. In the cases cited here, however, the next type of behavior shown cannot readily be predicted from the "consummatory" stimulus (see category V).

Granted that the consequences of the behavior produce a negative effect on the repetition or continuance of that behavior, there must be optimal values of the several consequences such that the effect will vary with the difference between the actual consequences and these optimal values. However, this does not necessarily imply a response to discrepancy such as that implied by Sokolov's suggestion that a response to novelty implies comparison with a "neuronal model" of the familiar, or as when the maintenance of different orientations to an external stimulus implies response to the deviation between the actual and required directions of that stimulus (see category III). Rather in the present case the response depends on a balance between positive and negative (consequential) factors, there being need to postulate a process of summation, but not of internal comparison. There is indeed more similarity to the hermit crab example cited in category III above, for we could regard the hermit crab's search for shells as depending on the resultant of positive factors, constantly present, and negative factors derived from the stimulation provided by the shell which it has at the moment. Nevertheless in some of these cases the animal may show one of a variety of responses suited to the particular situation in order to achieve the consummatory situation, and in such cases the behavior might well be labeled as goal-directed.

Another example is provided by Sevenster-Bol's (1962) analysis of the factors controlling the courtship of the male three-spined stickleback. The reduction in the readiness of the male to court females which occurs after fertilization is primarily a consequence of stimulation from a fresh clutch of eggs in the nest, and is almost independent of the previous performance of the courtship activities or of stimuli from the spawning female. In this case the influence of the consummatory stimuli is not due to a direct suppression of courtship activities, but to a change in the relative probability of the male showing attacking and fleeing responses, and thus to an increase in the probability of his attacking, rather than courting, a female who enters his territory (Wiepkema, 1961). This case thus forms a link with the next category, where the new stimulus acts by initiating a new response.

V. *The response produces a second stimulus, which elicits a second response.*

We now turn to cases in which the consequences of the response involve new stimuli that elicit new types of behavior. A well-analyzed case is the courtship of the three-spined stickleback, where each response by male or female can induce a new type of behavior in the partner. Although the chain is not so rigid as implied in Tinbergen's (1951) original scheme, a similar principle has been demonstrated in many other cases (e.g., Barlow, 1962).

In the context of nest-building by canaries similar principles apply in the control of responses with widely differing time constants. Thus in each sequence of building behavior the successive responses of gathering, carrying, and building in the nest form a chain, each of the first two bringing the female into the stimulus situation for the next. On a longer time scale, stimuli from the nest which the female builds induce endocrine changes that bring about both a decrease in nest-building and an increase in other types of behavior (Hinde, 1965).

In other cases the chain may involve a stimulus situation for a learned or operant response, which produces another stimulus, which in turn elicits a species-characteristic response not dependent on learning. Such a stimulus-response chain has been built up with Siamese fighting fish swimming through a ring to produce an image of another fish, which elicits a threat display (Thompson and Sturm, 1965); and with red gavillan roosters pecking a response key to produce an image of another cock, which similarly elicits threat display (Thompson, 1964).

Longer operant sequences have been built up on more conventional rewards such as food or water. The first step is to train the animal to approach the feeder and eat only when a signal is presented. If the signal and the approach and consummatory responses that follow it are made contingent on a response, such as a lever press, then the rate of lever-pressing will increase. If lever-pressing is then rewarded only when a signal for it is presented, and if another response is then trained to produce that signal, and so on, this step-by-step procedure can establish a stimulus-response chain of considerable length. For example, Pierrel and Sherman have used such a procedure to build up the following sequence performed by a food-deprived rat in a special chamber: The rat first "climbed to the top of a spiral staircase, 'bowed to the audience', pushed down a draw-bridge, crossed the bridge, climbed a ladder, used a chain to pull in a model railroad car, pedaled the car through a tunnel, climbed a flight of stairs, ran through a tube, and stepped into an elevator which descended to the base of the platform . . . At this point a buzzer sounded, and the rat pressed a lever and received a food pellet" (In Kelleher and Gollub, 1962, p. 544).

By common usage of the term, each stimulus in such an operant sequence could be called a subgoal. But it would be inappropriate to use this label in the case of species-characteristic responses. Most species-characteristic sequences are not built up in the way an operant chain is built up; and if such a sequence (e.g., nest material—picking up—material in beak—carrying—nest—placing) is broken, so that one of the stimuli (e.g., nest) is never produced, the preceding responses may persist for a long period. Unlike an operant chain, the sequence seems to be little influenced by achievement of the final stage (see Hinde and Stevenson, 1969).

SUMMARY

In this brief survey we have been concerned with the means by which responses are brought to an end. We have begun with those cases in which cessation does not result from the consequences of the response. Then, among the rest, we have attempted to describe different ways in which a response is ended by its consequences. Our initial aim was to assess criteria by which behavior can be labeled as goal-directed, and several points have arisen.

First, we would emphasize again that all our categories contain responses differing greatly in complexity: any criteria of goal-directedness based on them must not be taken to imply homogeneity of mechanism. Second, not all behavior is brought to an end by its consequences. Therefore, we should argue, not all behavior should be described as goal-directed: we do not believe that it assists analysis to label all stable systems as goal-directed. Third, the means by which consequences affect behavior are diverse, so that to categorize even all cases in which behavior is brought to an end by its consequences as "goal-directed" obscures important differences. Fourth, response to discrepancy, even discrepancy between present situation and an internal "standard," cannot by itself be used as an adequate criterion of goal-directedness.

Finally, most complex behavior involves several of the categories mentioned in the preceding pages. Thus maze-running is brought to an end by goal stimuli. It involves chain responses in so far as each turn accomplished or alley run brings the animal to a fresh stimulus situation. It involves response to discrepancy between the incident stimulus situation and a goal situation in so far as the direction of turning is such as to bring the animal nearer the goal and rate of running varies with the distance from the goal. Indeed, a given type of behavior may depend on error signals at one level of integration but not at others: the mechanisms controlling the individual movements of nest-building may be of a quite different kind from those controlling the sequence of behavior as a whole, and these in turn different from those controlling the incidence of the sequence from day to day.

We believe that the label "goal-directed" is useful only as long as it serves to discriminate categories of behavior. It can help to achieve this only if it is sufficiently flexible to cover behavior differing markedly in complexity, but it is then useful only if it is taken to imply questions about the mechanisms by which the goal is achieved. Perhaps the most useful function of the term "goal-directed" lies therefore in the questions it poses, such as, "In what sense?" and "By what means?"

ACKNOWLEDGMENTS

We are grateful to W. H. Thorpe, P. P. G. Bateson, and K. H. Pribram for their comments on the manuscript.

REFERENCES

Barlow, G. W. 1962. Ethology of the Asian teleost, *Badis badis:* IV. Sexual behavior. *Copeia* 2: 346–360.

Berlyne, D. E. 1960. *Conflict, arousal and curiosity.* New York: McGraw-Hill.

Birukow, G. 1954. Photo-geomenotaxis bei *Geotrupes silvaticus* Panz. und ihre zentralnervöse Koordination. *Z. Vergleich. Physiol.* 36: 176–211.

Blough, D. S. 1966. The reinforcement of least-frequent interresponse times. *J. Exp. Anal. Behav.* 9: 581–591.

Blough, D. S., and R. B. Millward. 1965. Learning: operant conditioning and verbal learning. *Ann. Rev. Psychol.* 16: 63–94.

Butz-Kuenzer, E. 1957. Optische und labyrinthäre Auslösung der Lagereaktionen bein Amphibien. *Z. Tierpsychol.* 14: 429–447.

Collias, N. E., and E. C. Collias. 1962. An experimental study of the mechanisms of nest building in a weaverbird. *Auk* 79: 568–595.

Craik, K. J. W. 1943. *The nature of explanation.* London: Cambridge Univ. Press.

Crook, J. H. 1960. Nest form and construction in certain West African weaverbirds. *Ibis* 102: 1–25.

Crook, J. H. 1964. Field experiments on the nest construction and repair behaviour of certain weaverbirds. *Proc. Zool. Soc. Lond.* 142: 217–255.

Doty, R. W., and J. F. Bosma. 1956. An electromyographic analysis of reflex deglutition. *J. Neurophysiol.* 19: 44–60.

Engen, T., and L. P. Lipsitt. 1965. Decrement and recovery of responses to olfactory stimuli in the human neonate. *J. Comp. Physiol. Psychol.* 59: 312–316.

Fraenkel, G. S., and D. L. Gunn. 1940. *The orientation of animals.* Oxford: Clarendon Press.

Frisch, K. von. 1954. *The dancing bees.* London: Methuen.

Hagiwara, S. 1961. Nervous activities of the heart in Crustacea. *Ergeb. Biol.* 24: 287–311.

Hagiwara, S., and A. Watanabe. 1956. Discharges in motorneurons of cicada. *J. Cell. Comp. Physiol.* 47: 415–428.

Harlow, H. F., and R. R. Zimmermann. 1959. Affectional responses in the infant monkey. *Science* 130: 421–432.

Heinroth, O. 1911. Beiträge zur Biologie, namentlich Ethologie und Psychologie der Anatiden. *Verh. 5 Int. Orn. Kong.,* pp. 589–702.

Henke, K. 1930. Die Lichtorientierung und die Bedingungen der Lichtstimmung bei der Rollassel *Armadillidium* cinereum Zenker. *Z. Vergleich. Physiol.* 13: 534–625.

Hinde, R. A. 1958. The nest-building behaviour of domesticated canaries. *Proc. Zool. Soc. Lond.* 131: 1–48.

Hinde, R. A. 1960. Factors governing the changes in strength of a partially inborn response, as shown by the mobbing behaviour of the Chaffinch (*Fringilla coelebs*): III. The interaction of short-term and long-term incremental and decremental effects. *Proc. Roy. Soc.* (London), B, 153: 398–420.

Hinde, R. A. 1965. Interaction of internal and external factors in integration of canary reproduction. In F. A. Beach, ed., *Sex and behavior*. New York: John Wiley.

Hinde, R. A., and J. G. Stevenson. 1969. Sequences of behavior. In D. S. Lehrman et al., eds., *Advances in the study of behavior,* vol. 2. New York: Academic Press. Pp. 267–296.

Holst, E. von. 1935. Über den Lichtrückenreflex bei Fischen. *Pubbl. Staz. Zool. Napoli.* 15: 143–158.

Horn, G. 1967. Neuronal mechanisms of habituation. *Nature* 215: 707–711.

Howard, E. 1935. *The nature of a bird's world.* London: Cambridge Univ. Press.

Jander, R. 1957. Die Optische Richctungsorientierung der Roten Waldameise (*Formica rufa L.*). *Z. Vergleich. Physiol.* 40: 162–238.

Kelleher, R. T., and L. R. Gollub. 1962. A review of positive conditioned reinforcement. *J. Exp. Anal. Behav.* 5: 543–597.

Konishi, M. 1963. The role of auditory feedback in the vocal behavior of the domestic fowl. *Z. Tierpsychol.* 20: 349–367.

Kuczka, H. 1956. Verhaltensphysiologische Untersuchungen uber die Wischhandlung der Erdkröte (*Bufo bufo L.*). *Z. Tierpsychol.* 13: 185–207.

Lipsitt, L. P., and Kaye, H. 1965. Change in neonatal response to optimizing and non-optimizing sucking stimulation. *Psychon. Sci.* 2: 221–222.

Lissmann, H. W. 1958. On the function and evolution of electric organs in fish. *J. Exp. Biol.* 35: 156–191.

Lorenz, K. 1941. Vergleichende Bewegungsstudien an Anatinen. *Suppl. J. Ornith.* 89: 194–294.

Lorenz, K., and N. Tinbergen. 1939. Taxis und Instinkthandlung in der Eirollbewegung der Graugans: I. *Z. Tierpsychol.* 2: 1–29.

MacCorquodale, K., and P. E. Meehl. 1954. Edward C. Tolman. In W. K. Estes et al., eds., *Modern learning theory*. New York: Appleton-Century-Crofts.

Marler, P. 1956. Behaviour of the Chaffinch (*Fringilla coelebs*). *Behaviour Suppl.* 5: 1–184.

Mast, S. O. 1912. Behavior of fireflies, with special reference to the problem of orientation. *J. Anim. Behav.* 2: 256–272.

Matthews, G. V. T. 1955. *Bird navigation.* London: Cambridge Univ. Press.

Maynard, D. M. 1955. Activity in a crustacean ganglion: II. Pattern and interaction in burst formation. *Biol. Bull.* 109: 420–436.

Miller, G. A., E. Galanter, and K. H. Pribram. 1960. *Plans and the structure of behavior.* New York: Holt, Rinehart and Winston.

Miller, N. E. 1957. Experiments on motivation. *Science.* 126: 1271–1278.

Miller, N. E. 1959. Liberalization of basic S-R concepts. In S. Koch, ed., *Psychology: A study of a science,* study I, vol. 2. New York: McGraw-Hill.

Mittelstaedt, H. 1961. Die Regelungstheorie als methodisches Werkzeug der Verhaltensanalyze. *Naturwissenschaften* 48: 246–254.

Mittelstaedt, H. 1962. Control systems of orientation in insects. *Ann. Rev. Entomol.* 7: 177–198.

Noirot, E. 1965. Changes in responsiveness to young in the adult mouse: III. The effect of immediately preceding performances. *Behaviour* 24: 318–325.

Osgood, C. E. 1953. *Method and theory in experimental psychology.* New York: Oxford Univ. Press.

Pantin, C. F. A. 1964. Learning, world-models and pre-adaptation. *Anim. Behav.,* Suppl. 1: 1–8.

Pennycuick, C. J. 1960. The physical basis of astro-navigation in birds: theoretical considerations. *J. Exp. Biol.* 37: 573–593.

Prechtl, H. F. R. 1958. The directed head turning response and allied movements of the human baby. *Behaviour* 13: 212–242.

Reese, E. S. 1963. The behavioral mechanisms underlying shell selection by hermit crabs. *Behaviour* 21: 78–126.

Rilling, S., H. Mittelstaedt, and K. D. Roeder. 1959. Prey recognition in the Praying Mantis. *Behaviour* 14: 164–184.

Rosenblueth, A., N. Wiener, and J. Bigelow. 1943. Behavior, purpose and teleology. *Phil. Sci.* 10: 18–24.

Schmidt-Koenig, K. 1965. Current problems in bird orientation. In D. S. Lehrman et al., eds., *Advances in the study of behavior,* vol. 1. New York: Academic Press.

Schneirla, T. C. 1933. Some important features of ant learning. *Z. Vergleich. Physiol.* 19: 439–452.

Schneirla, T. C. 1949. Levels in the psychological capacities of animals. In R. W. Sellars et al., eds., *Philosophy for the future.* New York: Macmillan.

Schneirla, T. C. 1959. An evolutionary and developmental theory of biphasic processes underlying approach and withdrawal. In M. R. Jones, ed., *Nebraska symposium on motivation.* Lincoln, Nebraska: Univ. Nebraska Press.

Sevenster-Bol, A. C. A. 1962. On the causation of drive reduction after a consummatory act. *Arch. Neerl. Zool.* 15: 175–236.

Skinner, B. F. 1959. *Cumulative record.* New York: Appleton-Century-Crofts.

Sokolov, E. N. 1960. Neuronal models and the orienting reflex. In M. A. B. Brazier, ed., *The central nervous system and behavior.* New York: Josiah Macy Jr. Foundation.

Stevenson, J. G. 1967. Reinforcing effects of Chaffinch song. *Anim. Behav.,* 14: 427–432.

Thompson, T. I. 1964. Visual reinforcement in fighting cocks. *J. Exp. Anal. Behav.* 7: 45–49.

Thompson, T., and T. Sturm. 1965. Visual-reinforcer color, and operant behavior in Siamese fighting fish. *J. Exp. Anal. Behav.* 8: 341–344.

Thorpe, W. H. 1963. *Learning and instinct in animals.* London: Methuen.

Tinbergen, N. 1942. An objectivistic study of the innate behaviour of animals. *Biblio. Biotheoreta.* 1: 39–98.

Tinbergen, N. 1951. *The study of instinct.* Oxford: Clarendon Press.

Tinbergen, N. 1953. *The herring gull's world.* London: Collins.

Tinbergen, N. 1959. Comparative studies of the behaviour of gulls (*Laridae*): a progress report. *Behaviour* 15: 1–70.

Tolman, E. C. 1932. *Purposive behavior in animals and men.* New York: Century.

van Iersel, J. J. A., and J. van den Assem. 1964. Aspects of orientation in the diggerwasp *Bembix rostrata. Anim. Behav.,* Suppl. 1: 145–162.

Welford, A. T. 1962. Experimental psychology and the study of social behaviour. In A. T. Welford et al., eds., *Society: Problems and methods of study.* London: Routledge and Kegan-Paul.

Whalen, R. E. 1961. Effects of mounting without intromission and intromission without ejaculation on sexual behavior and maze learning. *J. Comp. Physiol. Psychol.* 54: 409–415.

Whitman, C. O. 1919. The behavior of pigeons. *Publ. Carneg. Inst.* 257: 1–161.

Wiepkema, P. R. 1961. An ethological analysis of the reproductive behaviour of the bitterling. *Arch. Neerl. Zool.* 14: 103–199.

ETHEL TOBACH

Department of Animal Behavior
The American Museum of Natural History
New York

Some Guidelines to the Study of the Evolution and Development of Emotion

"The simple sensations of pleasure . . . are soft and gentle in their nature. The class of painful sensations is powerful . . . and, from the consciousness of its place or source, the efforts are directed to remove it."

Charles Bell (1844)

"When this child was about four months old, I made in his presence many odd noises, . . . but the noises, if not too loud, . . . were all taken as good jokes . . . accompanied by smiles. . . ."

Charles Darwin (1872)

"A significant fact is that approaches toward the source of stimulation . . . occur . . . to effectively weak stimulation, whereas withdrawals occur . . . to effectively intense . . . stimulation. . . . [Watson] elicited "love" . . . by mild stimulation such as patting, and the "fear-rage" . . . responses . . . by definitely intense stimulation."

T. C. Schneirla (1949)

Although there are many articles on the nature of emotional behavior, most are versions of the two dominant conceptualizations of emotion, that is, emotion as based on activation or arousal mechanisms and emotion as based on physiological expressions of some vaguely defined innate mechanism (Tobach, 1969). Both theories make excellent use of current research in neurophysiology, neuroendocrinology and comparative neuroanatomy, and they are not mutually exclusive. Both are accepted and integrated without any difficulty in the most widely held views of behavior —Freudian psychology (Greenfield & Lewis, 1965) and Skinnerian psychology (Skinner, 1966).

The experimental investigations resulting from these conceptualizations are primarily deductive, rather than inductive; they rarely test the assumptions basic to the theories; they are pragmatic and fact gathering. Further, this state of affairs in the scientific study of emotion has led to suggestions for avoiding wars among nations by the development of ritualized behavior (Lorenz, 1966) or by the use of hormones and electrophysiological techniques (Moyer, 1967).

Greater attention should be given to formulating the questions to be answered. The concepts offered by Schneirla may be useful in breaking the set in which modern-day experimentalists function. These concepts are not offered as theories to be tested by deductive experiments, and thus "proved" or "disproved"; they offer, instead, a means of organizing existing data, and asking questions about the data in such a way as to formulate further inductive types of studies. Admittedly, such studies are not popular today, as the results cannot be cast into the mold of support or refutation of a particular hypothesis.

Although it is generally agreed that the scientific method is based on both induction and deduction, in behavioral science, the inductive method is frequently derided as "armchair rumination."

Schneirla always emphasized the need and value of an evolutionary approach based on the material basis of natural phenomena, setting aside any explanations based on immaterial, *élan vitale* types of explanation. He rejected logical positivism and operationism as bases for scientific inquiry and opened the way to a dynamic, holistic approach based on process. This approach has not yet been completely documented or implemented, but his many students are attempting to do so.

In place of the predeterministic nonmaterial, nativistic concepts currently and widely in use in behavioral science, he derived two new concepts from an older concept which was dramatically and significantly improved by his use of it, namely, the concept of levels of organization and integration. The two concepts which he formulated are: (1) the biphasic approach-withdrawal process and (2) development as a fusion of experience, growth, differentiation, and maturation.

LEVELS OF ORGANIZATION AND INTEGRATION

As defined elsewhere (Schneirla, 1939, 1949, 1952; Tobach and Schneirla, 1968; Tobach, 1963, 1965, 1968), the concept of levels of organization and integration postulates that natural phenomena may be hierarchically ordered in respect to differentiation and complexity, beginning with the most fundamental aspect of matter, today described in terms of elementary particles, neutrons, protons, electrons, etc. Succeeding through ever increasing degrees of organization of the basic forms of matter, various levels are identifiable in such terms as physical phenomena and chemical phenomena within which are found other orders of organization, e.g., molecules, compounds, and crystals. A further division may be made in terms of inanimate and animate levels of organization, bearing in mind that mesolevels between the two exist, as in the case of viruses. Within the animal level of organization, a further subdivision is possible in terms of plant and animal life, again taking mesolevels into consideration (Novikoff, 1945).

From this ordering of matter, several corollaries are obtained. Each level of organization is subsumed in subsequently higher levels and is modified by such inclusion. Each level requires its own methods of investigation and yields its own laws and principles. Information about one level cannot be used to predict phenomena on the next higher level as it includes more and different variables and processes within it than the preceding levels. Although some types of explanation are possible within one level itself (concomitance), causal explanations are probably not possible without the consideration of preceding levels.

Although the concept of levels of organization has been described in many writings and has been modified by many, Schneirla added the important dimension of historicity to it in his application of the concept to three major problems in behavioral science: motivation, emotion, and development.

THE BIPHASIC APPROACH-WITHDRAWAL PROCESS

The process is described by Schneirla as follows:

> In all animals low stimulative intensities tend to elicit and maintain A-processes, the energy-conserving metabolic processes normal for the species and basic to approach responses, whereas high stimulative intensities tend to arouse W-processes, energy-expending metabolic changes underlying disturbed responses and withdrawal. (Schneirla, 1965, p. 3.)

The implications of this process for emotional behavior is given in detail in an earlier version of the paper printed in 1959 (Schneirla, 1959), and will be discussed below.

THE CONCEPT OF DEVELOPMENT AS A FUSION OF EXPERIENCE, GROWTH, DIFFERENTIATION, AND MATURATION

Schneirla's concept of the developmental process was clearly stated in his paper on development (1957) and repeated in two subsequent papers (1965; 1966). The following statement taken from his first detailed paper on the A-W process is most succinct: (1959)

> . . . behavioral development [is] . . . a program of progressive, changing relationships between organisms and environment in which the contributions of growth are always inseparably interrelated with those of the effects of energy changes in the environs. . . . the concept of "maturation" [is] redefined as the contributions of tissue growth and differentiation and their functional trace effects at all stages, and "experience" defined as the contributions of stimulation from the developmental medium and of related trace effects. . . .

LEVELS OF ORGANIZATION AND INTEGRATION OF EMOTION: FOUR ASPECTS

In accordance with the general theory of evolution, it may be hypothesized that emotional behavior evolved much as did other behavioral patterns, e.g., behavioral modifiability and social behavior.

The evolution of emotional behavior is studied in part by a comparison of individuals of different species. The investigation of the emotional behavior of the individual organism takes into account four levels of organization:

1. The phyletic status of the structural and functional capacities of the species;

2. The ontogenetic characteristics of the individual, that is, its developmental stage and experimental history;

3. The particular level of organismic integration to be studied, e.g., biophysical and biochemical level, as for example, neuronal membrane permeability to circulating endocrines; or the level of molar behavior; and

4. The level of social organization at which the individual is functioning. The question of the evolution of emotional behavior may be approached by any one of these orientations or by all. Our understanding of the process is enhanced in proportion to our information about all levels.

It should be possible, therefore, to define the process of emotion in such a way that comparisons and contrasts of all representative species are feasible. At least two requirements need to be satisfied by such a definition. The first is that some primary behavioral pattern observable on all phyletic levels can be reliably and validly studied so as to provide a continuum which might be called "emotional behavior." The criterion of validity to be satisfied in this instance is that the behavioral pattern chosen for study precisely reflect the processes responsible for the behavior. The second requirement is that the behavioral pattern denoting the process of emotion be differentiable from other response patterns by objectively verifiable techniques.

Every organism is at all times in a particular state of tensional adjustment to internal and external energy systems.[1] Such a state of tensional adjustment can be seen as a continuum from a minimal tensional state in which energy systems make the least demand on the organism as an integrated whole, to a maximal tensional state in which extensive and intensive adjustments are required. Both extremes are stressful and detrimental to the integrity of the organism if prolonged. At other points in the continuum the organism is operating at varying degrees of a tensional state approximating an optimum for maintenance of integrity.

Depending on the phyletic status of the individual, the structural and functional implementations of these adjustments are more or less limited. These adjustments vary also in relation to different stages of development, and are the derivative of all levels of organismic organization. The variety of behavioral patterns reflecting internal and external tensional adjustments is a function of the quantitative and qualitative nature of the stimuli to which the animal is responding. The organism may be responding to the physical, inanimate characteristics of the environment, to configurations of stimuli produced by other living organisms (social stimuli), or to the changes in its own organismic function.

Tensional adjustment is a behavioral continuum based on the concept of the biphasic approach-withdrawal process, as defined by Schneirla,

---

[1] The distinction between *internal* and *external* is of limited use, as it is obvious that the external stimulus is effective only when it is transduced to an internal process. The effective stimulus energy is observable on more than one level of organization: externally it is describable in terms of its physical characteristics; internally it is describable in terms of biophysical, biochemical, and physiological levels. The distinction between *internal* and *external* is helpful primarily in the analytic stages of investigation. The nature of the energy input needs to be dealt with in terms of a process in which the external characteristics are transduced, as a function of such factors as developmental stage, state of organism (e.g., receptor threshold), level of phyletic organization, and the experiential history of the individual. In such a context, the stimuli may be described as "predator," "nursing female," "conditioned stimulus," etc., and thus the external and internal levels of the process become integrated. When information is synthesized to determine causality of behavior, it is impossible to separate the two aspects.

which meets the requirements listed above for a definition of "emotion," and makes the study of emotion feasible.

APPROACH-WITHDRAWAL AS A FUNDAMENTAL PROCESS IN EMOTION

Schneirla's biphasic theory objectifies the study of so-called motivated behavior (see manuscript in preparation by Lehrman and Rosenblatt which evaluates current concepts of motivation). The biphasic theory does the same for the study of emotional behavior (Schneirla, 1959).

The Relationship Between Motivation and Emotion

The terms *motivation* and *emotion* reflect an ancient dichotomization of internal impulsion and external manifestation. Thus, motivation is traditionally discussed as related to *internal* "drive" states, and emotion is studied as *expression*. The study of emotion still suffers from the persistent apposition of the affective and effective aspects of emotion, with the affect being considered something internal and thus not available for objective study. Although much has been written on the mind-body "dilemma," the logical inconsistencies in such thinking continue to plague us in many places. Schneirla's emphasis on behavioral phenomena as processes resolves this seeming dilemma. Emotion is frequently considered the resultant or concomitant of interference with "motivated" behavior. What is motivated behavior? Is there a real distinction between motivation and emotion?

If one considers that the organism is in a constant state of tensional adjustment, one phenomenon that emerges and needs to be explained is the continuous change in quantitative and qualitative aspects of these adjustments. This phenomenon intrigues us: at some point, the animal does something other than what it had been doing. The animal may be lying on the substrate, perched on a limb, or maintaining its location in water and "suddenly" it moves, or changes color, or secretes or excretes a substance. Most behavioral theories that explain these changes by some version of "motivation" require the assumption of some innately determined drive.

The processes responsible for this change in tensional adjustment need to be analyzed. A first step in analysis may be the operational definition of the functional status of systems that we presume are relevant. These may include sensory systems in regard to environmentally based stimuli, sensory systems in regard to internal stimuli, and the relation between these sensory systems and other systems in the organism. Such a distinction between external and internal stimuli is necessary at this first analytical stage. In addition, one needs to know the structural and functional capacities of these systems, as indicated by the animal's stage of development. The data thus gathered permit the analysis of a particular behavior pattern as a process derived from many levels of organization. The fish does not change color because it has been "motivated to be aggressive" but because a behavior pattern emerges from the integration of a complex of processes, among which are those involving the melanophores. This pattern may be termed *combative* on the basis of relevant data.

Is anything gained by introducing the concept of motivation? Perhaps if motivation were defined in the narrow operational terms of the experimental manipulations of deprivation and resulting activity it would be defensible, but not as an explanatory construct. As currently defined, motivation is based on "drives." The biphasic process of approach-withdrawal outlined above goes far toward providing testable hypotheses about changes in ongoing behavior, without requiring a theory or concept of motivation based on innate drives.

Does the biphasic theory also eliminate the necessity for the concept of emotion? Before dealing with this question, it is necessary to consider the problem of emotion as "expression" or "process" or both.

The Relation Between Process and Expression

The traditional emphasis on the expression of emotion before and after Darwin's protean work (1872) is both a reflection of the theoretical positions of the experimenters and a consequence of the adaptive value of the process of emotion in interindividual relationships. Since we are continuously responding to the external evidences of emotion, these became predominant in our thinking. In addition, this emphasis on expression is derived from the fact that the operational approach is helpful in early stages of investigations, particularly in so complex a phenomenon as emotion.

Advances in the techniques and interests in neurophysiology have shifted the emphasis somewhat to process, but again in an operational sense. The questions asked are not usually in terms of process (i.e., the course of something happening, being done or working in a given temporal sequence; a continuing development involving change). Rather, reification of such constructs as "activation" or "arousal" (as originally proposed by Lindsley, 1951) has taken place, leading to a conceptualization of "arousal" as "drive." This tendency in theory construction about emotional behavior has received further support from the neuroanatomists and their emphasis on neural centers for different emotional patterns.

Along with the effects of improved neurophysiological techniques, the progress in applied physics, as in computer technology, has introduced another type of explanatory process, the comparator process (Pribram, 1967; Berlyne, 1969; Simonov, 1969). At times, the comparator function, as described in the literature, takes on the tone of a *deus ex machina,* another example of the unfortunate consequences of the reification of a construct.

The comparator process is related to the activation concept, in that it is generally thought that discrepancies of one kind or another resulting from the comparator process are responsible for the impulsion to behavior and for the expression of emotion. The "discrepancy" hypothesis tends to emphasize those aspects of emotional behavior that are episodic and ignores another characteristic of emotional behavior, which has been emphasized by other theories, namely, the continuous, or modulating nature of emotional behavior (see below).

In addition, the comparator process as the basic process leading to emotional behavior is ascribable only to higher phyla with nervous systems that are capable of the storage and integration of complex information. It is desirable to understand the evolution of emotional behavior, but the comparator process is only helpful in dealing with certain problems in the study of emotional behavior in higher vertebrates and does not pertain to the "emotional behavior" in simpler phyla.

The concept of emotion not only covers the phyletic range, but must deal with the complete range of "negative" and "positive" emotions (Leeper, 1948; Pribram, 1967). In some respects, theories that deal with emotion as a process of behavioral modulation, e.g., Gellhorn's and Loofbourrow's theory (1963), include the full range of emotion in their systems. Modulation is defined in terms of neural systems (primarily autonomic) and are related to mammalian behavior in adult stages, rarely dealing with development. I would like to propose that the concept of emotion as behavioral modulation is valid only when based on the concept of tensional adjustment as an evolutionary behavioral continuum, derived from the biphasic approach-withdrawal process, and placed in an appropriate ontogenetic and phylogenetic context.

The concept of emotion as behavioral modulation derived from tensional adjustment is applicable to all phyla at all stages of development. The proposed concept applies to the entire range of emotion in that tensional adjustment is tied to the definition of the animal's adjustment in terms of the maintenance of its organismic integrity. Tensional adjustment is an emergent continuum extending from passive dysphoria through euphoria to active dysphoria. Euphoria and dysphoria are defined in terms of their facilitation or obstruction of the maintenance of organismic integrity. At each end of the continuum, organismic integrity is obstructed (passive dysphoria and active dysphoria). It should be noted that euphoria and dysphoria are continuously changing into each other as they are contiguous on the continuum.

As the organism is adjusting to maintain its integrity (euphoria), it is also deteriorating and losing its integrity (dysphoria). In other words, they are the contradictories of the dynamic process of tensional adjustment (emotion).

Dysphoria and euphoria emerge as processes that affect other behavioral processes by modifying the valence, or force, of internal and external stimuli, and by modulating the responses of the organism. The system of "response-to-stimuli" that operates in a dysphoric organism is very different from that which operates in a euphoric organism, although the stimuli may appear to be equivalent when objectively measured.

Pribram's point about homeostasis (1967) is well taken and consonant with the discussion above: each homeostatic change does not bring the organism back to the same level, but to a new level, and in the context of tensional adjustment, to a new level of maintenance or organismic integrity. This concept of homeostatic change is also basic to Schneirla's formulation of the developmental process.

Tensional adjustment (emotional behavior), as a process, exhibits both negative and positive feedback phenomena, as do many processes. In negative feedback, the adjustment, which reflects the internal organization and its relation to the environment, results in a change that attenuates the action (internally or externally or both). This type of feedback has been overemphasized because of its obvious relationship to traditional concepts of motivation. In positive feedback, the act results in its further repetition so that the quantitative changes brought about by the repetitions bring about a qualitative change—"explosion"—as in ejaculation during mammalian sexual behavior; combat or fighting behavior; sudden escape or flight after immobility or "freezing." One aspect of positive feedback in emotional behavior has been studied in terms of "pleasure zones" in intracranial self-stimulation, but only insofar as continuous repetition of the act is concerned. It may be that the parameters of this action have not been sufficiently varied to enable us to determine whether it functions as a positive feedback phenomenon.

At all times, the actual performance of the acts that we call emotional behavior are in some type of feedback relationship, either negative or positive. The feedback aspect of actual performance of emotional behavior will affect the likelihood of its performance at another time—depending on the stage and nature of the feedback. However, many other factors will also affect the likelihood of performance. Among these are considerations of the biphasic process of approach-withdrawal.

Approach-Withdrawal Process in Different Levels
of Organization and Integration of Emotion

In the sense that Schneirla's formulation of the biphasic process objectifies the study of so-called motivated behavior, the same is true about emotional behavior. The approach or withdrawal of an organism from the source of stimulation is a function of the history of the organism (phyletic and ontogenetic); of the present state of the organism (for example, hormonal status; existing thresholds as a function of habituation or adaptation); and of the situation in which the animal is observed. All these factors together with the intensity characteristics of the stimuli determine the organism's response. The approach or withdrawal responses are part of the process of tensional adjustment. The modulation of the processes bringing about approach or withdrawal is the emotional process; also, the process of emotion modulates, filters, amplifies, and attenuates approach-withdrawal responses, always within the context of the phyletic characteristics of the organism and its developmental history.

Emotion (tensional adjustment) as a process has certain characteristics: (1) it is contemporaneous with all behavioral processes, e.g., the "vegetative" processes; the higher integrational processes such as reasoning, learning, or human creativity; (2) it is dependent on and correlated with the specialization of the neural and endocrine systems, rather than with other systems; and (3) it is correlated with the level of social organization of the species and the stage of social development of the individual. In addition, the process of emotion may be viewed from the levels of organization listed above (p. 241).

In the phyletic aspect, the process of emotion as defined above is to be contrasted with the traditional concept of emotion. Traditionally, emotion is not discussed on the invertebrate level. Does this mean that emotion, in contrast to behavioral modifiability or social organization, sprang into existence *de novo* in a saltatory fashion with the evolution of the chordate phylum? It is possible that the term *emotion* should not be applied to the invertebrates. However, it may be possible to postulate a behavioral continuum which is present on all phyletic levels, but is manifested in different ways by different species.

For example, the behavioral continuum of tensional adjustment is accompanied at different times and on different phyletic levels by the discharge of nematocysts and trichocysts; the giving off of chemicals; the expansion or contraction of melanophores; changes in the circulation in hemolymph or coelomic fluids. All of these functional modifications evolved along with increasing specialization of receptor and effector systems. They are the *anlage* of the more easily understood and empathized emotional expressions in vertebrates.

In the vertebrates, the striking color patterns or positions of specialized dermal organs and of the appendages lead quickly to anthropomorphic descriptions such as "nuptial dress," "defensive coloration," "threat displays" and so forth when they accompany the approach and seeking behavior of a reproductive or parental animal or when they accompany the approach and withdrawal responses of animals in combat. Yet, these functional modifications in the vertebrates are quite possibly related to the equivalent functional modifications in the invertebrates. In this connection, it is interesting to note that the only invertebrates that are frequently included in this "empathic" anthropomorphism are the social insects and the cephalopods.

In regard to the organismic levels of organization, we have indicated above the necessity for the consideration of neural and endocrine function.

Social organization is of paramount significance in the process of emotion (tensional adjustment). The concomitant evolution of social organization and emotion signifies the adaptive value of emotional expression in the communicative process. The feedback in the communicative process makes for high degrees of specialization and differentiation in emotional expression, dependent on the phyletic status of the organisms. The lower the level of social organization, the more restricted the expressive emotional repertoire.

At simpler phyletic levels, emotional processes are primarily evident in the context of adult-adult behavior, e.g., reproduction and its concomitant patterns such as spatially-referent behavior (so-called territoriality) during mating and parental behavior; aggression between males in relation to the female; inefficient mating attempts with nonreceptive females. At the higher levels of organization, emotion is primarily a social process involving all types of interaction: adult-adult, adult-young, and peer-peer, as most developed in the apes and man.

Intimately related to the aspect of social organization, therefore, and to all aspects of organization of levels within the process of emotion, is the process of development.

THE DEVELOPMENT OF EMOTION

The "Emotionless Organism"

The idea that immature organisms are incapable of all types of emotional behavior has been widely accepted primarily on the basis of work with altricial avian and mammalian organisms. Schneirla has discussed the data showing that it is likely that approach systems are operative in the neonatal bird and mammal in the earliest stages (1959) and has proposed further that the withdrawal systems develop at a later stage.

It is possible that we are dealing here with a technical problem rather than a substantive theoretical one. It may be possible that withdrawal patterns are evidenced on the gross motor level of behavior at more mature stages of development, and at the biophysical and biochemical levels of organization at earlier stages. These lower levels may respond to intense stimuli in such a way that they form the developmental nexus for the emergence of the molar patterns seen later. Early stimulation has a profound effect on the development of the organism (Tobach and Schneirla, 1962); yet these effects are not always seen at the time of the stimulation, but are revealed in later developmental and behavioral phenomena. In view of the adaptive significance of withdrawal behavior, it is not reasonable to believe that the immature organism is completely without mechanisms for responding to intense and noxious stimulation. It is possible that the repertoire for such responses is strictly limited in form and temporal characteristics. This is also true on the lower levels of phyletic organization for the adult organisms. On these levels, the withdrawal patterns are not as plastic as they are on higher levels, as evidenced in the prevalent failure of individuals to survive under adverse conditions.

The assumption of a lack of both aspects of the biphasic process in all organisms at all stages of development needs documentation by further inductive studies.

The Contexts of the Development of Emotion

In the ontogeny of the organism, the level of socialization achieved at different stages is concomitant with the level of emotional development. The socialization process is present in most species because of the intrinsic characteristics of reproduction, varying importantly in temporal and qualitative ways, depending on the phyletic status of the species (Tobach and Schneirla, 1968). The process is dependent not only on adult-young interaction but on peer-peer interaction, again differently organized in different species.

In the human, the socialization process is most complex and persistent from the earliest *ex utero* stages throughout the life of the individual. Concomitant with the stages of socialization, the level of emotional organization is biosocial or psychosocial (Tobach and Schneirla, 1968).[2] As the individual matures, emotion becomes integrated in the highest psychosocial levels, so that two consequences result.

---

[2] The prefixes *bio-* and *psycho-*, as in the words *biosocial* and *psychosocial*, are used by Schneirla as follows: *Bio-* refers to all phenomena that are shared to a greater or lesser extent by all living organisms. *Psycho-* is relevant to phenomena as they are dependent on the level of organization of nervous function, implying that the degree of plasticity of behavior and the complexity of integration with biological systems other than the nervous system places psychological phenomena on a hierarchically higher level than the other biological processes.

The first is that emotion is not only a derivative function of the various organic systems and their organizational integration, but becomes primarily and significantly a derivative function of social relationships, so that spoken and written communication supersede the role of intraorganismic, physiologically based stimuli. The second is that the effects of the emotional process derived from such social relationships are predominant and override the physiological relationships. This last consequence is exemplified by the phenomena of so-called psychosomatic (psychogenic) diseases, physiological manifestations of religious fervor, and martyrdom in the political, social context. The listing of these disparate phenomena in no way suggests their equivalence, for in this category of emotional behavior, there are also levels of organization with their own laws that require their own investigational methods and techniques. Their differences far outweigh their one important similarity: the high valence of ideas derived from psychosocial relationships in affecting physiological levels of organization.

The Development of Approach-Withdrawal Process at Different Levels of Integration of Emotion

The four traditional motivational-emotional behavior patterns are feeding, fear-flight, fighting, and reproduction. Each of these patterns is amenable to analysis in terms of levels. Thus, feeding may be a function of behaving to stimuli on the taxic level (Tobach and Schneirla, 1968), that is, responding to the inanimate and subanimal elements of the environment, such as plants and water. On the biosocial level, it is seen in predation and nursing, depending on phyletic level. On higher phyletic levels, nursing becomes psychosocial (as in delayed weaning) and feeding may also become psychosocial on the human level as in nonorganic obesity.

Fear may be analyzed similarly. On the taxic level, the behavioral phenomenon is withdrawal, flight, or escape. Such withdrawal, flight, or escape may be extremely intense, which may be the reason it is so frequently called "fear." On the biosocial level, the withdrawal process is also a function of intense stimuli, but these emanate from another organism, either as in predator-prey relations, or as in combative behavior, leading to "defense."

Attack on the biosocial level is an intense approach response; on the psychosocial level it is an integration of both approach and withdrawal. The withdrawal in this instance is complexly elaborated on the psychosocial level as it derives from anticipation of future events based on past experience,[3] which may be defined as fear.

---

[3] I am indebted to a discussion with Richard van Gelder of The American Museum of Natural History who brought out this aspect of aggressive behavior in man.

I propose that fear is specific to the psychological level as it implies some integration of stored information or stimulus equivalence in terms of past experience. Anxiety, defined as anticipation of fear, is similarly relevant to the psychological rather than the biological level as it requires a similar integration of stored information or stimulus equivalence (Schneirla, 1957). It is related, however, to experience both with the physical, inanimate environment (taxic) and with stimulation from other animals (biotaxic). Fighting and reproduction are only manifested as biosocial and psychosocial phenomena.

The apparent paradoxical character of aggression as both approach and withdrawal and the apparent ambiguity in the definition of anxiety, as pertinent to the psychological level in terms of both inanimate and animate aspects of the environment, are examples of mesolevels. It is the hallmark of the levels concept that each level contains within itself the possibility of change to a preceding or to a succeeding level. Where both tendencies are strongly in evidence a mesolevel is formed.

The process of emotion varies profoundly at different phyletic levels and at different stages of development. For the first, the phyletic aspect, it seems almost necessary to develop new words to denote the relationship between the emotional processes on lower phyletic levels with the processes on higher levels. Reproductive function in all species is accompanied by a greater or lesser euphoria. The greatest complexity of euphoria concomitant with the reproductive process is evidenced on the human level, and this euphoria is known as various kinds of "love" in human terms.

RECAPITULATION

Emotion is viewed as a process, which evolved as tensional adjustment based on the processes of approach-withdrawal formulated by Schneirla. Evolutionary and developmental considerations based on the concept of levels of organization and integration are cardinal in the view of emotion as a process.

ACKNOWLEDGMENTS

This paper was written while the author was the recipient of a Career Development Award from the National Institutes of Mental Health (K3–MH–21,867).

REFERENCES

Bell, Charles. 1844. *The anatomy and philosophy of expression.* London: Henry G. Bohn.

Berlyne, D. E. 1969. Arousal: Reward and learning. In E. Tobach, ed., *The experimental approaches to the study of emotional behavior.* Ann. N.Y. Acad. Sci. 159: 1059–1070.

Darwin, Charles. 1872. *The expression of the emotions in man and the animals.* London: John Murray.

Gellborn, E., and G. N. Loofbourrow. 1963. *Emotions and emotional disorders.* New York: Harper & Row.

Greenfield, N. S., and W. C. Lewis, eds. 1965. *Psychoanalysis and current biological thought.* Madison: Univ. Wisconsin Press.

Leeper, R. W. 1948. A motivational theory of emotion to replace "emotion as a disorganized response." *Psychol. Rev.* 55: 5–21.

Lindsley, D. B. 1951. Emotion. In S. S. Stevens, ed., *Handbook of experimental psychology.* New York: Wiley. Pp. 473–516.

Lorenz, K. 1966. *On aggression.* New York: Harcourt, Brace & World.

Moyer, K. E. 1967. *Kinds of aggression and their physiological basis,* Report No. 67–12; *Aggression as an internal drive state,* Report No. 67–23. Department of Psychology. Pittsburgh: Carnegie-Mellon University.

Novikoff, A. 1945. The concept of integrative levels and biology. *Science* 101: 209–215; *Science* 102: 405–406.

Pribram, K. H. 1967. The new neurology and the biology of emotion: A structural approach. *Am. Psychologist* 22: 830–838.

Schneirla, T. C. 1939. A theoretical consideration of the basis for approach-withdrawal adjustments in behavior. *Psychol. Bull.* 37: 501–502.

Schneirla, T. C. 1949. Levels in the psychological capacities of animals. In R. W. Sellars et al., eds., *Philosophy for the future.* New York: Macmillan. Pp. 243–286.

Schneirla, T. C. 1952. A consideration of some conceptual trends in comparative psychology. *Psychol. Bull.* 49: 559–597.

Schneirla, T. C. 1957. The concept of development in comparative psychology. In D. B. Harris, ed., *The concept of development.* Minneapolis: Univ. Minnesota Press. Pp. 78–108.

Schneirla, T. C. 1959. An evolutionary and developmental theory of biphasic process underlying approach and withdrawal. In M. R. Jones, ed., *Current theory and research in motivation.* Lincoln: Univ. Nebraska Press.

Schneirla, T. C. 1965. Aspects of stimulation and organization in approach/withdrawal processes underlying vertebrate behavioral development. In D. S. Lehrman, R. A. Hinde, and E. Shaw, eds., *Advances in the study of behavior.* New York: Academic Press. Pp. 1–71.

Schneirla, T. C. 1966. Behavioral development and comparative psychology. *Quart. Rev. Biol.* 41: 283–302.

Simonov, P. 1969. Studies of emotional behavior of humans and animals by Soviet physiologists. In E. Tobach, ed., *Experimental approaches to the study of emotional behavior*. Ann. N.Y. Acad. Sci. 159: 1112–1121.

Skinner, B. F. 1966. The phylogeny and ontogeny of behavior. *Science* 153: 1205–1213.

Tobach, E. 1963. The use of telemetry in the study of the social behavior of laboratory animals. In L. Slater, ed., *Bio-telemetry*. New York: Pergamon Press. Pp. 33–44.

Tobach, E. 1965. Comparative psychology, psychobiology, biopsychology, animal behavior, ethology: What's in a name? Animal Behavior Society Meeting, 1965, Berkeley, California. In ms.

Tobach, E. 1968. Introduction and discussion. In Margaret Mead, Theodosius Dobzhansky, Ethel Tobach, and R. E. Light, eds., *Science and the concept of race*. New York: Columbia Univ. Press.

Tobach, E. 1969. Introduction. In E. Tobach, ed., *Experimental approaches to the study of emotional behavior*. Ann. N.Y. Acad. Sci. 159.

Tobach, E., and Schneirla, T. C. 1962. Eliminative responses in mice and rats and the problem of "emotionality." In E. L. Bliss, ed., *Roots of behavior*. New York: Paul Hoeber.

Tobach, E., and Schneirla, T. C. 1968. The biopsychology of social behavior in animals. In R. E. Cooke, ed., *The biological basis of pediatric practice*. New York: McGraw-Hill.

ELLIOT S. VALENSTEIN
The Fels Research Institute
Yellow Springs, Ohio

Pavlovian Typology: Comparative Comments on the Development of a Scientific Theme

---

"Truth is never pure, and rarely simple."

Oscar Wilde, THE IMPORTANCE OF BEING EARNEST

The comparative approach may not be the only fruitful road to scientific discovery, but an awareness of diversities and similarities among related phenomena may provide the perspective necessary to avoid gross errors in interpretation of phenomenon, while increasing the likelihood of finding an appropriate level of analysis for revealing underlying mechanisms. The application of this approach to a very great range of problems is illustrated most brilliantly in the lifetime of work of the late Professor Schneirla. Schneirla's writings testify to his belief that the comparative approach is applicable to any discipline in which an appreciation of historical and development factors can add to our understanding of present events. With this thought in mind, it might not be inappropriate to trace the history of Pavlovian typology, as a theoretical scheme for organizing behavioral responses that developed in one experimental context, but has come to be applied to different problems and different species.

Pavlovian typology refers to the attempts initiated by Pavlov to order individual differences in behavior in terms of what are presumed to be "basic properties" of the nervous system. This topic seems especially timely because, in spite of little awareness of the similarities, recent theories in neurophysiology, physiological psychology and psychopharmacology seem to have much in common with some of the basic concepts of Pavlovian typology. In addition, selection of this topic was motivated in part by the fact that our increasing contacts with Soviet psychologists and physiologists have made it clear that much of the research in the behavioral sciences in the Soviet Union is influenced by Pavlovian typology and related explanatory concepts. Indeed one has difficulty fully comprehending the Soviet literature without knowledge of this theoretical background and specialized language.

Pavlov, in common with most other students of behavior from classical to modern times, was impressed with the great variety of individual response patterns. He attempted to group such individual differences according to a limited number of properties of what he regarded as the two basic neural processes, *inhibition* and *excitation*. From his extensive experimentation with dogs he derived three fundamental properties of these basic neural processes: *strength, mobility,* and *balance.* It is clear from Pavlov's writings that these properties were not a means of describing behavior but rather were postulated as basic, "genotypical" response characteristics of nerve cells and the nervous system. Nervous system type was derived from the particular configuration of strength, mobility, and balance.

STRENGTH

By nervous system strength, Pavlov meant those characteristics of the nervous system that enabled an animal to resist the disruption of performance caused by excessive excitation. Excessive excitation can be produced by novel, intense, or persistent stimuli, and Pavlov implied that there was some capacity of cells, particularly of the cerebral cortex, that enabled them to function normally while subjected to "strain." For example:

> The importance of the strength of nervous processes is clear from the fact that unusual, excessive events and stimuli of great strength are present more or less frequently in the external environment, so that the necessity often may arise of suppressing or restraining the effects of these stimuli to meet the requirements of other similar, or still more powerful external conditions. (Pavlov, as quoted in Krushinsky, 1960, p. 78.)

Pavlov maintained that suppression of the effects of intense stimulation was achieved by the growth of an inhibitory process that disrupted, diminished, or qualitatively changed the conditioned response. The limit of the working capacity of cells, therefore, was determined by their ability to respond in the face of intense stimulation without becoming inhibited. Understandably, Pavlov was never very specific about the process underlying the strength of cells, but according to Teplov (1964) he did speculate that cell weakness might result from "a small supply of excitatory substance," which is subject to "rapid functional destruction."

It is often instructive to learn how a concept developed. From the early stages of Pavlov's research with the conditioned reflexes, he noted that some dogs

> remained quiet in the new experimental conditions, both when they were placed in the stands mounted on tables, and when certain apparatuses were attached to their skin and even placed in their mouths. When food was given to them by means of an automatic device, they began to eat at once. (Pavlov, 1957, p. 316.)

In contrast, other animals had to be accustomed very gradually to the procedure, a process which often required many weeks. With every experimental modification the conditioned responses of such dogs tended to disappear. Perhaps some of Pavlov's own frustration was reflected in his choice of words as the former dogs with "strong nervous systems" were frequently described as "bold" or "heroic." In contrast, the latter were often referred to as "cowardly" or "timid" animals. In most of Pavlov's writings the strong nervous system is referred to in positive terms while the weak nervous system is referred to with negative connotations. Teplov (1964), however, makes it clear that at least some of Pavlov's students did not place a value judgment on this basic nervous system process. Investigators associated with Teplov have shown that while a strong nervous system may be better equipped to cope with intense stimuli, a weak nervous system may possess greater acuity as reflected in sensory threshold measures. A weak nervous system may be more vulnerable, but it is also more sensitive (Nebylitzyn, 1964). A strikingly similar view was expressed by the performer Barbra Streisand: "If my vulnerability goes in real life, it goes as a performer and an artist on stage too. I must retain the vulnerability or lose sensitivity as an artist" (*Life,* March 18, 1966, p. 96B).

The techniques for measuring the strength of the nervous system evolved from a very crude, preliminary era to a period of more precise quantitative methods. It is almost amusing to read descriptions of some of the early techniques. Teplov (1964), for example, describes the work of L. N. Fedorov in 1923:

> One of the workers in the laboratory would run into the experimental room dressed in a sheepskin coat, turned inside out so that the fur was on the outside, and wearing a mask, and would shake a rattle, bringing it closer and closer to the dog, or bang on a sheet of iron with a hammer, or produce a flash by setting fire to a heap of smokeless gun-powder next to the dog; a motor horn would be sounded just under the dog's stomach; the dog's flanks would suddenly be squeezed together by planks; or the dog would be suddenly raised in the air, while at the same instant an electric current was applied to its hind paws.

At times it appeared that a clinical judgment was used, as in Walter's (1953) description of the Russian physiologist, Rosenthal, during the latter's visit to Cambridge: ". . . he would snap his fingers till the dog came to him, then suddenly would yell at the top of his voice; if the dog paid more attention to the louder noise it was strong; if it reversed its behavior and ran at the yell it was weak."

Krushinsky (1960) describes similar tests for determining the strength of the nervous system. While the dogs were eating, the sound of a rattle or an automobile horn was presented. Classification was based on a judgment of the extent of interference with eating. Reactions of the animals ranged from refusal to eat when either stimuli was presented to unperturbance even by the automobile horn.

Throughout this same period a battery of more or less standardized tests for determining strength and the other nervous system properties was gradually developed. The more comprehensive battery is referred to in the Soviet literature as the "large standard" while the abridged version is called the "small standard." Both series of tests require a considerable period of time and Teplov (1964) estimated that determination of the type of nervous system in a dog requires a year and one-half with the "large standard" and even the "small standard" requires six months. It is probably because of this time factor that much of the relevant work in the Soviet Union today gives only a partial description of nervous system type.

Several types of tests are used in measuring the strength-weakness dimension of the nervous system. One test measures the ability of the organism to continue to respond with prolonged excitation. This method, which is often referred to as "extinction with reinforcement," determines the number of rapid presentations of a conditioned stimulus (CS) followed by reinforcement that can be made before the response magnitude is significantly changed as a result of what is said to be an exhaustion of the excitatory process. The large variability, and at times conflicting results with this method, account for the fact that it is not too frequently used today with animals. However, Teplov and Nebylitzyn (1966) report that this method, which they believe reflects the endurance of the nervous system as it relates to arousal, is most useful with human subjects. These authors have applied the "photochemical conditioned reflex" (PCR) method in combination with the "extinction with reinforcement" technique. The PCR refers to the fact that a previously neutral stimulus can elicit a decrease in visual sensitivity if it is paired with an unconditioned stimulus (US) such as a light flash to a subject's eye. Experiments with the pupils held constant indicate that part of this effect may result from pupillary conditioning, but it is believed that at least a portion of the decrease in sensitivity may be produced by a conditioned destruction of visual purple. In the procedure used by Teplov, Nebylitzyn, and their associates, the CS is followed by light-flash reinforcement a number of times in rapid succession. A measure of the persistence of the conditioned response is obtained by testing for absolute threshold after the series is completed. Details of the method for threshold determination are described by Gray (1964).

More commonly, strength is measured by determining the range of stimulus intensities that will continue to elicit an increasing magnitude of response before an inhibitory process alters this tendency.[1] This method is frequently employed in combination with drugs, such as caffeine and bromides, or food deprivation to alter the excitability of the nervous system. These "artificial" means of modifying nervous system excitability serve to exaggerate the differences between weak and strong nervous systems.

It would not be possible to describe in detail all of the methods that have been used to determine the strength of the nervous system. However, this report may take on more meaning if several of the more common techniques used in Soviet laboratories in the past and at present are described.

Individual differences in responsiveness to drugs are a problem faced in one way or another in most laboratories and clinics. The effort in the Soviet Union to correlate nervous system types with such differences is an approach to this problem not generally represented in the United States. It is of some historical interest to recall that the possibility of applying conditioned reflex technique to pharmacology was conceived at a relatively early date in Russia (Zavadsky, 1908). Today, in laboratories of pharmacology in the Soviet Union, the distinction between "strong" and "weak" nervous systems often provides some degree of predictability of the effect of a drug. Variations of the "caffeine test" have been useful for this purpose. In small doses caffeine will increase positive responses in all animals, but with higher dose levels a "transmarginal inhibition" (also referred to as "ultraboundary inhibition") develops. This is manifested by a decrease in the magnitude of response. This special inhibition is believed to be a

---

[1] The concept of inhibition growing out of excitation is an omnipresent theme in the Russian physiological school and was studied especially by Wedensky and his student, Ukhtomsky, under the name of "parabiosis." The term parabiosis as used by Wedensky indicated that the inhibition which developed was related to a basic protoplasmic reaction bordering on life and death, but reversible if the parabiotic factor was removed. Later, Pavlov (1928) described the changes which occur when a stimulus is made progressively more intense. Up to a point the response becomes increasingly greater until an intensity is reached where differentiation starts to break down and a positive stimulus may produce a response that is only equal to that produced by a negative stimulus (equivalent phase). At higher intensities a strong or positive stimulus may produce a weaker response than the negative stimulus (paradoxical phase). At still higher intensities, a positive stimulus may have no effect, but a previously elaborated negative stimulus may elicit a large response (ultraparadoxical phase).

reaction to heightened excitation and of a protective nature. The maximum dose of caffeine benzoate that can be tolerated before inhibition is produced may be determined. Dogs with strong nervous systems are said to tolerate up to 125 mg/kg of caffeine while those with weak nervous systems may show inhibition at 30 mg/kg. This caffeine test is said to be particularly discriminating with rabbits in which 240 mg/kg and 40 mg/kg are tolerated by animals with strong and weak nervous systems, respectively.

The strength of the nervous system may also be evaluated with intense stimuli. The procedure may involve pairing a positive stimulus with food presentation and a negative stimulus with the withholding of food. When responses (usually salivary) are stabilized, the intensity of the stimuli (usually tones of two frequencies) are increased in 10 db steps. The first reaction to the increased intensity is a disinhibition manifested by a salivary response to the negative stimulus, and after a number of trials, inhibition returns and responses are again stabilized. The intensity is increased again and the process is repeated until a "point of stable disinhibition" is reached, where the response to the positive stimulus may be two and one-half times the original base level and the response to the negative stimulus may equal the initial positive conditioned response. At this intensity the inhibition of the response to the negative stimulus does not return. Dogs with strong nervous systems reach stable disinhibition at intensities that may exceed 125 db, while animals with weak nervous systems may reach this point at 85 db or lower. It is concluded from these and similar results that the capacity to cope with the excitement caused by the intense stimuli is greater with a strong nervous system.

A mathematical model that attempts to depict the changes in strength of the inhibitory and excitatory processes as a function of stimulus intensity has recently been proposed (Savinov et al., 1963). The basic assumption of the model is that when the nervous system is excited by any stimulus, an inhibitory process is also brought into play through a negative feedback mechanism. With increasing stimulus intensities, the growth of the two processes is not parallel. With moderate increases in intensity, excitation grows proportionally larger than inhibition and this accounts for the observed enhancement of the response. At very high intensities, the inhibitory process increases exponentially and may dominate over excitation.

MOBILITY OR LABILITY

By the dimension mobility or lability, Pavlov meant the capacity of the nervous system to react rapidly to stimuli. In much work, particularly in the past, the emphasis has been placed on the ability of the nervous system to react to a change in the significance of stimuli. For example, when positive and negative conditioned stimuli are reversed, subjects with "labile nervous systems" reflect this change rapidly while subjects with "inert nervous systems" tend to perseverate, that is, to persist in responding in ways that are no longer appropriate to the new significance of the stimuli. This technique is often referred to in the Soviet literature as the "method of bilateral revision of the signal significance of an associated pair of conditioned stimuli." The mobility of the nervous system was commonly believed to be dependent on whether alternation between inhibition and excitation was capable of occurring quickly, but as we shall see, other interpretations have also been advanced.

One interesting method of evaluating the mobility of the nervous system employs the technique of "dynamic stereotype" formation. Positive and inhibitory ("differential" or nonreinforced) stimuli are presented in the same sequence over many trials. A stereotype is formed when the position in the sequence as well as the stimuli themselves takes on significance. This can be demonstrated by observing the change in response level when a neutral "test" or "indicator" stimulus is presented in the place of a positive stimulus. Skipin (1938) and Asratyian (1938) have used the stereotype to test the mobility of the nervous system by presenting intense and mild stimuli in different positions in the stereotype. It is known that within limits a strong or intense conditioned stimulus produces a greater response than a low intensity stimulus ("law of strength"). In animals with a high degree of mobility the direct effect of the stimulus is most important in determining the response; where there is a great amount of inertia of the nervous process the place in the stereotype is dominant.

A modification of the dynamic stereotype method is commonly used in Soviet laboratories to produce experimental neurosis. The so-called collision technique is believed to result from the strain produced by an attempt to substitute excitatory and inhibitory processes too rapidly. After a "dynamic stereotype" is stabilized an inhibitory stimulus is presented in the place of a positive stimulus. A few presentations of an inhibitory stimulus in the temporal position normally occupied by a positive stimulus may produce experimental neurosis characterized by a prolonged elevation of blood pressure (hypertension) and a decreased magnitude and increased variability of conditioned reflexes. Recently, Suvorov and Danilova (1965) reported an experiment using this technique with dogs of different nervous system types which are described characteristically by such phrases as "limiting dose of caffeine = 0.3 g" and "bilateral alternation occurred in the course of 15 trials." In addition to demonstrating that the collision method could produce prolonged hypertension, Suvorov and Danilova were able to produce "experimental neurosis" with stimuli in different sensory modalities ("analyzers"). They have concluded that the nervous system strain produced by this method does not take place in afferent pathways, but probably occurs in the "cortical representation of the unconditioned reflex."

Soviet conceptions of the mechanism(s) underlying mobility are not at all clear as individual investigators tend to stress very different aspects. In different articles, mobility is said to depend upon the speed with which the nervous process is initiated, the speed with which it is terminated, the rapidity of substitution of inhibition for excitation and vice versa the speed of spread ("irradiation") and concentration of inhibition and excitation (judged primarily by stimulus generalization tests), etc. With respect to the first two interpretations of mobility, Moldavskaya (1966) quotes Kupalov as having said: "If we draw an analogy with neuromuscular physiology, then by lability we would understand the highest frequency with which I could move my legs, and by this frequency measure the lability." It can be seen why such measures as critical flicker fusion (CFF) and duration of after-images have been found useful in studies of nervous system mobility. Soviet scientists have also utilized procedures similar to the Continuous Performance Test (Rosvold et al., 1956) used in this country with brain-damaged subjects. Several forms have been used. In one experiment a movie camera is used to present letters to subjects. The task is simply to press a button when the letter *S* appears, but not to press the button if this letter is preceded by an *N*. When the letters are presented at increasing speed, the task becomes more difficult especially when two *S*'s follow each other or an *S* follows an *NS*. It is believed that this task is tapping a component of mobility such as the temporal limit in substituting inhibitory and excitatory processes.

As with the strength dimension, Pavlov's writings seemed to imply that one end of the mobility continuum should be regarded as a positive and the other end a negative trait. A labile or mobile nervous system is good while an inert nervous system is bad. However, Teplov (1964) points out that there are positive aspects of an inert nervous system that are manifested in such traits as persistence, and in this context Teplov quotes Kantorvich: "The mobile type quickly forms new conditioned connections and quickly destroys them when they cease to be reinforced. . . . Inertness is typically shown in a relatively slow formation of conditioned connections, which, on the other hand, remain very stable. . . ."

BALANCE

The third dimension, balance or equilibrium, denotes the relative dominance of excitation and inhibition. Historically, the vigor or quality of the response to the positive signal as well as the latency indicated the force of the excitatory process, while the relative absence of response to negative (nonrewarded) signals reflected the degree of inhibition present. The ease with which differentiation between positive and negative stimuli was established and the steepness of the stimulus generalization curve also indicated the force of the inhibitory process. In a given animal, inhibition and excitation might be of equal (balanced or equilibrated) or unequal force and in the latter case, one or the other process might dominate.

In the past, a number of problems associated with the concept of equilibrium have not been faced directly. Teplov and Nebylitzyn (1966) have pointed out that no system of measuring the relative strength of the inhibitory and excitatory processes in comparable units has ever been established. Nebylitzyn (1966) has noted recently that equilibrium has descended to a role of secondary importance as a property of the nervous system. This is true because many of the measures of equilibrium are highly correlated with measures of nervous system strength and therefore do not seem to reflect an independent property of the nervous system. Lately, those Soviet investigators working primarily with humans have tended to use such indicators as the rate of formation of both inhibitory and excitatory reactions and the number of errors occurring during inhibitory and excitatory trials. Such measures are not correlated with strength of the nervous system, and it is proposed, therefore, that such tests are tapping a primary property of the nervous system and equilibrium should be considered a secondary property (Teplov and Nebylitzyn, 1966; Nebylitzyn, 1966). The term *dynamicity,* defined in terms of the ease and speed with which excitatory and inhibitory reactions are formed, would replace balance or equilibrium as the third component in the classification system.

PAVLOVIAN TYPOLOGY

According to Pavlov, temperament is the most general property of the nervous system which modulates all activity. Initially, Pavlov anticipated that the particular pattern of the three basic properties of the nervous system would enable one to describe the temperament of an animal or man rather fully. Pavlov always referred to four basic types of nervous systems but the selection of this number seemed to have been arbitrarily influenced by the Hippocratic classification of choleric, melancholic, sanguine, and phlegmatic. MacMillan (1963) has pointed out that of the many possible mathematical combinations of the three basic nervous system properties only four have been observed. Probably because of the great expenditure of time required, the general tendency has been to evaluate only one or two of the nervous system properties and to correlate these with performance under the influence of some experimental variable which had been singled out for study. Many investigations, for example, attempt to relate only nervous system strength with reaction to drugs without attempting a more complete description of the "personality" of the subject.

There are exceptions to this trend. Leites (1956), for example, maintains that a study of tenth grade male students in a Moscow school reveals individuals that seem to be almost "pure" cases of Pavlovian types.

> The mental work characteristic of Viktor M. indicates the fact that he can be categorized as belonging to the strong, balanced and mobile type of higher nervous activity, according to I. P. Pavlov's classification. The strength of his excitatory process is evidenced by a constant readiness for energetic intense work. . . . The strength of the inhibitory process comes out appreciably in Viktor's power of concentration, in his resistance to distracting stimuli. . . . (P. 42.)

It may be of some interest to learn how Pavlov's nervous system would be described by this system. Dr. F. P. Mayorov worked with Pavlov for many years and also edited the notes of his weekly talks to the staff ("Pavlov's Wednesdays"). In describing Pavlov to me, Dr. Mayorov volunteered the information that: "Pavlov's nervous system was strong and labile and excitation dominated over inhibition."

In addition to these three dimensions common to animals and man, Pavlov later elaborated a further distinction that applied only to man. The so-called first and second signal systems distinguished separate capacities for analyzing stimuli that were either primarily sensory or symbolic in nature, respectively. These two signal systems were said to be relatively balanced in the average person. In "artistic" individuals the first signal system dominated, while in the "intellectual" or "ideational" type the second system was most highly developed. Tests to determine the relative dominance of the first or second signal system involve various ways of determining the ease of transferring a conditioned response elaborated to an object to the verbal symbol for that object and the converse. For example, it is reported that in those individuals in which the second signal system dominates a conditioned response to a metronome stimulus is relatively easily transferred to the word "metronome." These procedures are referred to as "verbomotor" techniques and in the past have been particularly evident in the work of Ivanov-Smolensky and Krasnogorsky. A description of some of the testing procedures used to determine the balance of the signal systems is presented by MacMillan (1963) and by Brožek (1964) in his review of recent developments in Soviet psychology.

APPLICATIONS OF PAVLOVIAN TYPOLOGY

It would be impossible to summarize completely the present work in the behavioral sciences ("physiology of higher nervous activity") in the Soviet Union that is influenced by Pavlovian typology. During a five-month visit there in 1961, this theme seemed to me to be omnipresent in almost all laboratories except those concentrating on "pure" neurophysiology. Some appreciation of the variety of applications of typological thinking may be gleaned from a description of a few observations made by the author.

The reports from Roumania which have claimed promising results with procaine when used with the aged or patients suffering from bronchial asthma and allergies have aroused interest in several Soviet laboratories. The reaction to procaine in animal experiments appears to be highly correlated with type of nervous system. At dose levels used for minor surgery (2 mg/kg administered subcutaneously), excitation manifested by enhanced responding is seen in all dogs. However, the effect persists for several days in animals with strong nervous systems, but is limited to the day of injection with animals whose nervous system is classified as weak. With higher dose levels (5 mg/kg), animals with strong nervous systems continue to exhibit excitation while responses are depressed in weak animals. At very high levels (20 mg/kg) the responses of all animals are depressed, but this effect may last only one day with strong animals compared to several days with weak animals. There is a great amount of skepticism in the United States about the effectiveness of procaine treatment, but it is at least a logical possibility that averaging of results may conceal differences in responsivity.

It may be of some interest also to note that work with ginseng has been proceeding in a similar direction. The root of the ginseng plant, *Panax quinquefolium,* has been extensively used by the Chinese for its presumed medicinal and psychic properties (Lawrence, 1964). The Far Eastern Division of the Academy of Science of the Soviet Union, which sponsors a ginseng plantation, has extracted a substance from the roots of this plant. In 1961, prior to the political split between the Soviets and Chinese, it was claimed that both excitation and inhibition were strengthened by this substance as manifested by an increased response to positive stimuli and by improved performance in tests of differentiation. At that time, only preliminary correlations with nervous system typing had been obtained, but results suggested that response characteristics might be distinguished on this basis. It would be most interesting from the viewpoint of the sociology of science to learn if this work was being continued at all and if success was still being claimed.

In addition to work with procaine and ginseng, I observed a great number of pharmacological investigations correlating the effects of tranquilizers, stimulants, and sedatives on animals with different nervous system types. A typical application of Pavlovian typology to a pharmacological problem can be seen in the work of Dr. A. M. Nuzhina (1961). Her problem concerned the determination of the "optimal" dose level of the stimulant, phenamine. The criterion used for optimal level was that dosage which would produce maximal excitation (enhanced responding and shorter latencies) without disrupting the inhibitory process and thereby interfering with differentiation. Dr. Nuzhina concluded that the optimal dose level of phenamine was dependent upon type of nervous system.

In many laboratories there is a considerable investment in the problem of change in higher nervous activity after exposure to radiation. Typological thinking often dominates this research as exemplified in the work of Ayrapetyants (1963). Gravid dogs were subjected to 200 r and their offspring were studied up to 1.5 years after birth by the method of "bilateral revision of signal significance." Both conditioned salivary reflexes and the "shake-off" reflex (shaking of animal's head in response to ear stimulation) were employed. When the signal significance was reversed, complete revision of the responses was generally observed during the eighth training day with normal puppies, but complete revision did not occur at all with the irradiated animals. From these and similar results it was concluded that radiation had significantly reduced the mobility of the nervous system as a consequence of a weakening of the inhibitory process. A partial bibliography of other Soviet studies of behavioral change induced by radiation may be found in the review by Brožek (1964).

Considerable effort is also expended, at least in some laboratories, in developing new techniques for analyzing nervous system typology. A few examples from the Laboratory of the Genetics of Higher Nervous Activity (Pavlov Institute of Physiology) may illustrate the nature of these investigations. Besides measuring the quantity of saliva during conditioning, an analysis of its constituents may also be undertaken. During my visit to this laboratory, attention was being directed to the nitrogen content of the saliva. The nitrogen level of saliva is normally high while eating, but if acid is introduced into the mouth the level is significantly diminished. Dogs with strong nervous systems, however, show a resistance to acid effect; nitrogen concentration in the saliva may remain constant for about ten minutes. Often a procedure of alternating appetitive and defensive stimuli is employed whereby food and acid are introduced into the mouth on successive trials. Dogs with strong nervous systems show a seven- or eight-fold difference in nitrogen content of the saliva during the two conditions while animals with weak nervous systems may only show a four-fold difference. Apparently, the persistence of the influence of the acid trials is inversely related to nervous system strength. Dr. Krasussky viewed these findings as evidence that the capacity to change in this situation is much greater with strong than weak nervous systems, but hastened to add that results indicate that the nitrogen changes of the saliva are correlated only with the strength of the nervous system, not its mobility.

Another method being explored in this laboratory involves following the reaction to skin irritation induced with various substances. Studies of skin reaction (rating extent of redness and blister formation) indicate that with a strong nervous system the reaction is more marked, but recovery more rapid. Sterile skin abscesses are also produced and the eosinophilic change has been found to correlate with strength of nervous system. As with skin reaction, the eosinophilic change is of greater magnitude, but shorter duration, with the strong nervous system. These studies are being extended to include correlations between nervous system type and endocrine response (particularly ACTH) to various somatic stresses.

In the Laboratory of the Physiology of Nervous System Typology (Pavlov Institute of Physiology), Dr. V. K. Fedorov heads a program that is concerned with typology and resistance to disease. Carcinogenic substances are introduced into a strain of mice (C57) that show little tendency toward spontaneous tumors. It was found that labile animals are much less likely to develop tumors than animals with inert nervous systems, but if tumors are produced the survival time is much shorter. Similar conclusions were reached with another strain of mice in which there is a very high incidence of naturally occurring mammillary tumors.

In the Laboratory of Cortical-Visceral Physiology (Pavlov Institute of Physiology) I was reminded that Pavlov believed that the elaboration of trace conditioning was very useful in studying typology and had claimed that this method best illuminated the characteristics of the inhibitory and excitatory processes. In this laboratory the conditioned stimulus (CS), usually a tone, was presented for three minutes and followed by the immersing of the hand in 6° C cold water (US). A finger plethysmograph recorded the vascular response. In the initial conditioning stage, "normal" individuals exhibit a response to the tone during the entire period of presentation. Later, with the development of inhibition, the response is seen only at the end of the CS and to the US. With patients suffering from hypertension (200/110 mm Hg) and with ulcer patients (there are some distinctions between the two) inhibition does not develop and the response continues throughout the total CS presentation. If an interval between the CS and the onset of the US of one or two minutes is introduced, inhibition may be seen, but it tends to disappear with any additional disturbance. Psychiatric patients are also tested with this procedure. Although the results were only exploratory, some data suggested that schizophrenics exhibited a greater response to the CS than to the US. The results with the trace conditioning method are believed to be particularly sensitive to the influence of pathology on the balance between inhibitory and excitatory forces. It will be recalled that as early as 1925, Luria's research (1932) suggested that with humans, conflict interfered with the inhibitory process ("functional barrier") which normally blocked discharge from the affective into the motor areas.

In general, workers in this field have been rather slow in applying electrical recording methods to the problem of typology. This was probably attributable to Pavlov's skepticism about the usefulness of EEG techniques. In this connection, it may be of some historical interest to note that the late Professor Kupalov blamed himself for Pavlov's antagonistic attitude toward the EEG. In a conversation with the author in 1961, Kupalov remarked that he "tactlessly" suggested to Pavlov that the EEG might be a more sensitive measure than the conditioned salivary response. Pavlov apparently reacted very heatedly to this remark and did not give any encouragement to students who expressed an interest in the EEG. More recently, electrical recording techniques have been applied to the problem of typology. Suvorova (1966), for example, has reported a correlation between "persons with predominant inhibitory processes" and the presence of a high alpha index and a resistance to depression of the alpha rhythm in response to an auditory stimulus. The electrical activity of the cortex of such persons exhibits a two-peaked curve with peaks in alpha and delta rhythms. Persons with predominance of the excitatory process exhibit only a peak in the alpha rhythm range and a relatively low alpha rhythm index following an auditory stimulus.

Another example of an application of EEG technique to typology can be seen in the work of Golubeva (1966). This investigator has correlated nervous system strength and lability with the tendency of the brain to follow the rhythm of photic flashes (photo-driving). While results are somewhat dependent upon the frequency of stimulation employed, Golubeva has concluded that the correlation between the photo-driving effect and lability is positive, but a negative correlation exists with nervous system strength.

Another problem frequently discussed in the Soviet literature concerns the relative contribution of genetic and environmental factors to nervous system type. As many of the purebred strains of dogs were lost during the Siege of Leningrad, there is actually very little relevant genetic information available. It was quite surprising for the author to learn how little data had been collected on this problem in view of the many claims of experimental results issued during the "Lysenko era." Studies in the Laboratory of the Comparative Ontogenesis of Higher Nervous Activity (Pavlov Institute of Physiology) provide one approach to this problem. This Laboratory, headed by Dr. G. A. Obrastzova, is especially interested in the development of inhibition, and here, too, trace conditioning is employed. Salivary conditioning is used with new-born puppies in which the parotid ducts are externalized. The conditioning stimulus is presented for periods as long as three minutes prior to the US (food or 0.5% HCl). The US is introduced into the corner of the mouth via a tube clipped to the cheek. Young puppies give evidence of strong internal inhibition and frequently fall asleep after twenty seconds of tone (CS) presentation. According to Dr. Obrastzova, the work provides no evidence of the presence of different nervous system types in dogs prior to three to four months of age. Attempts to strengthen the nervous system through a special training regimen, which systematically introduces an animal to tasks of increasing difficulty, did not seem to have advanced beyond the pilot, experimental stage. On a conceptual level, this systematic scheduling of exposure to "frustration" seemed quite similar to Skinner's (1948) views on the training of children in an utopian society expressed in his novel *Walden Two*.

Leites (1966) has discussed the difficulty of categorizing the nervous system of young children. In general, it has been concluded that 8- to 12-year-old children exhibit a predominance of excitation over inhibition, but older children (13 to 16) no longer reveal this tendency. One procedure from which this conclusion is drawn involves measuring motor reaction time. With older children and adults the introduction of both positive and inhibitory conditioned stimuli between trials increases the reaction time on the subsequent trial. With the younger children this same procedure decreases reaction time. As a result of these and similar changes with age, Leites stresses the necessity of having reliable normative data before attempting any evaluation of nervous system type in children.

The ontogenetic approach is extended to work with the aged. At the Institute of Pathology and Therapy in Sukhumi, there existed in 1961 a program of testing the aged, who were in a special home in that resort city. Many of these elderly people are from Georgia or from Azerbaidzhan, people who are noted for their longevity (Young, 1961). While visiting this Institute I watched a 119-year-old Georgian being given an EEG examination. The not too surprising remark was made to me that the nervous system of older people like older monkeys becomes increasingly inert. Pavlov (1957) had also commented on the decline in the mobility of the nervous system in old age.

Work with the aged as well as psychiatric patients has been in progress at the Bogomolets Institute of Physiology in Kiev for a considerable period of time. In the Laboratory of the Physiology of Aging, studies of oxygen metabolism have been in progress for a number of years. This work was originated by the late Dr. V. P. Protopopov, who is considered by the people at this Institute to have been the first person to use conditioned reflex methods with psychiatric patients. There is the belief, which is supported by some evidence, that oxygen metabolism of the brain will influence the balance of inhibition and excitation. Dr. Kolshinskaya, who was writing a book on this work and was my informant, reported that arterial and venous studies have demonstrated a lowered oxygen level in the brain of schizophrenic, particularly catatonic, patients. These patients have been exposed to low pressure chambers to induce hypoxia. Metabolic adjustment to the oxygen deprivation will persist for some time when returned to normal pressure conditions, but it is difficult to use pressure chambers for prolonged exposure. To further these studies, medical stations have been established at different heights on Mt. Elbris in the Caucasus. Patients stay at these stations for several months. At a height of approximately 10,000 feet, internal inhibition is said to be decreased. At higher altitudes both excitation and inhibition may be depressed. At very high altitudes there is a loss of consciousness, but schizophrenics can stand higher altitudes than can normals. The aged and infants can also withstand these thin atmospheres. At 10,000 feet, catatonics may begin to move and patients with "psychogenic mutism" may start to talk. Improvement often is most evident after the patients have returned to the lower altitude of Kiev, but according to Kantorovich (in Akhmedov, Kalinicheva, and Lorents, 1963), remissions are not equally likely with all psychopathological syndromes. This author claims that in cases of manic states, recent paranoids, and in certain forms of schizophrenia, a sojourn at high altitudes produces excellent results.[2] However, no improvement was seen with various depressed, stuporous, or epileptic patients.

A recent report (Zvorykin, 1964) is typical of many experiments relating nervous system type to changes occurring under different barometric pressures. Nine dogs were studied with a somewhat modified "small standard" method. The tests included elaborating reflexes with caffeine, bromine, during starvation, and with intense stimuli. The methods of "alternation of the biological significance of the positive and inhibitory

---

[2] It may be of interest to note that long-term sojourns in the Andes mountains (6,200 meters) have been recommended for pilots (Klein, Bruner and Jovy, 1963). Results indicate that several weeks at these altitudes produce a marked adaptation in the unspecific hypophyseal-adrenocortical system and an improvement in the reaction of circulation, respiration and metabolism. A summary of recent Soviet physiological and psychological work in this area is reported in a book edited by Akhmedov et al., 1963.

conditional stimuli" and the dynamic stereotype with a "test" or "indicator" stimulus were also utilized. The experiments were performed in a pressure chamber with rarefaction of the air corresponding to altitudes ranging from 3,000–24,000 feet above sea level. It was concluded that the effects of exposure to rarefied air depend upon nervous system type. Among the findings were listed evidence that the reaction of a "differential stimulus" does not change with dogs with a strong inhibitory process; with dogs rated as possessing great mobility there is little change in this dimension, but with dogs whose nervous system was typed as somewhat inert, this inertness was exaggerated by high altitude simulation. The rate of recovery after descent also depended upon nervous system type.

CURRENT TRENDS

It is difficult to predict the future course of Pavlovian typology in the Soviet Union as this field is very much in flux at present. In the past, typological thinking has influenced studies which have encompassed such divergent fields as egg laying in chickens, lactation in cows, prognosis for improvement of schizophrenic patients, and the selection of suitable occupations and determinants of style of work (K. Gurevich, 1966; Merlin, 1964), and the classification and testing of brain injured individuals (Pevzner, 1959; Luria, 1963). In general there appears to be a trend away from conditioned salivary procedures in favor of conditioned motor responses both "classical" and what would be called "instrumental" in this country. Dr. B. Gurevich of the Pavlov Institute of Physiology, for example, described his studies of learning curves derived from food-searching behavior in an open and unfamiliar field. Initial searching patterns as well as adjustment to repositioning of the hidden food were correlated with nervous system type. However, particularly in Leningrad in 1961, one could still observe experimentation utilizing the salivary method in Pavlovian "cameras," which had almost a museum-like quality. My escorts were constantly commenting: "This is the same table Ivan Petrovitch used"; or "Here is the chair Ivan Petrovitch sat on while observing the experiments." Instrumentation consisting of hand operated pneumatic devices for turning food trays and manual switches for presenting stimuli were often the same equipment used by Pavlov's assistants.[3] The large "Opera-

---

[3] The reasons for this conservatism toward new instrumentation are complex and only in part to be explained by economic considerations. One other contributing factor is that animals tend to be studied for considerably longer periods than is customary in the United States. Any changes in the equipment, even those as minor as cleaning the dogs' harness straps, may weaken responses due to "external inhibition." Demonstrations frequently did not work during my visits and I became quite accustomed to the appellation, "external inhibitor." Also to be considered is the fact, referred to earlier, that Pavlov was not very favorably disposed toward electronic equipment.

tion Schedule Book" which I believe was still in use had entries by Pavlov and Podkopaev.

Newer devices, however, are being developed and will probably be used for salivary conditioning studies in the future. Although not yet in operation during my 1961 visit, I did see under construction new Pavlovian cameras which would incorporate automatic recording and programming features. Dr. A. I. Vasiliev of the Laboratory of Interoceptive Conditioned Reflexes (Pavlov Institute of Physiology) has designed equipment which would make possible the collection of human saliva separately from each side of the mouth. With Dr. Vasiliev's device, fluids may be introduced and withdrawn automatically and the mouth may be rinsed after each stimulus presentation. Absolute saliva level and rate of flow can be recorded continuously on a polygraph. It is possible that new instrumentation may revive interest in salivary conditioning. Judging from the comments of the younger investigators in the Soviet Union, however, there is a strong feeling that instrumental conditioning combined with electrical recording techniques are more promising. I detected little enthusiasm and some antipathy among the younger scientists for the salivary methods.

In the past, much of the antagonism to Pavlovian typology in this country stemmed from opposition to a theory which was considered to be based on two hypothetical central nervous system processes (Razran, 1965). There is little doubt that there were many good reasons for this antagonism. Even those only casually acquainted with the Soviet literature on higher nervous activity should be well aware of a kind of *deus ex machina* reasoning resulting from attempts to subsume such a great range of phenomena under a limited set of concepts. This tendency, exaggerated by political intrusions, frequently produced very poor science. Many of the shortcomings of Pavlovian typology are those common to any classification system. Phenomena are usually not divided up in discrete packages in nature and more specifically, it is certainly questionable whether all the potential of the nervous system can be summed up by activity of the basic processes of excitation and inhibition. Furthermore, one would certainly have to question the assumption that strength, mobility, and equilibrium (or "dynamicity") actually represent the fundamental properties of the nervous system. It does not seem likely that these are the only relevant dimensions that could be measured, nor is the evidence convincing that these are even unitary concepts, which are independent of each other. If strength, mobility and equilibrium were truly independent of each other, one would expect to find many more types than have actually been reported. Davidenkov (1947), for example, reported that dogs with strong, inert, and equilibrated nervous systems are extremely rare.

It would be unfair not to note that many of these criticisms have also been raised recently in the Soviet Union. The late Professor Teplov (1964) and his colleagues have stressed the need for different experimental techniques to measure nervous system properties and they have applied factor analysis methods to determine the relationships between different tests. This analysis has clearly indicated that many characteristics that were traditionally considered to be reflecting a single nervous system property were poorly correlated. For example, the speed of forming new conditioned reflexes and the speed of substituting excitatory and inhibitory processes ("bilateral revision of the signal significance") were both considered to be dependent on nervous system mobility, but it has been shown that the correlation between these measures is quite low or nonexistent (Teplov, 1959). Similarly, Moldavskaya (1966) concluded that lability as measured by the ability to respond correctly to rapidly presented stimuli was not correlated with the rate of reversing responses to differentiated stimuli.

Teplov and his colleagues at the Institute of Psychology in Moscow took the lead in pointing out that the value judgments placed by Pavlov on such terms as *strong* and *weak, mobile* and *inert* were often misleading. This same group of investigators has questioned the Pavlovian notion of four basic types and has pointed out that the typological descriptions of Pavlov do not emanate in any clear way from the really important contributions concerning the basic properties of the nervous system (Teplov and Nebylitzyn, 1966). These same authors have written: "We must never consider Pavlov's formulations as something immutable. Since Pavlov's death, many new experimental methods have been discovered and many new facts that establish new laws unknown to him have been accumulated. Moreover, we psychologists, working with people, must never forget that Pavlov worked only with dogs, and that his pronouncements on man were, as a rule, made only by analogy" (Teplov and Nebylitzyn, 1966).

Those familiar with the Soviet literature in this field from the end of World War II through the middle of the 1950s can immediately appreciate the extent of the change in attitude toward Pavlov's views expressed in the above quotation.

It seems clear that physiologists and psychologists in the Soviet Union are revising their thinking about the so-called basic properties of the nervous system such as "strength," "mobility," and "balance." New properties such as "dynamism" have been added and the list of tests to measure these properties are being changed at a rapid rate. In spite of the many changes, however, inhibition and excitation remain central to this approach to typology.

Outside the Soviet Union, current theories in several related fields reflect an increasing acceptance of the fundamental importance of the inhibitory and excitatory processes. It is generally agreed, for example, that interaction between neurons is primarily accomplished through the action of synaptic contacts which cover more than four-fifths of the surface of a neuron. Estimates of the total number of such contacts upon a single postsynaptic cell may exceed 2,000 (Wyckoff and Young, 1956). These contacts have their effect by liberating humoral transmitting substances of two types, either depolarizing (excitatory) or hyperpolarizing (inhibitory). As transmission can be effected only by the action of many contacts, the firing of a postsynaptic cell results from the differential effect of many excitatory and inhibitory influences. Eccles (1966) has written recently: "So extremely complex is the synaptic structure that there may be as many as 50,000 spine synapses on a single pyramidal cell, all of which are probably excitatory. In addition, there is now good evidence that the bodies of the cells are embraced by inhibitory synapses, just as the hippocampal pyramidal cells are." A recent book on synaptic transmission lists among factors which control nervous system action: "The balance between excitatory and depressant materials in the environment of the cell which may be influenced in turn by the over-all metabolism of the tissue" (McLennan, 1963). In view of such information it is not so fantastic to postulate a dominant body chemistry that would influence the action of excitation and inhibition,[4] and at least in this respect Pavlov's views ironically appear more modern today than they did thirty years ago.

The concept of inhibition growing out of excitation, which Pavlov took over from Wedensky and Ukhtomsky, also has support from neurophysiological research. Renshaw (1941) has described the way recurrent collaterals, which arise near the origin of an axon, may turn back and inhibit neighboring motor neurons, or activate inhibitory cells; and Granit (1955) has suggested that such recurrent inhibition may offset strong excitation effects on the spinal cord level. Eccles (1966) and others have extended the application of the inhibitory cell to the level of the brain. Earlier, Grundfest (1940) had described alternative explanations and noted that axons can alter the excitability of adjacent, but structurally independent, axons.

It is also interesting to note that many recent explanatory concepts relating brain function and behavior suggest a competition between excitatory and inhibitory processes. Only a few examples need be provided. Performance deficits in delayed-response tests produced by frontal lobe lesions have been attributed to a "disinhibition" or loss of inhibition by a number

---

[4] This could be brought about in many ways. It is possible that differences in the enzyme systems involved in the metabolism of either the neural transmitters or those substances which inactivate transmitters may be responsible. For example, differences in nervous system mobility could conceivably be related to the quantity or effectiveness of the inactivators which determine the persistence of the action of the transmitters. Sawitsky, Fitch, and Meyer (1948) have shown, for example, that the erythrocyte ChE was relatively constant in a given individual, but varied widely from individual to individual.

of investigators (Brutkowski, 1965). Motivational consequences of direct brain stimulation via implanted electrodes have been described in terms of activation of so-called positive and negative reinforcing systems which may have reciprocal excitatory or inhibitory effects (Olds and Olds, 1962; Valenstein, 1965). Stellar (1954) has postulated a general theory of motivation based upon the activity of hypothalamic excitatory and inhibitory centers. Although the reticular formation was originally described as an activating system, further studies have shown that descending influences on the spinal motor neurons may be either inhibitory or excitatory depending on the region of the brainstem stimulated (Magoun and Rhines, 1948). Bonvallet, Dell, and Hiebel (1954) and Dell, Bonvallet, and Hugelin (1961) have demonstrated the existence of ascending inhibitory influences on the cortex, elicited by stimulating the bulbar-reticular formation. Similarly, Tissot and Monnier (1959) and Monnier, Kalbere, and Krupp (1960) have described reciprocally antagonistic areas within the nonspecific regions of the thalamus. Depending upon the frequency of stimulation, either EEG arousal or inhibition could be produced.

Many other examples from the current literature could be provided. For instance, recent studies of "passive" and "active" avoidance responses has led to the postulation of neural areas within the limbic system concerned with motor-inhibition and other areas concerned with motor-facilitation (McCleary, 1961). Sterman and Fairchild (1966) have studied the effects produced by simultaneously stimulating the mesencephalic reticular formation, which normally elicits wakefulness in a sleeping animal, and the basal forebrain area, which has been shown to produce sleep and diffuse inhibition. These authors have concluded that these two brain regions represent antagonistic systems "which interact to determine the functional state of the organism at any given moment." Also relevant are the drug-behavior studies which have elicited speculation that cholinergic mechanisms may form part of a system in the brain that antagonizes a second system that activates behavior (Carlton, 1963). It is certainly true that these hypotheses often differ in significant ways from Pavlovian thinking, but the prevalence of formulations that are based upon competing excitatory and inhibitory forces illustrates that such ideas are very much a part of our *Zeitgeist*.

Pavlov assumed that the basic properties of the nervous system involved characteristics of inhibition and excitation which he believed to reside in each neuron, or at least each cortical neuron. Pavlov had no direct evidence that this was the case. Contemporary theories of inhibitory and excitatory processes tend to emphasize, on the one hand, the existence of neurons that are either inhibitory or excitatory or, on the other hand, regions of the brain that play an inhibitory or excitatory role with respect to a given function.

Psychopharmacological studies in this country have raised questions very similar to those that are stressed by people working within the framework of Pavlovian typology. In an anthology on drugs and behavior (Uhr and Miller, 1960) the problem of individual differences in reaction to drugs is raised many times. Correlations are presented between subscales on the Minnesota Multiphasic Personality Inventory (MMPI) and drug reaction and there seems to be a general acceptance that there are "fast" and "slow" reactors to drugs, that is, a tendency for some subjects to show a relatively greater response to all drugs. Subjects who are most impaired by chlorpromazine, for example, tend to be most stimulated by dextroamphetamine (Kornetsky, 1960). Unfortunately, little is offered in the way of a unifying theory relating "personality" and response to drugs.

Our understanding will quite naturally be increased when we have more complete knowledge of the cellular and even molecular changes underlying the action of specific drugs. However, where responses to different drugs, which presumably have their own sites of action and underlying mechanisms, appear to reflect characteristics of an individual, we may have to consider that some general integrating system is playing a significant role. Recently, Brodie (1962) has pointed out that the effectiveness of any drug is dependent upon absorption rates, penetration of barriers, and other physiochemical characteristics which permits a substance to reach its site of action in adequate concentrations. It is not impossible that some properties of the nervous system, the major integrating system of the body, may influence the response pattern to very dissimilar substances.

The postulation of an interaction between excitatory and inhibitory processes did not begin with Pavlov. It would not be difficult to cite views dating back to the classical Greek period that utilize concepts similar to Pavlov's physiological theories. In a more contemporary vein, it is well known that the Russian physiologists, Sechenov, Wedensky and Ukhtomsky, had a powerful influence on Pavlov's thinking about inhibition and excitation.[5] It seems evident that the German physiologist, Heidenhain, who trained a number of Russian students including Wedensky and Pavlov, must have provided direction to Pavlov's developing theories. Heidenhain's interest in hypnosis led him to experiment with morphine narcosis and he demonstrated that under light narcosis there was an actual increase in the magnitude of reflexes and in the tendency of epileptics to have seizures. A number of related observations led to the conclusion that inhibition was an active process not simply a cessation of excitation. Heidenhain theorized that excitation of the nervous system was always accompanied by an inhibitory process, which limited the magnitude and

---

[5] The reader is referred to an interesting historical review of the concept of inhibition by Diamond et al. (1963).

duration of the excitation. Pavlov was convinced that these parallel processes resided in individual cells, while Sherrington's idea of central excitatory and inhibitory states emphasized functional systems in the brain rather than neuronal properties.

It is clear, however, that until very recently theories of the relationship of excitation and inhibition were relatively neglected in this country. Hull's (1943) theory had a place for inhibition resulting from activation (reactive inhibition), but this construct was divorced from physiology. In contrast, the Pavlovian physiologically oriented school gave this theme special emphasis and study and, in addition, applied it to the area of individual differences. As a result of recent developments in neurophysiology and psychopharmacology, investigators in this country have once again found the processes of inhibition and excitation useful in their theory construction. Soviet investigators appear to be revising their views. It remains to be seen what new syntheses will emerge in the future and whether it will be possible to relate inhibition and excitation to the problem of individual differences in a more meaningful way.

ACKNOWLEDGMENTS

Preparation of this manuscript was supported by research grant M-4529 and Research Scientist Award MH-4947 from the National Institutes of Health, and research grant MsG-437 from the National Aeronautics and Space Administration.

This account is based in part on current Soviet literature and in part on experience gained during a five-month visit to laboratories in the USSR during 1961. This trip was sponsored by the National Academy of Science as part of the Lacy-Zaroubin Scientific and Cultural Exchange Program.

REFERENCES

Akhmedov, K Yu, I. G. Kalinicheva, and O. G. Lorents. 1963. *Problems of the physiology and pathology of the high mountains.* Dushanbe: Izdatel'stvo AN Tadzhikskoy SSR.
Asratyian, E. A. 1938. *Tr. Fiziol. Labor. im. I. P. Pavlova* 8: 1. Moscow-Leningrad.
Ayrapetyants, H. G. 1963. Mobility of nervous processes in dogs of different ages irradiated antenatally with ionizing radiation. In *Zhur. Vysskey*

Neronoy Deystel'nosti im I. P. Pavlov. Translated by Joint publications Research Service, Washington, D.C. Pp. 25–38.

Bonvallet, M., P. Dell, and G. Hiebel. 1954. Tonus sympathique et activité électrique corticale. *EEG Clin. Neurophysiol.* 6: 119–144.

Brodie, B. B. 1962. Difficulties in extrapolating data on metabolism of drugs from animal to man. *Clin. Pharmacol. Therap.* 3: 374–380.

Brožek, J. 1964. Recent developments in Soviet Psychology. In P. R. Farnsworth, ed., *Annual review of psychology,* vol. 15. Palo Alto: Annual Reviews. Pp. 493–594.

Brutkowski, S. 1965. Functions of prefrontal cortex in animals. *Physiol. Rev.* 45: 721–746.

Carlton, P. L. 1963. Cholinergic mechanisms in the control of behavior by the brain. *Psychol. Rev.* 70: 19–39.

Davidenkov, S. N. 1947. *Evolyutsinno-geneticheskiye Problemy v Neuropatologii.* Leningrad.

Dell, P., M. Bonvallet, and A. Hugelin. 1961. Mechanism of reticular deactivation. In G. Wolstenholme and M. O'Conner, eds., *The nature of sleep.* London: Churchill. Pp. 86–107.

Diamond, S., R. S. Balvin, and F. R. Diamond. 1963. *Inhibition and choice.* New York: Harper & Row.

Eccles, J. C. 1966. Conscious experience and memory. In J. Wortis, ed., *Recent advances in biological psychiatry.* New York: Plenum Press. Pp. 235–256.

Golubeva, E. A. 1966. Photo driving brain potentials and typological characteristics of the nervous system. *Symposium 9 Physiological Bases of Individual Psychological Differences.* XVIII Internat'l Cong. of Psychol., Moscow. Pp. 122–132.

Granit, R. 1955. *Receptors and sensory perception.* New Haven: Yale Univ. Press.

Gray, J. A. 1964. Strength of the nervous system as a dimension of personality in man. In J. A. Gray, ed., *Pavlov's typology.* New York: Pergamon Press. Pp. 157–287.

Grundfest, H. 1940. Bioletric potentials. In J. M. Luck and V. E. Hall, eds., *Annual review of physiology.* Palo Alto: Annual Reviews. Pp. 213–242.

Gurevich, K. M. 1966. Psychological manifestations of the basic characteristics of the nervous system and their significance for occupational fitness. *Symposium 9 Physiological Bases of Individual Psychological Differences.* XVIII Internat'l. Cong. of Psychol., Moscow. Pp. 72–91.

Hull, C. L. 1943. *Principles of behavior.* New York: Appleton-Century-Crofts.

Klein, K. E., H. Bruner, and D. Jovy. 1960–1961. *Presentation of coworkers of the Institute of Flight-Medicine of the DVL in London and Paris,* pp. 109–121. Scientific and Technical Aerospace Reports, NASA (N64-15034 07-16).

Kornetsky, C. 1960. Alterations in psychomotor function and individual differences in responses produced by psychoactive drugs. In L. Uhr and J. G. Miller, eds., *Drugs and behavior.* New York: Wiley. Pp. 297–312.

Krushinsky, L. V. 1960. *Animal behavior: Its normal and abnormal development.* New York: Consultants Bureau.

Lawrence, H. L. 1964. Man plant's return. *Natur. Hist.* 73: 34–37.

Leites, N. S. 1956. *Typological characteristics of human higher nervous activity.* Moscow. Translation of chapter entitled: "Experiment in the Psychological Characteristics of Temperament. (Office of Technical Services, U.S. Dept. of Commerce, Washington, D.C.)

Leites, N. S. 1966. The problem of the relationship between typological and age characteristics. *Symposium 9 Physiological Bases of Individual Psychological Differences.* XVIII Internat'l Cong. of Psychol., Moscow. Pp. 110–121.

Luria, A. R. 1932. *The nature of human conflicts.* (Translated from the Russian by W. Horsley Gantt.) New York: Liveright.

Luria, A. R., ed. 1963. *The mentally retarded child.* (Translated from the Russian by W. P. Robinson. English translation edited by Brian Kirman.) New York: Pergamon.

McCleary, R. A. 1961. Response specificity in the behavioral effects of limbic system lesions in the cat. *J. Comp. Physiol. Psychol.* 54: 605–613.

McLennan, H. 1963. *Synaptic transmission.* Philadelphia: W. B. Saunders.

MacMillan, M. 1963. Pavlov's typology. *J. Nerv. Ment. Dis.* 137: 447–454.

Magoun, H. W., and R. Rhines. 1948. *Spasticity: The stretch reflex and extrapyramidal system.* Springfield, Ill.: Charles Thomas.

Merlin, V. S., ed. 1964. Typologicheskie issledovaniya po psikhologu lichnosti i po psikhologu truda. [Typological investigation of the psychology of personality and the psychology of work.] Perm: Perm State Pedagogical Inst.

Moldavskaya, S. I. 1966. Correlation between the rate of conditioned reflex reversal and the lability of nervous processes in persons of different ages. *Soviet Psychol.* 5: 49–54 (translation).

Monnier, M., M. Kalbere, and P. Krupp. 1960. Functional antagonism between diffuse reticular and intralaminary recruiting projections in the medial thalamus. *Exp. Neurol.* 2: 271–289.

Nebylitzyn, V. D. 1964. An investigation of the connection between sensitivity and strength of nervous system. In J. A. Gray, ed., *Pavlov's typology.* New York: Pergamon Press. Pp. 402–445.

Nebylitzyn, V. D. 1966. Some questions relating to the theory of the properties of the nervous system. *Symposium 9 Physiological Bases of Individual Psychological Differences.* XVIII Internat'l Cong. of Psychol., Moscow. Pp. 14-32.

Nuzhina, A. M. 1961. Effect of various phenamine doses on higher nervous activity in animals, and determination of optimal doses in various types of nervous system. In V. S. Rusinov, ed., *Works of the institute of higher nervous activity,* vol. 6. Acad. Science USSR, Moscow. (Office of Technical Services, U.S. Dept. of Commerce, Washington, D.C.)

Olds, M. E., and J. Olds. 1962. Approach-escape interactions in rat brain. *Am. J. Physiol.* 203: 803–810.

Pavlov, I. P. 1957. *Experimental psychology and other essays.* New York: Philosophical Library.

Pavlov, I. P. 1928. *Lectures on conditioned reflexes,* vol. 1. New York: International Publishers.

Pevzner, M. S. 1959. *Oligophrenia, mental deficiency in children.* New York: Consultants Bureau.

Razran, G. 1965. Russian physiologists' psychology and American experimental psychology: A historical and a systematic collation and a look into the future. *Psychol. Bull.* 63: 42–64.

Renshaw, B. 1941. Influence of discharge of motoneurons upon excitation of neighboring motoneurons. *J. Neurophysiol.* 4: 167–183.

Rosvold, H. E., A. F. Mirsky, I. Sarason, E. B. Bransome, and L. H. Beck. 1956. A continuous performance test of brain damage. *J. Consult. Psychol.* 20: 343–350.

Savinov, G. V., L. V. Krushinsky, D. A. Pless, and R. A. Valershteyn. 1963. An experiment in the use of mathematical modelling in the study of the interrelation of the process of stimulation and inhibition. In S. M. Kuzin, ed., *Biological aspects of cybernetics.* Transl. by Joint Publications Research Service, Washington, D.C. Pp. 114–123.

Sawitsky, A., Fitch, H. H., and Meyer, L. M. 1948. A study of cholinesterase activity in the blood of normal subjects. *J. Lab. Clin. Med.* 25: 1325–1332.

Skipin, G. V. 1938. *Tr. Fiziol Labor im. I. P. Pavlova* 8: 16. Moscow-Leningrad.

Skinner, B. F. 1948. *Walden Two.* New York: Macmillan.

Stellar, E. 1954. The physiology of motivation. *Psychol. Rev.* 61: 5–22.

Sterman, M. B., and M. D. Fairchild. 1966. Modification of locomotor performance by reticular formation and basal forebrain stimulation in the cat: Evidence for reciprocal systems. *Brain Research* 2: 205–217.

Suvorov, N. F., and L. K. Danilova. 1965. Functional localization of overstrained neural processes due to their collision. *Soviet Psychol. Psychiat.* 4 (2): 3–8 (translation).

Suvorova, V. V. 1966. EEG correlates of individual differences in human behavior under stress conditions. *Symposium 9 Physiological Bases of Individual Psychological Differences.* XVIII Internat'l Cong. of Psychol., Moscow. Pp. 178–187.

Teplov, B. M. 1959. Tipologicheskie Osobennosti Vysshei Nervnoi Deyatel'nosti Cheloveka [Typological characteristics of higher nervous activity in man]. In B. M. Teplov, ed., *Academy of pedagogical sciences,* vol. 2, 1959, p. 228. Moscow.

Teplov, B. M. 1964. Problems in the study of general types of higher nervous activity in man and animals. In J. A. Gray, ed., *Pavlov's typology.* New York: Pergamon Press. Pp. 3–153.

Teplov, B. M., and V. D. Nebylitzyn. 1966. The study of the basic properties of the nervous system and their significance for the psychology of individual differences. *Soviet Psychol. Psychiat.* 4: 80–85 (translation).

Tissot, R., and M. Monnier. 1959. Dualité du système thalamique de projection diffuse. *EEG Clin. Neurophysiol.* 11: 675–686.

Uhr, L., and J. G. Miller, ed. 1960. *Drugs and behavior.* New York: Wiley.

Valenstein, E. S. 1965. Independence of approach and escape reactions to electrical stimulation of the brain. *J. Comp. Physiol. Psychol.* 60:20–30.

Walter, W. G. 1953. *The living brain.* London: Gerald Duckworth & Co.

Wyckoff, R. W. G., and Young, J. Z. 1965. The motoneuron surface. *Proc. Roy. Soc.* (London) 144: 440–450.

Young, P. 1961. Yes, death is afraid of us. *Life 61* (September 16): 123–127.

Zavadsky, I. V. 1908. An application of the method of conditioned reflexes to pharmacology. *Proc. Russian Med. Soc. in St. Petersburg,* vol. 75.

Zvorykin, V. N. 1964. The typology of higher nervous activity characteristics in dogs under barometric changes. In *The effect of the gas medium and pressure on body functions.* Acad. of Sciences, USSR, Institute of Evolutionary Physiology in Sechenov. (Translation available from the U.S. Dept. of Commerce, Springfield, Va., pp. 170–177.)

WILLIAM N. TAVOLGA

Department of Animal Behavior
The American Museum of Natural History, New York
and
Department of Biology
The City College of the City University of New York

Levels of Interaction in Animal Communication

Most animals influence each other's behavior by visual, acoustic, chemical, and other means. Some of these interactions have been called communication. In an objective and well-organized recent compilation, Frings and Frings (1964) recounted numerous instances of animal communication that have been observed and studied. The problem of definition immediately presented itself to these authors. What sorts of interacting behavior constitute communication? They stated that "communication between animals involves the giving off by one individual of some chemical or physical signal, that, on being received by another, influences its behavior." They set certain limits on this definition, namely that the sender must utilize some specialized structure in the production of the communicatory stimulus, and that both the sender and receiver must be members of the same species. In their text, however, the authors often strayed outside these limits. They were faced with two alternatives: a massive, unwieldy tome if they used a broad definition; or a concise readable text if some arbitrary restrictions were adopted. The very existence of this difficulty is of considerable interest because it points up the vast variety of animal interactions and the lack of a theoretical framework within which to place and relate them.

The study of animal communication and of human language tended to be separated for many years. This separation was, in a large part, the result of a strong reaction against anthropomorphism and teleology influenced by the principles stated more than sixty years ago by C. Lloyd Morgan and E. L. Thorndike. Nevertheless, the idea that both these phenomena can be united under the same set of generalizations was effectively proposed by Cherry (1957). He defined communication as: "The establishment of a social unit from individuals by the use of language or signs." Marler (1961) was strongly influenced by Cherry's approach, and he attempted to classify animal interactions and signals according to the analysis of human language made by Morris (1946). These primitive communicatory phenomena were fitted into the categories of "identifiors, designators, prescriptors and appraisors." In a similar attempt at synthesis, Hockett (1961) emphasized auditory phenomena and proposed a set of thirteen functional properties of communication, such as, broadcast, rapid fading, total feedback, semanticity, and interchangeability.

In further contrast to the zoological and analytical contribution by Frings and Frings, Sebeok (1965) attempted a synthesis from the viewpoint of an anthropologist and linguist. He placed animal interactions into the framework of semiotics, the theory of signs. Semiotics is conceived as the study of patterned communication and includes all modalities (Sebeok, Hayes, and Bateson, 1964), and the term *zoosemiotics* was coined by Sebeok to encompass much of the study of animal behavior.

A significant feature of some of the major theoretical approaches to animal communication that have been made in recent years is the tendency to make cross-phyletic generalizations with little regard for the differences in phyletic level among the organisms being compared. A comparative evolutionary approach is certainly appropriate, but to do this properly, the ontogenetic and evolutionary history of the organisms and phenomena must be considered. Aside from demonstrating a few aerodynamic principles in common, a comparison of the wings of birds and insects gives us little information or insight. Comparisons of behavior are further limited by the virtual lack of historical, i.e., paleontological data. Such comparisons demonstrate and emphasize that qualitatively different mechanisms operate at different phyletic levels, as, for example, a comparison of maze-learning in an insect and a mammal (Schneirla, 1962).

The social organization of the army ants is obviously different from human societies, and is based on entirely different endocrine, stimulative, neural, genetic, developmental, and other factors. It is, therefore, clearly inappropriate and meaningless to apply the complex of interacting factors that control the behavior pattern of these ants (see Figure 2 in Schneirla and Rosenblatt, 1961) to a theoretical analysis of human society. By the same token, is it at all appropriate to use the methods and theory developed for the study of human language toward the investigation of a form of communication found in another species at a different phyletic level?

The purpose of this article is to appraise the value of an approach to the study of communication that is based on the concept of qualitatively different phyletic levels, expressed as levels of behavioral organization, or levels of integration.

An excellent, brief statement of this concept was presented recently by Tobach and Schneirla (1968):

> The concept of *Levels of Integration* (Schneirla, 1953) postulates a hierarchical arrangement of energy (organization of matter) beginning with the simplest organized entities and increasing in the degree and elaborateness of organization from inanimate to animate forms. General categories of human knowledge which reflect such an ordering of natural phenomena are cosmology, physics, biology and sociology. Each of these levels of organization requires distinctive instrumentation, experimental operations and laws. For the wide range of existing animals, levels of integration are conceived as a series of progressive advances from the acellular (or "single-celled") animal through the multicellular. Each of these levels refers to those animal groups which have in common a set of distinctive capacities for behavior. Any one level of integration in behavior, although certain to be similar in some respects to levels judged "lower" in the scale to which it is related through evolution, differs from them sufficiently to warrant its separate categorization. For example, the analysis of the physiological and behavioral interactions that prevail among bees in a hive gives no adequate preparation for studying human societies. Human social organization is not only more complex than that of the honeybee but presents new aspects of development, stereotypy and plasticity in behavior, requiring special experimental techniques and the formulation of principles not applicable to bees. These are qualitatively different levels.

INTERACTIONS AMONG ANIMALS

Are all interactions equivalent to communication? Is the interaction between an animal and its physical environment also a form of communication? If such broad definitions are accepted, then all of animal behavior and ecology falls into this category and it is rendered meaningless. Which interactions among living organisms should be considered communication?

The stimuli that impinge upon any organism may be grouped into the following categories according to the stimulus source:

1. Stimuli from the physical environment, e.g., changes in light, temperature, and salinity.
2. Stimuli from living organisms
   a) from plants
   b) from animals
      i) from individuals of other species
      ii) from individuals of the same species
      iii) from the same individual, as in echolocation.

Even if item 1 is eliminated as a form of communication, the question arises whether the effect of an organism upon its physical environment is "communicated" to other organisms. The withdrawal of calcium carbonate from sea water by many organisms certainly has an effect on the behavior and physiology of various forms of marine life. Does a parasite communicate with its host by making it sick? When sparrows feed on seeds found in horse droppings, is this an example of communication between horse and bird? Is all mating behavior communication? Is there any communication in the reproductive act of the male salamander that leaves spermatophores scattered about for the female to pick up in his absence? Do predator-prey relations involve communication? Should communication as a concept be limited only to intraspecific interactions?

The above questions amply demonstrate the difficulty of where to draw the line, and most authors have done this arbitrarily and as a matter of convenience. But is there really a line to be drawn? The concept of levels can be used to resolve this dilemma. Communication does not exist as a single phenomenon; there are, instead, different levels of interactions involving different levels of integration. Some, but not all, of these levels can be called "communication." There are many kinds of communication, with each kind not only quantitatively but qualitatively different.

A comparison of the communicatory "dances" in the honeybee and the language of man is not meaningful without an appreciation of the anatomical, physiological, and psychological differences between these representatives of two extremes of evolutionary development. In spite of superficial resemblances that have led authors to use the same words to describe different phenomena, the fact remains that the underlying mechanisms and ontogenies are fundamentally different.

REQUIREMENTS FOR INTERACTION

For an interaction or communicatory event to take place, only three requirements are necessary.

1. The *Emitter*. An organism that produces an energy change in its immediate environment by means of some behavioral or physiological manifestation is an emitter. Various more or less teleologically loaded terms have been used for this entity in communication, e.g., addresser and sender. In order not to assume that the organism is in any way directing its energy output, I recommend the more objective term used by Moles (1963): emitter. In a broad sense, then, any organism is an emitter of stimuli, whether these be specialized, individualistic groupings of sounds

(as in the songs of birds) or the production of metabolic excretions. The emitting structure may be the entire organism, as in protozoa, or some specialized part of it, such as the tymbal of a cicada, or it may consist of some special movement, such as the "dances" of honeybees, in which no apparent morphological specialization is involved.

Many authors, including Sebeok (1965), conceive of the emitter as being an encoder in which some internal process is put in the form of a code or representation. To "encode" means to "put into a coded form," and this must involve the translation of one set of symbols into another. The concept is therefore teleological and metasymbolic, and applicable only to the higher levels of human communication. To extend this idea to interactions among animals is meaningless and misleading.

2. The *Energy Output* (*stimulus*). The available channels for a stimulus are photic, thermal, mechanical (including acoustic and tactual), chemical, and electrical (limited to certain fishes). In many instances more than one channel is involved in the output, and the energy output has to be more intense than the background noise for a communicatory event to take place.

A sign or signal is a specialized form of emitter output in which there is usually a single energy channel and only a narrow portion of its spectrum is involved, as in the production of a specific chemical odor or a sound of specific frequency and duration. From the viewpoint of the receiver, a sign or signal represents not only the discriminable difference between stimulus and background noise, but the perception of this stimulus in a manner determined by the receiver's morphology and development. At higher communicatory levels, signals are often patterned, as are the songs of birds, and this adds to the specificity of their function. The concept of a message or code belongs only to the highest of communicatory levels; possibly it should be restricted to the human species.

3. The *Receiver*. Any organism that is able, by virtue of its sensory equipment, to receive the energy change is potentially or actually a receiver. The most reliable way in which the receipt of the energy output can be determined is by some observable physiological or behavioral response, and therefore the receiver is not separable, in practice, from the responder. The response must be measurable either as a correlated physiological or behavioral change. Specialization of the receiver may occur at two levels: by the refinement of the sensory receptors, and by the development of specific central nervous mechanisms. The ultrasonic receptors of certain moths exemplify the first of these specializations, in which considerable signal analysis and filtering takes place in the receptor organ

(Roeder and Treat, 1961). The eye of a frog has also been shown to act as a peripheral analyzer (Lettvin et al., 1961). The abilities of birds to discriminate among a variety of complex song patterns, however, indicates that the analysis occurs at the level of the central nervous system (Marler, 1960). Symbols and codes are the property of the highest of communicatory levels. Context, defined as including all internal and external stimuli that impinge on the receiver in addition to the energy output of the emitter, is of profound importance in determining the quality and strength of the response (Smith, 1965).

LEVELS OF INTERACTION

The following is a tentative schema for the classification of levels of interaction (or communication) according to phyletic and psychological levels. Only a few examples in each category will be given, and the position of some of these may be doubtful, because of the lack of experimental study or because they may represent intermediate stages of development. The different levels are not always clearly separable, no more so than are phyletic levels, nevertheless they are discrete and not entirely artificial. Perhaps the more important point than the specific definitions of the various levels is the concept of levels of interaction itself.

Vegetative Level

An organism is an emitter simply by being. Its physical presence can affect the behavior of other organisms, and its growth pattern, tropisms and physical form can serve as a stimulus. This level of emitter is characteristic of plants whose influence upon the behavior of animals is on the level of that of the physical environment, with the one difference that plants are the basic source of organic food. Interactions among most plants are on this level, including the process of wind pollination. Exchange of DNA between bacterial cells and interactions between viruses and cells are vegetative and tropistic. The exact mechanism of DNA transfer among bacterial cells is still controversial, but the replication of DNA itself might provide the force for transfer (Gross and Caro, 1965). It might be argued that species that possess motile gametes could represent a higher level of interaction, but gamete production is clearly a primitive, vegetative function. Even the specializations of form, color, and odor that characterize many insect-pollinated flowers are vegetative in the sense of being the manifestations of growth, cellular differentiation, and development of form rather than of behavior. In this case, receivers, i.e., the insects, operate on a different level of interaction.

It is conceivable that under certain conditions, some animals interact on this level, although the presence of sense organs and organized behavioral responses would virtually eliminate all but some protozoa and some primitive metazoa, such as sponges, from this category. At early embryonic stages, vegetative interactions predominate in most, if not all, animals.

Tonic Level

Continuous, "on-going" processes that are basic to species-typical development and regular behavior are defined by Schneirla (1965) as *tonic;* e.g., homeostasis, excretion, cellular metabolism. Such processes exist at all levels of integration and contribute to the development of the organism. These processes result in certain physiological and behavioral manifestations that affect other organisms, and, therefore, they represent energy outputs, i.e., stimuli, in interactions among organisms. On this basis, Schneirla's definition can be expanded to consider interactions in which these processes predominate to be on the *tonic level.*

During the normal course of metabolism, animals produce a number of chemical exudates. Their normal locomotor patterns may also be considered as energy outputs in the photic and mechanical channels. All of these can serve as stimuli. Among protozoa and primitive metazoa, virtually all interactions are on this level. The output of the emitter is a direct and unspecialized result of its normal (tonic) state. In these forms, even reproductive activity consists simply of gamete production.

When acting as emitters, most animals interact at this level at some time in their lives. Symbiotic relationships probably belong in this category. The output of a prey organism is generally of this type, although the receiver, i.e., the predator, may be functioning at a higher level. Interactions in mutualism, commensalism, and parasitism are on this metabolic, physiological level. It is probable that some primitive types of trophallactic and trail-following behavior in insects are also on this level. It is probable that most animals pass through stages in their development during which they interact on this physiological level, as, for example, the biochemical interactions of a fetus with its mother.

Mimicry, in the sense of physical resemblance between members of different species, is the result of a specific developmental pattern in the morphology and a direct consequence of metabolic and physiological species-typical properties. Since much mimicry involves little, if any, behavior, the interactions based upon these properties might be considered to be on the vegetative level, however, even the classic case of the resemblance of the monarch and viceroy butterflies must contain similarities in flight and other movements. The amount of integration involved, therefore, would argue for placing such cases in the tonic level.

Many so-called flash colors and color changes, particularly among fishes, might be considered as intermediate between the tonic and phasic levels. These represent the lower level of interaction in that many of these color patterns and associated movements function as camouflage in either a cryptic or disruptive manner, although the receiver may be operating on a higher level.

The receiver can also operate on the tonic level. The kineses described for many protozoa and some higher invertebrates by Fraenkel and Gunn (1961) are rigidly controlled by the stimulus characteristics, as are the primordial approach-withdrawal responses discussed by Schneirla (1959, 1965). Primitive orientations, such as taxes, are also closely stimulus-bound, and involve low levels of integration on the part of the responding organism. Interactions that are on this tonic level can be identified as such on the basis of the phyletic level of the species, the degree of control by the stimulus quality and quantity, the low level of central integration, and the stage of development. It is quite possible that many instances, such as the optomotor response, may turn out to represent a higher level, but ontogenetic information, presently lacking, must be available for a proper analysis of these and other similar phenomena.

Phasic Level

In the sense used by Schneirla (1959, 1965), *phasic* processes refer to discontinuous, more or less regular stages or events in the development of an animal. As in the previous section, Schneirla's definition may be extended, and interactions in which these processes predominate would be allocated to the *phasic level*. Interactions at this level usually involve broad and multichannel energy outputs, with, however, some specialization and specificity on the part of the emitter. The receiver on this level responds in a discriminatory manner, e.g., in species and sex discrimination.

The majority of higher invertebrates (arthropods and cephalopod molluscs) interact on this level. The well-known hive dances of returning forager honeybees represent an effect of food stimulation, and the energy output of these dances covers a wide range of visual, olfactory, gustatory and, probably, acoustic and tactual stimuli. The response of the receiver bees in this situation results from the interplay of trophallactic factors, taxes, and experimental effects.

The case of the so-called language of bees deserves some careful scrutiny. The original studies of von Frisch (1950) seemed to indicate that honeybees interact in a highly specific manner, with the returning forager emitting signals as to the location of the feeding station, and the recruits receive this information and react to it in a remarkably precise manner. In recent years, data are accumulating to the effect that the energy output of the foragers is broad-band and multichanneled (Esch, 1967; Wenner, 1962). The precision of response by the recruit bees has also been opened to question by some carefully controlled experiments (Johnson, 1967; Wenner, 1967). This is in line with the evidence that a bee's response to a floral pattern is controlled by the flicker value of the pattern rather than any innate configurational schemata (Hertz, 1934; Wolf, 1937; see also review in Carthy, 1958). Unfortunately, there is no information as to the development of the bee orientations, and it is clear that the problem is far more complex than previously supposed. The type of study exemplified by Schneirla's analysis of army ant behavior (Schneirla, 1957) is needed here, and, until such data become available, it seems appropriate and parsimonious to classify these bee interactions on the phasic level.

Interactions among some of the lower vertebrates may be placed in this category. The simple mechanism of sex discrimination in the lamprey consists of males attaching themselves to other lampreys. If the second animal is a female, she remains attached to the substrate and spawning follows. If the second animal is another male, he then releases his hold on the substrate, and the first male also releases (Breder and Rosen, 1966).

At least a portion of sex discriminatory behavior in fishes and amphibians involves broad channel energy outputs and uncomplicated approach or withdrawal responses on the part of the receivers. The interplay of visual, chemical, and acoustic stimuli in the gobiid fish (*Bathygobius soporator*) (Tavolga, 1956), and the warning vibrations in many anuran amphibians (Bogert, 1961), may be cited as examples.

Many feeding responses by predators are on the phasic level of interaction, in that a simple discrimination is made between "food" and "nonfood" as in the case of the "bug-detector" mechanism of the frog (Lettvin et al., 1961). Many predators among insects, such as the praying mantis, fishes, and anuran amphibians depend on broad channel outputs from their prospective prey, e.g., their size and movement.

Among fishes the categorization of most sounds and color displays is difficult. The small amount of experimental evidence on the function of fish sounds indicates that, although the sonic mechanism may sometimes be quite specialized, the sounds represent phasic outputs rather than specific signals. These function in territoriality and sex discrimination. The fact that only a small portion of the acoustic spectrum is used may simply reflect the limitations of the hearing apparatus (Tavolga, 1964; Protasov, 1965).

In fishes, this phasic level of interaction seems to have attained some aspects of social organization. In studies on the schooling of fishes, Shaw (1960, 1961) has shown that the development of this form of an approach response begins at a very early stage in many species and may begin as a nonspecific optomotor reaction. Although the sensory cues used by schooling fishes are evidently broad and multichannel, the final result gives the appearance of a closely ordered, precisely spaced, organized moving mass of animals.

Signal Level

Emitters on this level are characterized by possessing specialized structures that produce their energy output generally along a single channel and, usually, within a narrow band of the spectrum of that channel. The production of specific chemical attractants, i.e., pheromones (Wilson, 1965), by some insects represents a channeling of their output into what may be termed a *signal*.

In a recent report, Ryan (1966) described the production of a chemical by premolt females of a crab (*Portunus*) that stimulates activity in males. Although the stimulating substance seems to be species-specific, it is probably premature to consider it a pheromone since its chemistry and physiological origin are unknown and the behavioral response of the males has not been exactly characterized. For the present, this example of interaction should be considered on the phasic level.

The specialization of sonic organs in many insects, especially among the Cicadidae and Orthoptera (Alexander, 1960; Dumortier, 1963), the specificity of their acoustic outputs, and their responses argue strongly for the inclusion of these phenomena on the signal level of interaction. Similarly, the waving behavior of male fiddler crabs can be called a signal (Salmon, 1965).

Among lower vertebrates, it is possible that at least some fishes will be shown to emit such specific signals, particularly where sonic mechanisms are highly evolved and specialized, as in the toadfish (Winn, 1967). The mating calls of most anuran amphibians are produced by specialized structures, are quite specific in their acoustic characteristics, and are produced under special circumstances of season, temperature, and internal endocrine condition (Bogert, 1961; Blair, 1963). Among reptiles, the highly specific courtship behavior of many lizards can be considered a combination of signals (Carpenter, 1963).

The call notes of most birds are signals in the sense of being produced by a special structure and under more or less specific circumstances. Social communication among herring gulls (Tinbergen, 1959) is based upon signals that are simple and involve narrow energy channels. The complex songs of many passerine species, however, are clearly on a higher level. The vocalizations of most mammals are signals in the sense of expressing some specific emotional state, for example, the bark of a dog, purr of a cat, and squeal of a porpoise. Many such vocalizations originate as phasic outputs, but as a result of conditioning they become associated with specific circumstances, and thereby develop into signals. In the case of porpoise sounds, despite much publicity to the contrary, it remains to be shown that the high-pitched whistles and squeals of porpoises represent anything more than emotional outbursts, although there is evidence that these sounds function in communication (Lang and Smith, 1965; Dreher and Evans, 1964).

Busnel, Moles, and Gilbert (1962) described a form of communication based on the use of whistles among some peoples of the Pyrenees and these authors compared this language to the whistle communication among dolphins. In this comparison, the authors have confused two levels of interaction. The human whistle language is based upon symbolization and develops as part of a culture like the "talking" drums of Africa. It is comparable to the symbolic gestures used for communication among skin divers. The symbols are derived from and convey in coded form messages that were originally in the normal spoken language. The standard Morse code is also such a representation.

In dolphins, or any other infrahuman animal for that matter, the communication signals are not a coded form of speech and are not derived from a culture. If further data demonstrate the specificity of signals among these whistles, it is possible that this form of interaction may be shown to be on the signal level, but it is clearly below the language level.

This example well illustrates the danger of comparing behaviors on the basis of superficial similarities and with no regard to ontogeny or to level of organization.

It is not unusual for the receiver in many cases of interaction to operate on a higher level than the emitter. The receivers of energy outputs from phasic or tonic level organisms may respond to these outputs in a highly specific and specialized manner, often with specialized receptor organs. The response of many predatory species, particularly among the vertebrates, is to specific color, behavior, scent, or auditory cues. In reproductive behavior, particularly in courtship, the approach of a female to a courting male often involves the detection of specific cues. This is probably true of a few fishes, most anurans, and many mammals and birds.

Echolocation is a special case of interaction on the signal level in which the emitter and receiver are the same animal. Many insectivorous bats and toothed whales possess elaborate mechanisms for the output of sound pulses and the reception and analysis of their echoes (see review by Vincent, 1963). In other species in which echolocation has been established, such as shrews, oilbirds and even in man, the level of specialization for echolocation appears to be low. Tentatively, at least, these echolocators would be considered at the phasic level of interaction.

The mechanism of electric location and prey detection by certain fishes (Lissmann, 1963) is, in one sense, similar to echolocation to the extent that the animal uses its own energy output to gain information about its environment. Although the electric organs are highly specialized, the energy channel is broad-band and nonspecific. Some electric fishes, however, use the information in a precise, specific way, and this instance of interaction probably should be allocated to the signal level.

At first glance, it might seem that the ethological concepts of "sign stimulus" and "releaser" are examples of the signal level of interaction. Many such studies are only cross-sectional, however, and the contribution of experience to the development of the interaction has not been analyzed. The ethological approach, with its rigid theory of instinct, is hampered by a priori assumptions as to the "innateness" of the responses. Moltz (1965), Schneirla (1966) and many other authors have pointed out that such assumptions are neither necessary nor useful. Experiential factors play a vital role in the development of all signal level interactions, particularly with regard to the receiver. Developmental studies, rather than cross-sectional ones, are needed.

Investigations on the development of the use of signals have been notably few. One study applicable to this problem is that of Schneirla (1962) in which the psychological levels of insects and mammals were compared with particular emphasis on the learning process. Another significant study (Schneirla and Rosenblatt, 1961) concerned the development of social behavior. In both insects and mammals, the development of specific responses to specific stimuli involves a process of *selective learning* (as defined by Maier and Schneirla, 1942), but the process of acquisition of this response is qualitatively and quantitatively different in the two groups. The insect level is *biosocial,* and dominated by organic processes, including maturation and experience, while the mammalian level is *psychosocial.* In mammals, learning controls the development of the social bond in addition to the organic processes. It is evident, therefore, that the signal level of interaction includes two subordinate, qualitatively different levels:

BIOSOCIAL SIGNAL LEVEL

This category includes the interactions of the more specialized forms of social insects, particularly members of the Order Hymenoptera. The development of pairing and other reproductive interactions among vertebrates (birds, reptiles, and possibly some amphibians and fishes) might be placed here, although it is quite probable that ontogenetic studies will show that even the lower vertebrates develop signal communication in a different manner from that of social insects.

PSYCHOSOCIAL SIGNAL LEVEL

As defined by Schneirla (1953), the development of a social bond in mammals involves communication at this level. The example of reciprocal stimulation between kittens and their mother serves as a model study for this type of communication (Schneirla and Rosenblatt, 1961).

As a special case, the development of the songs of higher passerine birds deserves mention. Lanyon (1961) compared the development of the songs of primitive and higher passerines. In the primitive forms the motifs are simple and relatively unaffected by motifs of other birds. According to the present schema, such songs would be considered to be on a primitive signal level. In higher passerines, i.e., the more typical songbirds, there are stages in development during which acoustic patterns from other birds, both of the same and different species, can be absorbed into the repertoire. The acquisition of song by the higher passerines is a function of the social situation, and, consequently, is on a qualitatively different level.

Symbolic Level

The numerous studies on social behavior in infrahuman primates, especially those of Yerkes and Yerkes (1929), and Köhler (1925), have shown that interactions among these animals are not only on the psychosocial level, but that primitive forms of symbolism can be observed. The "pointing" behavior and other gestures of a chimpanzee are clearly symbolic in an elementary fashion.

The large repertoire of facial expressions, vocalizations, and gestures of the anthropoids has been described by many authors, including Charles Darwin in his classic volume, *The Expressions of the Emotions in Man and Animals*. Studies on captive primates, especially chimpanzees, have demonstrated the specificity and complexity of these signals. Observations of wild colonies have substantiated the social basis for the development of communication. The complexity and variability of social communication in infrahuman primates has been emphasized recently in a review by Washburn, Jay, and Lancaster (1965) and in a compilation by Altmann (1967). These studies on free-ranging anthropoids have described some of the complex interactions that begin in infancy and develop into plastic social organizations that have been likened to primordial cultures.

Although the development of social communication among infrahuman primates has been studied only indirectly and sporadically, communication appears to emerge rapidly as part of social development. In contrast, in songbirds and lower mammals, the development of a social bond is based upon a concatenation of discrete signals and the gradual organization, through reciprocal stimulation, of complex patterns. For these reasons, and especially because of the primitive symbolism involved, the communication level in many infrahuman primates represents a distinct level.

Language Level

The term *language* has been used for several levels of communication, sometimes as distinct from and sometimes including "speech." For purposes of this thesis, I shall use language for that form of primarily vocal communication characteristic of man. Without reviewing the vast literature on linguistic theory, semiotics, and paralanguage (see, for example, Sebeok, Hayes, and Bateson, 1964; and Lenneberg, 1964), I should like to characterize this level of communication as one in which symbols are used in a teleological sense; in which abstract ideas are communicated; and in which not only present conditions, but past and future events are communicated. Yerkes and Yerkes (1935) pointed out that despite the complexity and primitive symbolism of intercommunication in many infrahuman primates, the property of communication of ideas is still exclusive to human speech.

The development of human speech is unique. Carmichael (1964) reviewed the emergence of speech in the human infant, and pointed out the difficulties in the study of prelinguistic utterances. This emergence does not, of course, follow through all the interaction levels in accordance with the Biogenetic Law, but the earliest sounds, including the so-called birth-cry, appear to be nonspecific energy outputs. Subsequently, the psychosocial level is reached and surpassed with the use of specific sounds representing objects, classes of objects, physiological and mental states, and, finally, abstract ideas. The qualitative differences between human language and other forms of animal communication clearly set this level apart.

Much has been said about the comparison of function of nerve networks and digital computers. Wiener (1950) pointed out that a nerve cell is essentially digital, i.e., on-off, in function. Although such comparisons may be useful in the technology of computers, this emphasis on superficial similarities masks the obviously profound differences in origin and development between brains and machines. "Communication theory," as expounded by cyberneticists and others, is applicable only to human communication since it is a set of generalizations derived from that level. Models and machines that demonstrate these generalizations are only representations or abstractions of human language. Communication theory, therefore, is not applicable to the other nonsymbolic levels.

DEVELOPMENT OF COMMUNICATION

In his recent exposition of the theory of biphasic (approach-withdrawal) mechanisms in the ontogeny of behavior, Schneirla (1965) began by saying: "In the evolution of behavior, operations which appropriately increase or decrease distance between organisms and stimulus sources must have been crucial for the survival of all animal types."

The above statement, as well as the major part of Schneirla's exposition, certainly applies to the evolution and development of communicatory mechanisms. At the vegetative level, tropisms predominate in interactions. At the tonic level, approach-withdrawal movements dominate, although their derivatives, the taxes and kineses, are also present. At the phasic level, interactions become more closely tied in with specific behavior patterns, although the underlying approach-withdrawal mechanisms can still be observed. The signal level involves the development of approach-fixation (i.e., imprinting) and the increasing role of recent experiential factors. At the psychosocial and language level, it is difficult, although possible at times, to detect the origins of these complex behavior patterns from primordial approach-withdrawal processes.

A developmental study is of basic importance in this levels approach to communication. Since there are virtually no paleontological data from which to draw conclusions on the evolutionary relationship of various forms of interaction, it is necessary to rely on morphological data for making phyletic comparisons. Although ontogeny does not recapitulate phylogeny, as clearly shown by de Beer (1958) and others, ontogenetic similarities may indicate relationships. In behavior, developmental analysis is an important tool in elucidating phyletic relationships.

In this connection, I might add that this necessity for developmental analysis is easily overlooked and masked by an urge to label and classify. In testing some of the ideas contained in this article on my colleagues and students, I found that the most common questions concerned the classification of this or that behavior in one or another of the proposed categories. Although perfectly valid, such questions are often unanswerable because there is a lack of information on the origin and development of the particular interactions. Indeed, many of the examples cited here are categorized without this information and are, therefore, only tentatively allocated. It is not the purpose of this schema to serve as a rigid taxonomy of interactions, but, rather, to assist in the analysis of interactions and point up the need for developmental data in such analyses.

SUMMARY AND CONCLUSIONS

An ontogenetic and evolutionary approach has been applied to the study of interactions among organisms. Such an approach requires the analysis of these interactions in accordance with their psychological and phyletic levels of integration. As a first approximation, the following schema of levels of interaction is offered:

1. Vegetative level: growth and tropism, physical presence. In plants, some protozoans and sponges.
2. Tonic level: metabolic processes and their by-products; unspecialized behavior; taxes and kineses. In many protozoa and primitive metazoa, but some higher forms may at times operate on this level.
3. Phasic level: discontinuous, but regular behavior; specialization of behavior of both emitter and receiver organism; primitive types of species and sex discriminatory behavior. In many invertebrates above the coelenterate level, and in some primitive vertebrates.
4. Signal level: specialized structures for production of specific, narrow channel signals; specific behaviors in response to cues. In some insects, and the majority of vertebrates.
Two subheadings of the signal level are defined:
4a. Biosocial signal level: communication within social situations in which organic processes control the development of behavior closely. In the social insects, and in the reproductive-parental behavior of birds, reptiles, and some amphibians and fishes.
4b. Psychosocial signal level: development of complex patterns of signals with an increasing role of experiential, especially social factors. In most lower mammals, possibly excluding only the higher primates. The songs of higher passerine birds represent a special form of signal developed on an essentially psychosocial level, but limited in scope and leading toward a more stereotyped form of communication.
5. Symbolic level: a highly plastic and variable form of psychosocial communication; strong dependence upon social interactions for its development; primitive utilization of symbolic sounds and gestures. In infrahuman primates, particularly the anthropoids.
6. Language level: communication of abstract ideas; speech; metalanguage and metasymbolism. Restricted to the human species.
It is evident that a given organism can communicate or interact with other organisms on more than one level at different times or simultaneously. For example, a chimpanzee sometimes exhibits symbolic behavior, but it often communicates its presence to other chimpanzees or other species on

a tonic, phasic, or signal level. In some cases, therefore, an animal that can achieve a given level of communication can also interact at lower levels, particularly during the course of its development.

Returning to the original problem of the definition of communication, communication could include any one or more levels in the above hierarchy. Even so, difficulties would arise in classifying doubtful or intermediate types. The difficulties of classification are further compounded by the fact that developmental information is often lacking or incomplete. My own opinion would be to include only the signal, symbolic, and language levels under the term communication. This is not very different from the definition used in practice by many animal behaviorists, including Frings and Frings (1964). The inclusion of the notion of a "social unit" (Cherry, 1957), seems to be unnecessarily restrictive, while the broad idea that communication includes any behavior that influences the behavior of another organism is too inclusive to be useful. Furthermore, this broad definition brings in the additional problem of defining "behavior." Although the definition and limitation of the concept of communication is frankly practical and arbitrary, levels of integration can be shown to exist as an objective fact. The utilization of the levels concept can make it possible to differentiate communication from other forms of interaction in a way that is biologically significant.

ADDENDUM

Since the above manuscript was submitted for this volume, there have been several major publications in the field of animal communication. I shall remark briefly upon three points of importance covered by these recent reports.

Sebeok (1968) produced an encyclopaedic compendium comprising 24 articles by 26 authors, covering virtually all aspects of animal communication, from the general and theoretical to the specific and specialized. The significance of qualitative differences at different levels of organization, and the importance of ontogenetic studies was touched upon by a few of the contributors, notably those dealing with human and other primate communication.

In the area of infrahuman primate behavior, it has now become clear, especially from field studies, that these animals communicate on a highly complex level. Recent volumes of collected papers, edited by Altmann

(1967), DeVore (1965), and Jay (1968), provide many examples of symbolic communication, particularly through gestures.

The problem of aggressive behavior has received considerable attention recently. A popular book by Lorenz (1966) was given laudatory reviews by many mass media contributors, and his rigid instinctivist views have been accepted as gospel by a significant segment of the lay public. Professionals in the field, however, have been strongly critical (cf. Montagu, 1968). It is interesting to note that even some of the critics treat aggression as if it were a distinctive behavior pattern found throughout the animal kingdom. Aside from the primitively anthropomorphic notions implicit in this broad concept of aggression, it should be clear that a behavior that looks like aggression in fish or rats must have very different ontogenetic antecedents than aggressive behavior among men. Examined from the point of view of levels of organization, aggressive behavior can logically be restricted only to some kinds of interactions in man, where motivations can be verbalized, however imperfectly. Such actions as biting, jostling, fluffing of feathers, spreading of fins, etc., among other animals are interactions on different levels of behavioral organization—qualitatively and ontogenetically different. To categorize all such behavior as aggressive is to ignore the profound differences that exist.

Altmann, S. A., ed. 1967. *Social communication among primates*. Chicago: Univ. Chicago Press.

DeVore, I., ed. 1965. *Primate behavior: Field studies of monkeys and apes*. New York: Holt, Rinehart and Winston.

Jay, P. C., ed. 1968. Primates. *Studies in adaptation and variability*. New York: Holt, Rinehart and Winston.

Lorenz, K. Z. 1966. *On aggression*. New York: Harcourt, Brace & World.

Montagu, M. F. A., ed. 1968. *Man and aggression*. Oxford: Oxford Univ. Press.

Sebeok, T. A., ed. 1968. *Animal communication*. Bloomington: Indiana Univ. Press.

ACKNOWLEDGMENTS

I wish to express my appreciation and gratitude to the late Dr. T. C. Schneirla, whose writings and comments strongly influenced and inspired the ideas contained in this essay. In addition, I am deeply grateful for the encouragement and criticisms of Drs. Lester R. Aronson, Ethel Tobach, James W. Atz, and David W. Jacobs, of The American Museum of Natural History. Thanks are also expressed to the many graduate students at the

City College who served as sounding boards for the ideas contained in this paper.

Support for this work was derived from Contract 552(06) with the Office of Naval Research, and Grant GB-4364 from the National Science Foundation.

REFERENCES

Alexander, R. D. 1960. Sound communication on Orthoptera and Cicadidae. In W. E. Lanyon and W. N. Tavolga, eds., *Animal sounds and communication*. Am. Inst. Biol. Sci., Publ. no. 7, pp. 38–92.

Altmann, S. A., ed. 1967. *Social communication among primates*. Chicago: Univ. Chicago Press.

Blair, W. F. 1963. Acoustic behaviour of amphibia. In R.-G. Busnel, ed., *Acoustic behaviour of animals*. Amsterdam: Elsevier. Pp. 694–708.

Bogert, C. M. 1961. The influence of sound on the behavior of amphibians and reptiles. In W. E. Lanyon and W. N. Tavolga, eds., *Animal sounds and communication*. Am. Inst. Biol. Sci., Publ. no. 7, pp. 137–320.

Breder, C. M., Jr., and D. E. Rosen. 1966. *Modes of reproduction in fishes*. New York: Natural History Press.

Busnel, R.-G., A. Moles, and M. Gilbert. 1962. Un cas de langue sifflée utilisée dans les Pyrénnées françaises. *Logos* 5: 76–91.

Carmichael, L. 1964. The early growth of language capacity in the individual. In E. H. Lenneberg, ed., *New directions in the study of language*. Cambridge: M.I.T. Press. Pp. 1–22.

Carpenter, C. C. 1963. Patterns of behavior in three forms of the fringe-toed lizards (*Uma*—Iguanidae). *Copeia*, pp. 406–412.

Carthy, J. D. 1958. *An introduction to the behaviour of invertebrates*. New York: Hafner.

Cherry, C. 1957. *On human communication*. New York: Wiley.

de Beer, G. 1958. *Embryos and ancestors*, 3rd ed. Oxford: Oxford Univ. Press.

Dreher, J. J., and W. E. Evans. 1964. Cetacean communication. In W. N. Tavolga, ed., *Marine bio-acoustics*. Oxford: Pergamon Press. Pp. 373–393.

Dumortier, B. 1963. Ethological and physiological study of sound emissions in Arthropoda. In R.-G. Busnel, ed., *Acoustic behaviour of animals*. Amsterdam: Elsevier. Pp. 583–654.

Esch, H. 1967. The evolution of the bee language. *Sci. Am.* 216 (Apr. 1967): 96–104.

Fraenkel, G. S., and D. L. Gunn. 1961. *The orientation of animals*. New York: Dover.

Frings, H., and M. Frings. 1964. *Animal communication*. New York: Blaisdell.

Frisch, K. von. 1950. *Bees, their vision, chemical senses and language*. New York: Ithaca.

Gross, J. D., and L. Caro. 1965. Genetic transfer in bacterial mating. *Science* 150: 1679–1684.

Hertz, M. 1934. Zur Physiologie des Formen und Bewegungssehens. III. Figurale Unterscheidung und reziproke Dressuren bei der Biene. *Z. Vergleich. Physiol.* 21:604–615.

Hockett, C. F. 1961. Logical considerations in the study of animal communication. In W. E. Lanyon and W. N. Tavolga, eds., *Animal sounds and communication.* Am. Inst. Biol. Sci., Publ. no. 7, pp. 392–430.

Johnson, D. L. 1967. Honey bees: do they use the direction information contained in their dance maneuver? *Science* 155: 844–847.

Köhler, W. 1925. *The mentality of apes.* New York: Harcourt, Brace.

Lang, T. G., and H. A. P. Smith. 1965. Communication between dolphins in separate tanks by way of an electronic acoustic link. *Science* 150: 1839–1844.

Lanyon, W. E. 1961. The ontogeny of vocalizations in birds. In W. E. Lanyon and W. N. Tavolga, eds., *Animal sounds and Communication.* Am. Inst. Biol. Sci., Publ. no. 7, pp. 321–347.

Lenneberg, E. H., ed. 1964. *New directions in the study of language.* Cambridge: M.I.T. Press.

Lettvin, J. Y., H. R. Maturana, W. H. Pitts, and W. S. McCullough. 1961. Two remarks on the visual system of the frog. In W. A. Rosenblith, ed., *Sensory communication.* New York: M.I.T. Press and Wiley. Pp. 757–776.

Lissmann, H. W. 1963. *Electric location by fishes. Sci. Am.* 208 (Mar. 1963): 50–59.

Maier, N. R. F., and T. C. Schneirla. 1942. Mechanisms in conditioning. *Psych. Rev.* 49: 117–134.

Marler, P. 1960. Bird songs and mate selection. In W. E. Lanyon and W. N. Tavolga, eds., *Animal sounds and communication.* Am. Inst. Biol. Sci., Publ. no. 7, pp. 348–367.

Marler, P. 1961. The logical analysis of animal communication. *J. Theoret. Biol.* 1: 295–317.

Moles, A. 1963. Animal language and information theory. In R.-G. Busnel, ed., *Acoustic behaviour of animals.* Amsterdam: Elsevier. Pp. 112–131.

Moltz, H. 1965. Contemporary instinct theory and the fixed action pattern. *Psychol. Rev.* 72: 27–47.

Morris, C. W. 1946. *Signs, language and behavior.* New York: Prentice-Hall.

Protasov, V. R. 1965. Bioakustika ryb. *Akademia Nauk, S.S.S.R.*

Roeder, K. D., and A. E. Treat. 1961. The reception of bat cries by the tympanic organ of noctuid moths. In W. A. Rosenblith, ed., *Sensory communication.* New York: M.I.T. Press and Wiley. Pp. 545–560.

Ryan, E. P. 1966. Pheromone: evidence in a decapod crustacean. *Science* 151: 340–341.

Salmon, M. 1965. Waving display and sound production in the courtship behavior of *Uca pugilator,* with comparisons to *U. minax* and *U. pugnax. Zoologica* 50: 123–149.

Schneirla, T. C. 1953. The concept of levels in the study of social phenomena.

In M. Sherif and C. Sherif, eds., *Groups in harmony and tension*. New York: Harper. Pp. 52–75.

Schneirla, T. C. 1957. Theoretical considerations of cyclic processes in doryline ants. *Proc. Am. Philos. Soc.* 101: 106–133.

Schneirla, T. C. 1959. An evolutionary and developmental theory of biphasic processes underlying approach and withdrawal. In M. R. Jones, ed., *Nebraska symposium on motivation*. Lincoln: Univ. Nebraska Press. Pp. 1–42.

Schneirla, T. C. 1962. Psychological comparison of insect and mammal. *Psychol. Beiträge* 6: 509–520.

Schneirla, T. C. 1965. Aspects of stimulation and organization in approach/withdrawal processes underlying vertebrate behavioral development. In D. S. Lehrman, R. Hinde and E. Shaw, eds., *Advances in the study of behavior*, vol. 1. New York: Academic Press. Pp. 1–74.

Schneirla, T. C. 1966. Behavioral development and comparative psychology. *Quart. Rev. Biol.* 41: 283–302.

Schneirla, T. C., and J. S. Rosenblatt. 1961. Behavioral organization and genesis of the social bond in insects and mammals. *Am. J. Orthopsychiat.* 31: 223–253.

Sebeok, T. A. 1965. Animal communication. *Science* 147: 1006–1014.

Sebeok, T. A., A. S. Hayes, and M. C. Bateson, eds. 1964. *Approaches to semiotics*. The Hague: Mouton & Co.

Shaw, E. 1960. The development of schooling behavior in fishes. *Physiol. Zool.* 33: 79–86.

Shaw, E. 1961. The development of schooling in fishes. II. *Physiol. Zool.* 34: 263–272.

Smith, W. J. 1965. Message, meaning, and context in ethology. *Am. Naturalist* 99: 405–409.

Tavolga, W. N. 1956. Visual, chemical and sound stimuli as cues in the sex discriminatory behavior of the gobiid fish, *Bathygobius soporator*. *Zoologica* 41: 49–64.

Tavolga, W. N., ed. 1964. *Marine bio-acoustics*. Oxford: Pergamon Press.

Tinbergen, N. 1959. Comparative studies of the behaviour of gulls. *Behaviour* 15: 1–70.

Tobach, E. and T. C. Schneirla. 1968. The biopsychology of social behavior in animals. In R. E. Cooke, ed., *Biological basis of pediatric practice*. New York: McGraw-Hill.

Vincent, F. 1963. Acoustic signals for auto-information or echolocation. In R.-G. Busnel, ed., *Acoustic behaviour of animals*. Amsterdam: Elsevier. Pp. 183–227.

Washburn, S. L., P. C. Jay, and J. B. Lancaster. 1965. Field studies of Old World monkeys and apes. *Science* 150: 1541–1547.

Wenner, A. M. 1962. Sound production during the waggle dance of the honey bee. *Anim. Behav.* 10: 79–95.

Wenner, A. M. 1967. Honey bees: do they use the distance information contained in their dance maneuver? *Science* 155: 847–849.

Wiener, N. 1950. *The human use of human beings*. Boston: Houghton Mifflin.

Wilson, E. O. 1965. Chemical communication in the social insects. *Science* 149: 1064–1071.

Winn, H. E. 1967. Vocal facilitation and the biological significance of toadfish sounds. In W. N. Tavolga, ed., *Marine bio-acoustics*, vol. 2. Oxford: Pergamon Press. Pp. 283–304.

Wolf, E. 1937. Flicker and the reactions of bees to flowers. *J. Gen. Physiol.* 20: 511–518.

Yerkes, R. M., and A. W. Yerkes. 1929. *The great apes: A study of anthropoid life*. New Haven: Yale Univ. Press.

Yerkes, R. M., and A. W. Yerkes. 1935. Social behavior in infrahuman primates. In C. Murchison, ed., *Handbook of social psychology*. Worcester, Mass.: Clark Univ. Press. Pp. 973–1033.

HELMUT E. ADLER

Yeshiva University
and
The American Museum of Natural History
New York

Ontogeny and Phylogeny
of Orientation

Ach! Zu des Geistes Flügeln wird so leicht
Kein körperlicher Flügel sich gesellen.
Doch ist es jedem eingeboren,
Dass sein Gefühl hinauf und vorwärts dringt,
Wenn über uns, im blauen Raum verloren,
Ihr schmetternd Lied die Lerche singt,
Wenn über schroffen Fichten höhen
Der Adler ausgebreitet schwebt,
Und über Flächen, über Seen
Der Kranich nach der Heimat strebt.

Goethe, FAUST

ANIMALS AS NAVIGATORS

Requirements of Navigation

Orientation is a selective process in which certain stimuli in the environment elicit a response sequence that results in a nonrandom pattern
of locomotion, direction of the body axis, or both. This process is limited
by the sensitivity of the sense organs involved, the steepness of the
selection gradient, and the accuracy of the mechanism relating sensory
(afferent) input to motor (efferent) output.

In most instances orientation can be looked upon as a closed feedback
system in which the consequences of the initial activity are taken into
account in the determination of subsequent response patterns. Oriented
activity continues in this way until the original stimulus pattern is removed

or superseded, adaptation lowers the effectiveness of the eliciting stimuli or stimulus differences, or the internal state of the organism changes.

FEEDBACK PROCESS

This point of view lends itself well to the analysis of specific orientation problems in terms of self-regulating systems. A number of successful models have been developed, using the techniques of systems analysis. Thus, for example, von Holst (1950) in a pioneering study, analyzed body position in fish as a function of the dorsal light reaction and statolith organs. Mittelstaedt (1962, 1963) applied systems analysis to the orientation of insects, working out, in particular, a model for prey-localization by the praying mantis. In another application, Hassenstein (1959) was able to develop a model to account for the optomotor turning of the beetle *Chlorophanus*.

These analyses owe their success to the fact that the specific stimuli to which orientation occurred were well known and the sensory mechanisms mediating the responses could be specified. In the case of migration and homing of birds and other animals, where the stimuli used for orientation cannot, as yet, be specified satisfactorily and where the mechanism itself still needs clarification, a simplified model can still serve a useful purpose. This model can be used as a basis for quantitative considerations and for investigating the components of the complex system of orientation that must be involved in navigation.

As Kalmus (1964) has pointed out, a bird must perform the functions of an analogue computer while navigating. It must take its cues from a reference source, code this input and compute direction and distance of its course from this information. In the parallel case of the bee, at least, the end result can then be displayed symbolically as the waggle dance on the vertical comb. The deviation of its pattern from the vertical represents the angle between the azimuth direction of the sun and the food source, while the number of waggles communicates the distance to be flown (Lindauer, 1967; but see also Johnson, 1967; Wenner, 1967).

In computer simulation of the migration of the whooping crane (Adler and Adler, 1965) and the homing of pigeons (Adler and Adler, 1966) a simplified model of orientation was employed. The flight path was broken down into a series of steps. This reiterative process was translated into a mathematical model, suitable for computer simulation, by assigning the position of the bird x and y coordinates in a system of coordinates that used the starting position as zero. The location of the bird after each step could be specified in terms of the distance traveled and the angle traversed. The degree of orientation, in such a system, was dependent on the departure from randomness of the distribution of angles or the inequalities of the distances traveled in each step.

In the simulation a bias in flight path was introduced by making the vector toward "home" proportionally larger than any other direction. This bias, which could be varied from zero to one, was determined by the eccentricity of an ellipse, the key part of the computer model. The longer axis of the ellipse was pointed toward the goal and each leg of the flight path consisted of the distance from one of its foci to the intersection with the boundary of the ellipse. The angle of each step was determined by a random number generator. At the end of each step the major axis of the ellipse was again aligned in the home direction and the process repeated. The flight terminated when the home region was reached or when the time allotted ran out.

The shape of the ellipse thus determined the degree of orientation and the total distance that had to be traversed in order to reach the goal. For example, if the eccentricity of the ellipse was zero, the distance would be equal in all directions and the flight path would be truly random. If, on the other hand, the eccentricity was one, the ellipse would in effect become a straight line and each step would be directly in the home direction and cover the maximum distance assigned to the model being tested.

In the simulation of pigeon homing it was assumed in addition that familiarity with the home territory would result in better orientation near home. The eccentricity of the ellipse was accordingly allowed to increase as a function of closeness to the "loft."

This model incorporated a "two-step" process, such as Kramer's (1953) "map and compass" concept, where the rotation of the axes of the co-ordinate system represented the "map" and the nonrandom distribution of distances the "compass" of the bird. A closed feedback loop was established by making each step in the flight pattern dependent on the position reached in the previous step. An additional advantage of the model was the fact that the uncertainty of the orientation process was represented by the degree of bias introduced at the time a new step was initiated. This uncertainty represented the margin of error due to the imperfections of the bird's navigational abilities, the limitations of its sensory discriminations, and the inaccuracy of its internal clock (Adler, 1963a).

Means Available for Orientation

Fundamental to any analysis of orientation is the nature of the stimuli that serve as cues in the orientation process. Discussion will be limited to long-distance orientation behavior, as found in navigation and homing. No clear differentiation can be made, however, between these types of orientation and orientation over shorter distances. In fact, some of the experiments in this area, such as those based on migratory restlessness, involve merely the alignment of the body-axis with respect to a fixed reference source of stimulation (e.g., Kramer, 1950). It is implicitly assumed that locomotion would continue in the same direction. If long-distance navigation is analyzed as consisting of a series of steps, the orientation process would be repeated many times during the course of a flight and the information obtained used to correct the flight-path.

A convenient distinction has been made between three types of orientation (Griffin, 1952). These can be restated in terms of the stimuli employed in the feedback loop. In random or patterned search an animal uses its own body as a reference source (Type I). If a single external reference is used, the animal can establish a direction or bearing (Type II). Two independent reference points are necessary and sufficient for true navigation or homing (Type III). (For examples, see reviews by Lindauer, 1963; Matthews, 1955; Schmidt-Koenig, 1964c).

Many different kinds of stimuli have been suggested as possible reference sources.

CELESTIAL BODIES

The use of the sun in orientation was first demonstrated by Santschi for ants (1911). The work of von Frisch (1946, 1950) indicated that bees not only oriented to the sun, but had the ability to set a steady course by it, compensating for its apparent movement. Kramer's findings (1949) established the same ability for birds. Kramer and St. Paul (1950) showed that starlings could be trained to a particular compass direction, using only the sun's position as a cue. Reptiles (Gould, 1957; Fischer and Birukow, 1960), amphibians (Ferguson, 1963), and fish (Hasler et al., 1958) have also been shown to have the capacity for sun-compass orientation. Many invertebrates, such as crustaceans (Pardi and Papi, 1953), spiders (Papi, 1955), and beetles (Birukow, 1953), also appear to have the same ability.

According to Matthews (1951a) and Pennycuick (1960) birds utilize not only the sun's azimuth angle but also some aspects of its altitude in navigation.

Besides the sun, other celestial cues may also be employed. Sauer and Sauer (1955) claimed to have demonstrated a complete navigational system in nocturnally migrating birds by use of the stars, either under the natural sky or under the artificial stars of a planetarium (Sauer, 1957; see also Wallraff, 1965b; Emlen 1967a,b). Papi and Pardi (1953) and Enright (1961) have shown evidence for orientation to the moon.

LANDMARKS

Orientation by means of landmarks is widely practiced. It has been worked out in detail for insects (see, for example, Chmurzynski, 1964). Aronson (1951) found that gobiid fish, living in tidal pools, used landmarks learned during high tide to locate their home pool during low tide. In a laboratory situation, fish (*Phoxinus laevis*) oriented to irregularities in the test tank ("landmarks") in preference to a simulated sun ("celestial mark"), when the two were in competition (Hasler, 1956b). Homing pigeons typically are trained by initial orientation flights to familiarize them with the vicinity of their loft and subsequent releases at increasing distances, usually in a single direction (Allen, 1959). Matthews (1952) found, however, that pigeons were relatively slow to learn "landmarks" in a laboratory choice experiment. A retest showed retention after periods of one and two years. Skinner (1950) reported that pigeons trained to peck at a specific spot on an aerial photograph were able to repeat their performance after a lapse of four years.

Migrating birds often follow well-known routes (flyways) defined by such geographical features as coastlines, river valleys and divisions between different ecological zones (Bellrose, 1967b). Often, but not always, their route is deflected at a coastline or an unfavorable landscape (Gruys-Casimir, 1965). Radar evidence has clearly shown, however, that some species of birds freely cross coastlines and that long nonstop flights are carried out over water, for example, from New England to South America (Lack, 1959a; Drury and Keith, 1962). According to Myres (1964), migrating Scandinavian thrushes, observed by radar while flying over water were seen to ascend around dawn, presumably stimulated by the sight of the sea beneath them, and reoriented their flight toward land. In many cases migrators seemed to follow an intercept course, flying over featureless or unknown terrain until a landmark, such as a coastline or familiar territory was reached.

METEOROLOGICAL FACTORS

The meteorological elements that make up the weather occur in complex sequential patterns. Wind direction and speed, temperature, humidity, cloud cover, precipitation and visibility depend on air masses, fronts, and barometric pressure patterns. Consequently the meteorological cues available for orientation are not independent of one another. For example, wind direction, temperature change and cloudiness would be highly correlated in the typical frontal passage.

The utilization of meteorological elements in orientation may be based on passive transport or on active recognition of their value as reference cues (Rabøl, 1967). Williamson (1955), for example, has stressed the effect of certain wind patterns in bringing about typical migration routes by down-wind drifting. Similarly, Lowery and Newman (1955) have interpreted data obtained by studying migrating birds silhouetted against the moon as favoring a down-wind drift. On the other hand, Lack (1959b) showed that birds were capable of correcting for wind drift.

Birds tracked by Drury and Nisbet (1964) and by Bellrose (1967a) compensated for wind drift, even though the whole sky was covered by clouds.

Cold front passages have long been noted as a stimulus for migrating birds. According to Hassler, Graber, and Bellrose (1963), migrators may have been responding to the wind shift, associated with the front, rather than any of the other elements involved. Mueller and Berger (1961) found that hawk migration correlated with frontal passages. The hawks may have been influenced by the occurrence of suitable updrafts for soaring and gliding associated with these weather conditions (Mueller and Berger, 1967). Ducks also may seek out favorable weather conditions, particularly of wind direction (Bergman and Donner, 1964). Radar studies of mass migration through central Illinois disclosed an elliptical migration pattern dependent on the general pattern of the atmospheric circulation (Graber, 1968). Not only is the evidence regarding specific meteorological factors still unsettled, but the complete absence of any relationship to the weather has also been found to be true in certain instances (Stevens, 1957).

MAGNETISM

Since first proposed in 1855 by von Middendorff (1859), the idea that birds gain some of their information on direction from the earth's magnetic field has given rise to many speculations and some experimentation. In its simplest form, this hypothesis assumes that birds follow the lines of magnetic force (Allen, 1948). Yeagley's well-known experiments (1947, 1951) are based on a hypothesis that birds navigate by means of a bi-coordinate system made up of the vertical component of the earth's magnetic field and lines of equal Coriolis force. These experiments must be considered to have had a negative outcome (Thorpe, 1949; Wilkinson, 1949; Matthews, 1955, pp. 79–91).

Direct approaches to this problem by trying to distort the near magnetic field by means of magnets attached to homing birds did not disturb orientation (Matthews, 1951b; Riper and Kalmbach, 1952). Attempts to obtain a threshold for magnetic sensitivity on pigeons by conditioning techniques have also failed (Orgel and Smith, 1955). Radio signals were at one time thought to exert a disturbing influence on the geomagnetic field and thus to affect birds using magnetic cues, but this possibility must be discounted according to the most recent survey (Eastwood and Rider, 1964).

Despite these negative results, two theories based on magnetic force have recently been proposed. Barnothy's (1964) twofold system assumes that a bird can gain information regarding its position either from an alternating current induced by the dynamo effect of the flapping wings or from a "ring current" generated, due to relativistic principles, by the interaction of the flying bird and the direction of its flight. The first of these effects had already been mentioned by Wilkinson (1949) as an alternative to the much simpler electrostatic field proposed by Yeagley (1947) and had been dismissed as being too small in magnitude to be perceptible. Talkington (1964) centers his theory on the pecten, a highly vascularized membrane extending into the eye of many species of birds. Electric potentials generated in the pecten by cutting the lines of magnetic force when birds are in flight cause current flow whenever changes in direction occur. Birds orient by keeping the current constant and navigate by matching current flow to that of some familiar flight directions. These theories suffer from the lack of a demonstrable receptor of sufficient sensitivity for magnetic or electric forces. The pecten has most recently been studied by Wingstrand and Munk (1965) and has been shown to be primarily a substitute for intraretinal blood vessels, without ruling out, however, additional functions.

Some evidence in support of magnetic field sensitivity has recently been presented. Robins (*Erithacus rubecula*) had been shown to orient indoors by Merkel and Fromme (1958) and by Fromme (1961) during migratory restlessness. Although Perdeck (1963) was not able to replicate these results, Merkel, Fromme, and Wiltschko (1964) answered Perdeck's objections and pointed out differences in their procedure (but see also Wallraff, 1965b). Merkel and Wiltschko (1965) have now reported that migratory restlessness was nondirectional in an all-steel chamber, unless the birds had first become adapted to the lowered intensity of the magnetic field in the shielded room. Orientation in a predictable direction could be brought about by introducing an artificial magnetic field, similar in strength to that of the earth, by means of Helmholtz coils. In a later experiment, Robins displayed directional migratory restlessness in an unshielded room in accordance with an artificially created magnetic field, when a second Helmholtz coil was used to cancel out the earth's magnetic field (Wiltschko and Merkel, 1965). In view of the previous discussion of recent theories of magnetic orientation, it should be pointed out that very little actual flying or even wing flapping was possible in these experiments.

A study by El'darov and Kholodov (1964) showing that a constant magnetic field increased motor activity of passerine birds may possibly be related to these findings. Conditioned responses to magnetic fields have also been obtained on fish. Pigeons failed to condition to a magnetic field, but an inhibitory action of a magnetic field on previously conditioned responses could be observed (Kholodov, 1964).

Some findings appear to support orientation to magnetic cues in invertebrates. According to Schneider (1963a,b), cockchafers (*Melolontha vulgaris*) responded to artificial magnetic fields. At least part of a population of house flies (*Musca domestica*) have been shown to orient to natural and artificial magnetic fields (Becker and Speck, 1964; Becker, 1965). In a series of experiments, mud snails (*Nassarius obsoletus*) (Brown, Webb, and Barnwell, 1964), planarians (*Dugesia dorotocephala*) (Brown and Park, 1965), and the protozoans *Paramecium* and *Volvox aureus* (Brown, 1962; Palmer, 1963) were found to give oriented responses to magnetic fields.

MISCELLANEOUS GEOPHYSICAL FORCES

The Coriolis force, an apparent effect on all bodies moving over the surface of the earth, was first suggested by Martorelli in 1899 as a possible factor passively deflecting birds on migration. Ising (1945) and Beecher (1952, 1954), as well as Yeagley (1947, 1951), have also advanced theories based on utilization of information from this source. However, Thorpe (1949), Wilkinson (1949), and Matthews (1955) have subjected this hypothesis to devastating criticism.

Because of the intimate relationship of magnetism and electricity, most theories involving magnetic cues imply also the presence of electrical effects. A theory, based specifically on the electrical charges accumulated on the bird's feathers in flight, has been proposed by Stewart (1957). Electrostatic fields have been found to influence the orientation of the body axis of cockchafers (Schneider, 1963a,b) and of planarians (Brown, 1962). On the other hand, Wiltschko and Merkel (1965) reported that the elimination of the electrostatic field did not interfere with directed migratory restlessness in European robins.

The gravitational field has been proposed as an explanation for experimental findings that cockchafers exhibited an oriented response to 40 kg blocks of lead placed adjacent to their containers, while they did not react to cardboard props of the same size and shape (Schneider, 1964).

INERTIAL NAVIGATION

Theories that birds navigate by dead reckoning date back to Darwin's proposal in 1873. Exner (1893) suggested the vestibular apparatus as the sensory basis of this ability. This theory has now been resurrected in modern guise by Barlow (1964) in the form of an inertial guidance system. Such a system, used today in the navigation of airplanes, missiles, and submarines, computes from information of the initial position and from the output of sensing elements, the present position, relative velocity and heading, without recourse to any external signals. The vestibular organs of vertebrates have been considered the locus of sensory input for such a system. Drury and Nisbet (1964) have discussed the possibility of inertial guidance for the somewhat simpler case of establishing direction only. An analogous system in insects (*Diptera*), based on the gyroscopic action of the halteres, has been hinted at by Pringle (1963). A major obstacle to such a theory, based on vestibular function, has been the results of experiments (Huizinga, 1935; Wallraff, 1965a) showing that cutting the semicircular canals of birds did not impair their homing ability (see also "Sensory Limitations," p. 320).

Ontogeny and Phylogeny

None of these orientation systems, either singly or in combination, have been found satisfactory to account for all observed phenomena in migration and navigation of birds and other animals. One problem has been our inadequate knowledge of the characteristics of the sensory receptors involved and their capabilities to make the extremely fine discriminations demanded by the various theories (Adler, 1963a).

Another serious deficiency has been the lack of attention paid to the development of orientation. Too many studies of orientation gloss over this fruitful avenue of approach by labeling the orientation process as innate or concluding their report of observations by the statement that the directed behavior observed was instinctive. These labels do not add to our understanding of the phenomenon involved.

One of the dangers of relying on the dichotomy between innate and acquired behavior lies in the tendency to be satisfied with classification of behavior into one or the other of these categories, to the detriment of analysis of the process. Schneirla has pointed out repeatedly, in connection with other problems (e.g., 1956), "the instinct problem is one of development." Examination of the ontogeny of orienting behavior therefore may supply us with valuable evidence from which the factors responsible for orientation may emerge.

Research on orientation has been deficient not only in its neglect of developmental aspects, but also in the lack of attention paid to phylogenetic comparisons. Since development differs for each phyletic level, ontogeny of behavior deserves to be studied in each type of organism. Valuable evidence is available in the comparison of both closely related and distantly related organisms with respect to the evolutionary history of orientation and phyletic similarities and differences (Schneirla, 1952, 1957). The study of the phylogeny of orientation, with due regard to the differing levels of functioning for different types of organisms (Schneirla, 1949), may make a valid and unique contribution to the understanding of the principles involved in orientation.

DEVELOPMENTAL VARIABLES

Orienting for the First Time

OBSERVATIONS UNDER NATURAL CONDITIONS

Typical are the kind of results obtained by the large-scale displacement experiments of Rüppell (1944) on hooded crows, Schüz (1949) on storks, and Perdeck (1958, 1967) on starlings. Young birds were transported to release points outside their normal range. Recoveries of marked birds indicated migration had taken place on a course parallel to the normal route, even when guidance by older, experienced birds could be excluded. Generally these birds continued to maintain their displacement in later years. For example, juvenile starlings displaced from Holland to Switzerland (Perdeck, 1958) flew to wintering grounds in southern France and the Iberian peninsula instead of wintering in their normal quarters in northwest Europe. Recoveries in later years showed that these same birds tended to keep on flying to their newly acquired wintering grounds, even though they may have returned to their old breeding range in Holland during the spring migration. Adult and juvenile starlings displaced from Holland to Spain (Perdeck, 1967) during autumn migration showed a clear-cut difference in their behavior. Juveniles were recovered southwest of the release point, adults to the north. Recoveries of juveniles in later seasons suggested parallel migration, rather than return to the area of hatching. It is possible that these starlings took to wintering in Spain. In the reverse situation, on the other hand, mallard ducks of English origin, incubator-hatched and raised in Finland, migrated with the local Finnish population and returned to the place in Finland where they had been raised, even though mallard ducks in England are nonmigratory (Välikangas, 1933).

McCabe (1947) found that wood ducks (*Aix sponsa*) transferred as flightless young from Illinois to Wisconsin returned to the location in Wisconsin where they had first learned to fly. An experiment by Vaught (1964), however, did not fit this pattern. On their first migratory flight, his blue-winged teals (*Querquedula discors*), displaced as flightless young from Minnesota to Missouri, took off in a northerly direction and were recovered in later years from north of the release site or from their wintering range in Texas, the West Indies and Venezuela. According to Löhrl (1959) only about 10 days' experience was necessary to establish the breeding range to which displaced young flycatchers (*Ficedula albicollis*) returned in later seasons. Russian attempts to introduce beneficial species of birds into new breeding territories by displacement of juveniles have met with some success in the case of flycatchers (*Muscicapa hypo-leuca*) and have also been carried out on starlings (Mauersberger, 1957). Schüz' (1964) intensive study of the migratory pathways of storks has established the existence of two populations, one taking a westerly route, the other an easterly route around the Mediterranean.

These observations indicate that young birds on their first migratory flight are already capable of choosing a direction. On their return to the breeding grounds, many birds appear to come back to the locality where they were hatched or first learned to fly. The flight direction on the initial flight seems to be typical of a given population.

The salmon run is an interesting parallel. Donaldson and Allen (1958) reported that silver salmon (*Oncorhynchus kisutch kisutch*) that had been displaced *in ovo* returned to the stream in which they had matured and developed into fry. The extensive studies of Johnson and Groot (1963) and Groot (1965) on sockeye salmon (*Oncorhynchus nerka*) smolts (i.e., first year juveniles) during their seaward migration through large, complex lake systems, appear to have demonstrated preferred directions of swimming corresponding to the most direct route to the outlet. Depending on their area of origin, each group of smolts shifted preferred directions in the sequence appropriate to their shortest pathway to the exit point. Since the smolts were tested in choice tanks outside the lake, current would not have been a cue. Unidirectionality of the orientation was indicated by the fact that displacement did not change the preferred direction. Groot (1965) concluded that celestial cues were used during clear sky conditions, polarized light when cloudy skies obscured the sun (>70%), and an unknown system under complete cloud cover, indoors, and when no other cues were available.

LABORATORY STUDIES

Kramer (1949) and his coworkers originated the technique of testing directional tendencies of birds during migratory restlessness in a circular choice cage (Kramer cage). It was possible to demonstrate consistent orientation of Silviid warblers and shrikes, when exposed to the clear night sky, although their fluttering did not necessarily point in the direction of their normal migratory route (Kramer, 1952; St. Paul, 1953). Although primarily nocturnal migrants, these species appeared to respond to the position of the sun as well (St. Paul, 1953).

A recent large-scale study, using the Kramer cage, by Shumakov (1967) on 250 birds of 36 species in the Soviet Union reported directed migratory restlessness that corresponded in general to the species' migration courses at that particular time and latitude. Juvenals showed the general orientation of their species. Nocturnal migrants were able to hold to a general direction not only by the stars, at night, but also by the sun, during daytime. Twenty-five nocturnal species showed two peaks of activity, in the evening and in the morning, whereas 11 diurnal migrators peaked only during the morning hours.

Two experiments employing modifications of Kramer's method seem to be especially relevant. Sauer (1963) obtained preferred directions for hand-raised golden plovers (*Pluvialis dominica*) during their migratory season. These birds had been collected at Boxer Bay on St. Lawrence Island in the Bering Sea. They were then flown by airplane to Madison, Wisconsin, where some were kept in light-controlled rooms on the day-night rhythm typical of their birthplace, and the rest kept on local time under the natural sky. About six months later, they were transported to San Francisco. When tested there during their period of nocturnal migratory restlessness in the spring, the mean vector of the birds that had been confined, approached most nearly a great circle route from San Francisco to Boxer Bay, whereas that of the birds that had been exposed to the Wisconsin sky appeared to match most closely a great circle route from Madison to Boxer Bay. Neither group showed the direction normally chosen by golden plovers during spring migration—from Hawaii to the Bering Strait. Hamilton (1962b,c) worked with two populations of bobo-links (*Dolichonyx oryzivorus*), one from North Dakota, the other from New York. When tested in San Francisco, the mean vector of the North Dakota birds during autumn migration appeared to parallel the great circle route from their breeding grounds to Florida, whereas the mean vector of the single bird of the New York population that yielded usable results indicated a response parallel to the route from its home territory to Florida.

These experiments confirm observations made under natural conditions that celestial orientation is used. The direction chosen during spring migration appears to be determined by early exposure to the sky, perhaps during some "critical period." Autumn migration seems to depend on the population being tested and displacement results in the choice of a direction parallel to the normal course.

The controlling factors in the directional behavior during the autumnal migration season are not known. The unidirectional orientation, specific to a population, appears to be in some ways parallel to the so-called nonsense orientation found in several species of waterfowl (Bellrose, 1958, 1963; Matthews, 1963b, 1967). Actual flight seems to be unnecessary for this kind of orientation, as it is found in flightless young Canada geese (Hamilton and Hammond, 1960) and Adelie penguins (Emlen and Penney, 1964; Penney and Emlen, 1967). Constant escape directions, dependent on astronomical cues, are also known for a number of inverte-brates, for example, the water skater *Velia currens* (Birukow, 1956) and wolf spiders of the genus *Arctosa* (Papi and Tongiorgi, 1963). For the latter, the learning of new escape directions has been demonstrated.

The nature of the experience necessary for birds to fixate on a breeding territory, to which they then return each season, is presently obscure. A set of experiments on homing pigeons, raised under various degrees of visual restriction constitutes the first significant attempt to investigate the ontogeny of this behavior. These experiments date back to an observation by Pratt (reported by Kramer, 1959) that homing pigeons raised in shielded cages without sight of the horizon, were lost without a trace when required to home. Wallraff (1966) reports on a series of experiments completing the work done by Kramer (1959). A total of 1,111 young homing pigeons were raised under nine different conditions, varying the amount of visual access to the environment. Birds raised in open aviaries showed a clearcut homeward orientation. Compared with this group, birds raised under two conditions completely eliminating sight of the horizon did not orient as well. A group that was brought up in shielded aviaries containing a gap in their north wall, another group with the gap in their south wall, and a third group with the upper part of the wall made of glass, also did not seem to be able to orient. On the other hand, birds raised under a wooden roof with only the lower part of the sky visible, birds raised in an aviary with a second floor from which the horizon could be seen, and birds raised in a screened enclosure provided with five gaps giving a noncontinuous view of their surroundings did not differ significantly from the group raised in open aviaries. At present these results raise more questions than they answer, but if confirmed and systematically extended, they could provide valuable information.

Improvement of Orientation with Practice

OBSERVATIONS ON NATURAL POPULATIONS

Learning clearly plays a part in both navigation and homing, as shown in experiments involving repeated releases. Matthews (1964) collected data on 144 Manx shearwaters transported over distances between 65 and 415 miles. Neither initial homeward tendency nor return rate were affected by practice, but homing speed improved markedly. Mewaldt (1964) studied golden and white crowned sparrows (*Zonotrichia atricapilla* and *leucophrys*) displaced from San Jose, California, their winter range, to Baton Rouge, Louisiana. Of the 26 birds that returned the next year, 22 were redisplaced to Laurel, Maryland. Six of these birds were recovered the next winter at San Jose. These birds apparently returned to their breeding grounds during spring migration, prior to coming back to their winter quarters. The comparatively high return ratio suggests that the first displacement had improved their ability to perform the same feat the second time.

There exists extensive evidence on homing pigeons from pigeon fanciers. Allen (1959) states that training of racing pigeons should start somewhere between six weeks and three months of age. Releases are made at increasing distances from the loft until ultimately 2- or 3-year old birds are ready for long distance races over a course of 500–600 miles or more. Systematic studies of homing pigeons are reported by Wallraff (1959) and by Schmidt-Koenig (1964c). Homing success increased rapidly at first, then gradually reached a maximum. Homing speed, but not necessarily initial orientation improved with practice. General experience appeared to be more important than training in a particular direction. As Schmidt-Koenig (1964c) has pointed out, however, some positive effect could be expected due to the selection caused by the nonreturn of poor homers. General habituation to being handled and familiarity with the expected task may also contribute to the improved performance (Matthews, 1964).

Initial orientation as a function of experience did show improvement in experiments by Matthews (1963a) and Graue (1965). The factor involved appeared to be distance from the loft. Matthews found that better performance was shown if the release points were less than 18 miles or more than 50 miles from home. Graue's release distance was approximately 10 miles from the loft. Schmidt-Koenig (1966), studying initial release directions of experienced birds, has also found a zone of roughly 12.5 miles from the home, where initial orientation was more nearly homeward directed than at distances between 12.5 miles and about 62–94 miles. Initial orientation between the latter two distances was marked by random scattering and large differences in performance between release days. Above 94 miles, the rate of initial choice of a homeward direction again reached the level found for pigeons released close to the loft.

Improvement of homing on repeated releases may be due to learning of landmarks surrounding the home site and to higher effectiveness of some other navigational system at greater distances. The "dead belt" found by both Matthews (1963a) and Schmidt-Koenig (1964a) would then be due to releases in unfamiliar territory, too close to benefit from the navigational cues available beyond a minimum distance.

LEARNING UNDER CONTROLLED CONDITIONS

The development of suitable methods has permitted experimentation aimed at elucidating the specific cues animals may need to learn to find a reward or to avoid punishment in a given compass direction. In the training experiments first developed by Kramer and St. Paul (1950), starlings learned to take food presented at any of twelve compass directions in a circular cage. Pigeons were trained in a situation requiring discrimination between four choice directions. Western meadowlarks (St. Paul, 1956) and various species of ducks (Hamilton, 1962a) also have been tested, the latter with water as a reward.

This ability to learn a compass direction is not confined to birds. Fish (Hasler et al., 1958) have been trained to a compass direction, using both food or escape as reward. Lizards (Fischer and Birukow, 1960) have learned to orient when rewarded by warmth. Newts (*Taricha granulosa*) learned to respond to goal boxes in relation to a stationary simulated sun (Landreth and Ferguson, 1969). And the pioneer work of von Frisch (1950) and his coworkers has demonstrated the ability of bees to learn to fly to a food source in a given compass direction.

Systematic exploration of the environmental variables these animals used to make their directional choices have yielded information pointing to celestial bodies as the main source of orientation. For example, Kramer and St. Paul (1950) were able to shift the search direction of their starlings by changing the apparent direction of the sun by mirrors, and Kramer (1952) used a 250 watt electric light bulb as a simulated sun to which orientation could be learned. Training to a constant direction took "about a fortnight."

Celestial orientation entails the simultaneous judgment of at least two factors, the azimuth position of the sun and the local time. Thus, for example, at local noon in the northern hemisphere the sun would be precisely above the southern horizon. The acquisition of the capacity to make these discriminations is essential to orientation. Evidence to throw some light on this matter is now becoming available.

A regularly changing stimulus can serve to pace the rate of key pecking in pigeons on fixed ratio or on fixed interval schedule (Ferster and Skinner, 1957). The sun's pathway could similarly become the discriminated stimulus providing the occasion for directed locomotion. Starlings allow for the movement of the sun when making a directional choice, and it is possible, but difficult, to train birds to make a response at a constant angle to the sun, as Kramer (1952) did with his starling "Heliotrope." Hoffmann (1953) raised a starling without sight of the sun beginning with the 12th day after hatching. This bird was able to use the sun as a compass, although its error was larger than that usually found in birds brought up with normal exposure to the sun. Shifts in the light-dark cycle resulted in predictable changes in choice direction of previously trained birds (Hoffmann, 1954). Thus, while it is not clear how or when birds learn to calculate the sun's path, its influence on directional choice behavior is well established not only in starlings, but in pigeons (Schmidt-Koenig, 1958).

Evidence is also available that lizards (Fischer and Birukow, 1962), brought up completely without view of the sun, learned the sun's pathway by being exposed to a minimum of three separate sun positions.

A parallel situation exists with regard to the use of the stars at night. In searching for the stellar patterns guiding Indigo Buntings (*Passerina cyanea*) Emlen (1967b) found, in a planetarium experiment, that star patterns were apparently used as cues for oriented migratory restlessness. No single star or constellation seemed to be essential, however, as could be shown by blocking out various portions of the artificial sky. Different buntings seemed to rely on different stellar information, although it appeared that features of the northern celestial hemisphere were especially predominant. The important fact is that no constant patterns were genetically fixed, each bird apparently forming its own guidance system.

Displacement experiments on birds, fish, and bees have added further information on the ability of some animals to learn to discriminate the movement pattern of the sun. These experiments were based on the fact that the sun's apparent motion varies as a function of latitude and with the seasons (see Braemer, 1960). If the ability to allow for the sun's movement were not subject to learning, animals in the tropics or moving across the equator would face an impossible task in attempting to use the sun for navigation without a Nautical Almanac. In the northern hemisphere, north of the Tropic of Cancer, they would have to base their orientation on a sun traveling clockwise from east to south to west; in the southern hemisphere, south of the Tropic of Capricorn, the sun would appear to them to move counterclockwise from east to north to west; and in the regions in between, the sun might culminate in the north (in summer) or the south (in winter) or directly overhead (at the equinoxes). Only inconclusive evidence is available on birds. Homing pigeons, trained in the northern hemisphere, did not compensate for the changed direction of the sun's movement when transported across the equator (Schmidt-Koenig, 1963b). Most likely they did not have an opportunity to learn during the short periods of exposure to local conditions. Fish, on the other hand (Braemer and Schwassmann, 1963; Schwassmann and Hasler, 1964), were able to adjust their behavior to changed patterns of the sun's movement. Bees (Lindauer, 1959, 1963) also demonstrated their ability to learn to orient to the apparent changes in the sun's path on displacement from the northern to the southern hemisphere. In a related experiment it could be shown that some 500 flights, distributed over five afternoons, was the minimum experience required by the bees to learn the local movement pattern of the sun.

Local time, the second factor essential in the utilization of the sun-compass, is derived from synchronizing the circadian rhythm with the alternation of day and night or other regular changes correlated with

24-hour periodicity. A circadian rhythm itself may be affected by changes in the course of ontogeny and is in turn capable of affecting ontogeny, as pointed out by Rensing (1965). In the absence of exposure to the day-night cycle, an animal's circadian frequencies will deviate from natural time. Birds trained to a compass direction will choose a direction differing from that previously learned, when tested under the natural sun after 11 or 12 days under constant conditions (Hoffmann, 1960). Ferguson, Landreth, and McKeown (1967) tested the sun-compass orientation of the northern cricket frog, *Acris crespitans*. When held in darkness for seven days, the frogs became unable to orient to the sun, but exposure to the normal light-dark regime or to daily temperature and humidity fluctuations reestablished orientation by resynchronizing the frog's circadian rhythm with local time. Shifts in time from the normal day resulted in the frog's making choices that took into account the real position of the sun, but they interpreted it in terms of the shifted time base. Similar findings in sunfish (Schwassmann, 1960) confirm the circadian rhythm as the "internal clock" by which the sun-compass operates.

Circadian rhythms are labile, as shown repeatedly by orientation experiments in which the animal's internal clock was subjected to entrainment out of phase with the natural day, leading to predictable changes in choice of direction (Hoffmann, 1954; Schmidt-Koenig, 1958; Braemer, 1959). Longitudinal displacement, such as Renner's (1959) transport of a colony of bees from Long Island, New York to California, similarly has shown the rapid adjustment of the time sense to new conditions. This sensitivity of the internal clock to environmental influences seems well fitted to celestial orientation by means of the sun compass, but raises serious questions about the use of celestial navigation in animals (Matthews, 1955), as this theory depends on the comparison of local time with a standard time in order to calculate east-west displacement.

EVOLUTION AND ORIENTATION

The Organic Basis

THE ORIENTING REFLEX

The physiological correlates of orientation, especially in birds and other vertebrates, are just beginning to be explored. Sokolov (1960) has described a model for orientation based on somatic and autonomic changes. Sensory input is mirrored by EEG activity, according to Sokolov. Hara, Ueda, and Gorbman (1965) describe the EEG patterns of homing salmon. Water from the home pond, infused into the nasal cavity, produced a vigorous response of high amplitude from the olfactory bulb, while water from other sources had no such effect. Since olfactory cues play a prominent part in salmon homing (Hasler, 1954), these physiological correlates provide confirmation for the behavioral observations. In the same study it was also found that adult spawning salmon showed consistently high EEG amplitudes from the olfactory bulb and the posterior cerebellum, but especially low amplitudes from the optic lobe. Young, hatchery-raised fish showed high activity from the olfactory and optic lobes. This evidence may be interpreted as indicative of orientation by both olfactory and optic modes in salmon fingerlings, as opposed to purely olfactory orientation in the adult salmon returning to the spawning beds and during spawning (see Aronson, p. 83, this volume).

SENSORY LIMITATIONS

Any attempt to account for orientation must take into account the sensory capacities of the species concerned. Many of the theories presently advanced either do not have a demonstrable basis in the animal's sensory capabilities, e.g., theories based on magnetic or electrical forces, or extrapolate from sensory threshold data that may be inappropriate (Wilkinson, 1949; Griffin, 1952; Adler, 1963a, 1963b).

An experiment specifically concerned with inertial navigation has recently been carried out by Wallraff (1965a). Nine homing pigeons with bisected horizontal semicircular canals showed no decrement in homing performance. In fact, three experimental pigeons demonstrated better than average homing ability. These findings effectively rule out Barlow's (1964) inertial navigation hypothesis.

IS THERE MORE THAN ONE KIND OF ORIENTATION?

There is no reason to assume that orientation must be confined to any one sensory modality or system of cues. A combination of stimuli appropriate to the sensory capacities of the animal and its past history may well play a role in complex migration and in homing. For example, salmon appear to be guided by celestial cues while descending to the ocean (Groot, 1965) and on returning to the shore (Hasler et al., 1958), while olfactory cues serve to guide them in their ascent to their breeding grounds (Hasler, 1956a).

Michener and Walcott (1966) followed homing pigeons equipped with radio transmitters by airplane. Their results suggested to them that the pigeons often used three methods to find their home: compass orientation, bi-coordinate navigation, and orientation by familiar landmarks. The pigeons appeared to start out originally in the trained compass direction, using landmarks as checkpoints. If the compass course led them into unknown territory, they generally stopped flying or made an abrupt turn to a well-directed homeward course. Landmarks probably played a major role within 10 miles of the loft.

Schmidt-Koenig's (1964a) discovery of a "dead belt" in pigeon homing (p. 316) points to a combination of learning of landmarks and the improvement of some other navigational system, operative at different distances from the loft. Vleugel (1952) has advanced the hypothesis that chaffinches first use the sun to establish a bearing, then keep to this direction by relating it to wind direction. Landmarks and celestial cues appear to function in the case of the bee (Lindauer, 1963). Indeed, there may exist a hierarchy of modes of orientation, depending on the development of sense organs, availability of cues, and previous experience of the animals concerned.

Random or systematic searching may also play an important role, when orientational cues are not available. Purely theoretical considerations, based on probability principles (Wilkinson, 1952), lead to conclusions that purely random searching can be used to calculate distributions that fit the evidence on percentage return, speed versus distance, maximum time of return, and return times for birds. Computer simulation of salmon homing (Saila and Shappy, 1963), the flight pattern of whooping cranes (Adler and Adler, 1965) and the return to the loft of homing pigeons (Adler and Adler, 1966) also indicates the possibility that an almost random movement pattern, corresponding to only a slight degree of orientation, suffices to account for the success rates found in real fish and birds (see pp. 304–305). In fact, the tracks of radio-equipped homing pigeons, shown by Michener and Walcott (1966) and the tracks of simulated pigeons displayed by a computer (Adler and Adler, 1966) showed a remarkable degree of similarity.

Transfer of the controlling factors in orientation from one set of cues to another appears to be relatively simple. Kramer (1952) obtained orientation to an "artificial sun," Sauer and Sauer (1955) found the simulated sky of a planetarium an effective stimulus. In both cases the cues were actually quite different from the real sun and stars and their acceptance as substitutes is remarkable and probably significant. St. Paul (1953) demonstrated that normally nocturnal migrants were able to orient by the sun.

These examples serve to indicate that a combination of orientation systems may be functioning in direction finding and that shifts from one system to another may be relatively easy. Two factor theories, such as Kramer's (1953) map-and-compass concept have assumed that such multiple processes do take place.

EVOLUTIONARY CONSIDERATIONS

There is a universal need for orientation. The organic basis for orientation must have evolved before any specific mechanisms of orientation could have been utilized. Presumably the earliest forms of orientation involved gradients of stimulation. The evolution of long-distance orientation demanded the perception of fixed environmental reference points. The use of celestial bodies as such reference cues may be quite ancient. As Ferguson and Landreth (1966) point out, "the physical prerequisites for celestial orientation, the presence of celestial phenomena, has gone unchanged for a longer time than other possible orientational cues" (but see "The Rate of Evolution," p. 324). Certainly, the widespread finding of celestial orientation in arthropod and vertebrate phyla (Lindauer, 1963) appears to strengthen this argument.

It should be possible, therefore, to relate the evolution of direction-finding over long distances to its physiological basis. Animals that show highly skilled navigational ability ought to have more highly developed organs, or other basis, for navigation, or vice versa.

In this connection, it should be noted that the domesticated pigeon is descended from the rock dove, a nonmigratory species. Yet the homing pigeon, a variety of the domesticated pigeon, has been bred specially for carrying messages and for racing competition, probably since the days of ancient Egypt (Allen, 1959). These pigeons are capable of great feats of navigation, even without training. It is interesting that a nonmigratory species has developed these capabilities. This fact appears to indicate that the capacity for long-distance orientation is present in nonmigratory, as well as in migratory species. It might be noted that swallows (migrants) and frigate birds (oceanic wanderers) have also been used to carry messages.

By the same token, the evolution of general mechanisms for long-distance orientation raises certain questions about theories of navigation based on special characteristics. If parallel methods of orientation have evolved in many phyla, then arguments in favor of theories demanding specialized organs are weakened. Thus, for example, theories based on the pecten (Talkington, 1964) or feathers (Stewart, 1957) of birds or the vestibular organ (Barlow, 1964) would not meet the criterion of universality.

Phylogenetic Characteristics

CROSS-PHYLETIC COMPARISONS

The many parallels between arthropod and vertebrate orientation must not be allowed to obscure phyletic differences that may exist. For example, the arthropod compound eye has evolved along different principles compared to the vertebrate visual organ. On the basis of this phyletic difference, it is quite possible for arthropods to utilize the polarization of the sky to locate the sun's position in a cloudy sky, while no assumption should be made that vertebrates without an eye demonstrably capable of discriminating the direction of polarization could accomplish the same thing, as claimed, for example, by Groot (1965) (cf. p. 313).

Sometimes differences in life-span ought to be considered. Displaced young birds migrate in a fixed direction, parallel to the "normal" route of the species (cf. p. 312). But if displaced *in ovo* they return to the location where they were hatched or first learned to fly. Bees reorient when transported across the equator (cf. p. 318). Both instances may serve as evidence against a genetically fixed mechanism of interpretation of directional cues. Yet it is important to realize that the individual bird changes its fixed direction, while in the bees, which live only a few weeks, the offspring of the original group must reinterpret the sun's apparent path.

Occasionally behavior patterns do not seem to fit into any phyletic categories. Laboratory experiments testing the angle of orientation to an artificial sun (cf. p. 317) at night have revealed two patterns (for summaries see Birukow, 1960; Schmidt-Koenig, 1964b). Starlings, fish, lizards, and bees behave as if the sun moved clockwise from west to north to east during the night. On the other hand, many arthropods, for example the sand flea, *Talitrus saltator,* and pigeons behave as if the sun moved counterclockwise from west to south to east at nighttime. Exceptions have also been found. Fish at the equator (Braemer and Schwassmann, 1963) and pigeons at Barrow, Alaska, north of the Arctic circle (Schmidt-Koenig, 1964b) individually followed either pattern.

INTRAPHYLETIC COMPARISONS

Differences within the various taxa also merit consideration. Some bird species, for example, are long-distance migrants, while other closely related species do not migrate. Even within the same species some populations may be migratory, while others are not, for example, in the starling. It has also been established that pigeons from various lofts vary in their homing ability (Hoffmann, 1959; Schmidt-Koenig, 1963a). The organic or environmental factors responsible for these differences are at present unknown.

The basic principles of orientation should be general enough to fit broad taxonomic groups. Both day-migrating and night-migrating birds probably are able to use the sun's position as a source of directional cues (St. Paul, 1953). Penguins (Emlen and Penney, 1964) showed consistent orientation, although being flightless, they would not be capable of generating the electrical current induced by wing flapping that Barnothy's (1964) hypothesis demands (cf. p. 308).

These considerations lead to the conclusion that direction of movement cannot be rigidly fixed for any species. On the contrary, considerable flexibility seems to be inherent in the basic method that animals use for orientation. Additional evidence for this point of view is provided by the nature of the physical cues that animals must use in orientation, considered in terms of evolutionary history.

THE RATE OF EVOLUTION

Any fixed reference system is subject to change, when viewed on the scale of evolutionary processes. Celestial navigation by means of star patterns would have to be modified to follow the differences introduced by the stars' proper motions and the precession of the earth's axis (Agron, 1962). Most of the constellations as we know them today will change their configuration so as to be unrecognizable over a period of some 100,000 years, due to the component of the actual motions of the individual stars that is perpendicular to the line of sight. Similar changes have occurred in the past. In addition to this nonrecurrent pattern of change, there is a cyclical change of the relationship between the constellations and the seasons, due to the precession of the equinoxes, caused by the wobble of the earth's axis. In a period of 25,800 years the celestial poles complete one circuit and the constellations return to the same seasonal relationship. Polaris, now the North Star, will be replaced by Vega in 13,000 years, just as Vega must have been the North Star in 1,000 B.C.

Animals relying on the earth's magnetic field would be no better off, if they were to base their navigation on a fixed inherited pattern. The position of the poles changes in an erratic course, so that the angle of declination, the difference between true north and magnetic north, is subject to constant adjustment. While the error introduced by this factor would be relatively small in the middle latitudes, it becomes important in the arctic and antarctic regions. Even more disturbing would be the periodic reversals of the earth's magnetic field shown by recent studies of rocks and sediments (see, for example, Opdyke et al., 1966). The present epoch, dating from about 700,000 years ago, was preceded by a period of reversed magnetic force beginning about 2.4 million years ago and several earlier reversals have been established.

CONCLUSIONS

The emphasis of much orientation research on fixed, innate patterns of behavior has obscured the possibilities inherent in studies of the orientation process itself. Its flexibility and dependency on the environment suggests that feedback from local cues and learning of recurrent patterns are likely to play an important role.

Many models of orientation have been proposed in the past. If learning of directions and the means of establishment of these directions provide a basis for orientation, then it should be possible to interfere with these processes by systematically excluding the availability of cues during the developmental sequence. Kramer's (1959) and Wallraff's (1966) visual restriction experiments represent a first, though unsystematic, step in this direction. In a parallel to these exclusion experiments, it would be valuable to study the capacity for learning to utilize the stimuli that potentially could serve as cues for orientation, such as the movement of celestial bodies.

Turning now from ontogeny to phylogeny, it seems apparent that theories based on the possession of special organs, such as wings (Barnothy, 1964) or a pecten (Talkington, 1964) do not have the universality that seems to be required by a successful theory of orientation. Cross-phyletic and intraphyletic differences, on the other hand, suggest that it might be fruitful to compare the sensory capacities and other factors of species known for their navigational abilities with those of species that do not have a great need for accurate navigation.

Consideration of these differences in orientational competence naturally raises the question of the existence of similarities. What means do good navigators have in common? Certainly, the accuracy of sensory discriminations must be a basic factor (Adler, 1963b), but it seems possible that investigation of other organic functions may yield useful information. In this connection, serious consideration should be given to the possible existence of more than one kind of orientation and to the likelihood that the establishment of a goal direction and the means of holding to a course in this direction may be due to two separate processes (Kramer, 1953).

Finally, the establishment of an organic basis for orientation would open the possibility of testing specific hypotheses. The development of organs involved in orientation should correspond to an increasing ability to navigate. Values obtained from investigating the capacities at various stages of development could then enter as parameters into a model of orientation as a self-regulating system.

SUMMARY

Orientation may be conceived as a feedback process in which selected environmental cues lead to directed locomotion or changes in the direction of the body axis. The resulting change in position leads to a new pattern of stimulation on which the next step in the oriented movement response is then based. This process is continued until a goal is reached or the stimulus pattern becomes ineffective.

A wide variety of stimuli has been proposed as cues for orientation, particularly in cases involving movement over considerable distances, as found in bird navigation and homing. These range from such well-established factors as the position of the sun and cues provided by landmarks and meteorological conditions to the more speculative role of magnetic forces, other geophysical factors, and inertial navigation.

None of these theories provides a completely satisfactory account of orientation, because of the complexity of the factors involved. It is proposed that greater attention be paid to the development of orientation and that valuable evidence may be obtained from the comparison of closely related and distantly related types of organisms.

Observations on natural populations indicate that young birds already find their direction on their first migratory flight, but that more than a fixed, innate flight pattern appears to be involved, since they return to the place where they were hatched or first learned to fly. Laboratory experiments, especially those using the Kramer cage technique, indicate that celestial cues are used for guidance and that visual restrictions during development, particularly in the direction of the horizon, lead to absence of orientation, at least in the case of homing pigeons.

Practice plays a role in the improvement of navigational skills, especially as far as speed of homing is concerned. Some evidence suggests that at least two systems, the learning of landmarks and the improvement of the ability to utilize navigational cues, are involved.

Training experiments have shown that both the reference system utilized in celestial navigation, the sun azimuth and local time, can be acquired by learning. Displacement experiments have confirmed the ability of organisms to adjust to novel patterns of motion of the celestial bodies and changed diurnal rhythms. This lability in turn raises some questions about navigational hypotheses that depend on an animal's comparison of these parameters after displacement with their value in the home territory as a cue to the amount and direction of displacement. If learning can be shown to play a role in orientation, it still leaves unsolved the question of how the direction of the goal is acquired.

Consideration of the evolutionary history of orientation weakens the case for theories based on the possession of special organs and capacities. The study of the evolution of the physiological basis of orientation indicates some apparent parallels among the basic modes of orientation at various phyletic levels, but due weight should be given to existing differences. Considerable flexibility as to method of orientation must be assumed, including the possibility of utilization of a combination of cues.

Even the physical framework that organisms must use for obtaining fixed reference points is subject to change on an evolutionary time scale.

Future investigations might profitably examine the effect of environmental influences on the orientation process by manipulation of the availability of environmental cues during ontogeny and by the study of how the relationships between cues and direction are learned. It might also be fruitful to compare the capacities of species with proven navigational abilities with those of species that do not have a great need for navigation. Ultimately these studies should lead to a better understanding of the organic basis of orientation and its modification during the developmental process.

ACKNOWLEDGMENTS

Support of the National Science Foundation and the Friedman Foundation is gratefully acknowledged.

REFERENCES

Adler, H. E. 1963a. Sensory factors in migration. *Anim. Behav.* 11: 566–577.
Adler, H. E. 1963b. Psychophysical limits of celestial navigation hypotheses. *Ergeb. Biol.* 26: 235–252.
Adler, H. E., and B. P. Adler. 1965. Computer simulation of bird migration. *Am. Zool.* 5: 387 (abstract).
Adler, H. E., and B. P. Adler. 1966. Computer simulation of pigeon orientation. Paper read at 14th Int. Ornith. Congress, Oxford, England.
Agron, S. L. 1962. Evolution of bird navigation and the earth's axial procession. *Evolution* 16: 524–527.
Allen, W. H. 1948. Bird migration and magnetic meridians. *Science* 108: 708.
Allen, W. H. Jr. 1959. *How to raise and train pigeons.* New York: Sterling.
Aronson, L. R. 1951. Orientation and jumping behavior in the gobiid fish *Bathygobius soporator. Am. Mus. Novitates* 1486: 1–22.
Barlow, J. S. 1964. Inertial navigation as a basis for animal navigation. *J. Theoret. Biol.* 6: 76–117.
Barnothy, J. M. 1964. Proposed mechanisms for the navigation of migrating birds. In M. F. Barnothy, ed., *Biological effects of magnetic fields.* New York: Plenum Press. Pp. 287–293.
Becker, G. 1965. Zur Magnetfeld-Orientierung von Dipteren. *Z. Vergleich. Physiol.* 51:135–150.

Becker, G., and U. Speck. 1964. Untersuchungen über die Magnetfeld-Orientierung von Dipteren. *Z. Vergleich. Physiol.* 49: 301–340.

Beecher, W. J. 1952. The unexplained direction sense of vertebrates. *Sci. Monthly* 75: 19–25.

Beecher, W. J. 1954. On coriolis force and bird navigation. *Sci. Monthly* 79: 27–31.

Bellrose, F. C. 1958. The orientation of displaced waterfowl in migration. *Wilson Bull.* 70: 20–40.

Bellrose, F. C. 1963. Orientation behavior of four species of waterfowl. *Auk* 80: 257–289.

Bellrose, F. C. 1967a. Radar in orientation research. *Proc. 14th Int. Ornith. Cong.*, 1966, pp. 281–309.

Bellrose, F. C. 1967b. Orientation in waterfowl migration. In R. M. Storm, ed., *Animal orientation and navigation.* Corvallis, Ore.: Oregon State Univ. Press. Pp. 73–99.

Bergman, G., and K. O. Donner. 1964. An analysis of the spring migration of the common scoter and the longtailed duck in southern Finland. *Acta Zool. Fenn.* 105: 1–59.

Birukow, G. 1953. Menotaxis im polarisierten Licht bei *Geotrupes silvaticus* Panz. *Naturwissenschaften* 40: 611.

Birukow, G. 1956. Lichtkompassorientierung beim Wasserläufer *Velia currens* F. (Heteroptera) am Tage und zur Nachtzeit. I. Herbst-und Winterversuche. *Z. Tierpsychol.* 13: 467–484.

Birukow, G. 1960. Innate types of chronometry in insect orientation. *Cold Spr. Harb. Symp. Quant. Biol.* 25: 403–412.

Braemer, W. 1959. Versuche zu der im Richtungsgehen der Fische enthaltenen Zeitschätzung. *Verh. Deut. Zool. Ges. Münster/Westf.*, pp. 276–288.

Braemer, W. 1960. A critical review of the sun-azimuth hypothesis. *Cold. Spr. Harb. Symp. Quant. Biol.* 25: 413–427.

Braemer, W., and H. O. Schwassmann. 1963. Vom Rhythmus der Sonnenorientierung am Äquator (bei Fischen). *Ergeb. Biol.* 26: 182–201.

Brown, F. A., Jr. 1962. Responses of the planarian, *Dugesia,* and the protozoan, *Paramecium,* to very weak horizontal magnetic fields. *Biol. Bull.* 123: 264–281.

Brown, F. A., Jr., and Y. H. Park. 1965. Phase shifting a lunar rhythm in planarians by altering the horizontal magnetic vector. *Biol. Bull.* 129: 79–86.

Brown, F. A., Jr., H. M. Webb, and F. H. Barnwell. 1964. A compass directional phenomenon in mud-snails and its relation to magnetism. *Biol. Bull.* 127: 206–220.

Chmurzynski, J. A. 1964. Studies on the stages of spatial orientation in female *Bembex rostrata* (Linné 1758) returning to their nests (Hymenoptera, Sphegidae). *Acta Biol. Exp.* (Warsaw) 24: 103–132.

Darwin, C. 1873. Origin of certain instincts. *Nature* 7:417–418.

Donaldson, L. R., and G. H. Allen. 1958. Return of silver salmon, *Oncorhynchus kisutch* (Walbaum) to point of release. *Trans. Am. Fish. Soc.* 87: 13–22.

Drury, W. H., and J. A. Keith. 1962. Radar studies of songbird migration in coastal New England. *Ibis* 104: 449–489.

Drury, W. H. Jr., and I. C. T. Nisbet. 1964. Radar studies of orientation of songbird migrants in southeastern New England. *Bird-Banding* 35: 69–119.

Eastwood, E., and G. C. Rider. 1964. The influence of radio waves upon birds. *Brit. Birds* 57: 445–458.

El'darov, A. L., and Y. A. Kholodov. 1964. Effect of constant magnetic field on motor activity of birds. *Zh. Obshchei Biol.* 25: 224–229. (*Fed. Proc.* 1965, 24: T 431-T 433).

Emlen, J. T., and R. L. Penney. 1964. Distance navigation in the Adelie penguin. *Ibis* 106: 417–431.

Emlen, S. T. 1967a. Migratory orientation in the Indigo Bunting, *Passerina cyanea*. Part I: Evidence for use of celestial cues. *Auk* 84: 309–342.

Emlen, S. T. 1967b. Migratory orientation in the Indigo Bunting, *Passerina cyanea*. Part II: Mechanism of celestial orientation. *Auk* 84: 463–489.

Enright, J. T. 1961. Lunar orientation of *Orchestoida corniculata* Stout (Amphipoda). *Biol. Bull.* 120: 148–156.

Exner, S. 1893. Negative Versuchsergebnisse über das Orientierungsvermögen der Brieftauben. *Sitzber. Akad. Wiss. Wien* 102: 318–331.

Ferguson, D. E. 1963. Orientation in three species of anuran amphibians. *Ergeb. Biol.* 26: 128–134.

Ferguson, D. E., and H. F. Landreth. 1966. Celestial orientation of Fowler's toad *Bufo fowleri*. *Behaviour* 26: 105–123.

Ferguson, D. E., H. F. Landreth, and J. P. McKeown. 1967. Sun compass orientation of the northern crikct frog, *Acris crespitans*. *Anim. Behav.* 15: 45–53.

Ferster, C. B., and B. F. Skinner. 1957. *Schedules of reinforcement.* New York: Appleton-Century-Crofts.

Fischer, K., and G. Birukow. 1960. Dressur von Smaragdeidechsen auf Kompassrichtungen. *Naturwissenschaften* 47: 93–94.

Fischer, K., and G. Birukow. 1962. Die Lichtkompassorientierung sonnenlos aufgezogener Smaragdeidechsen (*Lacerta virides* Laur.). *Verh. Deut. Zool. Ges. Wien*, pp. 316–321.

Frisch, K. von. 1946. Die Tänze der Bienen. *Österr. Zool. Z.* 1: 1–48.

Frisch, K. von 1950. Die Sonne als Kompass im Leben der Bienen. *Experientia* 6: 210–221.

Fromme, H. G. 1961. Untersuchungen über das Orientierungsvermögen nächtlich ziehender Kleinvögel (*Erithacus rubecula, Sylvia communis*). *Z. Tierpsychol.* 18: 205–220.

Gould, E. 1957. Orientation in box turtles *Terrapene c. carolina* (Linnaeus). *Biol. Bull.* 112: 336–348.

Graber, R. R. 1968. Nocturnal migration in Illinois-different points of view. *Wilson Bull.* 80: 36–71.

Graue, L. C. 1965. Experience effect on initial orientation in pigeon homing. *Anim. Behav.* 13: 149–153.

Griffin, D. R. 1952. Bird navigation. *Biol. Rev.* 27: 359–400.

Groot, C. 1965. On the orientation of young sockeye salmon (*Oncorhynchus nerka*) during their seaward migration out of lakes. *Behaviour,* Suppl. 14: 1–198.

Gruys-Casimir, E. M. 1965. On the influence of environmental factors on the autumn migration of chaffinch and starling: A field study. *Arch. Néerl. Zool.* 16: 175–279.

Hamilton, W. J., III. 1962a. Celestial orientation in juvenal waterfowl. *Condor* 64: 19–33.

Hamilton, W. J., III. 1962b. Bobolink migratory pathways and their experimental analysis under night skies. *Auk* 79: 208–233.

Hamilton, W. J., III. 1962c. Does the bobolink navigate? *Wilson Bull.* 74: 357–366.

Hamilton, W. J., III, and M. C. Hammond. 1960. Oriented overland migration of pinioned Canada geese. *Wilson Bull.* 72: 385–391.

Hara, T. J., K. Ueda, and A. Gorbman. 1965. Electroencephalic studies of homing salmon. *Science* 149: 884–885.

Hasler, A. D. 1954. Odour perception and orientation in fishes. *J. Fish. Res. Bd. Canada* 11: 107–129.

Hasler, A. D. 1956a. Perception of pathways by fishes in migration. *Quart. Rev. Biol.* 31: 200–209.

Hasler, A. D. 1956b. Influence of environmental reference points on learned orientation in fish (*Phoxinus*). *Z. Vergleich. Physiol.* 38: 303–310.

Hasler, A. D., R. M. Horrall, W. J. Wisby, and W. Braemer. 1958. Sun-orientation and homing in fishes. *Limnol. Oceanog.* 3: 353–361.

Hassenstein, B. 1959. Optokinetische Wirksamkeit bewegter periodischer Muster. *Z. Naturforschung* 14: 659–674.

Hassler, S. S., R. R. Graber, and F. C. Bellrose. 1963. Fall migration and weather, a radar study. *Wilson Bull.* 75: 56–77.

Hoffman, K. 1953. Die Einrechnung der Sonnenwanderung bei der Richtungsweisung des sonnenlos aufgezogenen Stares. *Naturwissenschaften* 40: 148.

Hoffmann, K. 1954. Versuche zu der im Richtungsfinden der Vögel enthaltenen Zeitschätzung. *Z. Tierpsychol.* 11: 453–475.

Hoffmann, K. 1959. Über den Einfluss verschiedener Faktoren auf die Heimkehrleistung von Brieftauban. *J. Ornith.* 100: 90–102.

Hoffmann, K. 1960. Experimental manipulation of the orientational clock in birds. *Cold Spr. Harb. Symp. Quant. Biol.* 25: 379–387.

Holst, E. von. 1950. Quantative Messung von Stimmungen im Verhalten der Fische. *Symp. Soc. Exp. Biol.* 4: 143–172.

Huizinga, E. 1935. Durchschneidung aller Bogengänge bei der Taube. *Pflüg. Arch. Ges. Physiol.* 236: 52–58.

Ising, G. 1945. Die physikalische Möglichkeit eines tierischen Orientierungssinnes auf Basis der Erdrotation. *Ark. Mat. Astr. Fys.* 32: 1–23.

Johnson, D. L. 1967. Honey bees: Do they use the direction information contained in their dance maneuver? *Science* 155: 844–847.

Johnson, W. E., and C. Groot. 1963. Observations on the migration of young sockeye salmon (*Oncorhynchus nerka*) through a large, complex lake system. *J. Fish. Res. Bd. Canada* 20: 919–938.

Kalmus, H. 1954. Sun navigation by animals. *Nature* 173: 657.

Kalmus, H. 1964. Animals as mathematicians. *Nature* 202: 1156–1160.

Kholodov, Y. A. 1964. Effects on the central nervous system. In M. F. Barnothy, ed., *Biological effects of magnetic fields*. New York: Plenum Press. Pp. 196–200.

Kramer, G. 1949. Über Richtungstendenzen bei der nächtlichen Zugunruhe gekäfigter Vögel. In E. Mayr and E. Schuz, eds., *Ornithologie als biologische Wissenschaft*. Heidelberg: Carl Winter Universitätsverlag. Pp. 269–283.

Kramer, G. 1950. Orientierte Zugaktivität gekäfigter Singvögel. *Naturwissenschaften* 37: 188.

Kramer, G. 1952. Experiments in bird orientation. *Ibis* 94: 265–285.

Kramer, G. 1953. Wird die Sonnenhöhe bei der Heimfindeorientierung verwendet? *J. Ornith.* 94: 201–219.

Kramer, G. 1959. Über die Heimfindeleistung unter Sichtbegrenzung aufgewachsener Brieftauben. *Verh. Deut. Zool. Ges. Frankfurt a. M.*, 1958, pp. 168–176.

Kramer, G., and U. von St. Paul. 1950. Stare (*Sturnus vulgaris*) lassen sich auf Himmelsrichtung dressieren. *Naturwissenschaften* 37: 526–527.

Lack, D. 1959a. Migration across the sea. *Ibis* 101: 374–399.

Lack, D. 1959b. Watching migration by radar. *Brit. Birds* 52: 258–267.

Landreth, H. F., and D. E. Ferguson. 1967. Newt orientation by sun-compass. *Nature* 215: 516–518.

Lindauer, M. 1959. Angeborene und erlernte Komponenten in der Sonnenorientierung der Bienen. *Z. Vergleich. Physiol.* 42: 43–62.

Lindauer, M. 1963. Allgemeine Sinnesphysiologie: Orientierung im Raum. *Fortschr. Zool.* 16: 58–140.

Lindauer, M. 1967. Recent advances in bee communication and orientation. *Ann. Rev. Entomol.* 12: 439–470.

Löhrl, H. 1959. Zur Frage des Zeitpunkts einer Prägung auf die Heimatregion beim Halsband-Schnäpper (*Ficedula albicollis*). *J. Ornith.* 100: 132–140.

Lowery, G. H., and R. J. Newman. 1955. Direct studies of nocturnal bird migration. In A. Wolfson, ed., *Recent studies in avian biology*. Urbana: Illinois Univ. Press. Pp. 238–263.

McCabe, R. A. 1947. The homing of transplanted young wood ducks. *Wilson Bull.* 59: 104–109.

Martorelli, G. 1899. Les apparitions des *Turdides sibiriens* en Europe. *Ornis* 10: 241–292.

Matthews, G. V. T. 1951a. The sensory basis of bird navigation. *J. Inst. Navig.* 4: 260–275.

Matthews, G. V. T. 1951b. The experimental investigation of navigation in homing pigeons. *J. Exp. Biol.* 28: 508–536.

Matthews, G. V. T. 1952. The relation of learning and memory to the orientation and homing of pigeons. *Behaviour* 4: 202–221.

Matthews, G. V. T. 1955. *Bird Navigation*. Cambridge, Eng.: Cambridge Univ. Press.

Matthews, G. V. T. 1963a. The orientation of pigeons as affected by the learning of landmarks and by the distance of displacement. *Anim. Behav.* 11: 310–317.

Matthews, G. V. T. 1963b. "Nonsense" orientation as a population variant. *Ibis* 105: 185–197.

Matthews, G. V. T. 1964. Individual experience as a factor in the navigation of Manx shearwaters. *Auk* 81: 132–146.

Matthews, G. V. T. 1967. Some parameters of 'nonsense' orientation in Mallard. *Wildfowl Trust 18th Annual Report*, pp. 88–97.

Mauersberger, G. 1957. Umsiedlungsversuche am Trauerschnäpper (*Muscicapa hypoleuca*) durchgeführt in der Sowjetunion—Ein Sammelreferat. *J. Ornith.* 98: 445–447.

Merkel, F. W., and H. G. Fromme. 1958. Untersuchungen über das Orientierungsvermögen nächtlich ziehender Rotkehlchen (*Erithacus rubecula*). *Naturwissenschaften* 45: 499–500.

Merkel, F. W., and W. Wiltschko. 1965. Magnetismus und Richtungsfinden zugunruhiger Rotkehlchen (*Erithacus rubecula*). *Vogelwarte* 23: 71–77.

Merkel, F. W., G. G. Fromme, and W. Wiltschko. 1964. Nicht-visuelles Orientierungsvermögen bei nächtlich zugunruhigen Rotkehlchen! *Vogelwarte* 22: 168–173.

Mewaldt, L. R. 1964. California sparrows return from displacement to Maryland. *Science* 146: 941–942.

Michener, M. C., and C. Walcott. 1966. Navigation of single homing pigeons: Airplane observations by radio tracking. *Science* 154: 410–413.

Middendorff, A. von. 1859. Die Isepiptesen Russlands. Grundlagen zur Erforschung der Zugzeiten und Zugrichtungen der Vögel Russlands. *Mém. L'Acad. Imp. Sci. St. Pétersbourg*, Sixth Series, vol. 10 (Nat. Sci., vol. 8) pp. 1–144.

Mittelstaedt, H. 1962. Control systems of orientation in insects. *Ann. Rev. Entomol.* 7: 177–198.

Mittelstaedt, H. 1963. Bikomponenten-Theorie der Orientierung. *Ergeb. Biol.* 26: 253–258.

Mueller, H. C., and D. D. Berger. 1961. Weather and fall migration of hawks at Cedar Grove, Wisconsin. *Wilson Bull.* 73: 171–192.

Mueller, H. C., and D. D. Berger. 1967. Fall migration of sharp-skinned hawks. *Wilson Bull.* 79: 397–415.

Myres, M. T. 1964. Dawn ascent and re-orientation of Scandinavian thrushes (*Turdus* spp.) migrating at night over the northeastern Atlantic ocean in autumn. *Ibis* 106: 7–51.

Opdyke, N. D., B. Glass, J. D. Hays, and J. Foster. 1966. Paleomagnetic study of antarctic deep-sea cores. *Science* 154: 349–357.

Orgel, A. A., and J. C. Smith. 1955. A test of the magnetic-vertical-coriolis theory of homing. *Proc. Ind. Acad. Sci.* 64: 241.

Palmer, J. D. 1963. Organismic spatial orientation in very weak magnetic fields. *Nature* 198: 1061–1062.

Papi, F. 1955. Astronomische Orientierung bei der Wolfspinne *Arctosa perita* (Latr.). *Z. Vergleich. Physiol.* 37: 230–233.

Papi, F., and L. Pardi. 1953. Ricerche sull'orientamento di *Talitrus saltator* (Montagu) (Crustacea-Amphipoda). II. Sui fattori che regolano la variazone dell'angolo di orientamento nel corso del giorno. L'orientamento di notte. L'orientamento diurno di altre popolazioni. *Z. Vergleich. Physiol.* 35: 490–518.

Papi, F., and P. Tongiorgi. 1963. Innate and learned components in the astronomical orientation of wolf spiders. *Ergeb. Biol.* 26: 259–280.

Pardi, L., and F. Papi. 1953. Ricerche sull'orientamento di *Talitrus saltator* (Montagu) (Crustacea-Amphipoda). I. L'orientamento durante il giorno in una popolazione del litorale Tirrenico. *Z. Vergleich. Physiol.* 35: 459–489.

Penney, R. L., and J. T. Emlen. 1967. Further experiments on distance navigation in the Adelie penguin *Pygoscelis adeliae. Ibis* 109: 99–109.

Pennycuick, C. J. 1960. The physical basis of astro-navigation in birds: Theoretical considerations. *J. Exp. Biol.* 37: 573–593.

Perdeck, A. C. 1958. Two types of orientation in migrating starlings, *Sturnus vulgaris* L., and chaffiches, *Fringilla coelebs* L., as revealed by displacement experiments. *Ardea* 46: 1–37.

Perdeck, A. C. 1963. Does navigation without visual clues exist in robins? *Ardea* 51: 91–104.

Perdeck, A. C. 1967. Orientation of starlings after displacement to Spain. *Ardea* 55: 194–202.

Pringle, J. W. S. 1963. The proprioceptive background to mechanisms of orientation. *Ergeb. Biol.* 26: 1–11.

Rabøl, J. 1967. Visual diurnal migratory movements. *Dansk Ornith. Foren. Tidsskr.* 61: 73–99.

Renner, M. 1959. Über ein weiteres Versetzungsexperiment zur Analyse des Zeitsinnes und der Sonnenorientierung der Honigbiene. *Z. Vergleich. Physiol.* 42: 449–483.

Rensing, L. 1965. Circadian rhythms in the course of ontogeny. In J. Aschoff, ed., *Circadian Clocks.* Amsterdam: North-Holland Publ. Co. Pp. 399–405.

Riper, W. von., and E. R. Kalmbach. 1952. Homing not hindered by wing magnets. *Science* 115: 577–578.

Rüppell, W. 1944. Versuche über Heimfinden ziehender Nebelkrähen nach Verfrachtung. *J. Ornith.* 92: 106–133.

Saila, S. B., and R. A. Shappy. 1963. Random movement and orientation in salmon migration. *J. Cons. Int. Explor. Mer* 28: 153–166.

St. Paul, U. von. 1953. Nachweis der Sonnenorientierung bei nächtlich ziehenden Vögeln. *Behaviour* 6: 1–7.

St. Paul, U. von. 1956. Compass directional training of western meadow larks (*Sturnella neglecta*). *Auk* 73: 203–210.

Santschi, F. 1911. Observations et remarques critiques sur le mécanisme de l'orientation chez les fourmis. *Rev. Suisse Zool.* 19: 303–338.

Sauer, F. 1957. Die Sternenorientierung nächtlich ziehender Grasmücken (*Sylvia atricapilla, borin* und *curruca*). *Z. Tierpsychol.* 14: 29–70.

Sauer, E. G. F. 1963. Migration habits ˅of golden plovers. *Proc. 13th Int. Ornith. Cong.*, pp. 454–467.

Sauer, F., and E. Sauer. 1955. Zur Frage der nächtlichen Zugorientierung von Grasmücken. *Rev. Suisse Zool.* 62: 250–259.

Schmidt-Koenig, K. 1958. Experimentelle Einflussnahme auf die 24-Stunden-Periodik bei Brieftauben und deren Auswirkungen unter besonderer Berücksichtigung des Heimfindevermögens. *Z. Tierpsychol.* 15: 301–331.

Schmidt-Koenig, K. 1963a. Neuere Aspekte über die Orientierungsleistungen von Brieftauben. *Ergeb. Biol.* 26: 286–297.

Schmidt-Koenig, K. 1963b. Sun compass orientation of pigeons upon equatorial and trans-equatorial displacement. *Biol. Bull.* 124: 311–321.

Schmidt-Koenig, K. 1964a. Initial orientation and distance of displacement in pigeon homing. *Nature* 201: 638.

Schmidt-Koenig, K. 1964b. Sun compass orientation of pigeons upon displacement north of the arctic circle. *Biol. Bull.* 127: 154–158.

Schmidt-Koenig, K. 1964c. Über die Orientierung der Vögel: Experimente und Probleme. *Naturwissenschaften* 51: 423–431.

Schmidt-Koenig, K. 1966. Über die Entfernung als Parameter bei der Anfangsorientierung der Brieftaube. *Z. Vergleich. Physiol.* 52: 33–55.

Schneider, F. 1963a. Ultraoptische Orientierung des Maikäfers (*Melolontha vulgaris* F.) in künstlichen elektrischen und magnetischen Feldern. *Ergeb. Biol.* 26: 147–157.

Schneider, F. 1963b. Systematische Variationen in der elektrischen, magnetischen und geographisch-ultraoptischen Orientierung des Maikäfers. *Vierteljahrsschr. Naturforsch. Ges. Zürich* 108: 373–416.

Schneider, F. 1964. Die Beeinflussung der ultraoptischen Orientierung des Maikäfers durch Veränderung des lokalen Massenverteilungsmusters. *Rev. Suisse Zool.* 71: 632–648.

Schneirla, T. C. 1949. Levels in the psychological capacities of animals. In R. W. Sellars, V. J. McGill and M. Farber, eds. *Philosophy for the future.* New York: Macmillan. Pp. 243–286.

Schneirla, T. C. 1952. A consideration of some conceptual trends in comparative psychology. *Psychol. Bull.* 49: 559–597.

Schneirla, T. C. 1956. Interrelationships of the "innate" and the "acquired" in instinctive behavior. In P.-P. Grassé, ed., *L'Instinct dans le comportment des animaux et de l'homme.* Paris: Masson. Pp. 387–452.

Schneirla, T. C. 1957. The concept of development in comparative psychology. In D. B. Harris, ed., *The concept of development.* Minneapolis: Univ. Minnesota Press. Pp. 78–108.

Schüz, E. 1949. Die Spätauflassung ostpreussischer Jungstörche in Westdeutschland 1933. *Vogelwarte* 15: 63–78.

Schüz, E. 1964. Zur Deutung der Zugscheiden des Weisstorchs. *Vogelwarte* 22: 194–223.

Schwassmann, H. O. 1960. Environmental cues in the orientation rhythm of fish. *Cold Spr. Harb. Symp. Quant. Biol.* 25: 443–450.

Schwassmann, H. O., and W. Braemer. 1961. The effect of experimentally changed photoperiod on the sun-orientation rhythm of fish. *Physiol. Zool.* 34: 273–286.

Schwassmann, H. O., and A. D. Hasler. 1964. The role of the sun's altitude in sun-orientation of fish. *Physiol. Zool.* 37: 163–178.

Shumakov, M. E. 1967. An investigation of the migratory orientation of passerine birds. *Vestnik Leningrad. Univ., Biol. Ser.*, 1967, pp. 106–118. (Engl. summary)

Skinner, B. F. 1950. Are theories of learning necessary? *Psychol. Rev.* 57: 193–216.

Sokolov, E. N. 1960. Neuronal models and the orienting reflex. In M. A. B. Brazier, ed., *The central nervous system and behavior.* New York: Josiah Macy, Jr. Foundation. Pp. 187–276.

Stevens, O. A. 1957. Fall migration and weather, with special reference to Harris' sparrow. *Wilson Bull.* 69: 352–359.

Stewart, O. J. 1957. A bird's inborn navigational device. *Trans. Ky. Acad. Sci.* 18: 78–84.

Talkington, L. 1964. On bird navigation. Paper read at A. A. A. S. meeting, 1964. Montreal, Canada.

Thorpe, W. H. 1949. Recent biological evidence for the methods of bird orientation. *Proc. Linn. Soc. London* 160: 85–94.

Välikangas, I. 1933. Finnische Zugvögel aus englischen Vogeleiern. *Vogelzug* 4: 159–166.

Vaught, R. W. 1964. Results of transplanting flightless young blue-winged teal. *J. Wildl. Mgmt.* 28: 208–212.

Vleugel, D. A. 1952. Über die Bedeutung des Windes für die Orientierung ziehender Buchfinken *Fringilla coelebs. Ornith. Beob.* 49: 45–53.

Wallraff, H. G. 1959. Über den Einfluss der Erfahrung auf das Heimfindevermögen von Brieftauben. *Z. Tierpsychol.* 16: 424–444.

Wallraff, H. G. 1965a. Über das Heimfindevermögen von Brieftauben mit durchtrennten Bogengängen. *Z. Vergleich. Physiol.* 50: 313–330.

Wallraff, H. G. 1965b. Versuche zur Frage der gerichteten Nachtzug-Aktivität von gekäfigten Singvögeln. *Verh. Deut. Zool. Ges. Jena,* pp. 338–356.

Wallraff, H. G. 1966. Über die Heimfindeleistung von Brieftauben nach Haltung in verschiedenartig abgeschirmten Volieren. *Z. Vergleich. Physiol.* 52: 215–259.

Wenner, A. M. 1967. Honey bees: Do they use the distance information contained in their dance maneuver? *Science* 155: 847–849.

Wilkinson, D. H. 1949. Some physical principles of bird orientation. *Proc. Linn. Soc. London* 160: 94–99.

Wilkinson, D. H. 1952. The random element in bird "navigation." *J. Exp. Biol.* 29: 532–560.

Williamson, K. 1955. Migrational drift. *Acta XI Congr. Orn. Basel,* 1954, pp. 179–186.

Wiltschko, W., and F. W. Merkel. 1965. Orientierung zugunruhiger Rotkehlchen im statischen Magnetfeld. *Verh. Deut. Zool. Ges. Jena.,* pp. 362–367.

Wingstrand, K. G., and O. Munk. 1965. The *pecten oculi* of the pigeon with

particular regard to its function. *Biol. Skr. Dan. Vid. Selsk.* 14 (3): 1–64.
Yeagley, H. L. 1947. A preliminary study of a physical basis of bird navigation. *J. Appl. Phys.* 18: 1035–1063.
Yeagley, H. L. 1951. A preliminary study of a physical basis of bird navigation. II. *J. Appl. Phys.* 22: 746–760.

KNUT LARSSON
Institute of Psychology
University of Göteborg,
Göteborg, Sweden

Mating Behavior of the Male Rat

Although the animal psychologist has been eager to acclaim the general application of his concepts, he has usually avoided testing the validity of his theorizing against behavior patterns that are normal and typical for the species. It is regrettable that this class of behavior patterns has remained unnoticed, because it is one that may be particularly suited for integrating behavioral phenomena into a larger biological context.

In this chapter a series of investigations will be reviewed, all devoted to mating behavior of male rats. Since this behavior depends on both the hormonal and neural systems, its biological foundation is obvious. At least in the case of the domesticated rat, it is a behavior that is easy to elicit in the laboratory and easy to observe and quantify. Although vital for the maintenance of the species, it is unnecessary for the existence of the individual organism thus allowing drastic experimental interference.

Figure 1 gives a diagrammatic representation of the mating pattern. Before ejaculation the male repeatedly mounts and penetrates the receptive female. The ejaculation is followed by a postejaculatory interval during which no sexual activity takes place. After some minutes' rest he may approach the female anew and initiate another series of copulations, again resulting in ejaculation. The alternating periods of activity and inactivity may continue for several hours until finally the male appears sexually exhausted and no longer responds to the female. By that time he may have produced ten or more ejaculations.

The various responses to the female are thus linked together in sequence forming an organized system of activities and processes. The dependence of the behavior on sensory stimulation, its detailed organization, its development in the maturing rat, and its neural basis represent four aspects of the mating behavior that will be dealt with in this chapter.

SENSORY STIMULATION OF MATING

Mating represents a response to the environment, and does not occur unless the male is properly stimulated. A problem of primary interest is, therefore, to determine the range of stimuli influencing the behavior, and to evaluate the relative importance of these stimuli.

Three classes of stimuli can be distinguished; each of which, in different ways, may affect mating behavior. One class is composed of stimuli, or rather patterns of stimuli, which are potential triggers of the mating response. These will be called *triggering* stimuli. Other stimuli, although not triggering a copulatory attempt, may yet influence the behavior by modifying the animal's excitability. They will be called *modulatory* stimuli. A third class is the *genital* stimuli produced by the tactile stimulation of the penis.

Triggering Stimuli

When presented with another animal, a sexually experienced male may, with equal vigor, mount another male or a female in diestrus the same way that he mounts a receptive female, showing that sexual receptivity is not a prerequisite for provoking sexual behavior in the male. This, however, cannot be taken as an indication that the behavior of the male is independent of that of the female. On the contrary, we have every reason to believe that the behavior of the receptive female, the vibration of her ears, her lordosis and quick spasmodic jumps with stiffly bent hind legs, her sniffing and presentations to the male, constitute important features of the sexual stimuli that contribute to the maintenance of excitability in the male.

A well-known study of Beach (1942) still remains the only systematic attempt to investigate the relative importance of different sensory cues in triggering mating. He deprived male rats of various senses including olfaction, vision, and cutaneous sensitivity in the head region, and did not find any specific sense modality or any particular pattern of senses essential for the elicitation of mating. Finding that removal of any two senses abolished mating in sexually inexperienced animals and destruction of all three senses eliminated mating in experienced rats, he concluded that sexual activity is a result of stimulation from several different sources.

Recently this problem was taken up for reinvestigation in our laboratory. In view of the great importance of olfactory stimuli in the female mouse reproductive cycle (Parkes and Bruce, 1961), we were particularly interested in studying the effects of surgical interference of the olfactory system.

The olfactory bulbs were destroyed bilaterally in a number of sexually vigorous males. Before and after the operations, the mating behavior was observed at regular intervals. Out of 11 operated animals, 4 did not show any changes at all in their behavior, while 7 males showed a considerable impairment of the mating performance. In no case did the operations entirely eliminate mating. The number of ejaculations achieved during the tests greatly diminished, however, and in many tests the animals did not attain any ejaculations. The intromissions were spaced at long intervals. Although some improvement took place during the four-month period of postoperative testing, the animals did not regain their preoperative level. These results indicate that the olfactory system exerts a powerful influence upon sexual behavior.

How does a stimulus acquire the property of eliciting mating responses? Beach's observation that sexually inexperienced rats were more susceptible to sensory deprivation than were experienced rats gives one indication of the importance of learning on this effect. Other evidence was recently reported by Carr, Loeb, and Dissinger (1965), who demonstrated that sexually experienced males, in contrast to unexperienced males, were able to discriminate a receptive female from a nonreceptive one on the basis of olfactory cues. In fact we may think of the "sexual stimulus" as a pattern of sensory cues to which the male has learned to respond.

Modulatory Stimuli

Relatively little attention has been paid to the possibility that stimuli other than those emanating from the mating partner influence the sexual behavior. Yet daily laboratory practice suggests that even minor variations in the experimental procedure may greatly influence the mating performance. In a study aimed at determining the onset of sexual maturation in male rats, it was observed that repeated handling of the male during the course of testing made him respond more readily to the female. In this way the onset of puberty was seemingly advanced by several days.

This problem was later taken up for a systematic study (Larsson, 1963). Twenty-five rats, about two years old, were allowed to mate under two conditions. On one occasion they copulated without any interference from the observer and on another, they were handled. The observer picked up the male from the floor of the mating cage approximately twice a minute, handled him for a few seconds, and dropped him to the female. Under these conditions the average intercopulatory intervals were shortened to approximately half of their normal length and the ejaculations showed an increase from 3.7 to 5.3 per hour.

Genital Stimuli

Beach and Holz (1946) observed that adult rats with infantile genital organs exhibited a high proportion of incomplete mounts and rarely ejaculated. Similar behavior was seen in males with fully developed copulatory organs from which a part of the penile bone was removed. These indications of the critical importance of genital stimulation to intromission were further amplified by Beach and Levinson (1950), who demonstrated changes in the histological appearance of the glans penis following castration.

In a recent experiment, a local anesthetic was applied to the glans penis of a number of sexually vigorous rats (Carlsson and Larsson, 1964). The treatment did not depress the males' readiness to initiate sexual activity but it completely abolished their ability to ejaculate, and greatly impaired their ability to penetrate the female. Close inspection of the animals during mating showed that the penis did not develop complete erection but remained semiflaccid.

Conclusion

It is tempting to think of sexual behavior as a pattern of responses elicited by a specific stimulus and unfolding according to a predetermined rule. The above discussion demonstrates how incomplete this view is. Not only is the female a complex pattern of stimuli to which the male may have to learn to respond, but many other factors in the environment may act to modify and regulate the rate of copulatory behavior at any point in the mating process.

FACTORS DETERMINING
THE OCCURRENCE OF EJACULATION

Once a male has mounted the female and intromission has taken place, he usually continues to copulate until ejaculation has been achieved. The number of intromissions before ejaculation varies in the subsequent series of copulations. The number is relatively high before the initial ejaculation, reaches a minimum in the second and third series, and thereafter slowly increases. The intromissions are separated by brief intercopulatory intervals which, after repeated ejaculations, increase considerably in length. Also the postejaculatory intervals vary. With each successive ejaculation they are lengthened, in an extremely regular manner by one or a few minutes.

Ejaculation sometimes occurs outside the vagina and may even take place without direct tactile stimulation of the penis. Normally, however, it is a response closely linked with the intromission, and does not occur unless the penis has penetrated the vagina. Sensory stimulation derived from the erect penis in the vagina thus seems to be prerequisite for the elicitation of the ejaculatory reflex, at least under normal mating conditions.

Why then does not the male ejaculate at each intromission? One possible answer is that ejaculation is a matter of chance. An increase in the number of intromissions therefore means an increased likelihood of ejaculation. The opposite hypothesis is that ejaculation does *not* occur at random but represents an occurrence strictly determined by preceding events.

Lindgren (1962), in our laboratory, tested these hypotheses by subjecting the data to statistical analysis. It was found that the very first intromissions in a series, in contrast to the following ones, rarely resulted in ejaculation, but that each of the succeeding ones was equally likely to be followed by an ejaculation. Thus the hypothesis of randomness applies to only part of the series.

A possible way of interpreting these findings is that the initial intromissions serve to build up a state of sexual excitation. During this phase, ejaculation naturally does not occur. After a certain number of intromissions, a level of excitation is attained, where the occurrence of ejaculation is unpredictable.

Focusing our attention upon those aspects of the behavior that can be kept under experimental control, we may ask what factors influence the occurrence of ejaculation. It might appear that the assumed rise in sexual excitation is exclusively determined by the quantity of sensory stimulation, i.e., by the number of preceding intromissions. The intromissions, however, are distributed in time, and their time relationship might be of no less importance than their frequency. Experimental evidence will be reviewed showing the importance of the time relationships within the series and between consecutive series.

Ejaculation as Determined by the Timing of
the Individual Intromissions

Within a single series of copulations, the intromissions are spaced at intervals varying between 10 and 60 seconds. No regular pattern in the timing of the intromissions can be seen except that the interval immediately preceding the intromission resulting in ejaculation is slightly prolonged. When the female is removed after an intromission, the behavior of the male changes drastically. Not only does he become very excited, running back and forth in the cage, but when again presented with the female, he very quickly ejaculates. The artificially prolonged intercopulatory interval thus produced an increased likelihood of ejaculation.

To investigate the effects of artificially prolonged intercopulatory intervals, the following experiment was performed. The male was placed in an open cage together with a receptive female. When an intromission had taken place, the female was removed for an interval, which, in different tests, varied from 0.15 to 2.00 minutes. The procedure was repeated after each intromission until ejaculation had occurred. The subjects were sexually active and highly experienced males, and immediately mounted the female when given access to her.

When the female was removed for only 0.15 minutes no change occurred in the intromission frequency as compared to the ad libitum behavior. By increasing the interval to 0.30 minutes there was a significant reduction in the number of intromissions required for ejaculation; and longer intervals reduced the number still further. The maximum effect was obtained with 1 and 2 minute intervals. The animals then ejaculated after a median of 3.4 intromissions compared to 7.4 under ad libitum conditions.

This experiment has been repeated in many versions and always with the same result. It is not necessary that all intercopulatory intervals be artificially prolonged. A single enforced interval will cause a marked reduction in the number of intromissions required before ejaculation is achieved (Beach and Whalen, 1959; Bermant, 1964). To be effective, however, the interval of enforced rest must exceed the average intercopulatory interval characterizing the individual rat under ad libitum conditions (Larsson, 1961a).

The excitatory effects of an intromission are of limited duration, and when the intercopulatory intervals are prolonged beyond a certain length, the occurrence of ejaculation can be wholly prevented. By interposing enforced rest of ten minutes between each intromission, many rats do not ejaculate even when allowed a relatively high number of intromissions (Larsson, 1956).

Ejaculation as Determined by Time Relationships
Between Subsequent Series of Copulations

As indicated above the second ejaculation is usually preceded by fewer and more closely spaced intromissions than the first one. The initial ejaculation, and/or the intromissions preceding it, thus facilitates the occurrence of subsequent ejaculations. What will occur when the normal time relations between complete series of copulations are changed experimentally?

To answer this question the following experiment was performed (Larsson, 1958b). The male was allowed to complete a series of copulations ending in ejaculation. The female was then removed and replaced after an interval varying from 6 to 120 minutes. The male was then allowed to achieve a second ejaculation.

It was found that when the interval of rest was prolonged from 6 to 90 minutes, a progressive increase occurred in the number of intromissions required for the second ejaculation. In fact, after 90 minutes the animal had to make as many intromissions to ejaculate a second time as he previously had made to attain the first ejaculation. The average intercopulatory interval, however, was still shorter in the second series than it was in the first one, even when the female was withdrawn for five hours (Beach and Whalen, 1959; Larsson, 1961b).

This facilitatory effect of the initial ejaculation has another effect upon the behavior. It was earlier mentioned that once an ejaculation has occurred, the male remains sexually inactive for a short period called the postejaculatory interval.

While looking for a systematic relationship between the length of the postejaculatory interval and the intromission frequency in the subsequent series, it was observed that animals exhibiting relatively brief postejaculatory intervals tended to use more intromissions to ejaculate than those having long postejaculatory intervals.

We therefore performed an experiment in which the postejaculatory intervals were slightly prolonged beyond their normal length (Larsson, 1959). Instead of starting to copulate ad libitum (e.g., 5 minutes after the initial ejaculation), the animals initiated the second series 8 minutes after the first ejaculation. As was expected, the number of intromissions preceding the second ejaculation was significantly decreased and the intercopulatory intervals shortened.

The results confirm the impression that the animals, particularly very excited and active males, tend to start to copulate before optimal conditions have been attained. This observation has another, perhaps still more interesting, aspect. The postejaculatory interval must be composed of two different phases. Immediately following ejaculation there is a relatively long period during which the male is completely incapable of mounting and penetrating the female. Following this there is another phase of only a few minutes' time, during which intromission may occur but its excitatory effects are lowered.

Using the information gathered above, we can arrange an experimental situation in which all responses are made to occur under optimal conditions. This means that within the series, the intromission must be spaced with intervals sufficiently long to cause a maximum reduction of the intromission frequency, and after an ejaculation has taken place the male must be allowed a period of rest so that he may completely recover from the depressive influence of the preceding ejaculation.

Such an experiment was actually carried out in our laboratory (Larsson, unpublished data). Twenty-five sexually experienced and highly active males were allowed to achieve 6 ejaculations. The intromissions within each of these series were spaced by intervals of 1½ minutes. Following the first, second, third, fourth, and fifth ejaculation, the female was removed for 6, 8, 10, 12, and 14 minutes respectively. It was found that under normal mating conditions, when the animals copulated ad libitum, they used a median number of 31.1 intromissions to ejaculate six times; but under the experimental conditions the same number of ejaculations was attained by 13.2 intromissions. The difference was highly significant ($p < .001$, Wilcoxon matched pairs sign-ranked test).

DEVELOPMENTAL CHANGES IN MATING

Changes in the mating behavior of the rat occur as a function of growth and development of the animal. When the complete behavior, including ejaculation, first appears at puberty (i.e., about 2–3 months of age), the male is capable of only a few ejaculations before becoming sexually exhausted. The coital pattern is, at that time, characterized by many intromissions preceding ejaculation, a long ejaculation latency, and long postejaculation intervals. This behavior rapidly changes, however. The capacity to achieve repeated ejaculations improves, the number of intromissions before ejaculation decreases, and the response latencies shorten. A month or two after sexual maturation, the behavior has largely assumed its adult shape, and a plateau occurs. At 15–18 months of age, changes in the behavior again begin to appear. First there is a slight prolongation of the ejaculation latencies which is followed by a significant prolongation of the average intercopulatory interval and the postejaculation interval. At 25–30 months of age a decrease is noted in the animal's ability to achieve repeated ejaculations.

Possible factors accounting for these variations in behavior include increasing sexual experience, hormonal changes, and modifications in the target tissues of the hormones.

The Role of Experience

The importance of contacts with other animals for the development of sexual behavior of the adult has been clearly established in the male guinea pig (Young, 1961). Experimental studies devoted to the rat are less conclusive, and the results obtained sometimes even contradictory. According to Beach (1958), adult rats that lived in separate cages beginning 14 days of age, when presented with a receptive female immediately mounted her, displaying a behavior that did not differ from group-reared animals. Zimbardo (1958), however, in a similar experiment reported results showing that isolated males were slow in initiating mating, and rarely achieved ejaculation.

Even in the hands of the same experimenter the rats may respond differently to what appears to be the same treatment. In our laboratory, Brusling (1964) compared the sexual behavior of adult males that had been isolated at 30 days of age, with that of group-reared litter mates. The isolated rats were found to be, in almost all respects, inferior to the group-reared animals in their behavior. They were slow in initiating mating; several males never ejaculated; and when a complete sexual behavior pattern was attained, the response latencies were relatively long. The experiment was repeated. Now, however, negative results were obtained. No differences could be detected between group-reared and isolated rats. It is obvious that there are factors involved in the sexual development of the rat that were not controlled in these experiments.

Hormonal Influences Upon the Development

EFFECTS OF PREPUBERTAL AND POSTPUBERTAL CASTRATION

The great decrease in the number of intromissions preceding ejaculation and the shortening of the ejaculation latency, which occur during the first months after sexual maturation, represent changes in the mating pattern. These changes take place whether or not the male has had heterosexual experience. Essentially, two factors may contribute to these alterations: age and gonadal hormones.

To decide which of these two factors is critical for producing these effects, the following experiment was performed (Larsson, 1967). Two groups of male rats were used. One group was castrated before maturity and the other after sexual maturation. They were then treated with the same amount of testosterone propionate either immediately after the operation or when they reached adulthood. Assuming that the behavioral changes were a factor of age, i.e., of nongonadal origin, the effect of the injected hormone would vary with the age of the animal at treatment, independently of his age at castration. If, however, the changes were related to secretions of the maturing testes, the effect of the exogenous hormone would vary with the age of the animal at the time of castration, independently of his age at treatment.

It was found that in the prepubertal castrated males, androgen induced all changes normally occurring during sexual development. The number of intromissions preceding ejaculation progressively diminished, ejaculation latency shortened, and ejaculation frequency increased. These alterations in the coital pattern took place whether the castrated male was treated by hormone immediately after the operation or the treatment was delayed until adulthood was reached. This shows that age is of no critical importance to the effect. When, by contrast, the hormone was injected into a male castrated after puberty, the adult mating pattern was immediately elicited.

The results of this experiment show that gonadal hormones and not age are responsible for the progressive modifications occurring in the coital pattern after puberty. Since the same dosage of androgen had different effects in animals castrated before puberty and in ones castrated after maturation, it may be concluded that there does not exist any direct relationship between the quantity of hormones in the body and the character of the coital pattern. The hormonal effect varies with the endocrine history of the animal, and the role of the hormones is twofold: to provoke the occurrence of the copulatory response and to produce changes in the sensitivity of the target tissues to androgenic stimulation.

EFFECTS OF NEONATAL CASTRATION

In the above experiment, treatment with exogenous androgen was found to restore the normal development of mating behavior when castration was performed immediately before puberty. At that time, however, the development of the genital organs is far advanced, and the sex differentiation long since established. It may therefore be asked if androgen still is able to induce normal development of sexual behavior if castration is performed at the neonatal stage when, in the male rat, differentiation is not yet completed.

To this aim the rats were castrated 4, 7, 10, 13, or 19 days after birth (Larsson, 1966). Beginning at 100 days of age the animals were treated with a daily dose of testosterone propionate (250 μg/100 grams body weight) for a period of 3 months. Regular observations of the mating performances showed that there were very marked differences between the groups. These differences did not disappear even when the animals were treated with very large quantities of hormones. While all animals ejaculated in the day 13 and day 19 groups, only 1 out of 12 ejaculated in the group of rats castrated at day 4. In the day 7 animals, 7 out of 13 ejaculated, and in day 9 animals, 9 out of 11 ejaculated. The day 7 animals, and to some extent even the day 10 animals, showed marked disturbances in their mating behavior. The frequency of mounts without intromissions was very high and the ejaculation latency and intercopulatory interval very much prolonged. The results of this experiment show that testosterone only induces normal sexual behavior when tissues sensitive to the hormone have been given time to develop normally after birth.

EFFECTS OF NEONATAL CASTRATION AND
EARLY HORMONE TREATMENT

It is apparent from the above experiment that sexual behavior does not develop normally unless the testes have been present during an early period of the sexual development. In the following investigation we tried to substitute the testes' own secretions by injecting the neonatally castrated male with testosterone propionate.

Three groups of rats were used in this experiment. One group was castrated at day 3 and given a daily injection of 100 μg testosterone propionate until day 13. A second group was castrated at day 3 and treated with sesame oil instead of hormone. A third group was castrated at day 13 and not treated at all. When the animals were 90 days old, they were treated with daily injections of 250 μg testosterone propionate for a 45-day period.

While none of the control animals ever ejaculated, all but a few of the day 3 castrated males were able to ejaculate. All of the animals castrated at 13 days attained ejaculation. Compared to the performance of the day 13 group, the behavior displayed by the day 3 castrated and hormone-treated animals was very much impaired. The mount frequency was high and the ejaculation latency very much prolonged.

The results of this experiment show that under these conditions of treatment, testosterone propionate cannot wholly compensate for the loss of the testes.

An observation of particular interest was that the penis and the accessory organs of the day 3 castrated and hormone-treated animals showed normal weight and measurement in contrast to the control rats. It is therefore possible that the behavioral impairments were not due to failure of the development of the genital organs but possibly to defects in the central nervous system.

Behavioral Changes in Old Age

There are striking differences between the mating behavior of a two-year-old rat and a one-year-old rat. The old male copulates at a low rate, and it takes him a long time to achieve ejaculation. One would expect that these alterations in the behavior merely signify a reduction in the animal's sexual potency, i.e., in his ability to achieve repeated ejaculations. This, however, does not seem to be the case. If the male is allowed sufficient time he may attain as many, and even more ejaculations than a younger animal (Larsson and Essberg, 1962). It may be that the threshold for sexual arousal is higher in old age. Thus he is more dependent upon

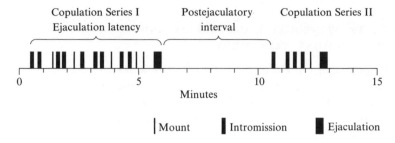

The mating pattern of the male rat. Definitions of measures of behavior. *Intromission latency:* time from the presentation of the female to the first intromission; *Ejaculation latency:* time from the first intromission until ejaculation; *Average intercopulatory interval (ICI):* average delay between each intromission computed by dividing the ejaculation latency by the intromission frequency; *Postejaculatory interval (PEI):* time from the ejaculation until the next intromission. *Series of copulations* means the sequence of mounts and intromissions culminating in ejaculation.

optimum sensory stimulation to produce maximum activity. A young male copulates at about the same rate during the light period of the diurnal cycle as during the dark period, whereas an old male shows increased activity during the dark period (Larsson, 1958a). Furthermore, by repeatedly handling the animal and thereby exposing him to increasing amounts of tactile and proprioceptive stimulation, the long response latencies can be shortened (Larsson, 1963).

It is not likely that the decline in sexual activity is related to a lowered level of sex hormones. In a carefully done experiment Jakubczak (1964) treated 30-month-old guinea pigs with large doses of testosterone propionate without being able to increase their sexual activity beyond the pretreatment level.

It is very probable that in old age modifications occur in the central nervous system making the animal less responsive to sensory stimulation. This would account for the long intercopulatory and postejaculatory intervals, which in turn explain why the old rat ejaculates after a relatively low number of intromissions.

THE NEURAL REGULATION OF MATING BEHAVIOR

While much knowledge has been obtained about mating behavior in different species, and some understanding of environmental and hormonal factors influencing the behavior has been achieved, surprisingly little is known of the neural mechanisms that mediate the behavior.

It is generally assumed that the basic genital reflexes are integrated at the spinal level, since after spinal cord transection mammals are still able to show penile erection and ejaculation (Bard, 1940). However, for integrating these reflexes into the broader context of sexual behavior, the brain is essential.

From the work of Beach and his collaborators, we know that the cerebral cortex is essential for mating in the male mammal. A decorticated male rat (Beach, 1940) or cat (Beach, Zitrin, and Jaynes, 1955; 1956) is no longer sexually active, nor is any activity seen in a rat subjected to a KCI induced cortical depression (Larsson, 1962). It was long thought that in the rat, in contrast to the cat, various parts of the cerebral cortex were equally important in maintaining sexual excitability. This, however, does not seem to be the case. In a recent series of studies, various parts of the cerebral cortex were destroyed (Larsson, 1962a,b; 1964). In no case did removal of the occipital or posterior temporal areas abolish mating. In fact, if the lesion was confined to the posterior portion of the cerebral cortex, as much as half of the entire cortical surface could be destroyed without causing any serious disturbances in the sexual behavior. If, however, the areas destroyed were located in the rostral end of the cerebral cortex, and particularly in the frontal lobes, even relatively small lesions, about 13% of the total surface, might be effective in eliminating sexual behavior permanently. No impairment of motor coordination was observed in the operated animals and thus the sexual deficiencies cannot be attributed to an inability in maintaining contact with the female. The possibility of a hormonal deficiency is not feasible in view of the fact that the testes remained normal and treatment with large amounts of testosterone failed to reactivate the animals.

It has been reported that lesions in the preoptic region and the anterior hypothalamus eliminate mating (Brookhart and Dey, 1941; Soulairac, 1963). To verify these observations an extensive investigation was initiated in our laboratory (Larsson and Heimer, 1964a; Heimer and Larsson, 1966/1967). Lesions were placed in almost every part of the hypothalamus and adjacent structures with the exception of the tuberal hypothalamic region. It was found that large lesions (about 2 mm²) in the medial preoptic-anterior hypothalamic continuum regularly abolished the mating behavior. Small lesions within this area were either wholly without effect or caused only a temporary impairment of the sexual behavior. In only a few cases did lesions in other regions including the septal areas, the lateral anterior hypothalamus, the medial dorsal hypothalamus, and the mamillary body (Heimer and Larsson, 1964) influence the sexual behavior.

These results suggest that the medial preoptic-anterior hypothalamic continuum functions as a mating center. The immediacy of the decline in sexual behavior, the normal appearance of the testes, and the failure of injected testosterone to restore mating indicate that the loss of sexual behavior following lesions in this region was not due to disruption of the pituitary-gonadal axis but to injury of a mechanism directly responsible for the integration of the behavioral responses.

While injury to the rostral end of the brain may eliminate or, in the case of olfactory bulb lesions, impair mating, lesions of more caudally located structures may have an opposite effect, causing a drastic increase in sexual activity (Larsson and Heimer, 1964b). In a series of 16 rats an extensive region in the junction of the diencephalon and mesencephalon was destroyed. Nine rats did not show any deviation in their sexual behavior but conspicuous changes were seen in 7 rats. They ejaculated after very few intromissions, the postejaculatory intervals were shortened to a few minutes, and the number of ejaculations was very much increased. Particularly striking changes appeared in 2 males. During the weeks immediately following the operation they displayed frequent penile erections. The erections occurred not only in the presence of the receptive female but also while being fed and weighed. The penile erections diminished in frequency and completely disappeared after about a month, but the other behavioral deviations remained until the animals were killed half a year after the operation.

These observations suggest the existence of an inhibitory mechanism which, when destroyed, causes a radical increase in sexual activity. It might be that there are two antagonistic systems controlling the behavior, one excitatory and another inhibitory in nature. The first system includes structures in the rostral end of the forebrain—the olfactory systems, the frontal lobes, and the medial preoptic-anterior hypothalamic region—all of these serve to receive and process information from the senses and the hormonal system and to establish a state of sexual arousal. The second system is inhibitory and may be primarily composed of structures in the brainstem.

ACKNOWLEDGMENTS

Support of the investigations reported in this chapter was provided by research grant HD 00344-04 from the Institute of Child Health and Human Development USA, by grant from the Swedish Council for Social Research, the Swedish Medical Research Council and the Magnus Bergwall Foundation.

REFERENCES

Bard, P. 1940. The hypothalamus and sexual behavior. In the hypothalamus and central levels of autonomic function. *Res. Publ., Assoc. Res. Nervous Mental Diseases* 20: 551–594.

Beach, F. A. 1940. Effects of cortical lesions upon the copulatory behavior of male rats. *J. Comp. Psychol.* 29: 193–244.

Beach, F. A. 1942. Analysis of the stimuli adequate to elicit mating behavior in the sexually inexperienced rat. *J. Comp. Psychol.* 33: 163–207.

Beach, F. A. 1958. Normal sexual behavior in male rats isolated at fourteen days of age. *J. Comp. Psychol.* 51: 37–38.

Beach, F. A., and Holz, A. M. 1946. Mating behavior in male rats castrated at various ages and injected with androgen. *J. Exp. Zool.* 101: 91–142.

Beach, F. A., and G. Levinson. 1950. Effects of androgen on the glans penis and mating behavior of castrated male rats. *J. Exp. Zool.* 114: 159–171.

Beach, F. A., and R. E. Whalen. 1959. Effects of intromission without ejaculation upon sexual behavior in male rats. *J. Comp. Physiol. Psychol.* 52: 476–481.

Beach, F. A., A. Zitrin, and J. Jaynes. 1955. Neural mediation of mating in male cats: II. Contribution of the frontal cortex. *J. Exp. Zool.* 130: 381–401.

Beach, F. A., A. Zitrin, and J. Jaynes. 1956. Neural mediation of mating in male cats: I. Effects of unilateral and bilateral removal of the neocortex. *J. Comp. Physiol. Psychol.* 49: 321–327.

Bermant, G. 1964. Effects of single and multiple enforced intercopulatory intervals on the sexual behavior of male rats. *J. Comp. Physiol. Psychol.* 57: 398–403.

Brookhart, J. M., and F. L. Dey. 1941. Reduction of sexual behavior in male guinea pigs by hypothalamic lesions. *Am. J. Physiol.* 133: 551–554.

Brusling, Ch. 1964. Social experience and sexual behavior in the male rat. Masters thesis, University of Göteborg.

Carlsson, S. G., and K. Larsson. 1964. Mating in male rats after local anesthetization of the glans penis. *Z. Tierpsychol.* 21: 854–856.

Carr, W. J., L. S. Loeb, and M. L. Dissinger, 1965. Responses of rats to sex odors. *J. Comp. Physiol. Psychol.* 59: 370–377.

Heimer, L., and K. Larsson. 1964. Mating behavior in male rats after destruction of the mamillary bodies. *Acta Neurol. Scand.* 40: 353–360.

Heimer, L., and K. Larsson. 1966/1967. Impairment of mating behavior in male rats following lesions in the preoptic-anterior hypothalamic continuum. *Brain Research* 3: 248–263.

Jakubczak, L. F. 1964. Effects of testosterone propionate on age differences in mating behavior. *J. Gerontol.* 19 (45): 458–461.

Larsson, K. 1964. *Conditioning and sexual behavior in the male rat.* Stockholm: Almqvist and Wiksell.

Larsson, K. 1958a. Age differences in the diurnal periodicity of male sexual behavior. *Gerontologia* 2: 64–72.

Larsson, K. 1958b. After effects in the copulatory activity of the male rat. 1. *J. Comp. Physiol. Psychol.* 51: 325–327.

Larsson, K. 1959. Effects of prolonged postejaculatory intervals in the mating behaviour of the male rat. *Z. Tierpsychol.* 16: 628–632.

Larsson, K. 1961a. The importance of time for the intromission frequency in the male rat mating behaviour. *Scand. J. Psychol.* 2: 149–152.

Larsson, K. 1961b. Duration of facilitatory effects of ejaculation on sexual behavior in the male rat. *J. Comp. Physiol. Psychol.* 54: 63–67.

Larsson, K. 1962a. Spreading cortical depression and the mating behaviour in male and female rats. *Z. Tierpsychol.* 19: 321–331.

Larsson, K. 1962b. Mating behavior in male rats after cerebral cortex ablation. I. Effects of lesions in the dorsolateral and the median cortex. *J. Exp. Zool.* 151: 167–176.

Larsson, K. 1963. Non-specific stimulation and sexual behaviour in the male rat. *Behaviour* 20: 110–114.

Larsson, K. 1964. Mating behavior in the male rats after cerebral cortex ablation. II. Effects of lesions in the frontal lobes compared to lesions in the posterior half of the hemispheres. *J. Exp. Zool.* 155: 203–214.

Larsson, K. 1966. Effects of neonatal castration upon the development of the mating behavior of the male rat. *Z. Tierpsychol.* 23: 867–873.

Larsson, K. 1967. Testicular hormone and developmental changes in mating behavior of the male rat. *J. Comp. Physiol. Psychol.* 63: 223–230.

Larsson, K., and L. Essberg. 1962. Effect of age on the sexual behaviour of the male rat. *Gerontologia* 6: 133–143.

Larsson, K., and L. Heimer. 1964a. Mating behaviour of male rats after lesions in the preoptic area. *Nature* 202: 413–414.

Larsson, K., and L. Heimer. 1964b. Drastic changes in the mating behavior of male rats following lesions in the junction of diencephalon and mesencephalon. *Experientia* 20: 460.

Lindgren, O. 1962. Statistical analysis of the male rat mating behavior. In ms. University of Göteborg.

Parkes, J. T., and H. M. Bruce. 1961. Olfactory stimuli in mammalian reproduction. *Science* 134: 1049–1054.

Soulairac, M.-L. 1963. Etude experimentale des regulations hormonnerveuses du comportement sexuel du rat male. *Ann. d'Endocrinol,* Suppl. 24, p. 98.

Young, W. C. 1961. The hormones and mating behavior. *Sex and Internal Secretions,* vol. 2. Baltimore: Williams & Wilkins. Pp. 1173–1239.

Zimbardo, P. G. 1958. The effects of early avoidance training and rearing conditions upon the sexual behavior of the male rat. *J. Comp. Physiol. Psychol.* 51: 764–769.

IV

SOCIAL BEHAVIOR

CARYL P. HASKINS
President
Carnegie Institution of Washington
Washington, D. C.

Researches in the Biology and Social Behavior of Primitive Ants

". . . *I will not here enter on these several causes, but will
confine myself to one special difficulty, which at first appeared to
me insuperable, and actually fatal to my whole theory. I allude to
the neuters or sterile females in insect-communities: for these
neuters often differ widely in instinct and in structure from both
the males and fertile females, and yet, from being sterile, they
cannot propagate their kind. . . . It may well be asked how is it
possible to reconcile this case with the theory of natural selection? . . .
This difficulty, though appearing insuperable, is lessened, or, as
I believe, disappears, when it is remembered that selection may be
applied to the family, as well as to the individual, and may thus
gain the desired end. . . . Consequently the fertile males and females
of the same community flourished, and transmitted to their fertile
offspring a tendency to produce sterile members having the same
modification. . . . It will indeed be thought that I have overweening
confidence in the principle of natural selection, when I do not
admit that such wonderful and well-established facts at once
annihilate my theory. . . .*"

Charles Darwin, THE ORIGIN OF SPECIES, Chapter VIII

Of all the Aculeate Hymenoptera, the ants are, to every appearance,
the longest established as social insects. To be sure, the fossil record of
the social Hymenoptera outside of the ants is meager and not very satisfactory. Certainly the social bees also have a geologically extensive history,
emphasized by the recent dramatic find of a worker of an obviously social
bee of apparently Meliponid affinities, complete with well-developed pollen-
baskets, in the Mexican amber of Chiapas, thought to date from Oligocene
times, perhaps thirty to forty million years ago. The earliest known ant,

however, is far older. Recently, two beautifully preserved fossil worker ants were found in the Magothy amber of New Jersey, which is of mid-Cretaceous age, and assigned an antiquity of about 100 million years (Wilson et al., 1967). These two individuals, named *Sphecomyrma,* are of undoubtedly archaic aspect, but, were they living today, they would instantly be recognized as ants by any passerby. We also have evidence that by Eocene times, there were ants that were far more "modern" in their physical structure than a vast number of ants living in our contemporary world. The Eocene fossil ant *Oecophylla brischkei,* for instance, finds its place, on morphological grounds, in one of the most highly evolved of modern Formicid subfamilies, albeit in one of the more conservative genera of that subfamily. Such evidence forcibly suggests that a long and complex social evolution had occurred among ants during pre-Eocene times, extending (it is not extravagant to guess) well back into Cretaceous, if not to Jurassic, times.

The abundant and well-preserved ant fauna of the Baltic amber, strikingly modern in its general aspect, emphasizes such a conclusion. Years ago, Wheeler (1913; 1928) noted that this amber contains numerous representatives of all living subfamilies except the Dorylinae, including some of the most specialized contemporary genera of the advanced subfamilies, such as *Aphaenogaster, Myrmica, Leptothorax,* and *Monomorium* of the Myrmicinae; *Dolichoderus, Tapinoma,* and *Iridomyrmex* of the Dolichoderinae; and *Prenolepis, Plagiolepis, Camponotus, Lasius,* and *Formica* of the Formicinae. There are amber specimens of *Ponera, Lasius,* and *Formica* so like living species of the same genera as to be virtually indistinguishable.

Only one fossil record of an entire ant colony is available to us—a nearly complete nest of *Oecophylla leakeyi,* well filled with brood and workers of both castes, found by Dr. and Mrs. L. B. Leakey in the Lower Miocene deposits of Mfwangano Island in Kenya, and recently described by Wilson and Taylor (1964). This nest shows that Formicid society in Miocene times was already, in a purely physical sense, structured almost exactly as it is today. Indeed, although this well-preserved nest is probably thirty million years old, it does not seem to differ significantly from the colonies of the contemporary *Oecophylla longinoda* of Africa or *Oecophylla smaragdina* of southeast Asia and northern Australia.

The fossil record will carry us only so far in interpreting the detailed social evolution of ants on a morphological basis, and, of course, it yields no evidence about the evolution of the "fine structure" of social behavior. I believe, however, that the primitive ant fauna living today offers great opportunities for the study of social evolution. To be sure, neither in the fossil nor the living record do we have the equivalent of such socially transitional forms among ants as are represented, for example, by the genera *Stenogaster* or *Zethus* among wasps, or of *Augochloropsis* or *Allodape* among primitive bees. But though all ants, living and fossil, appear fully social, the living record appears on morphological grounds, at least, to be fully as archaic as the fossil one.

Probably the most primitive and generalized fossil ant so far discovered is *Prionomyrmex longiceps* from the Baltic amber. Yet in 1934 two living ant workers, *Nothomyrmecia macrops,* that appear to be morphologically even more primitive than *Prionomyrmex,* were collected in southwest Australia somewhere between Esperance and Eucla inland of Israelite Bay (Brown, and Wilson, 1959). *Nothomyrmecia macrops* is an astonishingly generalized ant. Its social structure and ethology could be of the greatest interest and potential productivity for the student of social evolution in the Formicidae. Unluckily, *N. macrops* has never been found again, despite repeated and intensive search of the area by capable myrmecologists, both Australian and foreign. We are fortunate, however, that a related division of this most primitive known subfamily of ants, the Myrmeciinae, includes many highly accessible living forms. Ants of the genus *Myrmecia* are widely distributed on the Australian continent, where they occupy environments ranging from the soils of central and southwestern Australian deserts to the treetops of tropical Queensland forests. Several generalized species prove very amenable to culture in the artificial nest. They constitute an invaluable resource for behavioral studies of early forms of ant societies. The social life of *Myrmecia* may be a fairly faithful replica of that of some ants of early Eocene, late Paleocene, and, possibly, Cretaceous times. The genus *Myrmecia* may well be a rather late but morphologically and socially quite conservative offshoot of a truly archaic line. It is a fair guess that the evolution of its social ethology may have been as conservative as that of its morphology.

The resources for study of the ethology of primitive ants do not end here. Modern phylogenetic concepts of the evolutionary "tree" of the Formicidae (Brown, 1954; Cavill and Robertson, 1965) suggest that all modern subfamilies of ants may have evolved from one of two archaic groups, these derived in turn from some hypothetical wasp-like precursor of Tiphioid or perhaps Scoliid, affinity. In this scheme, one of these two major branches, the "Myrmecioid Complex," radiating from primitive Myrmeciine or pro-Myrmeciine type, gave rise to the modern subfamilies Pseudomyrmicinae, Dolichoderinae, and Formicinae. The other great evolutionary branch, the "Poneroid Complex," including the modern subfamilies Myrmicinae and Dorylinae, as well as the Cerapachyinae and the aberrant Leptanillines, is thought to have derived from primitive Ponerine-like ancestors, of which the archaic family Amblyoponini may be most representative. Living ants of the genus *Amblyopone* present some features of morphology, notably the structure of the petiole, that appear even more primitive than those of

Myrmecia, and are remarkably similar to those of such primitive genera of Tiphiid wasps as *Anthobosca* and *Diamma.*

Like *Myrmecia,* the genus *Amblyopone* offers a rich opportunity for the behavioral study of Formicid social structure at a level as close to the inception of ant social life as any accessible. And there are many contemporary species, which are global in distribution.

Myrmecia and *Amblyopone,* together with some of the higher but still remarkably primitive Ponerines, offer fascinating challenges for comparative social study. Following are certain observations of such ants, made in several social contexts over a long period of years, both in the artificial nest and in the field.

DISPERSION OF REPRODUCTIVES AND THE FORMATION OF NEW COLONIES AS PARADIGMS OF THE ORIGIN OF SOCIAL LIFE IN ANTS

The "Stem Pattern" in Higher Ants

It was René Antoine Ferchault de Réaumur who first observed and correctly interpreted a mating flight of ants, probably a species of *Lasius,* on the levee of the Loire near Tours, France, in September, 1731. His full description of the event, his capture of mating pairs, his distinction of a smaller from a larger form, and his interpretation of the former as males, the latter as queens, leave no doubt that he fully understood the significance of what he saw. (Somewhat later Réaumur recorded that he had observed flying and mating pairs of nearly every species of ant in France.) Réaumur reported that the larger members of the flying swarm, the females, dropped to the ground and cast off their wings after fertilization. He observed that females captured early in their flights and confined before fertilization had occurred were characteristically much more delayed in dealation than fertilized ones or even failed to dealate at all. At about the same time that these observations were made, M. Charpentier de Cossigny reported to Réaumur from the island of Martinique on the occurrence there of dramatic mass mating flights of tropical ants, similarly followed by dealation of the fertilized queens.

Réaumur confined some of his newly dealated queens in beakers. He describes their activity in excavating cavities and records that they sealed themselves within and did not again leave. He records the appearance of clusters of eggs, and the hatching of young larvae. He even speculates as to the subtle nature of sources of nourishment for the developing young. Contrary to popular concept, Réaumur clearly understood and described the typical mode of dispersion of reproductives and the foundation of new colonies in the higher ants as early as 1742 or 1743, anticipating the work of William Gould, published in England in 1747.

More than a hundred years would pass, however, before Sir John Lubbock in 1879 published his observations of the entire course of colony foundation—from the isolation of the queen through the bringing to maturity, without outside aid, of a first worker brood in the Myrmicine ant *Myrmica ruginodis*. More complete observations were recorded by Fritz Müller for the tropical twig-dwelling *Azteca mülleri* in 1880, by Potts in 1883 and by Blochmann in 1885 for *Camponotus pennsylvanicus*.

More recently, many apparently divergent modes of colony foundation in ants have been described. Pleometrosis, the founding of colonies by multiple queens rather than by a single one (an observation first made by Réaumur himself) has been repeatedly confirmed in a number of species. It has been observed that the founding queens often become mutually hostile as the common brood approaches maturity, leading to conflicts that end either with the separation of the females or, more usually, with the deaths of all but one of the foundresses. Secondary, as well as primary, pleometrosis occurs in a variety of ants, sometimes bringing large numbers of fertilized female reproductives together within a single colony. In some species, secondary pleometrosis is maintained by the return of young fecundated daughters to their home communities year after year, successively maintaining the reproductive capacity of the community over many years. In various species of the genus *Formica,* particularly, such young queens may be captured in considerable numbers by the workers at flight time; some may actually have been reared in different colonies from those where they finally take up residence. Indeed, sometimes these queens belong to a different subspecies from that of the colony into which they are inducted.

Many modifications of the dispersal pattern involving suppression of the mating flight in one or both sexes are familiar today. Most spectacular of all, perhaps, is the strange pattern of the Dorylinae (made known to us particularly through the long and beautiful contemporary research of Theodore C. Schneirla) in which dicthadiigyne (eyeless, wingless) females are fecundated within their columns by robust, winged males, which, uniquely among ants, may lose their own wings after their dispersal flight. As the life histories were worked out for slave-making species of Formicine ants, including members of the *Formica sanguinea* group and the genus *Polyergus,* or for members of such Myrmicine genera as *Leptothorax* or *Strongylognathus,* modifications of the colony-founding pattern involving early dependence of the queen on workers of other species became familiar. Also noted were more extreme cases of dependence involving workerless, wholly socially parasitic forms, such as *Anergates* and *Wheeleriella,* culminating in the situation of *Teleutomyrmex schneideri,* in which the minute fertile queen subsists essentially as an ectoparasite on the queen of the host species *Tetramorium caespitum.*

It seems more clear than ever today that all these varied and bizarre patterns of dispersal and colony multiplication in fact represent only radiating evolutionary derivatives from the basic course described by Réaumur and Lubbock and the many observers who followed them. For every species in which such bizarre modifications were discovered, many more exhibited essentially the straightforward pattern. Examples of that pattern have continued to multiply through the years, ranging through such genera as *Pachysima* and *Pseudomyrma* in the *Pseudomyrminae; Messor, Pogonomyrmex, Aphaenogaster, Myrmica, Leptothorax, Solenopsis, Monomorium, Tetramorium, Pheidole, Cremastogaster, Atta, Acromyrmex,* and many others in the Myrmicinae; *Liometopum, Monacis, Hypoclinea,* and *Dolichoderus* in the Dolochoderinae; *Polyrhacis, Melophorus, Notoncus, Dendromyrmex, Calomyrmex, Acropyga, Prenolepis, Myrmecocystus, Formica,* and many others among the Formicinae. Gradually it became clear that the straightforward course, in which both sexes participate in a mating and dispersal flight, followed by isolation of the newly dealated female in a cavity and the rearing of a first brood of young workers entirely claustrally, never leaving the cell or taking nourishment, must represent some approximation of a generalized type of colony propagation among the Formicidae.

In 1906 and 1907, Janet, in a series of histological investigations (Wheeler, 1913), demonstrated that in the isolated colony-founding queen of *Lasius,* the thoracic wing-musculature, which is highly developed in the newly-mated young queen when her wings are discarded, breaks down and disappears within a few weeks after dealation. Clearly these disintegrating muscles furnish an important nutritional reserve for the isolated young queen, sustaining it for many months and in addition permitting it to rear a first brood to maturity in isolation.

The striking uniformity of this pattern of the initiation of the colony in ontogeny and its widespread occurrence among many otherwise quite dissimilar ants suggest that it may not lie too far away from the primitive pattern in Formicid phylogeny. If this is true, one might conclude that the communal society of ants originated as a unit of a single female and young family and that the Formicidae, as a group, evolved from monophyletic beginnings.

There is, however, an important aspect of this hypothetical pattern of the origin and evolution of ant societies that may not have been sufficiently stressed: many species of social wasps, such as, notably, the tropical Polybiines, form large and aggressive and permanent communities; nevertheless they exhibit a very primitive and generalized level of colony organization. Caste is very little developed in such a community, as Richards, particularly, has shown (1953), and the ratio of queens to workers in a single community is high. The communities reproduce by swarming, and each new swarm contains a great number of females—all, apparently, in continuing reproductive competition, actual or potential, throughout the life history of the species.

It seems very possible that there may be an evolutionary connection between this constantly maintained polygyny in the Polybiines and the primitive stage of their communities. It seems plausible that considerable genetic variability may have existed, at least in the earlier stages of queen-worker caste differentiation in the social Hymenoptera in general, in the predisposition of different females of a given species toward caste plasticity in their offspring. That is to say, there may have been considerable variability among different females in the proportion of workers to perfect females that would appear among their daughters, given any particular set of environmental conditions. It seems quite possible that an evolutionary selection of this potentiality toward the two extremes of very low and very high worker production resulted in the many pairs of worker-rich host and workerless socially parasitic species, often closely related, of Aculeate Hymenoptera—the *Psithyrus-Bombus* pairs in the bumblebees, for instance; or *Vespula austriaca-V. rufa* or *Dolichovespula adulterina-D. arenaria* among the Vespine wasps; or the pairs of worker-less parasitic species of *Polistes* and their hosts; or the *Myrmecia vindex* or *M. nigriceps-M. inquilina* pair (Haskins and Haskins, 1964) among the Myrmeciine ants. It could well have been through a similar kind of selection that the true queen was lost in the "queenless" species of *Rhytidoponera* (Haskins and Whelden, 1965).

If such genetic variability for caste plasticity of offspring among females has existed since a socially primitive stage in the Aculeates, then we may think of the permanently polygynic community as, in effect, a population of females that are potentially highly competitive in this respect. Those socially "altruistic" mothers that produce a very high proportion of workers among their offspring in a given colony will benefit the community as a whole, including their laying sisters. But the penalty for this will be a relatively poor representation of their own genotypes in the following generation. Indeed, those genetic types that produce workers too exclusively might well be eliminated altogether, except as their genotypes might be conserved through their sons. Conversely, the females at the opposite end of the spectrum, producing predominantly queen-daughters, will be genetically "selfish." Their genotypes will be richly represented in the following generation. But they and their offspring will actually be supported by the progeny of their more "altruistic" sisters.

The result of such a situation in the evolution of queen populations of the species as a whole in respect of the caste ratios to be expected among their offspring might well be the dominance of intermediate types, where the ratio of reproductive to worker offspring was relatively high, the degree of queen-worker differentiation remained definite but fairly low, and colony reproduction regularly occurred by swarming. As long as such polygyny was maintained it is hard to see how the worker:queen ratio could rise to levels remotely approaching those in the Formicidae or the higher mono-gynic Vespidae or Apidae, or how the specialization of the community could nearly approximate that of higher forms. Viewed in this way, it may be proper to consider the colony of the Polybiine wasps, not as an early stage in an evolutionary progression toward the levels of greater specialization and perfection represented by the Formicid community, but rather as an evolutionary *cul de sac.*

Thus the pattern of colony foundation by initially isolated reproductives that is so general among the Formicidae takes on a special evolutionary interest. If in fact it was ancestral to the group, the Formicid queen was, at the very outset of social development, "freed," so to speak, to evolve a pattern of worker production adapted to optimize its own individual reproductive potential, while at the same time promoting maximal evolution of the society itself. In the polygynic Polybiines, the evolutionary "interests" of the individual reproductive and of the community must have been in one sense initially opposed so that a compromise had to be found between them, but in the Formicidae, those interests may have been identified from the beginning of social development.

The challenging question thus arises: is the pattern of dispersal flight followed by claustral colony foundation by isolated female reproductives really the primitive one? Or was it in turn derived from patterns yet more generalized? Do the patterns among the Myrmeciinae and the Ponerinae, so much more primitive in other respects, throw any light on a possibly yet earlier stage?

Dispersion of Reproductives in Primitive Ants

That dispersal and mating flights of winged males and females among the Myrmeciinae, undertaken individually or in large coordinated groups, may essentially resemble the basic pattern in the higher subfamilies was first (and vividly) suggested by Tepper's striking account (1882), of a massive nuptial swarm of one of the larger species of *Myrmecia,* perhaps *M. pyriformis.* Considering the inch-long size of its participants, it must have been an extraordinary sight. A similar massive flight was described for *Myrmecia gulosa* by W. W. Froggatt in 1915 (Wheeler, 1933), and John Clark (1934) recorded another for *Myrmecia forficata.* We have observed similar, though smaller, flights of *Myrmecia vindex,* and of its remarkable workerless social parasite *Myrmecia inquilina,* in each of which both sexes are winged (Haskins and Haskins, 1964).

Among the Ponerines, similar dispersal and mating flights have been reported repeatedly in several species of *Odontomachus* by a number of observers, including the author, in both the Old and New World tropics. They are typically executed as individual enterprises, reminiscent of the flights of the common carpenter ant, *Camponotus pennsylvanicus.* Among the more primitive Ponerines, dispersal and mating flights of both males and females of the American *Amblyopone pallipes* were witnessed over three successive years (Haskins, 1928), and flights of the related Australian *A. australis* were seen at Ferntree Gully, Victoria, Australia, in 1948. These too were characterized by the emergence of the winged forms, individually or in small groups. A huge mass flight of the common nearctic *Ponera pennsylvanica* was witnessed several years ago on the campus of Cornell University (Haskins and Enzmann, 1938). Thus, although modifications of the nuptial flight pattern seem to have taken place as often in living primitive groups as among living higher ants, the prevalence of essentially typical mating and dispersal flights among a great variety of Ponerines and Myrmeciines strongly suggests that this was indeed the ancestral pattern of the Formicidae. What, now, of the crucial point about the isolation of colony-founding females?

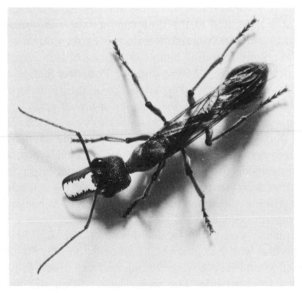

FIGURE 1
Fully winged virgin female of *Myrmecia vindex.*

Colony Foundation in Primitive Ants

Early in the present century W. M. Wheeler (1913) observed dealate females of two species of *Odontomachus* isolated with young brood in cells under stones in Texas and the West Indies. Later he observed similar incipient colonies of *Anochetus* and various species of *Neoponera*. In Australia in 1914 he discovered an isolated, dealate queen of the handsome *Myrmecia tricolor* tending young brood in a cell in the region of the Blue Mountains in New South Wales (Wheeler, 1933). John Clark recorded similar incipient colonies of *Myrmecia nigriscapa,* consisting of individual dealate queens with their broods (Wheeler, 1933). Even at this time it seemed clear that initial isolation of the young colony-founding queen was as general among the most primitive ants as among higher ones. Indeed, it was an assumption of the time that the dominant pattern of colony foundation among both the Myrmeciinae and the Ponerinae coincided exactly with that of higher ants.

There were some difficulties with this last inference, however. In the great majority of both Myrmeciine and Ponerine ants, morphological differentiation between queen and worker, while considerable by the standard of most Polybiine or Polistine wasps, or even of Bombine bees, is relatively slight in comparison with higher Formicids. In particular, queen and worker characteristically differ much less in stature than is usual among the higher subfamilies. It was unknown at the time whether the wing-musculature of the isolated Myrmeciine or Ponerine female was broken down as Janet had described for the colony-founding female of *Myrmica.* But even if this were true it seemed doubtful that a Myrmeciine or Ponerine female could bring a brood of relatively bulky worker larvae to maturity independently of outside sources of nourishment, in the manner of higher ants.

The first hints of an answer to this puzzle were again provided by Wheeler, when during a later visit to Australia he made extensive observations of *Myrmecia regularis.* In the great forests of eucalyptus, particularly karri and jarrah, outside the towns of Manjimup, Pemberton, and Albany in extreme southwestern Australia, *Myrmecia regularis* was a dominant and abundant and striking denizen. In those woodlands Wheeler unearthed some twenty incipient nests of *M. regularis,* each formed by a single dealate queen and frequently containing young brood. Some significant features were noted. The first was that in several nest cells, despite the fact that the walls were intact, young larvae were feeding on fresh insect prey, obviously recently obtained from outside. The second observation was that although the cell walls were indeed intact, many of them appeared to have been broken through and then rebuilt, as by a queen who periodically left and reentered the nest, and, upon reentry, repaired the wall. Third, Wheeler on several occasions encountered dealate queens wandering in the open, at a time when no young sexual forms were present in the mature colonies of the species. Conversely, several of the nests that he examined contained no queens, though healthy larvae were present. It looked very much as though, in the one case, he had intercepted young females foraging outside their nests and, in the other case, found nests whose queens had left to forage. Thus there was a strong suggestion that the isolated *Myrmecia* queen does not immure itself permanently, as higher ants do, but emerges from its cell at intervals to forage. It seemed a reasonable guess that the wholly claustral method of colony foundation of the higher ants had been preceded in evolution by a more generalized mode. Evolution to the modern pattern may have been made possible by increasing divergence in stature between queen and first-brood workers on the one hand, and increasing bulk of wing-musculature in the more specialized queen on the other.

We have had the pleasure of probing this situation in detail in two species of Myrmeciine ants, *Myrmecia forficata* and *M. regularis*. In Australia, near Melbourne, and on the lower slopes of Mt. Kosciusko in New South Wales, a number of typically isolated young queens of *Myrmecia forficata* dwelling in characteristic cells were collected. In the laboratory, they were transferred to glass earth-containing Lubbock formicaries of modified design. These, in turn, were placed within closed arenas, into which the females could emerge to forage. Dilute honey solution and fresh insect provender were provided daily in the arena. Within a few days the young queens, apparently so permanently immured, breached the walls of their cells and emerged, timidly and usually nocturnally, to forage for nectar in the arenas. After these excursions each female returned and once more closed itself securely into its cell. After a few weeks typical spherical eggs appeared, scattered loosely in the brood chambers. The females paid little attention to insect food until the eggs hatched. Then, as soon as young larvae appeared, insects were collected and carried into the nests. The larvae attached themselves to the provender and fed in the usual manner of Myrmeciine and Ponerine larvae. Thus nourished, they developed rapidly. At maturity, the larvae spun cocoons, and, during the process, they were covered with earth by the queens. When spinning was complete, all earth particles were scrupulously removed, in the pattern common to all ants having cocoon-spinning larvae. At that point the queens ceased to bring insect provender into the nests, but they continued for a time to feed on it outside, as though to restore a depleted protein balance. The whole process, from the excavation of the original chamber to the hatching of a second worker cocoon, required just four months for a typical female *Myrmecia forficata*. Even after the young workers had appeared, the queens continued to assist in foraging while the workers were still callow.

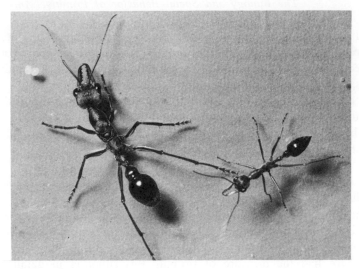

FIGURE 2
Fertile brood female of *Myrmecia vindex* with eggs and young worker.

Five years after this study, a similar one was made on a somewhat larger scale for *Myrmecia regularis,* the very species in which Wheeler had first suspected the pattern of progressive provisioning. Intensive search was made of an area of eucalyptus woodland near the town of Manjimup in extreme southern southwest Australia, not far from Margaret River and Pemberton, where twenty-one years earlier Wheeler had made his original observations. Ten closed earthen cells, each inhabited by a single colony-founding female, were uncovered in the course of the search. The occupants were brought to New York, where in mid-June they were isolated, again in earth-containing, modified glass Lubbock nests, enclosed, as with the queens of *M. forficata,* in foraging arenas similarly provisioned daily with dilute honey and fresh insect prey. All the young queens promptly excavated new nests, and within a month eggs were present in every cell. When larvae appeared, every female broke through the confines of its cell each evening and emerged into the arena, to feed avidly on the honey and to bring insect food back for the developing larvae. Of the nine queens permitted to found their colonies undisturbed, all reared larvae to the stage of cocoon spinning, and seven brought young workers to maturity, usually within a period of five to six months. The general pattern was identical with that of *M. forficata.* It strongly affirmed the speculation that the claustral mode of colony foundation so typical of higher ants may have been preceded in Formicid evolution by a more generalized mode dominated by periodic provisioning of the larvae from outside sources.

In view of this, it became of particular interest to determine whether a similar pattern of partial provisioning is the rule in colony founding among the less evolved species of Ponerine ants. In the same woodland near Manjimup where the queens of *M. regularis* were collected there also occur at the appropriate season abundant isolated young queens of the handsome and very primitive Ponerine *Amblyopone australis,* as Wheeler earlier reported for adjacent areas (1933). Seven such females, taken isolated without brood in cells which they had evidently constructed after recent fertilization, and three from cells containing brood, obviously at a slightly more advanced stage of colony foundation, were collected, and again housed in modified, earth-containing Lubbock nests. *Amblyopone* differs from *Myrmecia* in its ethology in two important respects. It is largely hypogeic in habit (living below the surface of the earth), and it is wholly entomophagous (insect-eating), accepting no nectar. Therefore the nests in which these queens were housed were not provided with external foraging arenas, and insect food was introduced directly into the formicaries. The colony-founding pattern of these females, though differing in detail, was essentially identical with that of *Myrmecia*. All constructed central cells, which they left at intervals to course through galleries that they excavated in the soil in search of arthropod prey. Some individuals closed the cells meticulously after every departure and reentry. Others left them permanently open. Unlike the *Myrmecia* female, those of *Amblyopone* collected provender from the very beginning, often well before brood was present. If the prey was large, the captor might not again leave the brood chamber until the food supply was entirely disintegrated, perhaps after two weeks

FIGURE 3
Colony-founding fertile female of *Amblyopone australis* in cell with first brood.

or more. This tolerance of decaying protein contrasted dramatically with the rapid and scrupulous manner in which all disused food, and all empty cocoons, are promptly removed from the chamber by the *Myrmecia* female. This behavioral difference appears to be correlated with a much greater resistance of the *Amblyopone* queen to the attacks of molds.

All ten females shortly laid eggs, hatched them, fed and reared their larvae as did *Myrmecia,* covered them with earth for spinning the cocoon and cleared them after spinning in similar fashion, tended the cocoons, and assisted the young adults in emerging as does *Myrmecia* (although in the Amblyoponini young workers are able to escape from the cocoon without adult assistance). The whole process of colony foundation required from three to six months. Long after the first workers had appeared, the queen continued to assist in foraging, this behavior even persisting in some colonies that were well established. The whole procedure in this primitive Ponerine had the same archaic aspect as in the Myrmeciines.

Have the higher Ponerines detectably evolved toward the claustral mode of colony foundation? Over a period of years we have observed young colony-founding females of a number of higher Ponerine species, belonging to several tribes, housed in modified, earth-containing Lubbock nests under conditions similar to those already described: *Bothroponera soror* from the African Congo; *Ectatomma ruidum, Odontomachus haematoda, Pachycondyla crassinodes,* and *Trachymesopus* (*Euponera*) *stigma* from Trinidad; *Pachycondyla impressa* from Panama; *Odontomachus ruginodis* from Bermuda; and *Proceratium croceum* from the southern United States. All constructed cells basically similar to those of the *Myrmecia* queens. In all, the founding females left the cells frequently to forage in the pattern of *Myrmecia* and *Amblyopone.*

In only a single species, *Brachyponera lutea,* was the fully claustral pattern found (Haskins and Haskins, 1950). This is a ubiquitous Australian Ponerine, so commonly associated with termite nests as to suggest that it may live as a thief in their galleries. Possibly in adaptation to this mode of life, the queen is of far greater stature in comparison with the worker than in any other Ponerine. Three isolated dealate females of this species were taken in well-formed, closed cells in an area of Hawkesbury sandstone near Sydney, New South Wales. In the artificial nest, all three formed cells, within which they sealed themselves. One shortly perished, but the remaining two each reared a brood of minute young workers to maturity without ever leaving their cells to forage, and thus without the benefit of any protein externally provided. Thus this one Ponerine species at least seems to have achieved the fully claustral pattern so widespread among higher ants.

It is not difficult to imagine some of the selective advantages provided by the specialized claustral mode of colony foundation over the diffuse

FIGURE 4
Colony-founding female of *Brachyponera lutea* in closed
cell with first brood.

pattern of progressive provisioning. The female of *Myrmecia* or of so many
Ponerines, forced to forage in the open every few days, is constantly exposed to the hazard of predators from which the cloistered colony-founding
queen of a higher ant is immune. Moreover, even if the young Myrmeciine
or Ponerine queen avoids or escapes the various perils of drought or excessive moisture or scarcity of food, or of reptilian, avian, or mammalian
predators, during those critical weeks of solitude, it is quite likely to become
lost in its wanderings and never regain the nest at all. Yet before the econ-
omies in the claustral mode of colony founding can be realized, critical
physiological specializations must have been achieved by the queen: the
capacity to administer ingluvial food to the larvae, apparently wanting in
the entomophagous Amblyoponini and poorly developed among the nectarivorous Myrmeciinae; the faculty of reconstituting the wing muscle tissue
as a nutritional reserve. Further, divergence in stature between queen and
first-brood worker must be great enough so that several young workers can
be produced wholly from the food-reserves of the queen. Among the
Ponerines investigated, only *Brachyponera lutea* appears to have met all
the conditions necessary for truly claustral colony foundation.

It appears that a perceptible evolutionary beginning toward the fulfillment of all these conditions has already taken place among the most
archaic living ants, as well as among the higher Ponerines. Histological
examination of the wing-muscle structure of mature fertile queens of
*Myrmecia vindex, Amblyopone pallipes, Euponera stigma, Tetraponera
laevigata, Ectatomma tuberculatum,* and *Ondontomachus haematoda* without exception indicated that those muscles had been as completely dis-

sipated as in the mature queen of any higher ant. Thus it appears that the pattern of muscle resorption antedated in evolution the adoption of the completely claustral mode of colony formation. Even for the isolated foraging *Myrmecia* queen and its first brood, the degenerating wing muscles evidently furnish valuable auxiliary sustenance. It was possible to demonstate this point rather vividly with the tenth founding female of *Myrmecia regularis* in the series already described. In contrast to the treatment given the other nine queens, all external supplies of insect food were denied this individual as soon as its eggs had hatched, though dilute honey continued to be supplied. For forty-five days this deprived foundress tended its larvae, feeding them ingluvially as was proven by incorporating methylene blue into the honey supply, but supplying no externally derived protein. Yet the larvae continued to live and grow. Toward the latter part of the period the female, obviously badly depleted of metabolites, abandoned its usual restriction to crepuscular hunting trips and foraged in the open almost constantly. On the forty-fifth day it was found dead outside the nest, leaving five healthy though by then partially emaciated larvae. The same experiment performed with an isolated female of *Myrmecia mandibularis* yielded a similar result, except that in this case some external protein was provided before the female perished.

The same situation has been demonstrated with more specialized Ponerinae. A young female of *Odontomachus haematoda,* captured immediately after the marriage flight, excavated a cell in a modified Lubbock nest and isolated itself in normal fashion. At this point all opportunities for obtaining external nourishment of any kind were withdrawn, leaving the young female entirely on its own metabolic resources. Eggs were nevertheless laid and hatched, and the young larvae developed rapidly. When about half grown, however, they began to decline and finally perished. The queen also ultimately perished, but only after having reared a good-sized brood through almost half its period of growth, evidently from its own bodily stores, and having sustained itself from the same sources for ninety days.

Thus in all the primitive ants investigated, marked evolutionary beginnings have already been made toward two of the conditions prerequisite for the claustral colony-founding pattern. The same is true of the third prerequisite situation—an adequate divergence in stature between queen and first-brood daughter workers. In both *Myrmecia* and *Amblyopone,* such daughters are usually abnormally small, because of meager nutrition. But this is not always true, at least in *Myrmecia.* Thus in the series of *M. regularis* founding females already described, insect food was available in plenty and the first workers were not of unusually small size—indeed they were considerably larger than smaller individuals normally appearing in mature colonies. Thus there does not seem to be any inner physiological mechanism to mediate the size of first-brood workers. The only adaptation at this stage of evolution appears to lie in an unusually highly developed capacity of half-starved larvae to transform into nanitic (abnormally small) adults.

In some of the higher Ponerines, however, an additional regulatory mechanism appears to have been introduced. Ledoux (1952) has found that the larvae that produce the small first-brood workers of *Odontomachus assiniensis* Latreille pupate after the third larval instar instead of after the fourth as do normal worker larvae in older colonies. In the Formicine ant *Oecophylla*, the larvae of the minor workers likewise pass through one less instar before becoming pupae than do those of major workers, and this mechanism is apparently fundamental to the differentiation of the subcastes of that species. It would be of great interest to learn whether this mechanism of subcaste regulation has evolved elsewhere among the Ponerines.

The essential query posed at the beginning appears to have been answered. The initial isolation of the colony-founding female from potentially reproductive competitors in the form of siblings or, even more menacing, of unrelated partners, appears to be an ancient and almost certainly the basic Formicid pattern. It may have contributed importantly from the beginning to the high degree of social organization ultimately achieved by the ants. But the fully claustral mode of colony foundation so characteristic of higher ants today is not primitive. It has been derived through a series of adaptive steps that are still to be seen in living Myrmeciines and Ponerines.

There is another point which, though subsidiary to the main one, is very interesting. If every element prerequisite to the development of the claustral mode of colony foundation is already primitively evident even in the most archaic ants, the very imperfection of their evolution has preserved for the Myrmeciinae and the Ponerinae an evolutionary plasticity in the development of other modes of colony formation which may have been lost to many higher ants. Thus one of the conspicuous evolutionary paths taken by many Myrmeciine and Ponerine species leads to the loss of wings and correlated simplification of the thorax in the female until it may closely approximate the structure of the worker—a trend evidently made possible in an evolutionary sense by the fact that though the young Myrmeciine or Ponerine queen is clearly aided in colony founding by the added nourishment derived from the deteriorating wing-muscles, she is by no means dependent upon it.

Some of these paths find dramatic illustration at very primitive levels. Brachypterous (short winged) and apterous (wingless) females occur surprisingly often within the genus *Myrmecia*. In some species they are the only functional type. In others they appear to coexist with normal forms, which appear to be little if any more successful in colony foundation. The virgin females of *Myrmecia regularis* are brachypterous, and, as we were able to demonstrate (Haskins and Haskins, 1955), they follow an unusual course of dispersion. Instead of embarking on a typical flight, the young females drop the short, delicate wings while still callow, and well before they leave the formicary. It seems likely that by the time they are fully pigmented and sufficiently hardened to leave the parent nest, degeneration of wing musculature may already have begun, though this remains to be proved. Mating takes place on the ground or on low herbage during a period of wandering, after which the female proceeds to found the new colony as already described. This evolution is carried further in *Myrmecia tarsata,* where the colony-founding female has a thorax so worker-like as to appear very similar to the ergate (worker). The patterns of fertilization and colony foundation seem to approximate those of *M. regularis.* With such nearly ergatoid forms, little if any auxiliary sustenance can be supplied from the thorax. Yet this lack evidently offers little impediment to colony-founding by the actively foraging queen.

The tendency toward development of ergatoid queens seems much more widespread among Myrmeciine and Ponerine ants than it is in the highest subfamilies. In the Ponerine genera *Paranomopone, Eusphinctus, Megaponera, Acanthostichus,* and *Plectroctena,* the queen approaches the worker form quite closely, reminiscent of the situation in, for example, *Myrmecia tarsata* or *M. aberrans.* Presumably their colonies must be

FIGURE 5
Brood female of *Myrmecia regularis* in nest (upper center) showing worker-like thorax. At lower left is a recently hatched Eucharid parasite.

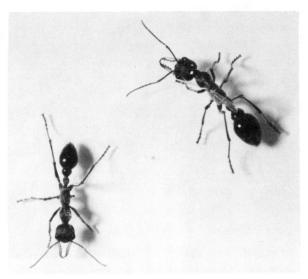

FIGURE 6
Young partially ergatoid female of *Myrmecia tarsata* (upper right) still in parent colony. Note the small left primary wing, not yet shed. In this species, as in *Myrmecia regularis*, the deciduous wings of the highly brachypterous females are usually shed before leaving the parent colony.

founded in virtually complete reliance on external food sources, unless colony fission takes place—the actual process does not seem to have been recorded. Finally, in such Ponerine genera as *Dinoponera* of South America and the related *Streblognathus* from South Africa, in *Diacamma,* and in the majority of species of the richly developed Australian genus *Rhytidoponera,* no queens distinguishable from the workers have ever been reported. It would appear that there has been an almost complete evolutionary convergence of the morphology of the two castes at worker level, save for the retention of functional spermathecae, and, possibly, of more fully developed ovarioles by the reproductives.

Nothing whatever is known of the method of colony foundation in any of these genera except *Rhytidoponera*. Several years of observation of *Rhytidoponera* have served to outline a most interesting situation (Haskins and Whelden, 1965). In a few species, such as *R. impressa, R. chalybeae,* and *R. purpurea,* mostly of tropical or semitropical distribution in northeastern Australia, a typical winged female is present, as in many related Ectatommine genera such as *Ectatomma, Gnaptogenys,* or the more closely related *Acanthoponera*. That this female fulfills the normal nest functions in a typical way is proven by long observation in the artificial nest. In the remaining species of *Rhytidoponera,* however, there seems to be no truly functional queen. A certain proportion of the "workers," however, retain functional spermathecae, and take the place of colony reproductives vacated

FIGURE 7
Rhytidoponera metallica colony composed entirely of workers.

by the queen. They have been seen to be fertilized by the low-flying males. A colony, even if isolated from others, has been shown able to maintain itself in the artificial nest for a decade, new workers being constantly produced by these older fertilized workers. In *Rhytidoponera convexa,* R. M. Whelden (1957) has found that an average of 8 percent of the workers in a sample of over 250 individuals carried sperm in their spermathecae, and the corresponding figures for the related *R. inornata* and *R. metallica* (Whelden, 1960) were respectively 2.3 percent and 5.4 percent. It seems probable that the percentage may vary widely with season and conditions, but it is clear that at all times a substantial fraction of the worker caste in the colony is able to produce worker offspring.

Rarely true queens may be produced under both natural and artificial conditions. This has been achieved for *R. metallica* in the artificial nest, and the behavior of such queens studied carefully. They appear wholly atavistic in behavior. Though fully winged, with bulky thoraces and well-developed eyes, and though they may fly actively, they have never shown the slightest tendency to found colonies, and when isolated artificially perished almost immediately. The only mode of genesis of new colonies that has been observed in this species is the detachment of small parties of workers, including one or more fertilized individuals, from older colonies. The whole pattern suggests that in *Rhytidoponera* the queen caste has simply been eliminated, and the reproductive niche so created reoccupied by the fertilized, laying workers.

It is possible that a similar course of evolution has been followed in *Dinoponera* and *Streblognathus*. These obviously ancient, monotypic forms, have been shown by Carpenter (1930) to be related to the beautifully preserved Miocene fossil *Archiponera wheeleri,* from the Florissant beds near Colorado Springs. Since no queen of either living species has ever been found, and since Carpenter was unable to discover a fossil queen of *A. wheeleri* in an extended and pointed search, we shall not be able to decide this question unless, as in *Rhytidoponera,* it proves possible to produce a true queen of either genus under artificial conditions.

Among the Myrmeciinae and the Ponerinae, we can find hints of virtually every development in colony structuring and in colony foundation that is collectively exhibited in much more specialized form by one or another species of the higher ants. Many species of *Myrmecia* and many Ponerines exhibit the beginnings of pleometrosis. The evolution of the ergatoid female is clearly foreshadowed among the Myrmeciinae, and may have been carried forward more luxuriantly in the Ponerinae than in any other ants. Even the dicthadiigyne is suggested in the queens of *Onychomyrmex,* whose affinities are with the most archaic tribe of the Ponerinae. In *Myrmecia inquilina* the pattern of social parasitism by a workerless inquiline species, so elaborately evolved in a whole series of higher ants, and culminating in *Teleutomyrmex schneideri* where the parasitic female is virtually an ectoparasite upon the host queen, is already well developed. One is left with the strong feeling that the Myrmeciines and the primitive Ponerines well typify early and rich reservoirs of evolutionary trends, from which irreversible specialization has occurred in many directions. In this connection, it is interesting to reflect that the well-developed wing muscles of the queens of so many of the higher ants may have been at least as adaptively important to the evolution of the fully claustral mode of colony foundation as to that of securing an effective dispersal flight.

THE EVOLUTION OF THE COLONY AS ORGANIZATION— THE BINDING TIES WITHIN THE COLONY

So much for the origin of the ant community. What now of its evolution as an organization?

The exquisite internal regulation of the colony of the higher ants, its high organization and the behavioral coordination of its elements, have been stressed many times, most recently, perhaps, by Wilson (1965) in his discussion of the social role of insect pheromones. Did the various behavioral modes by which this coordination and integration are achieved among higher ants appear contemporaneously in evolution? Or were older, cruder devices reinforced or successively replaced by finer, more discriminatory, and more powerful ones? In what order did such ties appear in evolution? Can behavioral studies of the Myrmeciines and the primitive Ponerines throw any light on these questions? There may be real possibilities here, though to date not a great deal has been done. What follows is a partial summary of our present knowledge of the evolution of ant colonies.

Relations Between Adult and Young

The long history of the ant colony as a family suggests that patterns of behavioral relationship between adults and brood may have been among the earliest perfected in insect social evolution. The Myrmeciinae and the Ponerinae seem to offer some evidence confirming this. Adult-brood relations in these subfamilies appear relatively more highly evolved than other patterns reinforcing colony solidarity, though even these relations, on the whole, are cruder than those in the higher ants.

The genus *Myrmecia* is unique among ants in its subspherical, pearl-like eggs, which in many species of the genus are entirely nonadherent, and so cannot be gathered into packets. They are typically arranged separately on the floor of the nest by the workers and are individually attended. Thus the newly hatched larvae cannot feed initially on younger eggs in the packet, a characteristic procedure among the Ponerinae and in many higher ants. The larvae appear, however, to be fed ingluvially by the workers, to a limited extent. Evidence of such feeding has been obtained for a number of species, notably *M. vindex* and *M. regularis,* by incorporating the dyes neutral red or methylene blue in the dilute honey with which the colony was supplied. Such dyes accumulate in the gut wall of the larva and can be seen through its integument, or made more clearly visible in sagittal section. As soon as the Myrmeciine larva is a few days old, whole or fragmentary arthropod prey is characteristically deposited beside it by a worker. Frequently the worker also carries the larva to the prey and places it nearby. If necessary, the larva can make its own way to this provender over considerable distances by a well-directed writhing. As soon as the prey is within reach, the larva commonly attaches itself, attacking the chitinous investment of the prey with its heavily sclerotized mandibles. In *Amblyopone australis* and *pallipes,* and very generally among the higher Ponerines, the care given the larvae by the adult workers is essentially similar, except that, as the eggs are adherent in packets, the newly hatched larva is able to feed on younger ova in the mass (a procedure, as Brian (1962) has shown, also characteristic of newly hatched brood in a higher ant genus such as *Myrmica*). The Amblyoponini are exclusively entomophagous, and there is no evidence that ingluvial feeding of the larvae by the adults is a possibility; this egg-feeding must substitute for it. In such higher Ponerine genera as *Odontomachus,* evidence from dye-feeding has indicated that ingluvial provisioning of the young larvae, in place of or in addition to egg-feeding, is a normal procedure. In many species of *Myrmecia,* a specialized mode of feeding worker-laid eggs to both adults and young has been developed. It was first observed by Freeland, and has been confirmed in a number of species by the author. There is suggestive evidence that in species such as *M. regularis,* in which a primitive type of ingluvial feeding of the larvae is most highly evolved, this pattern of egg-feeding may be slightly developed or absent, and that in such species as *M. vindex,* the converse is true. Further observation of this feeding process is needed.

All Myrmeciine and Ponerine larvae spin cocoons, which frequently are of considerable thickness and toughness. Typically the nurses cover the larvae with loose soil as they are about to spin, scrupulously clearing away the earth after the cocoon is formed. In some species, as, for example, *Myrmecia vindex* when nesting in loose and rather dry sand, a considerable proportion of the covering may be permanently incorporated in the silk of the cocoon, so that in extreme instances it may recall the silk-and-sand-grain cases of some caddis flies.

The behavior of Myrmeciine and Ponerine workers toward the larvae while spinning, and their subsequent care of the cocoons, is very similar to that of species of higher ants whose larvae also spin cocoons. When the young adults are about to be eclosed, however, worker behavior shows some interesting differences. When a pupa of one of the higher ants is ready for eclosion, adult workers typically cluster about the anterior pole of the cocoon, tearing at it until an adequate opening is made. The young imago is then pulled out, usually still enclosed in the pupal covering, its wings, if any, unexpanded. The pupal skin is stripped away under the constant pulling and rasping of the nurses; the wings, if present, are expanded; then the callow insect, feeble and lightly pigmented, is carried to a quiet corner. In the Myrmeciinae and the Ponerinae this nursing behavior is much less evolved. Indeed, a series of stages of increasing involvement of the adult nurses can be traced from these primitive ants to higher forms. In *Amblyopone australis* and *A. pallipes,* the young adult is able to escape from the cocoon even if entirely deprived of assistance, as isolation experiments show. The pupal skin has been shed within the cocoon, and the wings of the alate males and females are already expanded and hardened at emergence. The males are fully pigmented and capable of flight at once, although actually they rarely fly until a few days old. Though still somewhat callow, and of a light coloring which may be retained for several weeks, emerging females are able at once to take part in normal social duties, including the tending of brood.

Despite this capacity of the young insect to hatch by itself, it actually

FIGURE 8
Solicitation of larvae for exudates by adult workers
in *Myrmecia vindex* colony.

receives a great deal of attention and assistance from the adult workers
in the colony under natural conditions.

In *Myrmecia* an intermediate behavioral situation exists. Young imagoes
of *Myrmecia vindex, M. nigriceps,* and *M. pilosula* that have been observed
hatching in the artificial nest have all been greatly assisted by adult nurses.
As in *Amblyopone,* the pupal skin is shed before eclosion. But unlike that
genus, the young *Myrmecia* adults when first eclosed are characteristically
very feeble and lightly pigmented, like the hatching imagoes of higher ants.
Yet when cocoons of *M. vindex* and *M. pilosula* were isolated, a considerable proportion succeeded in emerging without assistance. Commonly,
however, such individuals were much delayed, and when they finally
escaped, they were almost as pigmented as adults. When cocoons of
M. forficata were similarly isolated, none succeeded in escaping. Evidently the dependence of young callows upon adult worker assistance in
hatching has been much increased in *Myrmecia.*

Though ingluvial feeding of the larvae is at best crudely developed among
the Ponerinae and the Myrmeciinae, the worker behavioral pattern of
soliciting the larvae for exudates and constantly licking them is conspicuous, so that the general impression of worker-larval trophyllaxis so typical
of higher ants is well sustained. In *Myrmecia vindex,* the rapid stroking
movements of the forelegs, which are so characteristic among higher ants
when one adult worker solicits ingluvial food from another, seem especially
adapted to palpation of the larval thorax and upper abdomen, which, accompanied by pinching of these parts with the mandibles, is presumably a
solicitation for exudates.

Relations Between Adults

The relationships among adults in colonies of the Myrmeciinae and the lower Ponerinae are considerably more elementary, and present a greater contrast to those among the higher ants in their crudeness and lack of variety than relationships between adults and brood. This again suggests that the adult-brood relations were, on the whole, pivotal in the evolution of early social life, while adult relations played a later role in reinforcing and elaborating the social structure.

DEPORTATION OF ADULT BY ADULT

Under certain conditions the adult workers of many higher ants characteristically carry one another from place to place. In the genus *Formica,* for instance, it is common for foraging workers that have become marooned outside the nest beyond normal hours of foraging to be picked up and carried home by passing sisters. Workers of many genera of ants, when changing the location of the nest, typically convey less aggressive sisters from the old to the new site. Not infrequently in such circumstances, the transported individual may return to the old site with its carrier to join with it in the process of removing further individuals. If an established nest is suddenly disturbed callow workers are often carried with the immature brood to safety. In all these situations, the deported individual usually assumes a rigid posture characteristic of the species, and remains passive while it is held.

In sharp contrast to this general situation among the higher ants, transportation of one adult by another seems very rare in all the species of *Myrmecia* that have been observed in the artificial nest. Moreover it nearly always involves moving males, callow workers or females, or old brood females to safety when the nest is disturbed. There is no conspicuous stereotyped pattern for this action, either by the individual being carried or its bearer. The deported individual is commonly grasped by the mandibles, or in the case of males by the antennae, and simply dragged, the deporting individual frequently proceeding backward. On rare occasions an old brood female of *Myrmecia* has been seen guided by workers to a fresh Lubbock nest in this way. The deported individual is rarely, if ever, lifted from the ground.

Occasional exceptions to this rule have been observed. One involved the handling of young queens of the workerless, social parasite *M. inquilina* by the workers of its host species *M. vindex* and *M. nigriceps*. On two occasions in the field, workers of *M. vindex* were seen to seize females of *inquilina* wandering in the open and drag them for considerable distances by the mandibles—in one case into a *vindex* nest. But in a third case, a worker of a vigorous and aggressive colony of *M. nigriceps* infested with *M. inquilina* seized and actively carried off a young queen of the parasite as though it were a host cocoon or larva when the nest was disturbed.

In *Amblyopone pallipes* this pattern is even more elementary. In the artificial nest an adult may occasionally seize another and drag it some distance along a gallery. In one case, thirteen such instances were observed within two minutes: workers seized sisters feeding on arthropod prey newly brought to the brood chamber and deported them from it. Any part of the body may be grasped, but most commonly the mandibles, margins of the head, pedicel, or gaster. There is no sign of tonic immobility on the part of the deported insect, and no other indication of any specialization in the process.

In more evolved Ponerines, characteristic manners of deportation approximate more closely those represented among the higher subfamilies of ants. Workers of *Odontomachus hastatus,* for example, regularly carry off their small, lightly pigmented males when the formicary is invaded, treating them like cocoons and handling them swiftly and efficiently. These males fold wings, legs, and antennae against the body during the process and remain immobile. Essentially the same procedure occurs within the genus *Lobopelta,* where the workers typically transport the immobile, folded males beneath their bodies when the nest is disturbed.

Fairly specialized deportation procedures have been recorded in the genus *Ectatomma.* In the related *Rhytidoponera,* deportation patterns are highly developed, closely resembling those in such a Myrmicine genus as *Leptothorax.* In *Rhytidoponera metallica,* for example, the deported individual is seized by the mandibles and held curled over the back of the carrier, its dorsal side forward, retaining a tonic immobility throughout. Thus processes of transportation of adults by adults, so crude and little specialized in the more archaic forms of the Ponerine evolutionary series, may among the more advanced members merge into the patterns of higher subfamilies.

THE EXCHANGE OF FOOD

It has been mentioned earlier that the workers of several species of *Myrmecia,* predominantly nectarivorous as they are, are capable to a limited extent of feeding liquid food to the larvae. Although among the Amblyoponini, with their specialized entomophagous habits, the capacity appears wanting, the process has been observed in many of the higher Ponerines, in particular in *Odontomachus, Ectatomma,* and, as already indicated, in *Brachyponera lutea.* On the whole, however, mutual feeding of adults by regurgitation is rarely observed in such species, and when seen, it is usually crude and unspecialized. This is a matter of considerable sociological interest. For it is precisely this process that is so vital to higher ants in securing colony-wide distribution of those social pheromones by which, to a great degree, community integrity and coordination are maintained. It is therefore of special interest that mutual ingluvial feeding of adults is conspicuous in such a higher Ponerine genus as *Rhytidoponera*. Here, despite the relative crudeness of the crop as a storage and pumping mechanism, worker-to-worker exchange of nectar occurs frequently.

In mutual feeding of adults, therefore, as in deportation, the genus *Rhytidoponera* approaches surprisingly close to the Myrmicinae to which, in Brown's view (1954), it may be related. Possibly linked to this habit is another remarkable behavior pattern in which *Rhytidoponera* also approaches the Myrmicinae. Though like other Ponerines, *Rhytidoponera* predominantly seeks insect prey for the larvae, it is, surprisingly enough, also an avid collector of seeds, like so many Myrmicinae. Such seeds are commonly thick-coated and hard, and they are characteristically stored intact within the nests. They may belong to a variety of indigenous plant genera. Thus seeds collected from nests of *Rhytidoponera metallica* near Sutherland, New South Wales, in December, 1963, were identified by Dr. Dorothy Johns, of the division of Plant Industry of the Commonwealth Scientific and Industrial Research Organization (Australia), as those of *Acacia baileyana,* while colonies of the same species of *Rhytidoponera,* examined in Ashton Park, near Sydney, yielded seeds of *Acacia longifolia* and *Hardenbergia violacea,* as well as of a species of *Bossiaea.* Colonies of *Rhytidoponera inornata* collected at Manjimup in southwestern Australia had stores of seeds of *Acacia pulchella* within their formicaries.

COMMUNICATION OF FOOD SOURCES; TRAIL-LAYING; AND THE ROLE OF PHEROMONES IN MYRMECIA

The communication of food sources by trail-laying, as well as by more direct means involving directive worker-worker stimulation, such as tandem running and similar patterns, has been extensively observed in higher ants. It would be interesting to know whether such communication patterns are developed to any extent in *Myrmecia,* whose habit of extensive and intensive foraging at long distances from the nest would make such patterns highly adaptive.

The author conducted a preliminary series of experiments on the potential for communication of food sources with two populous natural colonies of *Myrmecia gulosa* located near Sutherland, New South Wales, in the season when a large share of the brood was in the larval state and communal foraging was at its height (Haskins and Haskins, 1950a). Observations were made over two days, beginning in the morning between 10:00 and 11:00 A.M. and continuing until approximately 4:00 P.M. During these periods, foraging ants were leaving and reentering these large nests in a continuous stream.

The simple experiments involved placing plastic cups, filled with a mixture of honey and water, thirty inches from the nest craters. These were soon discovered (apparently by chance) by individual foraging workers, which drank from them avidly. The first six individuals to feed in each test were marked with spots of colored varnish, and were distinguished by a color- and position-code of the dots. After the individuals so identified had fed and left the cups, subsequent visitors were imprisoned. In all, 132 observations were made of the behavior of the marked ants subsequent to feeding. Their actions portrayed a singularly stereotyped picture. In somewhat over half the cases, the marked foragers returned to their nests within two minutes of leaving the plastic dishes. Occasionally they remained within the nests up to one hour before reemerging, but usually came out much sooner and returned to the cups, where they again fed. After a few such trips between cups and nests they ran more rapidly and found their way more directly than at the beginning. But never during the entire experiment was a forager returning from the nest crater to a cup accompanied by a nest mate. Nor was there the least behavioral evidence that an odor trail had been laid by the forager at any stage of the proceeding. The conclusion seemed unequivocal that the worker of *Myrmecia gulosa* hunts as a solitary individual, unable to communicate to its fellows the location of any food source that it finds. Subsequent experiments of a similar kind with *M. tarsata* and *M. nigrocincta* in the artificial nest reinforced this conclusion.

Cavill and Robertson (1965) reported in detail on the exocrine gland systems in *Myrmecia gulosa*. Such glands turn out to be similar in location and form to those so well known in the higher ants. They include pharyngeal glands, mandibular glands, salivary reservoirs and glands, metasternal glands, venom reservoirs and glands, accessory glands, and dorsal abdominal glands. If any of these systems were functional in trail-laying, some behavioral reflection of this use would surely be expected. Its decided absence in *M. gulosa* in the experiments cited suggests that here is a striking example of a physiological feature beautifully preadapted to social ends, but not yet, in this primitive social stage, actually put to such use.

It may be that the same should be said of the mandibular glands, which are well-developed in *M. gulosa,* and which, both in this and in other species, seem to be wholly functional in a physiological sense. Cavill and Robertson reported the odor of mandibular gland products, which is readily detectable at some distance from even a single captured worker, and we have repeatedly found the odor even more pronounced in such a species as *M. vindex,* and produced by the worker abundantly on the slightest handling. In higher species of ants the substances from this gland regularly evoke alarm frenzy in nest-mates, as Wilson has clearly demonstrated for *Pogonomyrmex badius* (1958a). Yet field tests made on some twenty colonies of *M. vindex* in Western Australia, in which mandibular glands, intact or contained in crushed and severed heads of workers, which to the human nostril reeked of the characteristic odor, were inserted into the entrances of otherwise undisturbed colonies and evoked no concerted alarm response whatever. The only reaction ever detected was that of a guard worker that seized and stung a severed head or head-fragment as it was dropped into the nest-entrance, as though it had been some bit of insect prey or of foreign material that had accidentally rolled in. Equally negative were similar field tests with *M. gulosa, M. tarsata, M. simillima* and *M. regularis.* Can it be that the secretions of the mandibular glands, though fully developed, have not yet been "matched" by a series of neural behavior patterns allowing their odor to be put to social use? This may be another example of social preadaptation, caught in evolution before it has become of social significance.

In at least one behavioral context, however, social reactions to odors are well developed even at the level of *Myrmecia.* From an adaptive standpoint, it is easy to see why this development should have occurred so early in social evolution. It involves the pattern of necrophoric behavior (burying the dead). Wilson (1958a,b) was able to demonstrate that when groups of *Pogonomyrmex badius* workers were allowed to come into direct contact with small amounts of formic acid, ethylamine, phenylethylamine, triethanolamine, phenol, n-butyric acid, n-valeric acid, n-caproic acid, or n-caprylic acid absorbed in 1 cm² pieces of filter paper, they exhibited weak to moderate alarm behavior, in time often passing into digging behavior, concentrated around the paper squares. The ants made no attempt to remove the squares, however. In the case of oleic acid alone, the ants removed the squares promptly to the kitchen middens—the refuse pile. This reaction could be expected, since oleic acid is a natural and prominent decomposition product of decaying insect material. A fatty acid component—possibly oleic acid—obtained from decaying bodies of *P. badius* was found to elicit necrophoric behavior in a species so far removed as *Solenopsis saevissima.*

The author has tested for necrophoric behavior in *Myrmecia.* In experiments to be reported more fully, an active nest of *M. vindex* was housed in a modified earth-containing Lubbock nest placed in a rectangular plastic arena (47″ × 23.3″ in inside dimensions, coated with clean brown paper) to which the ants had free access. A choice of chemically treated piles of brood from the nest was placed outside the nest on separate cardboard sheets, 6″ × 8″. It was quickly found that larvae were unsuitable for this work. Even when the larvae were heavily coated with the test chemicals, workers still often responded to the underlying larval odor. Such larvae were frequently carried back into the nest when discovered, whatever their chemical treatment. With cocoons, however, there was a different story.

Piles of six to ten cocoons were treated on the various cardboard rectangles respectively with caproic acid, methylbenzylamine, triethanolamine, n-valeric acid, formic acid, and oleic acid. Each cocoon in each group received three drops of the chemical. The cardboards with their cocoons were then placed in the arena at equal distances from the nest entrance in configurations differing in numerous tests. Each test included a control group. As with *Pogonomyrmex badius,* oleic acid consistently stimulated the *M. vindex* workers to carry objects (in this case the cocoons) treated with it to the middens. Digging and "burying" were associated with this reaction. Although no loose earth was available in the arena itself, sand grains were regularly brought from the interior of the Lubbock nest and placed on, or in a ring around, the objects treated with oleic acid, often almost completely covering them before they were finally carried away.

In all experiments control cocoons were carried back to the nest, as were those treated with formic acid once the acid had evaporated. There were other reactions of interest also. In contrast to Wilson's experience, caproic acid evoked a necrophoric reaction at least as strong as, and sometimes stronger than, oleic. Cocoons so treated were carried to kitchen middens even when these were situated as far as 44" from the nest entrance. On one occasion a single caproic-acid-treated cocoon was transported to a midden 40.5" from the nest and deposited; then, during the night, it was returned to the nest; then it was brought out again and carried 44" away and dropped; finally it was once more retrieved and brought back into the formicary. Similarly conflicting behavior patterns with caproic acid were observed in additional tests. Digging and "burying" reactions also appeared, similar to those evoked by oleic acid. Triethanolamine acted initially as a strong attractant. Cocoons treated with it were, on some occasions, carried into the nest promptly; on others they were transported as promptly to the middens.

Further experiments are needed, but a preliminary impression is clear that the necrophoric response to oleic acid is as strongly developed in *Myrmecia* as it is in those higher ants in which it has been investigated. Further, triethanolamine, which was found by Wilson to be without effect in *P. badius,* appears to markedly stimulate *Myrmecia,* though the nature of the response is somewhat indefinite; while n-caproic acid, which releases alarm and ultimately digging, but not necrophoric, behavior in *P. badius,* acts as a stimulant to necrophoric behavior in *Myrmecia.*

Few aspects of the development of social life are more fascinating than the evolution of "social signals." All sorts of questions arise. If necrophoric behavior is indeed one of the oldest of these social patterns, by what paths, and through what forms, did its response to a specific "signal" become "sharpened" from the apparently generalized and imprecise configuration in *Myrmecia* to a better defined and more specialized one in higher ants? If the glands that permit trail-laying in higher ants do not function in this capacity in *Myrmecia,* as appears to be true, did the capacity to respond in this way to their secretions develop at the mid-Ponerine stage of evolution—and if so, by what paths did such an evolution occur? Conceivably, it may have taken place very early. Schneirla's studies have given us a vivid picture of how developed trail-laying and trail-following have become at the Doryline level of social evolution. It would seem that it must have become a prominent behavioral feature at yet more primitive levels. It is hard to imagine that legionary Amblyoponine ants, like those of the genus *Onychomyrmex,* could maintain their migratory habits without such a capacity—though the matter has never, to my knowledge, been specifically tested—and the same must surely be true of such higher Ponerine "legionaries" as *Termitopone* and *Megaponera.*

Lastly, there is the matter of those behavioral mechanisms that must directly maintain the solidarity of the individual community, such as those conferring the remarkable capacity, so highly developed in many ants, to distinguish individuals of the same colony from those of other colonies, even if of the same species. This behavioral development, too, must have appeared very early in social evolution, since it is strongly manifested in all the species of *Myrmecia* and of *Amblyopone* in which we have investigated it. But there may be more to the story. Behavioral experiments con-

FIGURE 9
Hostile reaction of workers from different colonies in *Amblyopone pallipes.*

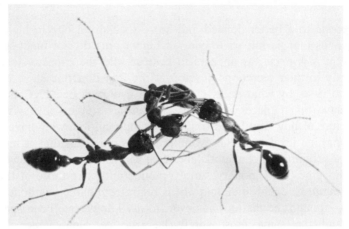

FIGURE 10

Hostile behavior between sister workers of *Myrmecia vindex,* one of which was eclosed from the cocoon and attained maturity in isolation from the main colony. Such failure to recognize nest-mates and siblings when eclosed and carried beyond the callow adult stage in isolation seems characteristic of some very primitive ants.

ducted over many years, though not yet ready for a full account, suggest that a rather sharp "break" in the specific means by which this recognition is accomplished may have intervened between evolutionary levels symbolized today by the Myrmeciinae and such lower tribes of the Ponerinae as the Amblyoponini, on the one hand, and by the higher Ponerines and higher ants in general on the other. As Wilson has pointed out (1965), quantitative research in this field is technically difficult to conduct, both because the substances that presumably mediate the behavior are present in such meager amounts that they are difficult to analyze by even the most sensitive available means, and because the behavior patterns themselves are readily subject to misinterpretation and demand most careful analysis. The experiments that support this guess about a "break" in the specific mode of colony recognition in the early social evolution of ants were based wholly on behavioral observation and analysis. But the consistency of the results obtained from many trials conducted over a long period of time suggests that more extensive investigation of this field might be rewarding.

CONCLUSION

Ants are probably the oldest of social insects, and ant fossils are abundant. Yet these circumstances do not make them the most attractive subjects for the study of social evolution. What is important is that throughout their history ants have primarily been dwellers within the earth. Under the mantle of the soil, with the manifold and varied protected environments that it provides, beneath the heaviest rocks, in decaying logs on the forest floor, in the sheltering cavities of plants themselves, ants have found refuge from predators, including higher ants that are characteristically ruthless exterminators of less adapted types wherever they come into direct competition. Thus there is today a rich assemblage of ancient forms, providing a living record of evolution, including near-facsimiles of types that were dominant on our planet sixty million years ago or more. In such "living fossils," we find in detail the specific evolution of many of the physiological and behavioral patterns that have cemented, reinforced, and perfected Formicid social life over the eons. The opportunity to study the development of that life at first hand offers one of the major challenges for the student of social evolution today. This is the opportunity that, above all, gives life and wings to adventures with primitive ants.

REFERENCES

Brian, M. V. 1962. Studies in caste differentiation in *Myrmica rubra* L. 5. Social conditions affecting early larval differentiation. *Insectes Sociaux* 4: 295–310.

Brown, W. L., Jr. 1954. Remarks on the internal phylogeny and subfamily classification of the family Formicidae. *Insectes Sociaux* 1: 21–31.

Brown, W. L., Jr., and Wilson, E. O. 1959. The search for *Nothomyrmecia*. *Western Australian Naturalist* 7: 25–30.

Carpenter, F. M. 1930. The fossil ants of North America. *Bull. Mus. Comp. Zool. Harvard* 70: 1–66.

Cavill, G. W. K., and P. L. Robertson. 1965. Ant venoms, attractants, and repellants. *Science* 149: 1337–1345.

Clark, J. 1934. Notes on Australian ants, with descriptions of new species and a new genus. *Mem. Nat. Mus. Melbourne* 8: 5–20.

Haskins, C. P. 1928. Notes on the behavior and habits of *Stigmatomma pallipes* Haldemann. *J. N. Y. Entomol. Soc.* 36: 179–184.

Haskins, C. P., and E. V. Enzmann. 1938. Studies of certain sociological and physiological features in the Formicidae. *Ann. N. Y. Acad. Sci.* 37: 97–162.

Haskins, C. P., and E. F. Haskins. 1950. Note on the method of colony foundation of the Ponerine ant *Brachyponera* (*Euponera*) *lutea* Mayr. *Psyche* 57: 1–9.

Haskins, C. P., and E. F. Haskins. 1950a. Notes on the biology and social behavior of the archaic ponerine ants of the genera Myrmecia and Promyrmedia. *Ann. Entomol. Soc. Am.* 43 (4): 461–491.

Haskins, C. P., and E. F. Haskins. 1955. The pattern of colony foundation in the archaic ant *Myrmecia regularis*. *Insectes Sociaux* 2: 115–126.

Haskins, C. P., and E. F. Haskins. 1964. Notes on the biology and social behavior of *Myrmecia inquilina*, the only known Myrmeciine social parasite. *Insectes Sociaux* 2: 267–282.

Haskins, C. P., and R. M. Whelden. 1965. "Queenlessness," worker sibship, and colony vs. population structure in the Formicid genus *Rhytidoponera*. *Psyche* 72: 87–112.

Ledoux, A. 1952. Recherches préliminaires sur quelques points de la biologie *d'Odontomachus assiniensis* Latr. (Hymen. Formicoidae). *Ann. Sci. Nat. Zool.* 14: 231–248.

Richards, O. W. 1953. *The social insects.* London: Macdonald & Co.

Tepper, J. G. O. 1882. Observations about the habits of some South Australian ants. *Trans. and Proc. Roy. Soc. South Australia* 5: 24–26; 106–107.

Wheeler, W. M. 1913. *Ants, their structure, development, and behavior.* New York: Columbia Univ. Press.

Wheeler, W. M. 1928. *The social insects.* New York: Harcourt, Brace.

Wheeler, W. M. 1933. *Colony founding among ants.* Cambridge, Mass.: Harvard Univ. Press.

Whelden, R. M. 1957. The anatomy of *Rhytidoponera convexa. Ann. Entomol. Soc. Am.* 50: 271–282.

Whelden, R. M. 1960. The anatomy of *Rhytidoponera metallica. Ann. Entomol. Soc. Am.* 53: 793–808.

Wilson, E. O. 1958a. A chemical releaser of alarm and digging behavior in the ant *Pogonomyrmex badius* (Latrielle). *Psyche* 65: 41–51.

Wilson, E. O. 1958b. Chemical releasers of necrophoric behavior in ants. *Psyche* 65: 108–114.

Wilson, E. O. 1965. Chemical communication in the social insects. *Science* 149: 1064–1072.

Wilson, E. O., F. M. Carpenter, and W. L. Brown. 1967. The first Mesozoic ants. *Science* 157: 1038–1040.

Wilson, E. O., and R. W. Taylor. 1964. A fossil ant colony: new evidence of social antiquity. *Psyche* 71: 93–103.

YOSIAKI ITÔ

National Institute of Agricultural Sciences, Tokyo

Groups and Family Bonds in Animals in Relation to Their Habitat

"Même à l'origine, la famille et la peuplade sont antagoniques; elles se développent en raison inverse l'une de l'autre."

Alfred Espinas (1877)

IS DOMINANCE HIERARCHY A NATURAL ORDER?

The concept of dominance hierarchy is one of the most important in animal sociology. The late W. C. Allee (1952), in his last and excellent review of social behavior among vertebrates, mentioned four kinds of relationships among organisms belonging to a single species as basic processes underlying the organization of animal groups. These are (1) sex pairing, (2) territoriality, (3) dominance hierarchy, and (4) leadership. Allee himself concentrated on the development of dominance hierarchy, the term being used here to mean "any social rank-order established through direct combat, threat, passive submission, or some combination of these behavior patterns" (Allee, 1952).

The existence of hierarchies among animals has long been known by cattlemen and fowl breeders, but modern studies on this problem are dated from Schjelderup-Ebbe's (1913) findings that the members of small flocks (up to about ten individuals) of common domestic fowl recognize one another as individuals (from his summary in Schjelderup-Ebbe, 1935). He made this fundamental observation when he distinguished individual hens by marking them with color spots. The order he found in bird flocks is now known as the "peck-right" type hierarchy. Remarkable advances along this line had been obtained by Allee and his collaborators since Masure and Allee (1934) first confirmed Schjelderup-Ebbe's findings.

Dominance relationships have hitherto been discovered in *experimental* groups of animals covering a wide range of species from arthropods to apes, but the implications of reported dominance are debatable. Howard (1955) put tenebrio beetles in a petri dish and found a peck-dominance type relationship in which larger beetles were consistently dominant over smaller ones. There is no evidence that such a dominance relationship plays a role in the natural life of this beetle, and the fact found here is not significantly different from a mere mechanical relationship as seen among metal balls. A dominance relationship is also found in naked hermit crabs competing for an empty shell (Allee and Douglis, 1945); lobsters, *Homarus americanus,* in aquaria (Douglis, 1946); and crayfish put into a round bowl (Bovbjerg, 1953). Lobsters and crayfish are often observed fighting in nature, but this fighting has nothing to do with the dominance hierarchy based on the recognition of individual positions but seems to be an expression of territorial behavior. If Bovbjerg had continued several days longer to rear those crayfish without food, he would have found cannibalism with the dominant animals eating subordinates.

Pardi (1948) described in detail a linear peck-right relationship among wasp females (*Polistes gallica*) in a nest during pleometrosis. He considered that this relationship may be an important factor integrating wasp colonies, especially useful in determining the division of labor. When two nests of another wasp, *Polistes fadwigae,* were artificially connected, Yoshikawa (1956) observed a similar hierarchical relationship. Deleurance (1957) considered, however, that such a hierarchical relationship among wasps would simply result from physiological heterogeneity among females. In this respect, Schneirla (1952a) said: "With reference to the ants, the fact is that attacks occur under many conditions. . . . There is no apparent evidence for any ant species, as far as I know, that such attacks are an expression of a regular colony organization of a dominance-subordination type. I believe that they are features of disorganization, rather than organization."

Etkin (1964), in a recent interview, said "An important aspect of dominance behavior in animal groups is the way it operates to make them closed societies." Experiments with caged animals show that animals can be arranged in order from the strongest to the weakest, but these findings cannot be considered evidence that, in their natural surroundings, the animals in question are organized in a social group as a result of the dominance hierarchy.

Allee (1952) suggested that the dominance hierarchy among vertebrates had been developed in animals phylogenetically higher than teleost fishes. He wrote that there was no demonstration of a social hierarchy among cyclostomes or elasmobranch fishes. From my viewpoint, however, there is no evidence in teleosts of the clear dominance-subordinate relationship which act to integrate social groups of poikilothermal animals, most birds, and lower mammals. Most of the evidence obtained to date from teleost fishes concerning the existence of a dominance hierarchy is based on experiments using confined animals.

For example, when Kawabata (1954) put some Medaca (*Aplocheilus latipes*) into a small basin, the dominance relationship of a peck-right type was observed. But, in their natural habitat, neither dominance hierarchy nor typical territorial behavior is seen among *Aplocheilus,* which usually live in unorganized groups. Occasionally, a large individual leaves the group and defends a small space around itself, but this is only a temporary and unstable phenomenon. Dominance hierarchies have also been found in many stream fishes such as the sword-tail (Noble, 1939a); sunfish, *Lepomis cyanellus* (Greenberg, 1947); Ayu, *Plecoglossus altivelis* (Kodama, 1960); and trout, *Salmo gairdneri* (Yamagishi, 1962), when these fishes were confined in an aquarium. But, in these cases, the dominance behavior resulted in disintegration rather than integration of the group, because all of these species show territoriality in nature. It was found in a group of Ayu entering a rapid that the dominant animals preferentially established their territories and, when all the good sites had been occupied, the remaining subordinate animals formed nonterritorial schools. No dominance behavior was observed in the school but if a territorial Ayu was caught by a fisherman, the most active schooling individual left the school and acquired the empty territory (Kawanabe, 1957). Thus, the dominance relations (though not expressed) certainly exist in the school of Ayu and act to disintegrate the school.

According to Uhrich (1938) and the authors, mice and rats cannot establish a linear dominance order even in cages but tend to show simple despotism, in which the alpha animal dominates over all other members of the group, the latter being all at one level in the hierarchy. Similar results were obtained for other small rodents, excepting a few gregarious species such as kangaroo rat. Most small rodents do not form social groups and are sparcely distributed over an area. Territoriality has not been well developed in these forms (Burt, 1940; Carpenter, 1958). Even among ungulates that develop typical dominance hierarchies in their natural herds, there is a notable case: "Among Indian antelope studied in the zoo, there is one dominant male who asserts his dominance over other males conspicuously while on the open range, but when the animals crowd around the feeding box, all evidence of this dominance disappears" (cited from Etkin, 1964). After citing this fact, Etkin wrote that dominance behavior is "situation specific."

In many passerine birds, among which typical linear dominance order (peck-right dominance, as a rule) has been reported,[1] aggressive behavior among males operates to separate migrating or wintering flocks into male-female pairs.

On the other hand, in some passerine and water birds (swans, geese, and cranes), the migrating groups consist of monogamous pairs plus their offspring, although family bonds usually break up during migration. Dominance based on parent-offspring relations seems to be limited in these groups (Kuroda, personal communication).

Lorenz (1952) described a very meaningful case: in a colony of jackdaws, *Corvus monedula,* the old individuals clearly reinforced the integration of the group, which is based on a parent-offspring relationship. The unique position of these birds in the evolution of social behavior will be discussed later.

In deer, *Cervus elaphus* (Darling, 1952), *C. nippon* (Kawamura, 1952) and *Odocoileus hemionus* (Dasman and Taber, 1956), dominance behavior among males in the breeding season operates to disintegrate loose male troops, which are formed during the nonbreeding seasons, while females have well-organized herds in which a clear-cut dominance order based on blood relations exists. In these cases, the dominance hierarchy is realized through leadership. Similar situations are seen in other ungulates. Finally, among higher primates, there are examples of social groups well organized by dominance and leadership. Further discussions of primate social behavior will be presented later.

To sum up, many of the dominance hierarchies reported so far seem to be artifacts rather than natural systems of social organization. Although aggressiveness among individuals of the same species is of importance in the social relations of animals, we must note two things: (1) aggressiveness generally functions to isolate individuals, and to disperse the group into individual pairs, rendering possible the evolutionary invasion of animals into new environments (discussed in the next section); and (2) in the most evolved groups, such as crows and large mammals, leadership based on dominance has appeared.

CONTRADICTION BETWEEN GROUPS
AND FAMILY BONDS

It is reasonable to believe that territoriality appeared earlier in the course of animal evolution than the dominance hierarchy. Territory, in the sense of "any defended area" (Noble, 1939) is reported in some

---

[1] For example, Kikkawa (1961) found in a winter flock of the white-eyed, *Zosteropis latelaris* that the linear peck-right type of dominance was found around the *supplied* food.

insects including dragonflies and ants, large crustacean species, some of teleost fishes, lizards, many birds and many mammals (Hinde, 1956; Aronson, 1957; Carpenter, 1958; Itô, 1959). Although spiders and some carnivorous insects (e.g., ant-lion) indiscriminately drive intruders out of their surroundings, this behavior seems to be a simple reaction to prey or enemies. Some supraorganismic insects like honey-bees defend their nests, but they do not defend a large area around the nest. There is no sure evidence to date showing the presence of territoriality among molluscs, elasmobranch and ganoid fishes, amphibia, and reptiles other than lizards.[2]

On the other hand, there are many cases in which a breeding pair in one species will defend a territory, while other closely related species live in groups in which neither consort-pair nor individual territoriality is observed. In the latter case, however, the group as a whole occasionally defends its common territory (e.g., monkeys).

How is territoriality developed during the course of social evolution? Why have some animals developed territoriality while others have not? In order to answer these questions we shall start with considerations of the origin of groups and family bonds among animals. Despite a great many descriptive papers on the social behavior of animals, there has been no attempt to explain the diversity of animal social relations.

In his papers on the evolution of reproductive rates of animals, Lack (1954a, 1954b, 1956) presented the idea that (1) the reproductive rate of animals is a product of natural selection tending to result in a maximum number of surviving offspring; hence (2) in animals that do not feed their young, the number of eggs laid is probably the maximum that the parent can produce, but there is an inverse correlation between egg-number and egg-size; and (3) in animals that feed their young, the clutch or litter size is limited by the number of young that the parents can feed.

I developed Lack's idea in my Japanese book *Comparative Ecology* (1959) and reached the following conclusions:

> The number of eggs or young produced is correlated with the availability of food for the young and inversely correlated with the degree of safety for the progeny. Animals living in environments where there is much food (eutrophic environment) that is easily available to the young, and where there are also many natural enemies, have evolved to produce many unprotected or weakly protected eggs or young. On the other hand, animals living in an oligotrophic environment or in an environment where

---

[2] The finding by Test (1954) of territoriality in the South American frog *Phyllobates trinitiatis* is an exceptional case. Grant's (1955) report on territoriality among *caged* North American salamanders, *Hemidactylum scatatum* and *Eurycea bislineata*, raised the possibility that this behavior was an artifact.

food materials are sparsely distributed (relatively oligotrophic environment) and are obtained with difficulty by the young have evolved to
produce a few well-protected eggs or young.

The following list contrasts oligotrophic environments with eutrophic
environments:

Fresh water	vs.	Sea
Land	vs.	Water
River	vs.	Lake
Upper stream	vs.	Estuary
Rocky & alpine area	vs.	Lowland
Forest	vs.	Grassland

Decrease in fecundity as described above is attended, at first, with
increased egg size and shortened larval period, and secondly, with the
development of parental care.

In the sea, a moderately eutrophic environment, parental care has been
evolved only exceptionally, because natural selection favors an increase
in fecundity. Family life of animals really evolved after the invasion of
the land.

Although there are some known cases of territoriality connected with
copulation and oviposition but not with the provisioning of young (see
Jacobs, 1957 and Itô, 1960 on dragonflies) or of territoriality connected
with the feeding of the adult itself (Miyadi, Kawabata, and Ueda, 1952 on
the Ayu), the main function of territoriality in most animals is undoubtedly the provisioning of offspring, that is, the defense of the family bond
and the nest.

Studying comparative reproductive behavior of darters (Percidae: North
American freshwater fish), Winn (1958) presented interesting suggestions
to explain the evolution of territorial behavior. In a primitive type, *Percina
caprodes,* an inhabitant of eutrophic lake shores and raceways, the female
scatters eggs over gravel. There is no male-female consortship, but the
male sits close by the female during the spawning, showing only weak or
no territorial behavior. The number of eggs in the ovaries of *P. caprodes*
females is the largest among the Percidae. *Hadropterus maculatus,* an
inhabitant of gravel raceways and stream pools, on the other hand, deposits eggs in sand or gravel. The male of this species shows weak territorial behavior around the moving female. Such a moving territory is also
seen in *Etheostoma caeruleum* and *E. blennioides* and *Ulocentra* spp., inhabiting rubble riffles and raceways, where the female deposits her eggs
on algae (*E. blennioides*) or into breaks in limestone shelves (*Ulocentra*
spp.). Few or single eggs are deposited in each spawning. The male of

these species defends a fixed territory in which the consorted female oviposits. In polygynous *Ulocentra* spp., a single male possesses several females in his territory. Lastly, in the most specialized species, *Etheostoma maculatum* and *E. flabellare,* which are inhabitants of rubble riffles, raceways, and the heads of riffles, a male and a female sit on the underside of a rock and the spawning of each egg is followed by ejaculation. The male strongly defends his "nest" and rubs the eggs with his pelvic and anal fins.

Parallel behavioral differences are seen in the Gobiidae and Cottidae in Japan. In Gobiidae, *Acanthogobius flavimanus* and other benthic species of the seashore and estuary produce more eggs of smaller size as compared with stream species. *Rhinogobius similis,* a downstream fish, the young of which spend their early larval life in the sea, produces more eggs of smaller size as compared with a closely related fluvial species, *Tukugobius flumineus,* that produces fewer eggs of larger size (Mizuno, 1961). Male parental care was reported in *T. flumineus,* whose newly hatched fry enter benthic life directly, while no parental care is observed in *R. similis* whose fry are, at first, pelagic. *Cottus pollux* is a typical benthic species in Japanese currents in which reproductive behavior is essentially similar to *E. flabellare.* The male guards his eggs until they hatch. Mizuno and Niwa (1961), however, described a variant of *Cottus pollux* that inhabits the midstream. Their young are carried down to the estuary. The female of this type produces more eggs of smaller size, which hatch at a less advanced developmental stage, than do other members of this species.

Many similar cases are known in fish where species living in oligotrophic water produce a smaller number of eggs of larger size or show more evolved parental care as compared with closely related species living in eutrophic water. Aronson (1957), in his review of reproductive and parental behavior of fishes, concluded that territoriality in fishes is developed in connection with sexual dimorphism and with decreased reproductive rate. It can be safely said that in fishes within a small systematic group (e.g., genera of family), integration of family relationships (parent-offspring and male-female ties) was developed as the animal became specialized for a relatively oligotrophic environment.

Birds offer another example of parental care and family bond that have developed in an oligotrophic environment or in an environment where food is obtained with difficulty by the young. The male-female relationship in gallinaceous birds, which take their food on the ground surface of open habitats, is polygynous (e.g., pheasants; see Collias and Taber, 1951) or promiscuous (e.g., sage grouse; see Scott, 1942, 1944). Parents do not feed the young: on hatching, young birds take their food without help. Territoriality among them is less developed as compared with that of perching birds (Collias and Taber, 1951; Blank and Ash, 1956) or completely absent (Howard and Emlen, 1942; Scott, 1944).[3]

On the other hand, most perching birds (tree-nesters) and marine birds are monogamous and feed their young until fledging. Most of these species (except for some that are considered to have regressed behaviorally) show typical territoriality in relation to parental care. Nice (1941) classified territoriality into six types, but type A (whole area for copulation, feeding and brooding is defended) and type D (the area for copulation and brooding is defended but feeding mainly takes place outside the territory) are especially important. Type A is highly developed in birds in temperate and subarctic forests (e.g., winter wren, willow warbler, nutcracker—passerine birds; horned owl, golden eagle—carnivorous birds), but type D is commonly seen in passerine birds in open habitat (e.g., reed warblers, house and tree sparrows, and swallows, *Hirundo* spp.), herons in open land, some carnivores (hawks, etc.), and in water birds. One cause of such divergence may be the variation in availability of food.

There is some evidence of polygyny (Kendeigh, 1941; Armstrong, 1956; Kikuchi, Sakagami, and Konishi, 1957) and polyandry (Hoffmann, 1949; Kobayashi, 1955) among birds that feed their young. Species showing such atypical behavior are often inhabitants of open land.

In 1877, Espinas wrote "Même à l'origine, la famille et la peuplade sont antagoniques; elles se développent en raison inverse l'une de l'autre." The above-mentioned facts on social relations among fishes and birds support his generalization. Parental care almost always brings about the establishment and isolation of male-female pairs, but when the group is formed, the pairs break up.

The contradiction between the family and the group is also seen among individuals of the same species. In many passerine birds, such as swallows and thrushes, the male-female pair breaks up before starting the migration in autumn. Neither a hierarchy nor a leadership is found in migrating passerine flocks. Offspring from different pairs are completely intermingled in the migrating group. On returning in spring, on the other hand, active individuals establish their own territories and form pairs, and the combinations of the previous year are renewed on many occasions. The grey starling, *Sturnus cineracus,* aggregates in the winter in a large group and feeds in a common feeding site, but as spring returns, the group breaks up into individual pairs (Kuroda, 1955).

The social structure is, of course, determined not only by the habitat but also by phylogeny and other factors. For example, the mandarine ducks,

---

[3] Among grouse, species living in open habitats are promiscuous "lek birds" (sage grouse, capercaillie, blue grouse, prairie chicken, etc.) but species living in forests and alpine are territorial (and monogamous and polygamous) during the breeding season (hazel grouse, ptarmigan, red grouse, spruce grouse, etc.) (Dr. M. Uramoto's pers. com.).

Aix galericulata, build their nests in trees but the reproductive rate remains as high as in surface-nesting water birds. The skylarks, *Alauda arvensis,* retain monogamy and territoriality despite their adaptation to plains habitats.

In tropical birds, on the other hand, monogamy is often concomitant with unorganized feeding groups (e.g., parakeets, *Melopsittacus undulatus* and *Psittacus erithacus*). Territoriality is not well developed in tropical birds, even in monogamous species. In addition, many species are "lek birds," i.e., they form promiscuous breeding flocks (e.g., *Manacus v. vitellinus;* see Chapman, 1935).[4] These tendencies in tropical birds are also observed among monkeys, as mentioned later. The tropical rain forest is characterized by the extreme diversity of trees, and as a result, the food for birds (especially seeds and fruits) is abundant. This may be the main reason for the lack of territoriality in many tropical, arboreal animals.

As noted in the previous section, some water birds, such as geese, swans, and cranes, and the social Corvidae (especially the jackdaw) hold a unique position in the range of relationships between the family and the group. In these birds, leadership based on blood relation serves to integrate groups during migration.

In short, in the process of adaptation to arboreal life and life on rocky shelves overlooking the sea, parental care and reinforced family bonds evolved; and monogamy, which offers the most efficient form of parental care, arose at the last stage of this adaptation.

Espinas' generalization on the contradiction between groups and families is seen also in mammals. In ungulates, species living on the prairie are known to be more gregarious than the systematically related species living in forests and in rocky alpine areas, the latter being solitary or living in small family bands. Some examples are shown in Table 1.

It should be noted that antelopes (including many species omitted from Table 1) that inhabit grasslands form large bisexual bands, while those adapted to life in forests or marshy areas are solitary or form small polygynous bands. Similar differentiation in social habits among closely related species is shown in wild Bovidae and wild pigs.

There are contradictory reports on social relationships among deer. In general, these relationships are characterized by the existence of two kinds of groupings, namely male troops and matriarchal female herds, but there are many exceptions to this general picture. For example, caribous in arctic tundra form large bisexual bands throughout the year. In reviewing the literature on social relationships among deer, Dasman and Taber

---

[4] Other examples of lek birds are many species of Birds-of-Paradise, some species of hummingbirds, and some Cocks-of-the-Rock.

TABLE 1
Social groupings of ungulates in relation to their habitats.

Grassland and tundra	Intermediate zone	Forest	Alpine or rocky area
Burchel's zebra	Mountain zebra		
Rhinoceros spp.	Somali wild ass		
Wart hog	Peccary (*Tayassa*)	Giant forest hog	
	Bush-pig		
	(*Potamocoerus*)		
Hippopotamus		Pigmy hippopotamus	
			Vicuña
	Giraffe	Okapi	
Caribou	Roe deer	Red deer	
	Elk (*Cervus*	Muntjac	
	canadensis)	(*Muntiacus*)	
	Blacktailed deer	Moose (*Alces*)	
American bison	Banteng	Forest buffalo	
		(*Syncerus nanus*)	
Cape buffalo			
(*Syncerus cafer*)			
Eland		Bushbuck	
Roan antelope	Waterbuck	Pigmy antelope	Klipspringer
Gnu	Sitatuunga		
Impala			
Uganda kob			Chamois
	Ibex		Bighorn sheep
			Argali

Species in bold-face type are gregarious species living in large bisexual bands. Species in standard type are species living in small (less than 20) bisexual bands or in polygamous bands (in some species such as deer, only females and young form small matriarchal bands). Species in small standard type are those living solitarily or in small (2–3) loose groups.

(1956) concluded that species in the open land (grassland and tundra) are characterized by large herds while species in a luxuriant environment live in small groups or solitarily and show territorial defense.

Among the blacktailed deer, *Odocoileus hemionus columbianus,* Lindsdale and Tomich[5] and Dasman and Taber (1956), who made observations in forests and chaparrals, found small mother-offspring groups that defended territorial borders, while McLean[5] and Palmer[5] observed deer in savannas and found that the typical social structure was a large female herd with a matriarchal hierarchy. Such a female herd was described in detail by Darling (1937) for the red deer. According to him, there are two types of groupings in wild red deer, unorganized male troops and matriarchical female herds in which a clear dominance hierarchy based on blood relationship exists. According to Cameron,[5] however, the red deer in Scotland are thought to have lived formerly in small groups consisting of a mother and her young offspring, but as the virgin forests were destroyed, the deer began to form larger (about ten individuals) female herds. The sika deer, *Cervus n. nippon,* are considered to be originally solitary animals in forest areas, but when they had been artificially fed

---

[5] Cited from Dasman and Taber (1956).

in open land, Nara Park, they formed female herds. These herds consist of some dozen individuals with a dominance hierarchy and they do not defend territory.

That species of mammals living on the prairie are more gregarious than species in forest and alpine areas is, of course, a rough generalization, which is modified by many other factors, and is realized only *within small phylogenetic groups*. Among these factors, phylogeny seems to be most influential, that is, the phylogenetically primitive species tends to live solitarily or in a small group as compared with more evolved species (e.g., pigmy hippopotamus, many species of rhinoceros and a group of small primitive antelopes—duikers).[6]

Like ungulates, carnivores, such as wolves and wild dogs (e.g., *Lycaon pictus* in Africa and *Canis dingo* in Australia), living in grasslands and semideserts, form bisexual groups while forest inhabitants, such as tigers and lynx, are solitary hunters. The lions have their place between these extremes because they hunt in polygynous family groups. Another terrestrial cat, the cheetah, lives in small groups. There are, however, some small cats living solitarily on the steppe (e.g., *Felis ocreata, F. manul, Lynx caracal*).

Basing his description on Murie's (1944) observations of wolves on Mt. McKinley, Etkin (1964) wrote about the interesting social habits of this pack-hunting animal:

> They not only attack in a pack but carry through such pack stratagems as tiring out their prey by spelling each other off in keeping up the pursuit. The pack appears to be derived from the family unit, consisting of father and mother and offspring of both sexes and of several successive years. Apparently on occasion several such family packs may join forces, at least temporarily, to form large packs. Whether the subgroups in a large pack are blood relatives . . . is not known. . . . Male and female remain together apparently for life. . . . After the pups are born in the spring, food is brought back to the den by all the adult wolves. . . .

Despite the high level of group coordination, the monogamous family of wolves does not break up during group activities, in contrast to that of birds and ungulates.

With the evolutionary advance of animals to the relatively oligotrophic environment, social evolution became possible. In animals with a lower mental capacity, this usually resulted in disorganization of groups and the establishment of family bonds. Only in animals with higher mental capacities, such as crows, the wolf and probably some apes, did the family bond

---

[6] Rowell (1966) recently studied baboons living in forest. The life-type was semiarboreal. The composition of two groups (mentioned in Table 2) shows a similarity with that of grassland-living baboons.

remain in group life and play its role in the establishment of leadership; in other words, there is no contradiction here between the group and the family.

SOCIAL RELATIONS AMONG PRIMATES IN RELATION TO ARBOREAL AND TERRESTRIAL HABITS

Since Japanese ecologists (e.g., Imanishi, 1957, 1960) broke the long pause in the study of primate socioecology that followed Carpenter's extensive works, increasing results of field observations on the social behavior of primates are being accumulated. We now have information on the basic social structure of more than ten species of primates from the lemur (Petter, 1962) to the chimpanzee (Goodall, 1965; Itani, 1967; Kortlandt, 1962; Reynolds and Reynolds, 1965; Sugiyama, 1968a).

Carpenter (1942c) recognized that "the characteristic group scatter or spatial distribution varies from species to species under different environmental conditions." He considered that, since the basic behavioral root of primate social groupings is individual dominance, the "relative individual dominance" or "dominance gradients" differ from species to species. According to him, baboons and langurs show a very sharp dominance gradient while the howler monkey shows a dominance gradient of very low slope. Between these extremes other species of primates have their places. Carpenter did not consider, however, the relationship which may exist between differences in dominance gradients or types of social groupings (size, hierarchy, type of consortship, etc.) and the ecological and phylogenetic conditions. There have been only a few attempts to study primate social organization comparatively "in terms of similarities and differences" (Schneirla, personal communication).

In one of these studies, DeVore (1962) stated "When the size of the social group is compared to the degree of arboreal or terrestrial adaptation of the species, a trend toward larger groups in the terrestrial species is apparent," and also "there is a clear trend toward increasing sexual dimorphism from the arboreal gibbon . . . to the gorilla. . . ."

Aside from polygynous bands of ungulates living in alpine area (e.g., vicuña) and bisexual groups of wolves, primates may be the only mammals that integrate males and females into *permanent* social groups based on the dominance-subordination relationship. In Lemuroidea, however, the social grouping is not permanent; some species form nocturnal groups that disperse again with the sunrise, and there are frequent changes of members among groups. Other species are solitary and form only temporary combinations (Petter, 1962). Petter found somewhat permanent groups which consisted of several males, several females and their young only in some diurnal or semidiurnal species (active at dawn and twilight). In *Lemur macaco,* the number of females in such a group was, as a rule, larger than that of males and the groups may be led by a female (Petter, 1962). This leadership by a female is different from groups of the other primates.

Group size of primates other than Lemuroidea varies from several individuals, as in the redtail monkey, to several hundreds, as in baboons. On the other hand, there is a great divergence in social relationship from the gibbon's monogamous "family" to the macaque's hierarchical organization. Table 2 is a summary of field studies known to me.

It is clear from this table that the social relations of almost all the terrestrial and semiterrestrial species of monkeys and apes are characterized by their permanent large bands, in which several hierarchical divisions (classes) can be distinguished according to ages and sex, but, the arboreal species tend to form small groups consisting of a single male, many females and young, or even monogamous pairs.

Since this hierarchy was investigated for the first time in the Japanese macaque, *Macaca fuscata* (Itani, 1954; Imanishi, 1957, 1960), a primate group exhibiting such a hierarchy is called a "*Macaca*-type" group. Basic social classes in the *Macaca*-type group are:

> Leaders (adult males)
> Subleaders (adult males) ⎫
> Young (subordinated males) ⎬ or Peripheral males
> Adult females
> Juveniles (males and females)
> Infants (males and females)
> Solitary (males)

In wild Japanese macaque bands, leaders and females (with or without babies) sit and feed at the center of the group, while subleaders and young males are located at the periphery of the group (Fig. 1). When a band moves, some young go first, the subleaders follow, leaders with females and juveniles come next, followed by other subleaders and finally by the remaining young. When a natural enemy approaches, young males give warning calls, but, if the danger continues, subleaders and leaders take turns in the young male's position. Neither a bisexual subgroup nor a polygynous harem is observed in groups of Japanese macaques, or in the rhesus monkey, *M. mulatta* (Carpenter, 1942a, 1942b), the crab-eating monkey, *M. irus* (Furuya, 1963a), the bonnet monkey, *M. radiata* (Nolte, 1955), langurs, *Presbytis cristatus* (Furuya, 1963b), or in baboons (Washburn and DeVore, 1961; DeVore and Hall, 1965). According to Sugiyama (1968b), *M. silenus* and *M. nemestrina* also form *Macaca*-type groups. Sugiyama (1965) reported, on the other hand, that the basic

TABLE 2
Social organization of fifteen species of primates.

Species	Habitats	Group size (mean)	Typical (average) composition of group	Subgroup or class-division	Literature[b]
Lemur macaco	A	4–13	4.8♂♂ + 3.6♀♀ + (1–2)I,J	obscure	(95)
Alouatta palliata	A	4–29 (17)	3♂♂ + 8♀♀ + 3I + 4J + S(♂)	obscure	(15)
Callicebus moloch	A	2–4 (3.3)	1♂ + 1♀ + 0.5I + 0.8J	monogamy	(82)
Ateles geoffroyi	A	30–100 (50)	(1♂ + 4♀ + 2I + 4J) + ♂troop	sub-group (3–17)	(16)
Cercopithecus ascanius	A	2–11 (5)	1♂ + (5–14)♀♀ + 3I + 4J + ?	see p. 404	(13)
Presbytis johni	A-T	10–25 (14)	3♂♂ + (5–14)♀♀ + 6I + 5J + ♂troop	polygyny	(114)
Presbytis entellus	A-T	12–61 (29.1)	(2–6)♂♂ + (4–19)♀♀ + I,J + ♂troop	Macaca-type (?)	(111)
	A-T	5–120	1♂ + 9♀♀ + I + J + ♂troop	near Macaca-type	(60)
Presbytis cristatus	A	9–24 (15.1)	2♂♂ + 10♀♀ + 3I + 7J + S(♂)	polygyny	(111)
	A	22	3♂♂ + 13♀♀ + 6I + 26J + ?	polygyny	(7)
	A	48		Macaca-type	(36)
Macaca mulatta	A-T	13–147 (70)	n♂♂ + 2n♀♀ + nI + nJ + S(♂)	Macaca-type	(36)
	A-T	13–74 (42)	13♂♂ + 28♀♀ + 13I + 30J + S(♂)	Macaca-type	(19, 20)
Macaca fuscata	A-T	3–600 (30–50)	n♂♂ + 2n♀♀ + nI + nJ + S(♂)	Macaca-type	(108, 109)
Macaca irus	A-T	30–73	n♂♂ + (2–6)n♀♀ + I,J + ?	Macaca-type	(50)
Macaca radiata	A-T		4♂♂ + 6.5♀♀ + 7.5I + 14.5J + ?	Macaca-type	(35)
	A-T		(3–5)♂♂ + (7–13)♀♀ + I,J + ?	Macaca-type	(92)
Baboons	T	10–100 (30–50)	16♂♂ + 13♀♀ + 9I + 7J + ?	near Macaca-type	(92)
	A-T	32	5♂♂ + 12♀♀ + 2I + 14J + S(♂,♀)	Macaca-type	(108)
	A-T	58	5♂♂ + 5♀♀ + 6I + 16J	see footnote 6	(117)
Papio hamadryas	A-T		14♂♂ + 16♀♀ + 13I + 15J	see footnote 6	(98)
	R	2–13 (6)	1♂ + (1–4)♀♀ + I + J + ♂troop	polygyny	(98)
	R	8–12	1♂ + (4–6)♀♀ + I + J + ♂troop	polygyny	(72)
Hylobates lar	A	2–6 (4)	1♂ + 1♀ + 1I + 3J + S(♂,♀)	monogamy	(106)
Pongo pygmaeus	A	< 5	solitary	see p. 404	(18)
Gorilla g. beringei	A-T	5–27 (16.9)	1.7♂♂ + 1.5♂'♂' + 6.2♀♀ + 4.6I + 2.9J + S(♂)	polygyny (?)	(99)
	A-T	5–18 (12)	1♂ + 0.3♂' + 3♀♀ + 2I + 2J + S(♂)	polygyny	(100)
Pan troglodytes	A-T	2–62 (20–50)	No stable composition (5.3♂♂ + 0.3♂' + 13.8♀♀ + 6I + 4.3J)[a]	see p. 407	(62)
					(6, 37, 54)
					(54)

Abbreviations: A: arboreal, A-T: semi-terrestrial, T: terrestrial, R: rocky area
  ♂: adult male, ♀: adult female, ♂': blackback or immature male in Gorilla and Chimpanzee.
  I: infants, J: juveniles, S: solitary.
a Examples of composition of "temporal group."
b See reference at end of chapter.

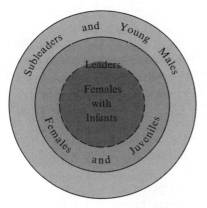

FIGURE 1
Schematic representation of the spatial arrangement of the classes in Japanese monkeys. (After Itani, 1954.)

group composition of a semiterrestrial langur, *Presbytis entellus,* in a monsoon forest of west India was a polygynous one-male group (average group size was 15.1), and only two out of 39 bisexual groups observed contained more than two adult males). But he also found in a drier area of central India, that this langur formed larger (average group size was 29.1) bisexual groups, of which 40 percent included more than two adult males. With the same species, Jay (1963, 1965) found, in central India, many groups (group size was 5–50) including more than two males among which a relatively rigid dominance hierarchy was shown. For an arboreal langur, *Presbytis johni,* Tanaka (1965) reported that the groups usually consisted of a single male, several females and young. Groups of another arboreal langur, *Presbytis cristatus,* also consisted of "a single male, multiple females and associated young," except one case (Bernstein, 1966). Furuya (1963b) found, however, two groups that consisted of multiple males, females (more females than males), and young in the latter species. Finally, Hall (1964) said of the terrestrial patas monkey, *Erythrocebus patas,* "In each group so far observed, there has been only one large male with several adult females and young." It must be mentioned that Colobinae, in which *Presbytes* and *Erythrocebus* are included, is the typical forest-living foliage-eater whose morphological characteristics show strong specialization for arboreal life. It seems that *P. entellus* and *E. patas* have acquired their semiterrestrial or terrestrial habits in rather recent geological time. Further studies of these species are needed and attention must be paid to the relation between social structure and phylogenetic history of the species in question.

As also seen in birds and ungulates, the family relation breaks up in large groups of these monkeys. Consort pairs are formed within groups, but are of brief duration, lasting from two hours to one day (in *Presbytes entellus;* see Jay, 1963) or from several days to one month (*Macaca fuscata,* see, e.g., Itani, 1954). Nevertheless there is a marked difference here from the social relationship of ungulates in that the primate bisexual groups are well integrated by the dominance hierarchy and leadership.

Blood relations also deeply affect the organization of the Japanese macaque's social groups (Tokuda, 1957; Itani and Tokuda, 1958).

The gibbon and the orangutan represent another extreme in social relations among primates. The gibbon, *Hylobates lar,* lives in a permanent monogamous pair; territory is defended by females as well as males. Mature offspring of both sexes leave their family and, after a solitary life for a certain time, a male and a female unite in a new pair. According to McClure (1965), *Hylobates syndactylus* may form monogamous family-like groups, too.

As for the orangutan, *Pongo pygmaeus,* Carpenter (1938) wrote, "The group consisted of two adult females with young babies and a large adult male, the adult male was widely separated from the females." Recent observation by Schaller (1961) showed, however, that the orangutan is an extremely solitary animal. According to Schaller, the most frequent combinations of the orangutan were (1) lone males, (2) females with one or more subadults of various ages, (3) pairs, and (4) small groups of subadults. Bisexual groups were rather rare (only two cases were hitherto found). In a group of four animals consisting of a male, female, subadult, and a small infant, he found one night that all four animals nested in adjoining trees; another night the male slept about 400 feet away from the female and juveniles and the third night there was no sign of the male in the vicinity. Thus, observations by Carpenter and Schaller suggest that there may be no constant male-female grouping. In other words, the orangutan is considered to be a solitary animal.

Discovery of monogamy in gibbons led some anthropologists to assume this to be evidence of the monogamous origin of human male-female relations. However, as the Hylobatinae are the only present-day true brachiators (Napier, 1963), monogamy in gibbons is considered to be a result of extreme adaptation to the life in the forest canopy such as in the case of passerine birds mentioned above. It must be noted that Mason (1966) found monogamous pairs in all of nine observed groups of the South American monkey, *Callicebus moloch,* which is well adapted to the arboreal life. The male-female bond is said to be strong and enduring.

Between these extremes, namely, life in the large band and in the isolated familylike small group, other primates have their intermediate positions.

The howler (*Alouatta*) and the redtail monkey (*Cercopithecus ascanius*) form small bands, in which one or a few males live together with many females. In the latter species, it was found that night-resting bands consisting of two to four individuals united in daytime into a large feeding group (Buxton, 1951, 1952; Haddow, 1952). Such a feeding group occasionally consisted of more than 100 individuals and sometimes of several species (Itani, 1958).

According to Lumsden (1951), the social organization of other monkeys adapted to the upper layer of the tropical rain forest, e.g., *Cercopithecus mitis stuhlmani* and *Cercocebus albigena johnstoni,* is similar to that of the redtail monkey. On the other hand, the semiterrestrial African species, *Cercopithecus ascanius centralis* and *Cercopithecus l'hoesti,* form constant large bands in which no polygynous sub-units are seen (Buxton, 1952).

According to Carpenter (1935), about 200 spider monkeys, *Ateles geoffroyi,* living in the Coto area of the Panama Canal Zone, were divided into four groups. But the individuals in these groups were rather sparsely distributed and subdivided into some permanent polygynous bands (Carpenter's "subgroup"), which jointly possess the group's nomadic range. As the monkey species inhabiting the upper layer of the tropical rain forest often lack territoriality (probably because of the concentration of a huge amount of food), the subgroup of the spider monkey and the night-resting band of the redtail monkey should probably be regarded as their basic social unit, corresponding to the gibbon's monogamous "family." Thus, our generalization that arboreal life develops family relations while terrestrial life develops large groups can generally be applied to the primates, too.

In *Comparative Ecology* (Itô, 1959), where my first attempt to present the comparative study of social relations among primates was made, I expressed some doubt as to Zuckerman's (1932) descriptions of the "harems" in baboon troops, and suggested that the baboon (*Papio cynocephalus*) should have a *Macaca*-type social structure. The results of extensive field studies conducted by Washburn and DeVore (Washburn and DeVore, 1961; DeVore and Washburn, 1963; DeVore and Hall, 1965) supported my expectation.[6]

They wrote:

> The baboon troop is subdivided by age, sex, and individual preference. For example, as a large troop moves along, the lesser adult males are first, then come the juveniles and females, then the most dominant males and females with small babies, followed by more females and juveniles, and finally by the other adult males. . . . Some individuals [male, female] prefer each other. . . . Groups of this kind may have given rise to the stories of "harems," but these preference groups are temporary associations and limit sexual activities in no way. (Washburn and DeVore, 1961.)

More recently, Kummer and Kurt (1963) found, in the hamadryas baboon, *Papio hamadryas,* living on rocky mountains in Ethiopia, "the

---

[6] Rowell (1966) recently studied baboons living in forest. The life-type was semiarboreal. The composition of two groups (mentioned in Table 2) shows a similarity with that of grassland-living baboons.

small one-male-groups to be the solid unit, the larger associations becoming less constant with increasing number of individuals." In "troops" that are formed on the shelf of precipitous rock of the Ethiopian highland for sleeping at night, they could not find a dominance order nor any leaders (see also Starck and Frick, 1958 for the same species). A similar social structure has been also found in another species, *Papio gelada,* living on rocky mountains (Crook, 1966).

In previous sections, I suggested that the adaptation of animals to alpine or rocky habitats is equivalent to the adaptation to arboreal life. Both are invasions of a relatively oligotrophic environment where it is hardly possible for the young to grow without highly developed parental care. Contrasting social structures seen in baboons and in hamadryas seem to present proof of my assumption.

Social relations among mountain gorillas present another interesting case. All of the six natural groups carefully observed by Kawai and Mizuhara (1962) were led by a single silverbacked male (mature male). The number of mature females in the groups varied from two to four. About two-thirds of several dozen groups described by Schaller (1963) were also led by a single silverbacked male. It may be safe to say that the basic social tendency in the mountain gorilla is polygyny.

However, there were several large groups (consisting of more than ten individuals) in which several silverbacked males coexisted with numerous females. Is the polygynous subgroup retained even in such a large group? Basing his theory on the observation that gorillas in a group did not defend their territory and seemed to tolerate participation of solitary animals in their group, Imanishi (1961) considered that the large group with two or more silverbacked males might be a combination of polygynous bands. There are observations opposing this assumption, however. Schaller (1963) noted that in a group consisting of 4 silverbacked males, 1 blackbacked (immature) male, 10 adult females, 3 juveniles and 6 infants, there was a dominance hierarchy among silverbacked males, and the subordinate silverbacked males were found at the periphery. Despite this, when a subordinate silverback attempted to copulate with a female near the dominant male, the dominant was not interested in the proceedings. During these observations which continued over a year, 2 silverbacked males joined this group, 3 other silverbacked males left, and 2 mature females joined, each at different times. Schaller also noted that 2 silverbacked and 2 blackbacked males *not followed by females* joined another group, which consisted of 2 silverbacked males, 1 blackbacked male, 3 adult females, 3 juveniles and 2 infants.

We cannot, therefore, conclude whether the basic male-female relationship among gorillas is polygynous or promiscuous (with temporary consortship). If the mountain gorillas actually were polygynous, they would hold a position equivalent to that of the jackdaw and the wolf, and, owing to their higher phylogenetic position, should be of major importance in understanding the evolution of human society.

Social relations among the chimpanzee *Pan troglodytes* are more loose and colorful than those of the mountain gorilla. Although Nissen (1931–1932) having found groups of chimpanzees composed of 4 to 13 individuals, believed that the chimpanzee lives in polygynous bands, there is no reliable evidence for this assumption. Kortlandt (1962) wrote that chimpanzees are usually organized into two types of groups, namely, the bisexual group of more than 20 individuals and the mother-infant group (nursery group) of about 15 individuals. He found that the mother-infant group (in which several adult females were usually found) occasionally had 1 or 2 adult males, but there were many chances for interchange between groups.

On the other hand, Azuma and Toyoshima (1963) found the following four types of groupings: (1) mother-infant group; (2) group of preadult individuals, usually consisting of 3 to 5 individuals with or without a few adolescent females; (3) polygynous familioid; and (4) large bisexual group.

Azuma and Toyoshima considered the basic type of chimpanzee's bisexual group to be polygynous (one-male group: $1\,\delta + 1{-}2\,\female\,\female + I + J$), but they observed a group having 2 adult males and another group consisted of 10 to 15 individuals including 2 or more males. There is no proof for Azuma and Toyoshima's supposition that this group was derived from a combination of two or more one-male groups.

Reynolds and Reynolds (1965), who made observations of chimpanzees at Budongo Forest in Uganda, stated that their groups "were not closed social groups. Groups were constantly changing membership, splitting apart, meeting others and joining them, congregating or dispersing." They also said that "there was no evidence of any 'family' or 'harem' groups," and their observation suggests "some degree of sexual promiscuity."

For chimpanzees living in the savanna-deciduous forest area of Tanzania, Goodall (1965) reported that "there are no stable groups (other than mother/infant)." She found only temporary associations of individuals and also promiscuous sex relations.

Contrary to these authors, Itani (1966) assumed the existence of a polygynous "preband" (which was considered to be the prototype of human family) in chimpanzee groups. But he later (1967) came to consider that the chimpanzee's social organization is characterized by "changeability in grouping, and no lower social unit." Sugiyama (1968), who studied the chimpanzees of the deciduous forest at Budongo, also concluded that, although chimpanzees form temporary groups of many kinds according to the situation, there is no permanent membership. He could not find any permanent familylike social unit but did find interchange of members between "temporary groups."

In short, aside from unorganized lemur groups, chimpanzee social groupings appear to be the most loose and open among primates (Bourlière, 1964; Sugiyama, 1968a, 1968b). Reynolds (1966) and Sugiyama (1968a) stressed that such an open group organization as seen in the chimpanzee would play an important role in the evolution of the human society.

Without doubt, the *Australopithecus*-type men, who started their life as bipedal gatherer-hunters in African grassland, formed groups of more than ten individuals (Oakley, 1961). But whether these men had established monogamous or polygynous families in their groups is not yet known. The results of extensive studies of the chimpanzee's social organization suggests a relatively loose male-female relationship in the groups of early man. More studies are needed to answer this question about primates and primitive men, especially studies of the social relationship of African apes in different habitats, such as savanna, gallery forests and tropical rain forests.

We have discussed in the first section of this paper the limitations of the experimental study of dominance hierarchy among animals. Dominance relationships based on aggressiveness have long been recognized as a central mechanism in the integration of natural groups of primates. However, it has been known that direct aggression is rarely found in primate groups, even in the baboon. Also, under natural conditions, there is scanty evidence of an actual precedence of dominant individuals around natural food. All the observations of a precedence of dominants in taking food are cases in which food is supplied. On the other hand, division of labor and cooperation, both being based on a *Macaca*-type hierarchy, are other important factors that integrate groups.

Schneirla wrote in 1952:

> The "dominance-hierarchy" concept, moreover, has influenced the attention and thought of students at the expense of concepts relating to the more positive and central aspects of group structure, so that the potentialities of concepts such as "cooperation" seem largely unrealized. . . . It is now clear that unless dominance behavior is viewed in its relation to other social factors, investigation and theory in this field are certain to be seriously limited and distorted.

The present examination seems to support Schneirla's statements.

Any generalization about such a complicated problem as the reasons for the differences in basic social structures among animals can only be described as an "impression," before much more quantitative data (especially on several species within the same genus) have been accumulated

(Hall, 1965). It is, however, the author's opinion that efforts like the present discussion will serve as a reference in programming future studies.

REFERENCES

1. Allee, W. C. 1952. Dominance and hierarchy in societies of vertebrates. *Colloques Int. Centre Natl. Rech. Sci.* XXXIV, "Structure et Physiologie des Sociétés Animales," Paris. Pp. 157–181.
2. Allee, W. C., and M. B. Douglis. 1945. A dominance order in the hermit crab, *Pagurus longicarpus* Say. *Ecology* 26: 411–412.
3. Altmann, S. A. 1959. Field observations on a howling monkey society. *J. Mammal.* 40: 317–330.
4. Armstrong, E. A. 1956. Territory in the wren *Troglodytes troglodytes*. *Ibis* 98: 430–437.
5. Aronson, L. R. 1957. Reproductive and parental behavior. In M. E. Brown, ed., *The physiology of fishes,* vol. 2. New York: Academic Press. Pp. 271–304.
6. Azuma, S., and A. Toyoshima. 1963. Progress report of the survey of chimpanzees in their natural habitat, Kabogo Point Area, Tanganyika. *Primates* 3: 61–70.
7. Bernstein, I. S. 1966. *A field study of the male role in Presbytis cristatus* (lutong). Paper read at the Symp No. 2 of XIth Pacific Congr., Tokyo.
8. Blank, T. H., and J. S. Ash. 1956. The concept of territory in the partridge, *Perdix p. perdix. Ibis* 98: 379–89.
9. Bourlière, F. 1961. Patterns of social grouping among primates. In S. L. Washburn, ed., *Social life of early man.* New York: Viking Fund. Pp. 1–10.
10. Bourlière, F. 1964. *The natural history of mammals,* rev. 3rd ed. New York: A. A. Knopf.
11. Bovbjerg, R. V. 1953. Dominance order in the crayfish *Orconectes virilis* (Hagen). *Physiol. Zool.* 26: 173–178.
12. Burt, W. H. 1940. Territorial behavior and populations of some small mammals in Southern Michigan, *Misc. Publ. Mus. Zool., Univ. Michigan,* no. 45, pp. 1–56.
13. Buxton, A. P. 1951. Further observations of the night-resting habits of monkeys in a small area on the edge of the Smliki Forest, Uganda. *J. Anim. Ecol.* 20: 31–32.
14. Buxton, A. P. 1952. Observations on the diurnal behaviour of the redtail monkey (*Cercopithecus ascanius schmidti* Matchie) in a small forest in Uganda. *J. Anim. Ecol.* 21: 25–58.
15. Carpenter, C. R. 1934. A field study of the behavior and social relations of howling monkeys. *Comp. Psychol. Monogr.* 10 (2): 1–168.

16. Carpenter, C. R. 1935. Behavior of red spider monkeys in Panama. *J. Mammal.* 16: 171–80.

17. Carpenter, C. R. 1938. A survey of wild life conditions in Atjeh of North Sumatra, with special reference to the Orang-utan. *Netherland Committee for International Nature Protection, Communications,* no. 12, pp. 1–34.

18. Carpenter, C. R. 1940. A field study in Siam of the behavior and social relations of the gibbon (*Hylobates lar*) *Comp. Psychol. Monogr.* 16 (5): 1–212.

19. Carpenter, C. R. 1942a. Sexual behavior of free ranging rhesus monkeys (*Macaca mulatta*). I. Specimens, procedures and behavioral characteristics of estrus. *J. Comp. Psychol.* 33: 113–142.

20. Carpenter, C. R. 1942b. Sexual behavior of free ranging rhesus monkeys (*Macaca mulatta*). II. Periodicity of estrus, homosexual, autoerotic and nonconformist behavior. *J. Comp. Psychol.* 33: 143–162.

21. Carpenter, C. R. 1942c. Societies of monkeys and apes. Levels of integration in biological and social systems. *Biol. Symp.* 8: 177–204.

22. Carpenter, C. R. 1958. Territoriality: A review of concepts and problems. In A. Roe, and G. G. Simpson, eds., *Behavior and evolution.* New Haven: Yale Univ. Press. Pp. 224–250.

23. Chapman, F. M. 1935. The courtship of Gould's Manakin (*Manacus vitellinus vitellinus*) on Barro Colorado Island, Canal Zone. *Bull. Am. Mus. Nat. Hist.* 68: 471–525.

24. Collias, N. E., and R. D. Taber. 1951. A field study of some grouping and dominance relations in ring-necked pheasants. *Condor* 53: 265–275.

24a. Crook, J. H. 1966. Gelada baboon herd structure—A comparative report. *Symp. Zool. Soc. Lond.* 18: 237–258.

25. Darling, F. 1937. *A herd of red deer.* London: Oxford Univ. Press.

26. Darling, F. 1952. Social life in ungulates. *Colloques Int. Centre Natl. Rech. Sci.* XXXIV, "Structure et Physiologie des Sociétés Animales," Paris. Pp. 221–226.

27. Dasman, R. F., and R. D. Taber. 1956. Behavior of Columbian black-tailed deer with reference to population ecology. *J. Mammal.* 37: 143–164.

28. Deleurance, E. P. 1957. Contribution à l'étude biologique des *Polistes* (Hyménoptères Vespidés) I. L'activité de construction. *Ann. Sci. Nat. Zool. Biol. Anim.* 19: 91–222.

29. DeVore, I. 1962. A comparison of the ecology and behavior of monkeys and apes. In S. L. Washburn, ed., *Classification and human evolution.* New York: Viking Fund. Pp. 301–319.

30. DeVore, I., and K. R. L. Hall. 1965. Baboon ecology. In I. DeVore, ed., *Primate behavior.* New York: Holt, Rinehart and Winston. Pp. 20–52.

31. DeVore, I. and S. L. Washburn. 1963. Baboon ecology and human evolution. In F. C. Howell, and F. Bourliere, eds., *African ecology and human evolution.* Chicago: Aldine. Pp. 335–367.

32. Douglis, M. B. 1946. Some evidence of a dominance-subordination relationship among lobsters, *Homarus americanus. Anat. Rec.* 94: 57 (Abstr.).

33. Espinas, A. V. 1877. *Des Sociétés Animales.* Paris.

34. Etkin, W., ed. 1964. *Social behavior and organization among vertebrates.* Chicago: Univ. Chicago Press.
35. Furuya, Y. 1963a. On the ecological survey of the wild crab-eating monkeys in Malaya. *Primates* 3: 75–76.
36. Furuya, Y. 1963b. The social life of silvered leaf monkeys (*Trachypithecus cristatus*). *Primates* 3: 41–60.
37. Goodall, J. 1965. Chimpanzees of the Gombe Stream Reserve. In I. DeVore, ed., *Primate behavior.* New York: Holt, Rinehart and Winston. Pp. 425–473.
38. Grant, W. C. Jr. 1955. Territorialism in two species of salamanders. *Science* 121: 137–138.
39. Greenberg, B. 1947. Some relations between territory, social hierarchy, and leadership in the green sunfish (*Lepomis cyanellus*). *Physiol. Zoöl.* 20: 267–99.
40. Haddow, A. J. 1952. Field and laboratory studies on an African monkey, *Cercopithecus ascanius schmidti* Matchie. *Proc. Zool. Soc. Lond.* 122: 297–394.
41. Hall, K. R. L. 1962. Variations in the ecology of the chacma baboon, *Papio ursinus. Symp. Zool. Soc. Lond.,* no. 10, pp. 1–28.
42. Hall, K. R. L. 1964. Aggression in monkey and ape societies. In J. D. Carthy, and F. J. Ebling, eds., *The natural history of aggression.* New York: Academic Press. Pp. 51–64.
43. Hall, K. R. L. 1965. Social organization of the old-world monkeys and apes. "Social Organization of Animal Communities." *Symp. Zool. Soc. Lond.,* no. 14, pp. 265–289.
44. Hinde, R. A. 1956. The biological significance of the territories of birds. *Ibis* 98: 340–369.
45. Hoffmann, A. 1949. Über die Brutpflege des polyandrischen Wasserfasans *Hydrophasianus chirargus. Zool. Jahrb. Abt. Syst.* 78: 367–403.
46. Howard, R. S. 1955. The occurrence of fighting behavior in the grain beetle *Tenebrio molitor* with the possible formation of a dominance hierarchy. *Ecology* 36: 281–284.
47. Howard, W. E., and J. T. Emlen, Jr. 1942. Intercovey social relations in the valley quail. *Wilson Bull.* 54: 162–170.
48. Imanishi, K. 1957. Social behavior in Japanese monkeys, *Macaca fuscata. Psychologia* 1: 47–54.
49. Imanishi, K. 1960. Social organization of subhuman primates in their natural habitat. *Current Anthropol.* 1: 393–407.
50. Imanishi, K. 1961. The origin of human family—A primatological approach. *Jap. J. Ethnol.* 25: 119–138 [in Japanese with English summary].
51. Itani, J. 1954. *Japanese monkeys in Takasaki-Yama* [in Japanese]. Tokyo: Kobun-sha.
52. Itani, J. 1958. Afrika. *Shizen* 13(1): 58–65 [in Japanese].
53. Itani, J. 1966. Social organization of chimpanzees. *Shizen* 21 (8): 17–30 [in Japanese].
54. Itani, J. 1967. The social unit of chimpanzees. *Primates* 8: 355–381.

412   YOSIAKI ITÔ

55. Itani, J., and K. Tokuda. 1958. *Japanese monkeys in Ko-Shima, with reference to their sexual behavior* [in Japanese]. Tokyo: Kobun-sha.
56. Itô, Y. 1959. *Comparative ecology* [in Japanese]. Tokyo: Iwanami Shoten. (Rev. 2nd ed., 1966.)
57. Itô, Y. 1960. Territorialism and residentiality in a dragonfly, *Orthetrum albistylum speciosum* Uhler (Odonata: Anisoptera). *Ann. Entomol. Soc. Am.* 53: 851–853.
58. Jacobs, M. E. 1957. Studies on territorialism and sexual selection in dragonflies. *Ecology* 36: 566–586.
59. Jay, P. 1963. The Indian langur monkey (*Presbytis entellus*). In C. H. Southwick, ed., *Primate social behavior.* Princeton, N.J.: Van Nostrand. Pp. 114–123.
60. Jay, P. 1965. The common langur of north India. In I. DeVore, ed., *Primate behavior.* New York: Holt, Rinehart and Winston. Pp. 197–249.
61. Kawabata, M. 1954. Socio-ecological studies on the killi-fish, *Aplocheilus latipes.* I. General remarks on the social behaviour. *Jap. J. Ecol.* 4: 109–113 [in Japanese with English summary].
62. Kawai, M., and H. Mizuhara. 1962. An ecological study on the wild mountain gorilla (*Gorilla gorilla beringei*). *Primates* 2: 1–42 (1959).
63. Kawamura, T. 1952. *Sika-deers in Nara Park* [in Japanese]. Tokyo: Kobun-sha.
64. Kawanabe, H. 1957. Social behaviour and reproduction of a salmon-like fish, *Plecoglossus altivelis,* or ayu, with reference to its population density. *Jap. J. Ecol.* 7: 131–137 [in Japanese with English summary].
65. Kendeigh, S. C. 1941. Territorial and mating behavior of the house wren. *Ill. Biol. Monogr.* 18: 1–120.
66. Kikkawa, J. 1961. Social behaviour of the white-eye, *Zosterops lateralis* in winter flocks. *Ibis* 103a: 428–442.
67. Kikuchi, H., S. F. Sakagami, and M. Konishi. 1957. Ethology of the eastern great reed warbler, *Acrocephalus stentoreus orientalis* with special reference to the life history of a polygyneous family (Preliminary Report). *Jap. J. Ecol.* 7: 155–160 [in Japanese with English summary].
68. Kobayashi, H. 1955. Observations on a shark, *Rostratula bengalensis.* (2). *Tori* 14 (66): 1–13 [in Japanese].
69. Kodama, Y. 1960. Studies on the specific growth of freshwater fishes refferring to the weight increase of individual and population—III. *Bull. Freshw. Fish. Res. Lab.* 10 (2): 23–40.
70. Koford, C. B. 1963. Group relations in an island colony of rhesus monkeys. In C. H. Southwick, ed., *Primate social behavior.* Princeton, N. J.: Van Nostrand. Pp. 136–152.
71. Kortlandt, A. 1962. Chimpanzees in the wild. *Sci. Am.* 206 (May, 1962): 128–138.
72. Kummer, H., and F. Kurt. 1963. Social unit of a free-living population of hamadryas baboons. *Folia Primatol.* 1: 4–19.
73. Kuroda, N. H. 1955. Field studies on the Grey Starling, *Sturnus cineraceus* Temminck. 1. *Misc. Rep. Yam. Inst.,* no. 7, pp. 7–19 [in Japanese with English summary].

74. Kuroda, N. 1962a. Avis. In T. Utida, ed., *Dôbutsu Keitô-Bunruigaku.* 10 (Part 2) [in Japanese]. Tokyo.
75. Kuroda, N. 1962b. Mammalia. In T. Utida, ed., *Dôbutsu Keitô-Bunruigaku.* 10 (Part 1) [in Japanese]. Tokyo.
76. Lack, D. 1954a. The evolution of reproductive rates. In J. Huxley, A. C. Hardy, and E. B. Ford, eds., *Evolution as a process.* London: George Allen and Unwin. Pp. 143–156.
77. Lack, D. 1954b. *The natural regulation of animal numbers.* Oxford: Clarendon Press.
78. Lack, D. 1956. Variations in the reproductive rate of birds. *Proc. Roy. Soc.* (London), B, 145: 329–33.
79. Lorenz, K. 1952. *King Solomon's ring.* London: Methuen.
80. Lumsden, W. H. R. 1951. The night resting habits of monkeys in a small area on the edge of the Semliki Forest, Uganda. A study in relation to the epidemiology of sylvan yellow fever. *J. Anim. Ecol.* 20: 11–30.
81. McClure, H. E. 1965. Some observations on primates in climax Diptocarp forest near Kuala Lampur, Malaya. *Primates* 5: 39–58.
82. Mason, W. A. 1966. Social organization of the South American monkey, *Callicebus moloch:* A preliminary report. *Tulane St. Zool.* 13: 23–28.
83. Masure, R. and W. C. Allee. 1934. The social order in flocks of the common chicken and the pigeon. *Auk* 51: 306–327.
84. Miyadi, D., M. Kawabata, and K. Ueda, 1952. Standard density of "Ayu" on the basis of its behaviour and grazing unit area. *Misc. Rept. Physiol. & Ecol., Faculty of Sci., Kyoto Univ.,* no. 75 [in Japanese].
85. Mizuno, N. 1961. Study on the gobioid fish, "Yoshinobori." *Rhinogobius similis* Gill. I. Comparison of life histories of three ecological types. *Jap. Soc. Sci. Fish.* 27: 6–11 [in Japanese with English summary].
86. Mizuno, N. and H. Niwa, 1961. Two ecological types of the freshwater scorpion, *Cottus cottus* Günther. *Dôbutsugaku Zasshi* 70: 267–275 [in Japanese with English summary].
87. Napier, J. 1963. Brachiation and brachiators. *The Primates, Symp. Zool. Soc. Lond.,* no. 10, pp. 183–195.
88. Nice, M. M. 1941. The role of territory in bird life. *Am. Midland Naturalist* 26: 441–487.
89. Nissen, H. W. 1931–1932. A field study of the chimpanzee: Observations of chimpanzee behavior and environment in western French Guinea. *Comp. Psychol. Monogr.* 8 (36): 1–122.
90. Noble, G. K. 1939a. The experimental animal from the naturalist's point of view. *Am. Naturalist* 73: 113–126.
91. Noble, G. K. 1939b. The role of dominance in the social life of birds. *Auk* 56: 264–273.
92. Nolte, A. 1955. Field observations on the daily routine and social behavior of common Indian monkeys. *J. Bombay Nat. Hist. Soc.* 53: 177–184.
93. Oakley, K. P. 1961. On man's use of fire, with comments on toolmaking and hunting. In S. L. Washburn, ed., *Social life of early man.* New York: Viking Fund. 176–193.

94. Pardi, L. 1948. Dominance order in *Polistes* wasps. *Physiol. Zoöl.* 21: 1–13.
95. Petter, J. J. 1962. Recherches sur l'écologie des l'émuriens Malgaches. *Mém. Mus. Nat. d'Hist. Natur. Série A*, pp. 1–146.
96. Reynolds, V. 1966. Open groups in hominid evolution. *Man* (N.S.) 1: 441–452. (Cited from J. Itani, and A. Suzuki, 1967. The social unit of chimpanzees. *Primates* 8:355–381.)
97. Reynolds, V., and F. Reynolds. 1965. Chimpanzees of the Budongo forest. In I. DeVore, ed., *Primate behavior.* New York: Holt, Rinehart and Winston. Pp. 368–424.
98. Rowell, T. E. 1966. Forest living baboons in Uganda. *J. Zool.* 149: 544–564.
99. Schaller, G. B. 1961. The orang-utan in Sarawak. *Zoologica* 46: 73–82.
100. Schaller, G. B. 1963. *The mountain gorilla: Ecology and behavior.* Chicago: Univ. Chicago Press.
101. Schjelderup-Ebbe, T. 1913. Hönsenes stemme. Bidragtil hönsenes psykologi. *Naturen* 37: 262–276.
102. Schjelderup-Ebbe, T. 1935. Social behavior of birds. In C. Murchison, ed., *A handbook of social psychology.* Worcester, Mass.: Clark Univ. Press. Pp. 947–972.
103. Schneirla, T. C. 1952a. Discussion to L. Pardi "Dominazione e gerarchia in alcuni invertebrati." *Colloques Int. Centre Natl. Rech. Sci.* XXXXV, "Structure et Physiologie des Sociétés Animales," Paris. P. 197.
104. Schneirla, T. C. 1952b. A consideration of some conceptual trends in comparative psychology. *Psychol. Bull.* 49: 559–597.
105. Scott, J. W. 1942. Mating behavior of the sage grouse. *Auk* 59: 477–498.
106. Scott, J. W. 1944. Additional observations on mating behavior of the sage grouse. *Anat. Rec.,* Suppl. 89: 24 (Abstr.).
107. Simmonds, P. E. 1965. The bonnet macaque in south India. In I. DeVore, ed., *Primate behavior.* New York: Holt, Rinehart and Winston. Pp. 175–196.
108. Southwick, C. H., M. A. Beg, and M. R. Siddiqi. 1961. A population survey of rhesus monkeys in villages, towns, and temples of northern India. *Ecology* 42: 538–547.
109. Southwick, C. H., M. A. Beg, and M. R. Siddiqi. 1965. Rhesus monkeys in north India. In I. DeVore, ed., *Primate behavior.* New York: Holt, Rinehart and Winston. Pp. 111–159.
110. Starck, D., and H. Frick. 1958. Beobachtungen an äthiopischen primaten. *Zool. Jahrb., Abt. Syst. Oekol. Geogr. Tiere* 86: 41–70.
111. Sugiyama, Y. 1965. Group composition, population density, and some siciological observations of hanuman langurs (*Presbytis entellus*). *Primates* 5: 7–38 (1964).
112. Sugiyama, Y. 1918a. Social organization of chimpanzees at the Budongo forest, Uganda. *Primates* 9 (in press).
113. Sugiyama, Y. 1968b. A comparative study of primate societies. *Seibutsu Kagaku (Biol. Sci. Tokyo)* 20: 113–120 [in Japanese].
114. Tanaka, J. 1965. Social behavior of nilgiri langurs. *Primates* 6: 107–122.

115. Test, F. H. 1954. Social aggressiveness in an amphibian. *Science* 120: 140–141.
116. Tokuda, K. 1957. Macaques in a Zoo. In K. Imanishi, ed., *Nihon Dôbutsu-ki* 4: 168–268 [in Japanese].
117. Uhrich, Jacob. 1938. The social hierarchy in albino mice. *J. Comp. Psych.* 25: 373–413.
118. Washburn, S. L. and DeVore, I. 1961. Social behavior of baboons and early man. In S. L. Washburn, ed., *Social life of early man.* New York: Viking Fund. Pp. 91–105.
119. Winn, H. E. 1958. Comparative reproductive behavior and ecology of fourteen species of darters (Pisces-Percidae). *Ecol. Monogr.* 28: 155–191.
120. Yamagishi, H. 1962. Growth relation in some small experimental populations of rainbow trout fry, *Salmo gairdneri* Richardson with special reference to social relations among individuals. *Jap. J. Ecol.* 12: 43–53.
121. Yoshikawa, K. 1956. Compound nest experiments in *Polistes fadwigae* Dalla Torre. Ecological studies of *Polistes* wasps, IV. *J. Inst. Polytechnics, Osaka City Univ.* D,7: 229–243.
122. Zuckerman, S. 1932. *The social life of monkeys and apes.* London: Kegan Paul, Trench Trubner & Co.

PAULO NOGUEIRA-NETO
Department of Zoology, University of São Paulo
São Paulo, Brazil

Behavior Problems Related to the Pillages Made by Some Parasitic Stingless Bees (Meliponinae, Apidae)

EVOLUTION OF PARASITISM AMONG BEES

Parasitism may be considered the complete or partial living of an organism at the expense of another one, without direct benefit to the latter. Pillage, therefore, is a type of parasitism.

The more primitive parasitic bees lay their eggs on the larval food of other bees. In this way, the food stored by one species is appropriated by another to rear its own brood. This kind of parasitism belongs to the broad category of pillage. However, the adults are free living. Only their brood are dependent on food previously gathered by other bees. (Perhaps the oldest detailed discussion of parasitism in bees was written by Lepeletier, 1841, pp. 409–423.)

Among the bees of the family Apidae, *Psithyrus* alone (the parasitic bumble bee) has a similar behavior. The *Psithyrus* queen, however, does not take possession of larval food already present in a brood cell. She merely lays her eggs in a *Bombus* nest and the resulting larvae are fed daily by the workers of this colony. It should be pointed out that the *Bombinae,* to which *Psithyrus* also belong, are one of the most primitive Apidae subfamilies (the habits of *Psithyrus* were discussed by Sladen, 1912, pp. 59–72; Plath, 1934, pp. 34–54; Free and Butler, 1959, pp. 71–79, and others).

Among the Apinae (honey bees) and the Meliponinae (stingless bees), pillage evolved in lines quite different from those of *Psithyrus*. When they pillage another colony, they transport to their own nests the materials that are stolen. Only when their swarms conquer another nest does the booty remain there. *Bombus* (the common bumble bees) may also pillage other nests (Plath, 1934, p. 22).

In all the Apinae, and in most Meliponinae, robbery is incidental. However, among the meliponins some true parasites evolved. What probably started as facultative, later became obligatory. Perhaps *Melipona flavipennis* Smith is presently in a transitional stage, since it frequently robs other bees. Its ravages were studied in Bolivia by Kempff Mercado (1952, p. 5), and by Moure and Michener, in Belém do Pará, Brazil (Moure, Nogueira-Neto, and Kerr, 1958, p. 488).

Only two genera of stingless bees, *Lestrimelitta* (Neotropical) and *Cleptotrigona* (African), are true parasites. However, they differ from the parasitic solitary bees, and from *Psithyrus* bumble bees, in three major aspects: (1) they do not invade foreign nests to lay their eggs inside brood cells built by other bees; (2) they do not gather food from flowers; (3) they work in their nests, as the nonparasitic members of the subfamily do.

The transportation to the thieves' nest of the materials that were stolen had a marked influence on the evolution of the parasitic Meliponinae. Contrary to what happened with bees that rear their young in alien nests, *Cleptotrigona* and *Lestrimelitta* were able to develop a nest architecture of their own. Their nests, it is certain, do not differ greatly from those of the other members of the subfamily. Nevertheless, in several ways they are very characteristic. The enormous entrance tube of *Lestrimelitta,* for instance, is a unique structure. *Cleptotrigona* also has some distinctive features in their nests (Portugal Araújo, 1958, pp. 204–206).

The swarms of the parasitic meliponins lodge themselves in nests belonging to other members of the subfamily (such nest conquests were first reported by Muller, 1874, p. 103). However, these nests are evidently reworked to suit the needs and habits of their new owners. Never was a nest of these parasites found to remain the same as when it belonged to the dispossessed species.

The existing knowledge of *Lestrimelitta ehrhardti* Friese and *Cleptotrigona cubiceps* (Friese) is small. There is a nest description for *L. ehrhardti* made by Sakagami and Larocca (1963, pp. 332–336); on *C. cubiceps* the only existing paper is the work of Portugal-Araújo (1958). I was never able to observe the activities of either of these parasites. That is why this chapter deals mostly with *L. limao* (Smith), the commonest parasitic stingless bee in the neotropical region.

WORK INSIDE THE NESTS

It is a unique character of both *Cleptotrigona* and *Lestrimelitta* that in spite of being obligatory parasites they work and apparently live the usual meliponin life in their nests, except for the absence of wax secretion and related behavior. This means that while these groups were specializing in the exploitation of other Meliponinae, they retained almost all their working capacity at home. And not only at home. F. Muller (1921, p. 229) wrote that when pillaging a nest of another meliponin, the *L. limao* bees made their distinctive entrance tube over the entrance of the hive that they were invading. I have often seen this, too. Muller (op. cit.) also observed that on one occasion these robbers stacked the dead members of a *Tetragonisca jaty* (Smith) colony, which they were pillaging, and fastened to the hive wall a loose comb of these bees, which Muller later displaced during his observations. Portugal-Araújo (1958, p. 209) saw *C. cubiceps* removing "trash" that had obstructed the entrance of the invaded colony.

In two cases I observed *L. limao* bees throwing dead and moribund *Plebeia droryana* (Friese) out of the nests the *L. limao* had invaded. In spite of doing this work, the parasites soon afterwards abandoned both hives.

As Muller (1921, p. 229), Michener (1946, pp. 196–197), Nogueira-Neto (1950, p. 329), Kerr (1951, p. 301), Moure, Kerr, and Michener (Moure, Nogueira-Neto, and Kerr, 1958, p. 489), Sakagami and Larocca (1963, pp. 329–330), and Lucas de Oliveira (Sakagami and Larocca, 1963, p. 331) reported, the *L. limao* guard the entrance of the colonies that they pillage during the duration of their occupation.

Some kinds of work that *L. limao* undertake while occupying an alien nest may be easily understood because they are necessary for defense or for their pillaging robbing activities. Such is the case of the building of *L. limao* type entrance tubes in invaded colonies. However, it is difficult to interpret the significance of fixing a loose brood comb and stacking or throwing out dead and moribund bees.

Muller (1921, p. 229) thought that the building of the entrance tube and the stacking of dead bees meant that the invaders were establishing a new colony in the nest they occupied. However, he destroyed the robbers and so the outcome of the invasion could not be observed. In the observations made by me, the bees soon abandoned—at most after 3 days—the hives that they had invaded. During their occupation, the *L. limao* simply acted as if they were in their own nest. The presence of dead or disabled bees seemed to initiate the action of stacking or throwing them outside, and this probably occurs anywhere these bees happen to be. The same could be said about the process of fixing an old brood comb, although new combs of invaded colonies are destroyed by *L. limao* for the sake of their larval food (Muller, 1921, p. 238). Wilson, Durlach, and Roth (1958, pp. 113–114) reported that in ants, dead specimens are removed from the nest because oleic acid and possibly other substances, also present in the corpses, bring about such behavior. However, in the cases that I observed, the *L. limao* removed not only dead bees but also many that were still alive.

THE FLOWER-VISITING QUESTION

Some authors think that *Lestrimelitta* bees also work visiting flowers, as most other bees do. Ihering (1930) saw no other meliponins near a nest of the parasites, and Fiebrig (1908, p. 378) reported seeing these bees return home with pollen. They concluded, therefore, that *Lestrimelitta limao* can forage in flowers, too.

One may suspect that these opinions were based on the fact that the *Lestrimelitta* bees work inside their hives. Therefore, it could be expected that they would also work with flowers. However, in the course of time, the opposite reasoning of Muller (1874, p. 103; 1921, pp. 235, 267), Michener (1946, p. 197), and others prevailed. Ihering could have easily missed nearby subterranean or concealed nests. Fiebrig possibly thought that returning *Lestrimelitta,* loaded with stolen cerumen, were bringing in pollen. At any rate, no observer has ever actually seen these bees on flowers. They are, in fact, obligatory parasites, and no good evidence was ever produced to prove the contrary.

WHAT IS STOLEN?

Muller (1874, p. 103) said that *L. limao* "sometimes takes possession" of the nests of other bees. Certainly, in doing so, they also use the food and other materials found there. He was the first to observe (1921, pp. 238, 267, 271, 311) that these parasites steal larval food and cerumen from the hives of other bees, and transport this to their own nests. He said that they take everything they can find in the provision pots and in the brood combs. Often I saw pots that the *L. limao* did not loot, but I never saw a new brood comb that was spared by them (because of its larval food). *Cleptotrigona* also destroys brood combs containing larvae, in order to obtain their food (Portugal-Araújo, 1958, p. 209).

Kerr (1951, p. 292) discovered how *Lestrimelitta limao* and other stingless bees steal pollen from the alien nests. He noticed that the robbers first make a suspension of the pollen in a liquid they bring themselves. This suspension of pollen is then carried inside their honey crops to the parasites' hive.

Strangely, the pillage of propolis by *L. limao* is reported here for the first time. However, it was already known that this material is stolen by stingless bees (Nogueira-Neto, 1949, p. 23; Moure, Nogueira-Neto, and Kerr, 1958, p. 488). Fiebrig did not see any pillage made by *L. limao,* but he noticed (1908, pp. 380, 385) some sticky lumps around their nest entrance. Such lumps almost certainly were of propolis, although Fiebrig spoke of them as made of "wax."

In another observation I noted that *L. limao* invaders licked the tongue region and the abdomens of many dead or dying *P. droryana* in the hives they attacked. In one hive, 46 dead or dying *P. droryana* bees were examined. Eighteen of them had cuts or signs of having been mauled in the abdomen and also in the head; 16 had been bitten only in the head region; 6 were attacked solely in the abdomen; 1 had a cut in the protorax, and 5 showed no signs of violence. In this colony, several *L. limao* licked the abdomen and tongue region of the bees they had mauled.

In another nest, all 23 dead or dying *P. droryana* examined had marks showing that they had been severely bitten in the abdomen. One of them was a drone. Several others still had recently secreted wax plates over their tergites. In this nest, no *L. limao* bees were seen licking the tongue region of their victims.

During many of my observations of *L. limao* bees pillaging nests of *Nannotrigona testaceicornis* Lepeletier and species of *Plebeia,* I noticed something that seems to have escaped the attention of other observers. Many members of the colonies that were attacked apparently filled themselves with liquid food. In this way they protected part of their reserve stores. This will be discussed later in this chapter.

The *L. limao* attacks on the two *P. droryana* colonies took place in the cold season; at that time the colonies had no brood combs. As Muller (1921, p. 247) also reported, the absence of such combs is common in the cold season. Both hives had honey, which could have been easily carried away, but the parasites did not take it. It seems that they were looking for other booty. Not finding brood combs with larval food, the *L. limao* possibly tried to obtain it from the abdomens or from the tongue region of the bees they had mauled. By springtime, when the weather is warmer, the building of new combs starts and proceeds very rapidly. This may be due to an accumulation of larval food inside the honey crop of the bees of a colony during the cold season. This would explain why many *P. droryana* had their abdomens bitten and their tongue regions licked.

Possibly there is also a relation between an attack in which the heads of the bees are dislocated and the licking of the tongue region of dead or disabled bees. More observations, however, are needed before such a relationship can be established.

SMALL-SCALE VIOLENCE AND
THE "SUPERSEDING ODOR" HYPOTHESIS

The way in which the *Lestrimelitta* attack the nests of other bees is not well understood. Evidence of violence in the typical "harvesting" invasions of these bees has been reported as occurring on a small scale. Few dead bees were seen in those attacks.

Castello Branco (1845, p. 63) more than a century ago stated that the first attacking *L. limao* robbers are killed, but after a while, they take possession of the nest, ". . . because the owners, not being able to withstand their smell, yield their house to them." It is remarkable that Kerr's recent hypothesis (see below) had a precursor so long ago.

Muller (1921, p. 235) noticed that some members of a gurupú colony, invaded by *L. limao,* carried several dead parasites attached to their bodies. They died when biting the gurupú (*M. nigra* Gribodo). Muller was the first to note that the bites of *L. limao* are a weapon in their assaults. Schwarz (1948, pp. 88–89) observed what seemed to be two colonies of *L. limao* fighting each other. However, he could not verify this. Kerr (1951, pp. 300–301) reported that in the pillages made by *L. limao* there were very few deaths on either side. Portugal-Araújo (1958, p. 207) stated that in robberies for food and cerumen, the *Cleptotrigona* ". . . only kill the guards and a few other bees willing to fight." This, he reported, is what occurs in the usual forays made by these bees. However, when they start a new colony, all bees of the invaded nest are killed. Wille (1961, p. 24) said that "one or few *Lestrimelitta* . . . invade by force the nest they are going to pillage and are sacrificed. . . ." Sakagami and Larocca (1963, p. 331) wrote that it was "highly probable that this professional robber can fight vigorously if necessary. . . ." However, they did not actually witness such fights. Summing up, it is thought that relatively few bees are killed during the common or "harvesting" invasions of the parasitic meliponins.

To explain this and some other things observed by him during the pillages made by *Lestrimelitta limao,* Kerr (1951, pp. 300–301) formulated an attractive hypothesis. He postulated that: (a) because of their strong smell, the invaders can easily enter the hive they attack, since the odor of the colony assaulted is superseded; and (b) apparently this smell is a strong repellent to other bees. Later, Moure, Nogueira-Neto, and Kerr (1958, pp. 488–489) extended this theory, stating that once the odor of the pillaged colony is no longer effective, the *Lestrimelitta* workers can enter and leave freely. *Plebeia* apparently become terrorized and stop all work inside the colony. The odor seems to be the important factor in orienting the robbers, guiding them to the correct entrance, even if other nests are close by, as was observed in Curitiba.

Kerr's theory, in its new version, states that the *L. limao* smell: (a) gives the invaders free access to the pillaged colony; (b) seems to terrorize *Plebeia;* and (c) guides latecomers to the nest attacked. Sakagami and Larocca (1963, p. 320) spoke of Kerr's views as "new observations," considering them to be facts. Wille (1961, p. 24) supported the conclusions expressed above in items (a) and (c), but did not mention any terrorizing activities. Kerr's postulations on this matter could be called the "superseding odor" hypothesis. However, as appealing as such an interpretation may be, numerous observations suggest that the odor is not necessarily a terrorizing agent (though there is evidence to support the theory that the odor is a guide to arriving invaders).

Portugal-Araújo (1958, p. 210), writing about his observations on the African robber meliponin (*Cleptotrigona cubiceps*), said that "during the assaults . . . I noticed the odor common to this species, especially when bees were fighting, but I cannot say that there was such an odor inside the hives." This statement implies that a superseding odor is not essential to the invasion of other colonies by the parasitic stingless bees, at least not to all of them.

My observations on *Lestrimelitta limao* show that its odor is far from being the repellent that Kerr (1951, p. 301) suspected. On three different days, six specimens of *L. limao,* one each time, were taken with my fingers and placed before the entrances of 4 hives of *Plebeia droryana,* 1 of *P. remota* Holmberg, 2 of *Nannotrigona testaceicornis,* and 3 of *Melipona quadrifasciata* Lepeletier. The *N. testaceicornis* guards nearly always retreated immediately when the *L. limao* were presented to them, but they did the same thing even when no bee was put near their hives. Twice, however, a guard attacked the *L. limao.* Every time that a parasite was placed near the entrance of the *P. remota* hive, several of these bees came out of it. They touched and sometimes even seemed to lick the imprisoned *L. limao.* One of the parasites was daubed with a clear liquid. Some *P. remota* fanned their wings near the *L. limao.* At times they also attacked the robber bee. In all the *P. droryana* hives, some guards went out to attack the *L. limao* bees, often also assuming a defensive posture, with forelegs raised and mandibles open. It is interesting to note that this reaction was decidedly milder in the case of a specimen of *L. limao* which produced very little or no perceptible odor. A comparison could be easily made with the next *L. limao* that I presented to the *P. droryana;* this individual had a pronounced lemon-like smell and evoked a far stronger response.

In similar experiments, the *M. quadrifasciata* and *M. rufiventris* Lepeletier bees rushed outside their nests aggressively, flying in great numbers.

Time and again, during other observations, an *M. quadrifasciata* colony attacked the entrance of a neighboring *L. limao* hive, fighting its guards. Also, in several instances other bees were seen stealing cerumen from the entrance tube of the parasites. Such robbing behavior was reported before by Sakagami and Larocca (1963, pp. 327–331). These authors also mentioned fights (in Curitiba) between the *L. limao* that were pillaging a *Plebeia* hive, and the *Tetragonisca jaty* bees of a neighboring colony. Such fights were also confirmed by me, during several invasions of a *Nannotrigona testaceicornis* nest.

At the entrance of a hive that was being pillaged I observed that some *L. limao* guards kept the returning bees outside of the hive. Michener (1946, p. 196) described this situation in a nest of *Nannotrigona testaceicornis perilampoides (Cresson)* (= *Trigona t. perilampoides* in his paper), in Panama. He said that "at the approach of a *Trigona,* one or more of the *Lestrimelitta* guards rears up and opens its mandibles in a threatening manner. The *Trigona* flies away and hovers near the nest, often trying several times, but rarely successfully, to enter the nest. As a result, a considerable swarm of workers, many of them carrying pollen, usually develops outside a nest being robbed." This clearly shows that the *L. limao* odor, which in one case was perceived at approximately 150 cm from the invaded nest (Kerr, 1951, p. 300), did not prevent the members of the attacked colonies from coming back. Apparently, however, after a few unsuccessful trials, the incoming bees just fly nearby, forming the "swarm" mentioned by Michener. The presence of such "swarms," resulting from the activities of the *L. limao* entrance guards, was observed also by Kerr (1951, p. 301), Moure, Kerr, and Michener (Moure, Nogueira-Neto, and Kerr, 1958, p. 489), Sakagami and Larocca (1963, pp. 329–330), Lucas de Oliveira (in Sakagami and Larocca, 1963, p. 331), and by myself. Wille (1961, p. 25) saw the same behavior of the *L. limao* sentries, but did not mention a "swarm" of repulsed bees.

All these observations question the possibility of the *L. limao* odor acting as a repellent. The avocates of this hypothesis would have to explain why the same odor that sometimes "infuriates" bees, or makes them attack the source of the odor (1) does not, on other occasions, prevent them from stealing cerumen from the entrance tubes of *L. limao* hives, (2) does not prevent returning bees from trying to enter their own nest during a pillage, when they first arrive, and (3) why, in different circumstances, it apparently "terrorizes" the victims of the *L. limao* robberies.

The only way to explain all these events in accordance with the superseding odor hypothesis is to postulate that different concentrations of *L. limao* odor have different effects on other bees. This is consistent with several reports summarized by Wilson (1963, p. 351) in his discussion of ant alarm pheromone substances. Pheromones are biochemical inhibitors or stimulators of behavior and some of them evoke different responses in accordance with their concentrations. Conceivably, the *L. limao* odor could have a "terrorizing" action when present in a relatively high concentration, as inside the nests of other bees.

In spite of these possibilities, the superseding odor hypothesis fails to explain several facts. During an invasion of a nest of *N. testaceicornis* by *L. limao,* bees of a neighboring *T. jaty* colony fought the *L. limao* attackers. This not only occurred outside the nest (as mentioned above and noted by Sakagami and Larocca, 1963, pp. 327–331), but also inside the *N. testaceicornis* hive being attacked. Although the *T. jaty* fighters were near the entrance, they displayed aggressive behavior in the same atmosphere in which the *N. testaceicornis* fled from the *L. limao* attackers.

Moreover, Weyrauch (see Schwarz, 1948, p. 178) reported that a nest of *L. limao* was occupied by *T. jaty* ". . . after a battle lasting two days." This clearly shows that the *L. limao* bees did not "terrorize" their opponents even in their own nest, the very place where their odor presumably would be strongest. However, nests of *T. jaty* may also be invaded by *L. limao.* I saw a case in São Simão, São Paulo State, Brazil. Wille (1961, p. 25) noted the same thing in Costa Rica.

Also significant, as evidence that the *L. limao* odor does not "terrorize" other bees, is a situation I observed in two *Plebeia droryana* hives that had been attacked by *L. limao.* The robbers had just abandoned one hive, and the characteristic lemon-like odor was still strong inside the nests. However, in spite of this smell, the *P. droryana* were excitedly walking again, and fanning with their wings. In the other hive, the parasites were still there, but retreating; here, the *P. droryana* walked only in that part of the nest where no *L. limao* bees were present, although the strength of the odor must have been practically uniform throughout the entire small hive. Often there are many *L. limao* inside the hive of a "terrorized" *P. droryana* colony. However, in one case the invaders were as few as 10 but the *P. droryana* still seemed "terrorized" (Nogueira-Neto, 1950, p. 329).

All these observations show that whatever advantage the odor of *L. limao* may give them, it does not work as a "terrorizing" agent. However, we agree that probably the strong lemon-like smell of *Lestrimelitta* bees serves to attract new recruits to the exact location of the nest being pillaged (Moure, Michener, and Kerr, in Moure, Nogueira-Neto, and Kerr, 1958, pp. 488–489; Wille, 1961, p. 24).

LARGE-SCALE VIOLENCE DURING PILLAGING

As already mentioned here, several authors observed violence on a small scale during the usual pillages by *Lestrimelitta limao.* At times, however, violence also occurs on a large scale. Perhaps this was not considered important during the formulation of the superseding hypothesis.

Although he did not mention how they died, Muller (1921, p. 229) said that in an invaded colony "most *jaty* were already dead, including the very robust queen." This colony subsequently perished, because of such an attack. Muller thought that the invaders were trying to establish a new nest of their own. Erhardt (see Friese, 1931, pp. 7–8) reported that *L. limao* annihilated the inhabitants of the nests they conquered, but perhaps he did not actually see such attacks. At any rate, his theory of cyclic nest conquests was disproved by the facts. Portugal-Araújo (1958, p. 207) stated that all the other bees are killed when the *Cleptotrigona* invade a nest to establish a new colony of their species. However, he said that in typical robberies few bees die. Recently, a new report came from Venezuela. Stejskal (1962, p. 27) wrote that honey bees were bitten by *L. limao* and sprayed with a pleasant smelling liquid, until they became very sticky. In his opinion this liquid was probably poisonous.

In the meliponin colonies that I observed during an attack, the *L. limao* invaders never made their victims sticky. Perhaps the honey bees observed by Stejskal were sprayed with the liquid that the *L. limao* use to steal pollen. When one takes an *L. limao* from its nest with one's fingers, these may become slightly smeared with a very strong smelling and somewhat sticky liquid. Although he did not say it expressly, apparently Stejskal thought that the honey bees he saw were attacked with a product of the mandibular glands. However, it seems that the secretion of such glands is not so plentiful. Possibly the pheromone produced by the mandibular glands was mixed with another more abundant liquid, but this is only a guess.

Another very interesting observation by Stejskal (op. cit.) was the great persistence of the strong odor of *L. limao*. This smell remained in the empty hives of *Apis* for a month. This is in accordance with the characteristics of some pheromones. Wilson and Bossert (1963, p. 680) wrote that the amount of "the sex attractant contained in one female gypsy moth, 10^{-2} μg, would require about 8×10^7 seconds or 930 days to diffuse below threshold level; i.e., fade out, if it were released in one puff in perfectly still air."

A few observations that I recently made may throw some light on the difficult question of how the *L. limao* bees apparently "terrorize" members of the *P. droryana* colonies and of other species.

Often the *L. limao* intruders were seen chasing the inhabitants of the invaded colonies. The parasites ran over their victims, with their mandibles quite open, frequently biting them. In some of the colonies attacked, few or no bees were killed. However, as already told here, on two occasions the effects of the invasions were particularly severe. This happened in two colonies of *P. droryana*. In both cases, the *L. limao* were seen repeatedly biting many *P. droryana* individuals. Immediately after a bite, the bee that received it remained motionless for a while, but soon recovered and made frantic movements. Meeting another *L. limao,* the victim was bitten again, until, after repeated assaults, it became motionless. In this condition, the bees that were mauled often survived a relatively long time. The floor of both hives became littered with hundreds of dead and dying *droryana.* The bodies of 69 such bees were examined under a stereoscopic microscope. Recognizable marks of violence were found in all but 5 of them. It was very clearly the force of the parasites' mandibles—not a liquid viscous poison—that killed the *P. droryana.* In both these cases, the *L. limao* bees were not trying to establish new nests. The mother queen was killed in one of the invasions, as Muller (1921, p. 229) also saw, but the parasites abandoned the two hives after remaining there 2 and 3 days respectively.

In subsequent pillages of these same *P. droryana* colonies, there was no large-scale slaughter. Moreover, at the same time the *L. limao* had been killing *P. droryana* in one of these violent attacks, a neighbor hive of *P. droryana* was also robbed by bees from the same *L. limao* colony, but no deaths were seen. Why in some cases the invading *L. limao* slaughter their victims and in other pillages they kill few or no bees is not known. Probably few *P. droryana* would have been killed if the colonies so violently attacked had had brood combs. Muller (1921, p. 271) wrote that the *Plebeia* (mirins, in his words) of a colony he saw being assaulted took refuge between their brood combs. The larger *L. limao* could not reach there. However, the two violent attacks mentioned above were made in the winter, when *Plebeia droryana* nests have no brood combs. Therefore, the *P. droryana* could not hide efficiently from the *L. limao.* Nevertheless, it should be remembered that in several other pillages that occurred at the same time of the year and apparently in the same circumstances, even in the same nests, and also in the absence of brood combs, no slaughter was observed.

THE DURATION OF INVASIONS

L. limao invaded a nest of *P. droryana* (which, at that time, I called *T. (P.) mosquito*) and were forced to remain because their mother colony was destroyed. The attackers lived there for 24 days before they were removed (Nogueira-Neto, 1950, p. 329). In other colonies the stay of *L. limao* bees was not so prolonged. In 23 pillages that I saw in 1965, only 5 times did the robbers stay in the nest either overnight or for 2 or 3 nights and the corresponding days. This occurred in nests of *Nannotrigona testaceicornis* and *Plebeia droryana.* Muller (1921, p. 234), too, observed the presence of *L. limao* overnight in a colony they had invaded. Sakagami and Larocca (1963, pp. 329–330) reported the same in a colony of *P. emerina*. Wille (1961, p. 25) noticed that a nest of *T. jaty* was invaded by *L. limao* bees for several hours every day for nearly a week. Lucas de Oliveira observed that attacks generally continued for 2–3 days and could have lasted longer if not stopped artificially (in Sakagami and Larocca, 1963, p. 331). In the great majortiy of *L. limao* pillages (18 out of 23) observed by me in 1965, the parasites did not remain overnight in the nests they attacked. The relatively short stay of the robbers was also noted by two other authors (Muller, 1921, p. 273; Michener, 1946, p. 196).

LEARNING DURING PILLAGING

In one case, it was amazing to see how relatively few bees it took to maintain large numbers of their victims in a defenseless state (Nogueira-Neto, 1950, p. 329). At the end of the raid in which the *L. limao* remained in the *Plebeia droryana* hive for 24 days, only 10 *L. limao* bees were seen in the hive, some guarding the entrance and the rest in other parts of the hive, while some 200 *droryana* bees were living in a restricted area of their nest.

Schneirla (1960, p. 33) wrote that in ants there is ". . . a certain stability and a certain persistence of habits in time. . . ." In this case, there seems to have been what may be considered the persistence of a retreating behavior learned by the *P. droryana* as a consequence of being bitten by *L. limao* bees. Possibly the many *P. droryana* of the nest had become so habituated to retreat from the *L. limao,* that they continued to do so even when the number of the parasites seems to have become too low to dominate the colony.

The retreating behavior of the *P. droryana* was so strong that they apparently did no work during the 24 days. The food reserves in the hive spoiled because neither the *droryana* nor the *L. limao* did the work necessary to perserve them. *L. limao* bees, as was explained, may do some kinds of work in the colonies they invade, but, in this case, they did not tend the food stores. Many of the parasites and *P. droryana* died of starvation, as the food supplies were depleted.

Perhaps a minimum number of *L. limao* is necessary to mount a successful invasion. When there are only a few dozen *L. limao* in a colony, they no longer leave their nest to pillage. Then, the parasites' colony is doomed.

Once I saw only two *L. limao* enter a nest of *P. droryana,* but no pillage developed. However, no more invasions were seen during the same day. After 141 days without being attacked by the *L. limao,* two parasites from the colony that had invaded it earlier were placed near the entrance of the same *P. droryana* nest. Immediately, 3 workers of *P. droryana* attacked and bit both the *L. limao* bees.

Possibly, the entry made by only two *L. limao* succeeded because the *P. droryana* had learned to retreat from the parasites, as a consequence of their recent experience of suffering mass invasions of *L. limao*. However, 141 days afterwards the guards of the *P. droryana* nest were almost certainly new bees that had never suffered an attack of the parasites and, consequently, had not learned to retreat from them. In two other species of stingless bees, guards are young bees just beginning their flying activities (Kerr and Santos Neto, 1956, p. 421; Hebling, Kerr and Kerr, 1964, pp. 122–123). Apparently as a result of the presence of new guards, both specimens of *L. limao* were immediately assailed by the *P. droryana*.

Kerr (1951, pp. 300–301) also saw only two *L. limao* bees invading a colony of *Melipona quadrifasciata*. Possibly this, too, happened because of previous attacks of the parasites. Former invasions were not mentioned, but they could have occurred, since Kerr kept a *L. limao* colony (Kerr, 1951, p. 292).

The formation of a "swarm" of bees near a nest occupied by *L. limao* seems to be due to a learning process. As Michener (1946, p. 196) noted, the bees that go back to their nest are received aggressively by the *L. limao* entrance guards. What probably happens is that these bees learn to avoid the guards. Therefore, after perhaps a few such encounters, instead of making successive tries to enter their nest, they simply continue to fly nearby. As many bees are thus flying together, they may form what seems to be a swarm.

The "certain stability of habits in time," which Schneirla (op. cit.) described, may offer also an explanation for the intriguing fact that while 23 recorded pillages occurred during a 3 month period in 5 colonies of *P. droryana,* 1 colony of *P. remota* and 2 colonies of *N. testaceicornis,* no robberies were seen in 2 colonies of *P. droryana,* 1 of *N. testaceicornis,* 3 of *M. quadrifasciata anthidioides,* 2 of *M. compressipes* Fabricius, 2 of *Tetragonisca jaty,* and 1 of *Scaptotrigona postica* Latreille, all of them present in one of my apiaries (in S. Paulo). These data revealed a remarkable preference of *L. limao* for pillaging certain hives, while others were spared, although they were all in the same garden. The sparing of some colonies belonging to the species *P. droryana* and *N. testaceicornis,* the ones most attacked during those 3 months, is very remarkable indeed.

A possible explanation is that *L. limao* chiefly invades the hives in which former pillages occurred. Such colonies, because their bees had learned to retreat from *L. limao,* would be easier prey for the parasites.

The possibility that learning plays a very important role in the behavior of bees belonging to the invaded nests still needs more investigation. At present it is a hypothesis that explains facts otherwise not clearly interpreted.

THE "RETREAT MESSAGE" HYPOTHESIS

Why *Lestrimelitta limao* bees attack some colonies so violently on certain occasions and not on others is difficult to understand. However, the data now available are perhaps sufficient to formulate a more precise explanation of what happens during an attack of these parasites.

Contrary to the superseding odor hypothesis, it was demonstrated here that the success of *Lestrimelitta limao* invasions does not seem to depend on odor, except as a "trail" pheromone for guiding new robbers. The capacity of *L. limao* for using violence was also examined. These bees, I believe, invade and rob the nests of other bees, based chiefly on the power of their mandibles.

The initial phase of an assault by *L. limao* was never recorded in detail. Only Lucas de Oliveira (in Sakagami and Larocca, 1963, p. 331) reported that in pillages made by *L. limao,* "at first *P. emerina* showed a little resistance to the robbers, but it soon ceased." Because there was some resistance, and also because the entrance guards of the nests of several stingless bees responded aggressively to the *L. limao* odor, it can be presumed that the robbers make their way into other colonies with their mandibles. In fact, during an invasion the parasites are often seen with their mandibles wide open, when biting, "menacing" or running over the members of the colony that they are assaulting.

When a colony is attacked, several *L. limao* bees guard the entrance of the invaded nest. In Michener's words (1946, p. 196) . . . "the *Lestrimelitta* guard rears up and opens its mandibles in a threatening manner." In this case, again, the power of the invaders' mandibles seems to be the primary factor that keeps the nest inhabitants out of the parasites way. The smell of *L. limao* may be perceived before the returning bees arrive at the entrance (Kerr, 1951, pp. 300–301), but it does not prevent them from trying to enter the nest, at least when they first come home. Afterwards, possibly the odor may become associated with the presence of *L. limao.*

What happens inside the nest, when an invasion starts, can be easily surmised. In the interior of the hive, some or many of the defenders may be killed. Others are only bitten. As explained here, several authors noticed that few bees die in the typical *L. limao* pillages. However, other observations showed that sometimes large numbers perish.

The bees inside the hive that is assaulted retreat to the walls, combs, or to other places of their nests (Muller, 1921; pp. 229, 235, 271). As the invaded colony has hundreds or thousands of bees, individual "retreat learning" (through the bites of *L. limao*) by all members of a colony seems practically impossible. Also, it was already demonstrated in this chapter that the odor of *L. limao* alone has no repellent or "terrorizing" properties. Therefore, it must be assumed that when some bees are attacked, somehow a retreat message is issued by them and received by the other members of the colony.

The retreat or distress message could be a vibration or a smell, as among other animals. Schneirla (1957, p. 110) wrote that in colonies of army ants certain "excitatory effects are transmitted widely through the colony", as long as the cause of the ". . . colony excitation is maintained at a high level. . . ." He was referring to the "active brood" as the source of excitation. In the case of the assaulted nests of stingless bees, this source must be the presence and the aggressive behavior of the parasitic invaders.

If the retreat message is an odor, it should be classed as an alarm pheromone, a category of substances capable of evoking a retreat or an aggressive reaction under certain concentrations (Wilson and Bossert, 1963, p. 690). In this chapter it was already shown that the *L. limao* smell makes other bees attack them. However, aggressive alarm behavior in stingless bees is not provoked only by odor. When one strikes the hives of certain meliponins, many bees may rush outside aggressively. This was first noticed by H. v. Ihering (1930, p. 651). Lindauer and Kerr (1960, pp. 34–36) stated that the Meliponinae ". . . field bees gave another alarm signal, a characteristic buzzing sound." These authors showed that "buzzes" were "very effective in alerting," in an experiment with colonies of *M. scutellaris* Latreille. Esch, Esch, and Kerr (1965, pp. 320–321) recently demonstrated that not only smell but sound, too, is important in the communication of food sources in stingless bees. Several times when I placed a *L. limao* before the entrance of *P. remota,* some bees came out of the hive and started vibrating (fanning) their wings near the parasite. All these facts show that sound vibrations may possibly communicate a retreat message among the Meliponinae.

In ants, Wilson (1963, p. 352) stated that "tactile and auditory stimuli probably play some role in alarm communication," although a secondary one. Schneirla (1952, p. 259) wrote that "in ants there is a wide range of transmissive media differing according to the species. . . ." In bees, probably the same situation exists. Therefore, until the nature of the retreat communication of meliponins is better known, one should simply call it a "message," an all-inclusive word that may cover either pheromones or sound vibrations.

The distress "information" also makes many bees of the invaded nests fill their honey crops with liquid food. As mentioned in this chapter, the filling of the honey crop is a very interesting method of protecting some of the food stores. In any colony of meliponins, some individuals normally keep larval food in their honey crops. However, during an *L. limao* invasion many more bees can be seen with full crops as they are hiding or gathering together in places out of the parasites' way. The bees of the invaded colonies were not actually seen collecting food before fleeing. However, obviously this must have been done before they sought refuge, not after, at least when they retreated to places with no food stores. Honey bees also display a strikingly similar reaction when their hives are pounded or smoked (Root, Root and Root, 1959, p. 593).

As was told here, a colony of *P. droryana* was observed when being evacuated by the *L. limao* robbers. In parts of the hive already abandoned by the parasites, there were plently of *P. droryana* bees fanning with their wings and walking excitedly, in spite of the *L. limao* odor still present. In another *P. droryana* hive, the same behavior was observed soon after the *L. limao* left. Obviously, in both cases the resident bees somehow knew that the *L. limao* invaders were leaving the place or had gone. Then, apparently, the retreating message was no longer used. Perhaps a kind of a "go ahead" signal was substituted for it. At any rate, both cases show that the *L. limao* odor during pillage does not block or overpower the *P. droryana* actions.

Schneirla (1960a, p. 315) stated that ". . . the army-ant cyclical pattern does not pre-exist in the genes of any one type of individual. . . . Rather, organic factors basic to the species pattern have evolved in close relationship with extrinsic conditions which in the evolved pattern, supply key factors essential for integrating processes from all sources into a functional system."

It could be said, also, that the retreat behavior that *P. droryana* and other bees display when they are attacked by *L. limao* is not simply an "innate impulse," as a statement of Moure, Nogueira-Neto and Kerr (1958, pp. 488–489) may perhaps imply. These authors wrote that "*Plebeia* becomes apparently terrorized," as a consequence of the parasite's smell, which supersedes the odor of the colony attacked. However, the facts discussed in this chapter show that there is no "inborn impulse" brought about by the presence of a *L. limao* smell.

The "retreat message" issued by *P. droryana* and the resulting behavior seem to be fundamentally stereotypic. Its effects are very similar in different colonies and even in different species of stingless bees. However, as it was explained here, the use of a retreat message, in connection with the attacks of *L. limao,* apparently occurs only after some bees of an invaded nest are bitten. One may say, therefore, that the generalized retreat behavior observed in nests of *P. droryana* and other bees attacked by the parasitic Meliponinae (and also the extinction of such behavior when the parasites leave the nest) is the final result of an interaction of extrinsic factors and of stereotyped patterns that have an organic base. In fact, "certain basic organic factors may . . . be postulated as centrally involved in the species pattern" (Schneirla, 1956, p. 397). As in the case of the army-ant cyclical pattern, here also there seems to be a "close relationship" of "organic factors" and "extrinsic conditions." In connection with this whole situation, one should always remember that distinctions of the "native" and "acquired" are ambiguous (Schneirla, 1956, p. 391).

The study of the behavior related to pillages made by Meliponinae robber bees was begun more than a century ago. Nevertheless, good harvests will certainly reward those who further plow this fruitful field.

REFERENCES

Castello Branco, L. S. D. 1845. Memoria acerca das abelhas da Provincia do Piauhy no Imperio do Brasil. *O Auxiliador da Industria Animal Nacional* 2–3: 49–72.

Esch, H., I. Esch, and W. E. Kerr. 1965. Sound: an element common to communication of stingless bees and to dances of the honey bee. *Science* 149: 320–321.

Fiebrig, K. 1908. Skizzen aus dem Leben einer Melipone aus Paraguay. *Z. Wiss. Insektenbiol.* n.s. 3, 12: 374–386.

Free, J. B., and C. G. Butler. 1959. *Bumblebees.* New York: Macmillan. Pp. 1–208.

Friese, H. 1931. Wie honnen Schmarotzerbienen aus Sammelbienen entstehen? II. *Zool. Jahrb., Abt. Syst.* 62: 1–14.

Hebling, N. J., W. E. Kerr, and F. S. Kerr. 1964. Divisão de trabalho entre operárias de *Trigona (Scaptotrigona) xanthotricha* Moure. *Pap. Avul. Dept. Zool. Secr. Agr. S. Paulo* 16: 115–127.

Ihering, H. von. 1930. Biologia das abelhas melíferas do Brasil [a translation of a work published in 1903]. *Bol. Agr.*, S. Paulo, 31:435–506, 649–714.

Kempff Mercado, N. 1952. Enemigos del apiario: algunas abejas indigenas. *Campo* (Bolivia) 5: 5–7.

Kerr, W. E. 1951. Bases para o estudo da genética de populações dos Hymenoptera em geral e dos Apinae em particular. *An. Esc. Sup. Agric. L. Queiroz,* Univ. S. Paulo, 8: 219–354.

Kerr, W. E., and G. R. dos Santos Neto. 1956. Contribuição para o conhecimento da bionomia dos Meliponini. 5. Divisão de trabalho entre operárias de *Melipona quadrifasciata quadrifasciata* Lep. *Ins. Soc.* 3 (3): 423–430.

Lepeletier de Saint-Fargeau, A. 1841. Histoire naturelle des insectes. Hymenopteres. vol. II. *Lib. Enc. Roret,* Paris.

Lindauer, M., and W. E. Kerr. 1960. Communication between the workers of stingless bees. *Bee World* 41:29–41, 65–71.

Michener, C. D. 1946. Notes on the habits of some Panamanian stingless bees (Hymenoptera, Apidae). *J. N.Y. Entomol. Soc.* 54: 179–197.

Moure, J. S., P. Nogueira-Neto, W. E. Kerr. 1958. Evolutionary problems among Meliponinae (Hymenoptera, Apidae). *Proc. Tenth. Int. Congr. Entomol.* 2: 481–493.

Muller, F. 1874. The habits of various insects. *Nature* (London) 10: 102–103.

Muller, F. 1921. Letters published by A. Moller in *Fritz Muller Werke, Briefe und Leben,* vol. II. Jena: G. Fischer.

Nogueira-Neto, P. 1949. Notas bionômicas sôbre meliponineos. II. Sôbre a pilhagem. *Pap. Avul. Dept. Zool. Secr. Agr. S. Paulo* 9(2): 13–32.

Nogueira-Neto, P. 1950. Notas bionômicas sôbre meliponineos (Hymenoptera, Apoidea). IV. Colônias mistas e questões relacionadas. *Rev. Entomol.* 8 (1–2): 305–367.

Plath, O. E. 1934. *Bumblebees and their ways.* New York: Macmillan.

Portugal-Araújo, V. 1958. A contribution to the bionomics of *Lestrimelitta cubiceps* (Hymenoptera, Apidae). *J. Kansas Entomol. Soc.* 31 (3): 203–211.

Root, A. I., E. R. Root, H. H. Root. 1959. *The ABC and XYZ of bee culture.* Medina, Ohio: A. I. Root Co.

Sakagami, S. F., and S. Larocca. 1963. Additional observations on the habits of the cleptobiotic stingless bees, the genus *Lestrimelitta* Friese (Hymenoptera, Apoidea). *J. Fac. Sci. Hokkaido Univ.* 15 (2): 319–339.

Schneirla, T. C. 1952. Basic correlations and coordinations in insect societies, with special reference to ants. *Colloques Int. Centre Natl. Rech. Sci.* XXXIV, "Structure et Physiologie des Sociétés Animales," Paris. Pp. 247–269.

Schneirla, T. C. 1956. Interrelationships of the "innate" and the "acquired" in instinctive behavior. In P.-P. Grassé, ed., *L'Instinct dans le comportement des animaux et de l'homme.* Paris: Masson. Pp. 387–452.

Schneirla, T. C. 1957. Theoretical consideration of cyclic process in Doryline ants. *Proc. Am. Phil. Soc.* 101 (1): 106–133.

Schneirla, T. C. 1960a. Instinctive behavior, maturation-experience and development. In B. Kaplan and S. Wapner, eds., *Perspectives in psychological theory.* New York: International Universities Press. Pp. 303–334.

Schneirla, T. C. 1960b. L'Apprentissage chez la fourmi et le rat. *J. Psychol. Norm. Path.* no. 1, pp. 11–44.

Schwarz, H. F. 1948. Stingless bees (Meliponidae) of the western hemisphere. *Bull. Am. Mus. Nat. Hist.* 90: 1–546.

Sladen, F. W. L. 1912. *The bumble-bee.* New York: Macmillan.

Stejskal, M. 1962. Duft als "Sprache" der tropischen Bienen. *Suedwestdeut. Imker* 14 (9): 271.

Wille, A. 1961. Las abejas jicotes de Costa Rica. *Rev. Univ. Costa Rica,* no. 22, pp. 1–30.

Wilson, E. O. 1963. The social biology of ants. *Ann. Rev. Entomol.* 8: 345–368.

Wilson, E. O., and W. H. Bossert. 1963. Chemical communication among animals. In G. Pincus, ed., *Recent progress in hormone research,* vol. 19. New York: Academic Press. Pp. 673–716.

Wilson, E. O., N. I. Durlach, and L. M. Roth. 1958. Chemical releasers of necrophoric behavior in ants. *Psyche* 65 (4): 108–114.

GASTON RICHARD
Department of Zoobiology
Faculty of Sciences
University of Rennes
Rennes, France

New Aspects of the Regulation of Predatory Behavior of Odonata Nymphs

A very abundant literature attracts the investigator who wishes to review the feeding habits of insects. The recent texts of Jourdheuil (1963), Dadd (1963), or Cathy (1965) provide surveys of the principal problems posed.

Since the works of Huber (1960), Vowles (1961), Roeder (1963), and others, the tendency has been to concentrate on the problem of eliciting behavior by direct stimulation of neural centers. This approach has provided us with interesting rudiments of the levels and processes of integration in the insect. Most of these studies, however, are concerned with locomotor behavior (in the broad sense) or sexual behavior. Very few provide information on any aspect of feeding behavior.

Under the influence of cyberneticians, a complementary tendency is expressed in the theoretical investigation of behavioral models and the structuring of the nervous system. Following von Holst (1950), Mittelstaedt (1957) and Hassenstein (1960) became the champions of this tendency and we owe them much for our understanding of these phenomena. The recent Ojai symposium (1964) exemplifies this approach. Nonetheless, relatively general aspects of behavior have been studied.

It is undoubtedly surprising that few studies of the feeding behavior of insects followed on the pathway opened by Schneirla's work with ants. It would seem useful to know the possibilities of insect behavioral modifiability as a function of alimentary motivation; such research would offer another route to an understanding of higher levels of behavioral integration.

GENERAL CONSIDERATIONS OF
THE FEEDING BEHAVIOR OF INSECTS

Let us remember that feeding behavior in general consists of various phases: localization, contact, ingestion, and then digestion of the food. These phases depend on different stimulus-inputs in the insect, through the cephalic organs (eyes, ocelli, antennae, mouth parts, etc.), through the organs of the thoracic appendages (particularly the legs), through different organs distributed over the entire body (as in the case of tactile receptors), or through the sense organs related to the digestive tract itself (control of ingestion). One must not forget that the peripheral *sensitivity* of the insect is very *diffuse*: for example, the distribution of chemoreceptors has little in common with that of vertebrates.

On the other hand, it is equally important to remember that the situation stimulating feeding is always a *situation made complex by the possibility of the heterogeneous summation of many factors*: actual nutritive factors, repellent factors, and finally, "token" or phago-stimulants, that is, substances which play a role in the choice of food without having any nutritional value themselves (Thornsteinson, 1960; Atwall, 1963; etc.). The order in which stimuli are presented is not always irrelevant: *Philanthus apivorus,* the hymenopterous predator of bees, first recognizes its prey by visual cues (the general shape of the body, the flight pattern), but this does not always enable it to distinguish a bee from a bee-like fly, e.g., *Eristalis*. Therefore, when the prey is near and registering in another sensory modality, it is the chemical cues that will maintain or breakdown the alerted predatory set, leading toward or away from an actual capture.

But, the correct sequence of stimuli arranged by the external environment of the animal is not sufficient. Each stimulus must be present when a specific internal state has developed (the law of maximum coincidence). Willington (1948) and Fleschner (1950) are among those who have emphasized this aspect by showing that observed deviations in feeding behavior cannot be explained by environmental changes, but by more or less abrupt variations in the internal state. For example, the bug *Triatoma* is attracted by chemical stimuli when fasting, whereas photic and thermal factors play the fundamental roles in its behavior after satiation. We consider this dynamic aspect of factors mediating approach and withdrawal (Schneirla, 1966) of great importance.

Too few authors have studied analytically the movements of insects, particularly of the mouth parts, during ingestion. We have been limited to a compendious classification of the mouth parts of adult insects according to their presumed mode of action (crushing, sucking, piercing, etc.) but the mechanisms of these functions have not been investigated sufficiently. Yet, Chilova (1954), Popham (1959, 1961), Kraus (1958), Dethier (1960), Pick (1962), and Richard (1965b), among others, have attempted to understand how the relevant neuromuscular components operate by specifying the innervation of the mouth parts. This is extremely important and we will return to this later.

Finally, we will emphasize one point; that is, that progressive entrainment of feeding behavior with other behavioral patterns in the insect, e.g., particularly wiping face or mouth parts, is very often observed as a compulsory activity after feeding.

FEEDING BEHAVIOR OF ODONATA NYMPHS

General Considerations

For several years, part of the activity in our laboratory has been devoted to a study of the neurophysiological aspects of the control of feeding behavior in various nymphs. These investigations complement the early studies of Amans (1881), Baldus (1926), Brocher (1917), and Koëhler (1924), among others, as well as the more recent works of Bücholtz (1961, 1964), Caillère (1965), Satija (1958), Mill (1964), and others. In all our studies, the Odonata nymphs have proved to be the preferred subjects.

The Sequences of Predatory Behavior

Typically, the predatory behavior of an Odonata nymph occurs in the following sequence, as described in whole or in part by Vasserot (1957), Richard (1960–1965), Caillère (1965) and Vogt (1964):

1. Resting position
2. Attention set
3. Location change (this may not occur)
4. Investigation of prey
5. Capture of prey
6. Ingestion of prey
7. Grooming

Important differences between the nymphs of various species are evident in both the form and the quantitative aspects of activities 2 to 5, as we have already stressed elsewhere (Richard, 1965b). When sequence 3 takes place, it may appear as very slow locomotion (walking on the substrate) or as a rapid movement (propulsive reaction of the large Anisoptera nymphs). The investigatory phase can manifest many different degrees of equilibrium between sensory receptors. We will now consider some of these problems.

Stimuli That Initiate Capture

THE STUDIES BY BALDUS

The investigations of Baldus (1926), devoted to the feeding behavior of aeschnid nymphs, have particularly emphasized the releaser role of visual stimuli presented by the prey. The image of the prey must be situated on particular symmetrical ommatidia: those whose optical axes cross at the tip of the mask,[1] and between the movable pincers on the mask. It should be sufficient, in theory, for an animal about to effect a capture to move until the prey is situated at the intersection of the appropriate optic axes. Regulation of positioning would be simpler than that invoked by Mittelstaedt (1957) for the praying mantis. The regulatory mechanism in the mantis resembles that of the aeschnid nymph in regard to proximal sensation. The aeschnid nymph has a head that is relatively immobile in relation to the thorax, and, in contrast with the praying mantis, capture is accomplished directly by mouth parts, as is true also for *Notonecta,* as Ludtke described (1950).

STUDIES RELEVANT TO THE WORK OF BALDUS

Old aeschnid nymphs.   Let us stress that, if Baldus (1926) is correct, it should be possible to elicit projection of the mask of an aeschnid nymph toward an appropriate prey situated at the proper distance, but separated from the nymph by a thin glass plate (Koëhler's [1924] technique). But, this produces a variable response. Under these conditions, the behavior of the nymph becomes irregular and there is a considerable increase in the sequence of prey investigation (at least tenfold). The aeschnid remains at the normal fixation distance a long time and the projection of the mask is delayed. When the nymph does project its mask, the tip of the mask hits the glass barrier and the nymph is thrust back forcefully. After several unsuccessful attempts, the insect abandons its efforts. All of this is confirmed by the experiments of Etienne and Howland (1964) or by Vogt (1964) working with moving points of light.

Young aeschnid nymphs.   We have shown (Richard, 1960) that the behavior described above closely resembles that of aeschnid nymphs whose antennae have been removed. This enables us to attribute to these organs

---

[1] The "mask" refers to the specialized labium which is thrust forward (projected) when the nymph is about to seize its prey.

a role in the dynamics of capture. Moreover, if we study the behavior of very young individuals soon after hatching, instead of that of older aeschnid nymphs as all the authors have done, we find (apart from all the quantitative sensory aspects which we will discuss later) that for the young nymphs, the eyes play a very limited role, while the antennae have the dominant role. These young nymphs, which are very mobile, move about very frequently on their substrate. They rarely dart in the water; the only jumps we have observed were sudden retreats. When a prey of appropriate size enters its vicinity, and is at least 2 mm away, the nymph shoots forth its mask, but only after having lightly brought its antennae together and after raising its head. We have never observed a distal attention set by headturning as in old nymphs. When a glass partition is placed between a young nymph and a potential prey located at a distance less than the maximum for capture, the predatory response is not elicited in the least.

Experiments with models (very fine glass fibers, metal filings, tips of vegetable fibers, etc.) have ascertained that stimulation of the tactile receptors of the antennae of the young nymphs is sufficient to initiate capture behavior. Antennal stimulation in old nymphs is sufficient to bring about a strong attention set, but capture follows rarely unless visual stimulation takes place simultaneously. The logical complement of the experiment cited above and an improvement of Koëhler's (1924) study is the representation of prey behind a glass partition while stimulating the antennae of the nymphs with a fine fiber at the same time. In all cases (young or old nymphs), the mask is projected as under natural conditions.

OTHER SPECIES

We stress this mechanism of antecedent sensation in predatory behavior because the balance between visual and antennal stimulation is highly specific. In other species of Odonata, there is also a possible complementary cue originating in tarsal tangoreception as Vasserot (1957), Büccholtz (1961, 1964) and Caillère (1965) have shown in *Agrion*. All intermediate degrees of intersensory relationships in predatory behavior are possible, from a visually-based capture assisted by a slight qualitative and quantitative intervention by the mechanoreceptors of antennae and legs (*Aeschna*) to the antennally-based capture assisted by a slight qualitative and quantitative intervention by the eyes and the mechanoreceptors of the legs (*Agrion*).

The current research in our laboratory (Urvoy, Guillou) will elucidate certain aspects of this sensory equilibrium because these investigators are studying the recovery of sensory activity during the regeneration of antennae in *Agrion Virgo*. This recovery is progressive and is accompanied by a decrease in the importance of the anterior tarsi in predatory behavior.

Approach and Withdrawal

MODALITIES OF MOVEMENTS

Having elsewhere (Richard, 1960) specified the dynamic sensory equilibrium involved in the recognition and localization of prey by the nymphs of *Libellula,* we wish now to stress a problem that seems to us of the greatest importance: the effect of sensory feedback on movement (sequence 3; see preceding section) of the predator in relation to its prey. The observed cases can be related to an approach or to a withdrawal, in the sense of Schneirla's theory (Maier and Schneirla, 1964, 1966); furthermore, each can be accomplished slowly (normal gait) or quickly (swimming reaction of Anisoptera, undulating swimming of Zygoptera). The direction and intensity of the reaction are determined by the quantitative aspects of the stimulus. A prey of a certain size and amount of movement *proportional to the predator* establishes the choice of approach or withdrawal, but above and below that value, the animal regulates its speed of approach or withdrawal as shown by the data in Table 1 from an experiment with aeschnid nymphs.

TABLE 1

Approach and withdrawal responses of aeschnid nymph (last stage) to a stimulus of varying intensity.

Stimulus	Response
Midge larva (*Chironomus*), living, 15 mm long	
Presented 10 cm from nymph	Approach: swimming reaction
Presented 2 cm from nymph	Slow locomotion and capture
Presented with facial contact	Slow backward movement and capture
Experimenter	
Finger held 1 cm from nymph	Withdrawal: swimming reaction
Standing 1.5–2 m from nymph	Approach: slow locomotion oriented to experimenter
Sudden movement 50 cm from nymph	Withdrawal: backward movement with locomotion or swimming reaction

In nymphs whose predatory behavior is regulated to a great extent by mechanical stimuli, the speed or intensity of stimulus contact with the mechanoreceptors determines the form of the locomotor response, but in many cases we find that it is possible to disrupt the balance between different sense organs. Thus, it is possible to stop the phase of visual investigation of prey by an old aeschnid nymph by a sudden rap on the antennae or tarsi. This halt may even be accompanied by a slow backward movement (on rare occasions by a rapid withdrawal).

TYPES OF MOVEMENT COORDINATIONS

Let us stress another equally important factor in the movements of our nymphs. The slow walk is always carried out with the normal hexapodal rhythm with tarsal contact on the substrate (Wilson, 1965). In sharp contrast, the rapid swimming pattern is accompanied by a very specific positioning of the legs which stretch out from both sides of the body for the entire period of this type of locomotion (Amans, 1881; Baldus, 1926). This is always true for the swimming reaction of Anisoptera and usually true for the undulating swimming of Zygoptera. We see here a first-rate method for analyzing the total coordination of the movements of the animal, because the control of this postural set is contemporaneous with changes in the excitation of the cephalic sensory receptors. At the level of this approach mechanism, a loop is formed between the sensory afferents and the posterior abdominal neuromotor units (for jumping) or the thoracic neuromotor units (for walking). The locomotion thus effected modifies the form and intensity of the stimulus until it places the predator in the proper position for aiming at the prey.

Projection of the Mask

GENERAL CONSIDERATIONS

A new stage of coordination becomes necessary for the projection of the mask. Indeed (and this has not failed to intrigue morphologists for a long time), the mask of the Odonata nymphs does not seem to have sufficient musculature to carry out its movements. Certain morphologists believe that an increase in blood pressure in the head region is necessary for extension of the mask. The hemolymph involved in this response could come only from the postcephalic region. In fact, if one presses strongly and quickly on the abdomen of an anesthetized nymph, the mask extends completely. If the neck of an active nymph is tied off, the total extension of the mask is prevented. If the ligature is not too tight, however, a little hemolymph enters the head region and succeeds in progressively extending the mask after several attempts at predation, but the return of the mask and its folding under the head are prevented. Let us make clear that the neck ligature does not modify the elements initiating capture and does not prevent feeding (holding and chewing prey), but it does make it necessary for the experimenter to assist in the predation by placing the prey between the jaws when they are open.

THE TWO PHASES OF PROJECTION

All of this suggests that the extension of the mask is accomplished in two sequential phases, which is confirmed by analysis of cinematographic records of capture. The first phase consists of a release of the labial base brought about by muscles innervated by the suboesophageal ganglion. These are the muscles that contract and thereby prevent mask extension when pressure is applied to the abdomen of an unanesthetized nymph. The second phase consists of complete extension brought about by the increase in pressure resulting from the closing of the dorsal thoracico-abdominal blood vessel by synergistic contraction of various abdominal and diaphragmatic muscles. The retraction of the mask is effected in part by elastic rebound after the drop in blood pressure, and in part by active muscle contraction.

The total enactment of these phenomena by *Anax* or *Aeschnia* nymphs takes place in the following manner as shown by the analysis of a film (64 frames per second): The nymph about to make a capture fixates on the prey for about 1/20 of a second (3 frames of film), with the tip of the mask open, then in about 1/100 of a second the rear of the mask is loosened, followed immediately in the next 1/100 of a second by the total extension with capture. Finally, during eight frames of film (approximately 1/8 of a second), the mask returns to its original position. The entire sequence lasts about 1/5 of a second.

ANATOMY OF THE PROJECTION APPARATUS

A detailed account of the control of this behavior must take into consideration the anatomical apparatus described by Snodgrass (1954) and reviewed by Richard (1965b). Our experiments involving the destruction of muscles, of their cuticular attachments, or muscular denervation, show clearly the integrated roles of the rigidity of the labial base (hypopharyngeal apodem), of the coordinated contraction of the retractor muscles of the hypopharynx, as well as of the extensors and flexors of the prementum—all of which bring about the first phase of mask release. The second phase depends on the coordinated contraction of the abdominal and thoracic diaphragms, permitting the hemolymph to flow into the head.

Our morphological studies (Richard and LeBoëdec, in press) complement certain aspects of Büccholtz's (1961, 1964) research, permitting us to specify the zones of the suboesophageal ganglion involved in the control of muscles of the labial area for aeschnid nymphs, as Dethier (1960) did for *Phormia*. Other studies (Richard and Kouma, in press) following the work of Mill (1964) define the areas of neural innervation in the abdominal and thoracic diaphragms, ascertaining that the median nerve plays a role in this innervation, just as certain parts of the sympathetic system in the neck are involved in the innervation of the labial base.

THE ROLE OF THE ABDOMINAL NERVOUS SYSTEM

All these morphological signposts enable us to trace more clearly the pathway of physiological research related to the behavioral patterns we are discussing. We will be more specific after we have related some further complementary experimental results (see Figure 1).

A complete transection of two connections of the chain at a level posterior to the suboesophageal ganglia suppresses a full extension of the mask. Nevertheless, partial movements, especially the opening of the distal jaws in response to any stimulation (prey or model) may still be carried out. Lesions that isolate the thoracic region by transecting connections anterior to the suboesophageal ganglia result in disorganized motor activity (kicks, sudden extensions, incoordination of the legs). Lesions in the abdominal ganglia (particularly the fourth ganglion), such as the transection of the connectives, cause a severe loss in the capacity to

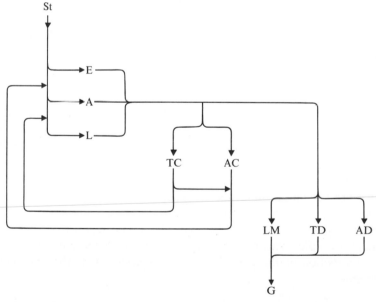

FIGURE 1

Control pattern of grasping behavior in Odonata: A chain including a loop. A: Antennae; AC: Abdominal locomotor coordination; AD: Abdominal diaphragms and muscles; E: Eyes; G: Grasping; L: Legs (particularly tarsus); LM: Leg muscles; St: Stimulus; TC: Thoracic locomotor coordination; TD: Thoracic diaphragms and muscles. (See pp. 443–444.)

project the mask, though some capacity may be slowly recovered (Richard, unpublished data). These findings are in perfect agreement with other morphological studies (Satija, 1958; Mill, 1964) and with electrophysiological studies. Burtt and Catton (1956), among others, have shown that in various insects (*Calliphora, Locusta, Aeschna*) illumination of the optic lobe is followed by, and coordinated with, the appearance of action potentials in the abdominal nerve cord. We think that these potentials directly or indirectly assume control of the contraction of the diaphragms involved in the projection of the mask. If this is indeed the case (and current research in the electrophysiology laboratory of our Institute should clarify this) we could discover some new aspects of the control of these behaviors.

OTHER PROBLEMS

In fact, these remarks raise other problems: one of these is related to the ontogeny of the nymphs. It seems to us that in Anisoptera, the entire neuromotor schema of control of the mask is present from the first nymphal instar, for in this stage we find all the labial elements and all the muscles cited above.

The other problem is related to the phylogeny of the Odonata. The mask apparatus, with some minor modifications, is found in all Anisoptera. The Zygoptera have no muscles or abdominal diaphragms that could be compared to those of the Anisoptera. Yet, the masks seem to be projected in the same way, mechanically speaking. Perhaps the thoracic diaphragms (a little more powerful than in the Anisoptera?) studied so thoroughly by Brocher (1917) in the context of respiratory functioning, play a role? The influence of the abdominal contraction itself could be involved. The analysis of the fast speed films that we are currently using in our laboratory will perhaps provide us with some clarification. In any case, we are keeping in mind the work of Mill (1964) which shows that the orientation of stretch receptors in the abdominal muscles of aeschnid nymphs changes at the level of the sixth abdominal segment. This suggests that there are differences between the contraction of the anterior part and in the contraction of the posterior part (respiratory) of the abdomen of the nymphs.

Regulation of Feeding Behavior as a Whole

TRANSECTION EXPERIMENTS

We will not stress the possible relationship between certain ganglionic structures of the head and the feeding behavior of our nymph since our studies are at present far from being as precise as those of Huber (1960). Simply to show the direction of our research we will point out that any important lesion at any level of the optic lobes of the aeschnid nymphs results in a severe deficit in the predatory capacity of these nymphs. Various lesions of the protocerebrum have increased the general activity level of the nymphs, stimulated strong but poorly directed releases of the mask at potential prey, and increased the duration of postfeeding antennal wiping.

But transections of the entire width of the cerebral ganglia at the level of the pars intercerebralis had extreme effects on the degree of excitation of the nymphs. Such individuals manifested an increase in the intensity of all their reactions: the legs exhibited certain partially uncoordinated jerking movements; stimulation of the antennae initiated a violent capture; stimulation of the tarsi a violent recoil. We may see this as a demonstration of the importance of the cerebral component in the control of the coordination of predatory behavior.

EFFECTS OF THE LOCATION OF CENTERS

We will review several other aspects of research on the components of control. The works of electrophysiologists (Autrum and Gallwitz, 1951; Burtt and Catton, 1956) and of morphologists (Viallanes, 1884; Richard, 1959) have shown all the changes in position and relative importance of the cephalic centers (particularly the optic lobe) in the course of the ontogeny of the Odonata nymphs. We will stress only two specific points on this subject:

1. The relative sizes and positions of various cellular and fibrous zones of the optic lobe are important in the structuring of visual control. We should note that the young nymph of *Aeschna* does not have the three classical optic zones until its third molt.

2. The relative importance of the protocerebrum (visual afferents) and of the deutocerebrum (antennal afferents) on the one hand, and the importance of the corpora pedunculata and the central body, on the other hand. These relationships vary considerably in the course of the maturation of the nymphs. Thus the antennal lobe of the Zygoptera nymphs is more developed than the optic lobe until quite an advanced stage of development, while in the Anisoptera nymphs, the growth of the optic lobe is greatly in advance of the antennal lobe.

All of this enables us to catch a glimpse of the morphological and physiological substrate of the behavioral patterns described above. For, as I have emphasized elsewhere, one cannot fail to be impressed by the parallelism between the variations in sensory control of predation and the variations in the morphology of the nervous system.

Alimentary Conditioning

At this point, we would like to describe certain other aspects of the integration of predatory behavior in the general life of the nymph. As we indicated above, and as it is easy to observe, the *Aeschna* or *Anax* nymphs are amenable to "being tamed." Their vision is very precise at great distances. The appearance of a human form in front of the tank in which we raise our nymphs in the laboratory elicits a quick orienting movement on their part and locomotion toward the form.

We wanted to understand certain aspects of this behavior better. We placed the nymphs in adjoining rectangular boxes on a table, taking the precaution of making the lateral walls of the box opaque, so that the nymphs could not see each other nor the experimenter approaching from above. Once a day, at the same time of day, the food (*Chironomus*) was placed in the same corner of the box by the assistant who always stationed herself in the same position outside the box. One person carried out these experiments.

At first the results were irregular. In a second phase of the experiment, we marked the floor of the corner of the box with a cross; in a third phase, we reinforced the marking with a series of vertical lines for several

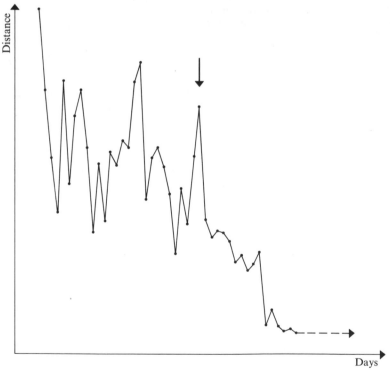

FIGURE 2

Variation of distance from stationary nymphs to marked corner in experimental boxes. The arrow indicates the day on which the corner was marked by straight vertical lines.

TABLE 2
Changes in the localization in the experimental box at time of molting: aeschnid nymph #27 (10th instar).

December 7:	
9 A.M.	Nymph in the cross[a]
10 A.M.	Nymph 5 cm away from cross. Does not move at time of feeding. Does not eat worms left in box.
2 P.M.	Molt. Nymph immobile on its exuvia 5 cm from the cross.
6 P.M.	Same position.
December 8:	
9 A.M.	Nymph in the cross.
10 A.M.	Feeding response: Captures worm.

[a] Nymphs were maintained in a box in which one corner was marked with a cross. Food was always placed in that corner.

centimeters around the corner. The curve drawn in Figure 2 shows that all the nymphs were soon found in the feeding corner.

A comparison of the activity cycles of the experimental and control nymphs shows great differences between them. The experimental nymphs are less active; the inactive period of the 24-hour cycle is longer; when they do move, they return to the marked area. For a few days before and after a molt, the nymphs exhibit a disinterest in feeding. At that time, as Table 2 shows, there is a loss and recovery of the localization response of the nymphs. Certain of our nymphs (those which always showed a sharp visual localization of the prey) molted three or four times, and the data for these animals were the same for each molt.

These experiments are in progress now and we will develop all the aspects of the results in another publication, but we wish to stress their value now as an analytical study of behavior. This type of conditioning allows us to ascertain easily two aspects of the biology of our insects:

1. The role of alimentary motivation in regulating the form and intensity of different modalities of predatory behavior as we have described above; and

2. The types of relationships between different behavioral patterns, as we have seen in the interaction of habituation of investigatory and wiping behavior. Perhaps we could even determine the effect of such relationships on the thresholds of different sensory receptors; e.g., the tactile organs on the tarsi which regulate contact with the substrate.

CONCLUSION

Throughout our approach, the precepts of Schneirla are invaluable. Like Schneirla, we refuse to study an isolated behavior item in an animal that is considered to be stabilized at the end of somatic development. On the contrary, we deem it necessary to continually compare the animal to itself at all stages of development. We consider all dynamic aspects equally worthy of study in our research into the causal mechanisms of behavioral regulation, but we consider the following to be the first and foremost among them: The reciprocal transition of approach ⇄ withdrawal in relation to the external environment; and the plasticity of these relationships in the course of conditioning or learning.

In all these investigations, the cooperation of the morphologist, ethologist, and electrophysiologist is of the greatest importance, and we have found it so in all our daily research activities. It is this alliance alone which will enable us to attain an objective understanding of the animals that surround us.

Translated by
MARTHA T. COLE AND ETHEL TOBACH

ACKNOWLEDGMENTS

I wish to thank Mme. Bernard, Assistant Technician, who carried out the tedious task described in the alimentary conditioning experiment with devotion and regularity and made precise protocols.

I am very grateful to Martha T. Cole and Ethel Tobach for the time and consideration they put into their perfect translation of my manuscript, and to Michael Boshes and Howard R. Topoff who assisted them in the vernacularization of certain technical terms.

REFERENCES

Amans, P. 1881. Recherches anatomiques et physiologiques sur la larve de l' *Aeschna grandis. Rev. Sci. Nat. Montpellier* 3 (1): 63–74.
Atwall, A. 1963. Biochemical basis for the food preference of a predatory beetle. *Current Science* 32 (11): 511–512.
Autrum, H., and U. Gallwitz. 1951. Zur Analyse der Belichtungspotentiale des Insektenauges. *Z. Vergleich. Physiol.* 33: 407–435.
Baldus, K. 1926. Experimentelle Untersuchungen über die Entfernungs Lokalisation der Libellen (*Aeschna cyanea*). *Z. Vergleich. Physiol.* 3: 475–505.
Brocher, F. 1917. Etude expérimentale sur le fonctionnement du vaisseau

dorsal et sur la circulation du sang chez les Insectes: les larves des Odonates. *Arch. Zool. Exp. Gén.* 56: 445–490.

Büccholtz, C. 1961. Eine Verhaltensphysiologische Analyse der Beutefanghandlung von *Calopteryx splendens* unter besonderer Berücksichtigung des optischen AAM nach partiellen Röntgenbestrahlungen des Protocerebrums. *Verhandl. Deut. Zool. Ges. Saarbrücken, Zool. Anj.* 25: 402–412.

Büccholtz, C. 1964. Elektronenmikroskopische Befunde am bestrahlten Oberschlundganglion von Odonaten-Larven (*Calopteryx splendens* Haar). *Z. Zellforsch.* 63: 1–21.

Burtt, E. T., and N. T. Catton. 1956. Electrical responses to visual stimulation in the optic lobes of the locust and certain other insects. *J. Physiol.* 133: 68–88.

Caillère, L. 1965. Description du reflexe de capture chez la larve d'*Agrion splenders* Haar. *Bull. Soc. Lin. Lyon* 34(10): 424–434.

Carthy, J. D. 1965. *The behavior of Arthropods.* San Francisco: W. H. Freeman and Co.

Chilova, An N. 1954. L'alimentation des larves de *Cricotopus silvestris* en rapport avec la structure de l'apparail buccal. *Compt. Rend. Acad. Sci. URSS Zool.* 100: 1191–1193.

Dadd, R. H. 1963. Feeding behaviour and nutrition in grasshoppers and locusts. *Advan. Insect Physiol.* 1: 47–110.

Dethier, V. G. 1960. The nerves and muscles of the proboscis of the bowfly *Phormia regina* in relation to feeding response. *Smithsonian Inst. Misc. Collections* 137: 157–173.

Etienne, A., and H. Howland. 1964. Elicitation of strikes of predatory insects to projected images and light spots. *Experientia* 20: 1–4.

Fleschner, C. A. 1950. Studies on searching capacity of the larvae of three predators of the citrus red mite. *Hilgardia* 20: 233–266.

Hassenstein, B. 1960. Die bisherige Rolle der Kybernetik in der Biologischen Forschung. *Naturwissenschaften* 13: 349–355.

Holst, E. von. 1950: Quantitative Messung von Stimmungen im Verhalten der Fische. *Symp. Soc. Exp. Biol.* no. 4, pp. 143–174.

Huber, F. 1960. Untersuchungen über die Funktion des Zentralnervensystems und insbesondere des Gehirns bei der Fortbewegung und der Lauterzeugung der Grillen. *Z. Vergleich. Physiol.* 44: 60–132.

Jourdheuil, P. 1963. L'appetence chez les Insectes. *Ann. Nutr. Aliment.* 17: 283–304.

Kraus, C. 1958. Die Stechakte von *Rhodnius prolixus. Acta Tropica* 15: 127–141.

Koëhler, O. 1924. Sinnesphysiologische Untersuchungen an Libellenlarven. *Verhandl. Deut. Zool. Ges.* 29: 83–91.

Ludtke, H. 1950. Die Funktion waagerecht liegender Augenteile der Rückenschwimmer. *Z. Vergleich. Physiol.* 22: 67–118.

Maier, N. R. F., and T. C. Schneirla. 1964. *Principles of animal psychology.* New York: Dover.

Mill, P. J. 1964. The structure of the abdominal nervous system of Aeschnid nymphs. *J. Comp. Neurol.* 122 (2): 157–158.

Mill, P. J. 1965. An anatomical study of the abdominal nervous and muscular systems of dragonfly nymphs. *Proc. Zool. Soc. London* 145 (1): 57–73.

Mittelstaedt, H. 1957. Control systems of orientation in insects. *Ann. Rev. Entomol.* 7: 177–198.

Ojai Symposium. 1964. *Neural theory and modeling.* R. F. Reiss, ed. Stanford, California: Stanford U. Press.

Pick, F. 1962. Sur la mise en évidence expérimentale du jeu des stylets des Réduvides hématophages, permettant la reconstitution du mécanisme de leur piqûre. *Ann. Parasitol.* 37: 326–337.

Popham, E. J. 1959. The anatomy in relation to feeding habits of *Forficula auricularia* and other Dermoptera. *Proc. Zool. Soc. Lond.* 133: 251–300.

Popham, E. J. 1961. The functional morphology of the mouth parts of the cockroach *Periplaneta americana. Entomologist* 94: 185–192.

Richard, G. 1960. Les bases sensorielles du comportement de capture des proies par diverses larves d'Odonates. *J. Psychol. Norm. Path.* 57: 95–107.

Richard, G. 1961. Ontogenèse du comportement chez diverses larves d'Odonates. *Congr. Psychol. Arcachon,* pp. 1–6.

Richard, G. 1965a. Contribution à l'étude des régulations du comportement alimentaire des larves d'Odonates. *Proc. XII Internatl. Congr. Entomol.,* London, pp. 321–323.

Richard, G. 1965b. La régulation nerveuse de la prise d'aliment chez l'Insecte. *J. Psychol. Norm. Path.* 62: 33–55.

Richard, G., and G. Gaudin. 1959. La morphologie du développement du système nerveux chez divers Insectes. Cas plus particulier des centres et des voies optiques. *The ontogeny of insects.* Acta Symposii de Evolutione Insectorum, Praha.

Richard, G., and Kouma, P. Le nerf médian chez quelques Insectes primitifs. *Bull. Soc. Sci. Bret.* (in press)

Richard G., and LeBoëdec, C. Les centres nerveux sousoesophagiens de commande du labium chez les Odonates. *Bull. Soc. Zool., France* (in press).

Roeder, K. D. 1963. *Nerve cells and insect behavior.* Cambridge, Mass.: Harvard Univ. Press.

Satija, R. C. 1958. A histological study of the brain and thoracic nerve cord of Aeschna nymph with special reference to the descending nervous pathways. *Res. Bull. Panjab Univ.* 138: 33–47.

Schneirla, T. C. 1957. The concept of development in comparative psychology. In D. B. Harris, ed., *The concept of development.* Minneapolis: Univ. Minnesota Press.

Schneirla, T. C. 1959. An evolutionary and developmental theory of biphasic processes underlying approach and withdrawal. In. M. R. Jones, ed., *Nebraska symposium on motivation.* Lincoln: Univ. Nebraska Press.

Schneirla, T. C. 1960. L'apprentissage chez la fourmi. Comparison avec le rat. *J. Psychol. Norm. Path.* 57: 47–94.

Schneirla, T. C., 1966. Aspects of stimulation and organization in approach/withdrawal processes underlying vertebrate behavioral development. In

D. S. Lehrman, R. A. Hinde, and E. Shaw, eds., *Advances in the study of behavior*. New York: Academic Press.

Snodgrass, R. E. 1954. The anatomy of Odonata. *Smithsonian Inst. Misc. Collections* 123: 1–36.

Thorsteinson, A. J. 1960. Host selection in phytophagous insects. *Ann. Rev. Entomol.* 5: 193–218.

Vasserot, J. 1957. Contribution à l'étude du comportement de capture des larves de l'Odonate *Calopteryx splendens*. *Vie et milieu* 8: 127–172.

Viallanes, H. 1884. Etudes biologiques et organologiques sur les centres nerveux et les organes de sens des animaux articulés: le ganglion optique de la Libellule *Aeschna maculatissima*. *Ann. Sci. Nat. Zool.* 18: 1–33.

Vogt, P. 1964. Ueber die optischen Schlusselreize beim Beuteerwerb der Larven der Libelle *Aeschna cyanea* Müll. *Zool. Jahr. Physiol.* 71: 171–180.

Vowles, D. M. 1961. Neural mechanisms in insect behavior. In W. H. Thorpe and O. L. Zangwill, eds., *Current problems in animal behavior*. New York: Cambridge Univ. Press, Pp. 5–29.

Vowles, D. M. 1964. Models and the insect brain. In R. F. Reiss, ed., *Neural theory and modeling*. Stanford, California: Stanford Univ. Press.

Willington, W. G. 1948. The light reaction of the spruce budworm *Choristoneura fumiferana*. *Can. Entomol.* 80: 56–82.

Wilson, D. M. 1965. The nervous coordination of insect locomotion. In J. E. Treherne and J. W. L. Beament, eds., *The physiology of the insect central nervous system*. New York: Academic Press.

EVELYN SHAW

Department of Animal Behavior
The American Museum of Natural History
New York

Schooling in Fishes: Critique and Review

"and the thousands of fishes moved as a huge beast,
piercing the water. They appeared united, inexorably
bound to a common fate. How comes this unity?"

Anonymous, 17th century

For about 40 years now the phenomenon of schooling in fishes has held the attention of investigators. Schooling has obvious importance for world food-supply problems as some four thousand species live in schools. But simply as a behavioral problem, schooling stands out as one of the few examples of vertebrate animal groupings having apparent synchrony of action among large numbers of its members. The fish appear to act in concert, moving forward simultaneously, keeping equal distances apart, all turning to swim in another direction at apparently the same moment.

This, at least, defines schooling for some investigators. But unfortunately there is no definition of schooling that is acceptable to all students of schooling, though all will agree that schooling is a social grouping. To some, defining a school is a trivial problem and is dismissed casually. To others, the definition itself serves as an approach to the experimental analysis of schooling behavior. Thus, an investigator who considers a school to be any fish grouping will ask questions different from those asked by the investigator who considers a school to be a fish grouping with specific qualities of orientation and organization. Therefore, because the definition of a school determines the direction of research, I wish first to present the views of a number of authors on "what is a school." Afterwards, I wish to discuss some of the current research into schooling behavior.

WHAT IS A SCHOOL?

The first to tackle the definition problem, albeit indirectly, was Parr (1927), whose main aim was to take the mysticism out of schooling and to explain the behavior on a firm biological basis. Even now, his paper, with its numerous ideas neither affirmed nor rejected, continues to stimulate readers. Parr pointed out that schooling had been interpreted as a social instinct "in an anthropomorphic way assuming the school to be the result of deliberate activities of the single individuals seeking the protection of great numbers, thus each acting with the school as a whole as their aim." Based essentially on the visual system, schooling behavior came about through "mutual attraction and adjustment of direction, with the latter checking the approach." These he considered "as automatic responses originating from a specialized tropism." However, with prescience for present problems in defining a school he pointed out that there were differences among certain fish groupings. Although the end result was the same, that is, a group formed, the forces which brought it about were based on significantly different biological factors. Parr essentially separated those fish that only occasionally school and those that school all the time. The first group was identified by their reaction to stimuli created by a temporary situation; that is, the gathering and dispersing of those schools seemed to be determined by immediate environmental factors. Parr's other group (which was the only one he considered in his paper) is the true schooler "characterized by great stability through the most varied of environmental conditions." Such constant stability suggested to Parr that the maintenance of these schools, e.g., mackerel, herring, is dominated by internal factors.

Actually Parr's researches were the first of a wave of studies that emanated from the laboratory of Allee and his students. Allee's group was generally investigating the bases of social behavior and, particularly, animal communities into which fish groupings comfortably fitted. Allee (1931) differentiated two main types of groupings. One type was the animal grouping that forms because each member was attracted to the same place because of its physical environmental conditions, e.g., light, temperature, to which each commonly responded. At the same time, they tolerated the proximity of the others. In this type, mutual attraction between species members played an insignificant role, if any at all. The second type of grouping was based on mutual attraction among species mates. This attraction had different degrees of intensity, some animals were strongly, moderately, or weakly attracted. In this social grouping, the physical environmental conditions played a secondary role, if any. Allee proposed an elaborate system of classification, listing more than 50 separate categories for almost any social situation and biological change during the life of the animal. Unfortunately, the categories proved unwieldy and unworkable.

Allee's work stimulated many analyses about the physiological advantages of community life, but up until 1946, schooling, as such, was not examined except for two papers, one by Spooner (1931) and the other by Shlaifer (1942). In 1946, Breder and Halpern presented some thoughts and experiments on aggregations in fishes. They pointed out that schooling is an extreme form of the aggregating tendency and that the distinguishing features of a school are that (a) all individuals are similarly oriented, (b) uniformly spaced, and (c) moving at a uniform pace. They recognized, however, that a definition with exactly defined limits was not possible because in many of the species studied there was considerable flux in orientation, spacing, and speed. At one time members of a group showed uniformity of spacing and orientation, and at another they showed randomness in their orientation, though remaining clustered. The point at which an aggregation became a school and a school changed into an aggregation remained yet to be established. Their important contribution was the statement that aggregation as a social grouping was based on mutual attraction of fish. They eliminated from their study of fish as social groupings any aggregations that occurred because the animals were attracted to a common region as a result of an external source of stimulation, other than their species types. And schooling, as an extreme form of aggregation, was given a characteristic geometry. Unfortunately, the word aggregation was never clearly defined.

In his review, Morrow (1948) slipped away from Breder's definition and considered aggregation to be "a chance grouping of individuals brought into a given locality by some external factor or factors *not concerned* with relationships between individuals, and schools as a closely knit cohesive group, in which there appears to be a definite centripetal influence existing between fish and fish." He eliminated specific spatial orientation as a principal criterion for schools. Atz (1953) defined schools in Breder's and Halpern's (1946) terms and agreed that the word aggregation should be reserved for fish showing mutual social attraction, but not polarized orientation. Keenleyside (1955) called a school "an aggregation formed when one fish reacts to one or more fish by staying near them. The chief factor . . . is a definite mutual attraction between individuals." Thus, Morrow considered an aggregation as an incidental coming together of fish while Keenleyside considered a school to be an aggregation based on mutual attraction of one fish for another.

Breder (1959) in a broad survey of social groupings retained his

earlier definition of schooling and aggregation. In each type of organization, aggregation or schooling, the fish are brought together on the basis
of mutual fish-to-fish attraction, essentially independent of environmental
circumstances. In aggregation, the geometric orientation is nonpolarized.
In schooling, it is polarized.

In 1964, Williams, in a paper on the evolution of schooling, embraced
Morrow's attitude when he referred to an aggregation as a group of
individuals independently seeking the same localized condition. The term
school he designated "to mean any contagious distribution that owes its
persistence to social (but not sexual) forces." A school may show no,
little, or high polarization. The only point of agreement among authors
is that a school is based on mutual attraction of fish.

In their incisive discussion of the biopsychology of social behavior,
Tobach and Schneirla (1968) made a number of comments that are of
pertinent interest here. Although they were not concerned with schooling,
per se, their categories are appropriate to this discussion. They subdivided
associations, the collection of two or more individuals, into two types of
aggregations—passive or active. Only the active aggregation concerns us
here. It is formed "as the result of responses of individual organisms to
stimulation from an external source." This association is generally the
result of taxis involving the movement of the organisms to a particular
place. The "external source," however, is defined to include "another animal" and thus could be called a biotaxic reaction. "Aggregations based
upon biotaxic responses may provide the setting for the reciprocal stimulation of the collected animals." Such animal group bonds Tobach and
Schneirla termed "biosocial." These animals are "capable of behavioral
modification within distinct limits: and their activities are determined by
structural and physiological factors." Schooling behavior, with its biotaxic
responses, therefore falls under their definition of aggregation.

Breder (1967) in his most recent paper makes another attempt to
define schooling. In many ways his earlier thinking (1946, 1959) on
schooling is reflected, only now he separates schooling fishes into facultative and obligate schoolers. The obligate schoolers "must be coherently
polarized; can only be forced to stop schooling momentarily, and then
only by means of considerable violence; and will not maintain a state of
random orientation." Facultative schoolers are not so clearly defined and
in effect are those schools that form as a result of environmental changes
and fright, but when the stimulus is removed the fish tend to become
nonpolarized and/or to disperse. According to Breder, in obligate schoolers, "the drive to associate with others in a body of great unanimity of
orientation is clearly a positive matter of great strength . . . unlike the
fragile schools of fright or other temporary mutual orientation. . . ."

Breder's definition makes a positive contribution to this apparently semantic argument about schooling by providing a means to distinguish fish that school only occasionally from those that school virtually all the time. Such a definition is useful especially for the phenomenon of fish that suddenly switch back and forth from a polarized school to a non-polarized, aggregated group.

In sum, all authors agree that schooling is a social grouping that is based on biosocial mutual attraction. Disagreements revolve around two main points: (1) should the term aggregation be used for fish that are mutually attracted, or should it be restricted only to associations that are formed when an external stimulus (other than conspecifics) attracts animals to a common location; and (2) should a distinction be made between biosocial groupings that are polarized as opposed to others that are nonpolarized. If no distinction is made then perhaps all fish groupings founded on a biologically based mutual attraction can be called a school. Our inability to find a universally acceptable definition of schooling for the many thousands of species makes us hesitate in assuming biological homogeneity for so many diverse groups. After all, the end product, the school, may have similar appearances but possibly quite different causal factors. Breder, in fact, essentially made a distinction between different causal factors when he discussed obligate and facultative schoolers.

By examining the possible biological causal mechanisms underlying aggregation and schooling we see why a distinction must be drawn between these two forms of behavior. Biosocial attraction I assign to a first order of importance, and polarization or nonpolarization, which occurs after biosocial attraction, to a second order of importance. I think, in time, as more work is carried out, we will find that these two orders are real, and in fact are correlated with different brain mechanisms. Certainly, the researches into forebrain function in fishes point in this direction, although the descriptions of behavior are too vague to permit drawing positive conclusions. When Janzen (1933) removed the forebrain in *Phoxinus,* he stated that "initiative" was lost. Berwein (1941) found that forebrainless schooling minnows readily accepted a strange fish, quite the opposite of their behavior with the forebrain intact. Aronson and Kaplan (1968) in a review of forebrain function emphasize the idea that the forebrain acts as a regulatory or arousal mechanism, which, by neural processes of excitation or inhibition, facilitates responses in the lower centers. Many ablation studies reveal a decline or an altering of behavioral patterns. Such motor patterns, however, are not eliminated but, on the contrary, may even be strengthened. Janzen demonstrated this by testing forebrainless fish with an optomotor apparatus (optomotor response to be discussed later) in which they showed constant and consistent swimming behavior. Evidently it was almost as if the fish and the rotating drum were synchronized in speed and direction of movement. He pointed out that intact fish often carried out flight reactions and showed erratic behavior.

Similar behavioral results were obtained in our laboratory after *Tilapia macrocephala* were treated with a tranquilizer known as Frenquel. Prior to treatment, the fish darted wildly around the test chamber and it took considerable time before they began to swim in the direction of moving stripes. After treatment, they showed no erratic darting behavior, but responded directly to the moving stripes of the optomotor apparatus by swimming smoothly and at the same speed as the stripes. One can see that the forebrain, serving as an excitatory or arousal mechanism, could function to sensitize the animal to many stimuli, including those stimuli concerned with species discrimination and subsequent attraction. With the loss of the forebrain the fish could lose its sensitivity to these stimuli, but not its orientative capacity, nor its visuomotor coordination, which is located in the mid- and the hindbrain. The tranquilizer eliminated sensitivity to external cues (the aquarium conditions may not have been "comfortable") and depressed the erratic, darting "escape" reactions. Lower center coordination could then be displayed in its almost reflex-like characteristics.

By evaluating the various studies of schooling according to this two-order system we see that many investigators are concerned with the first order, biosocial attraction, and others primarily with the second, polarization, which is parallel orientation. Breder's recent subdivision of schooling fish into facultative and obligate schoolers fits nicely into a two-stage neural system: the *facultative schoolers function mostly at the first order* and occasionally move into the second, whereas the *obligate schoolers function at the first and second orders* at all times.

SENSORY COMPONENTS

Parr (1927) suggested that the entire schooling mechanism, both approach and adjustment of position in parallel orientation, could be mediated through the visual system. His own studies, and those of other researchers, with blinded fish showed that the visual system was unequivocally linked to schooling, for without vision no schooling occurred (see reviews by Morrow, 1948; Atz, 1953; Breder, 1959; and Shaw, 1962). Parr said that when visual perception of a companion is the stimulus for an approach by a fish to its companion, this very same visual stimulus could also check the approach as the fish come closer to each other. However, Parr recognized that the approach could also be checked by other stimuli, for example, the smells or sounds produced by the fish. If we examine schooling as the suggested two-order system, we see that Parr was concerned here with the first order, that of biosocial attraction with a subsequent approach.

It is difficult to separate the attraction response from the approach response. Indeed, in our present state of knowledge we know that an interaction occurred only through manifestation of the approach. Furthermore, attraction as such may be a poor designation for a phenomenon involving species discrimination based on a number of sensory cues. Visual stimuli give rise to an approach, and as the fish swim closer together these same stimuli, now of increasing intensity, force a withdrawal or checking of the approach. In some studies (Keenleyside, 1955), vision plays a role in attraction and approach without, however, sustaining the approach. In his comprehensive paper dealing with several species that fit into Breder's category of facultative schoolers, he attempted to define visual cues that attract schooling fish. Visual stimuli alone were shown to be essential to the approach but were not enough to keep a single fish near its group. The fish—separated by glass from the group—at first responded and approached quickly but then "moved away more and more frequently." Even though larger numbers and larger size provided a stronger approach stimulus, they, too, were eventually inadequate to keep the individual fish nearby. From results on other experiments he suggested that olfaction may be an additional essential, at least in *Scardinius*.

Hemmings (1966b) in a recent study examined the responses of freely swimming *Rutilis rutilus, Mugil chelo,* and *Chromis chromis* to three visual stimuli: (1) the presence of a conspecific separated by a plastic barrier (caged fish), (2) a mirror image of itself, and (3) a freely swimming species mate. He found positive attraction under all three conditions although the mirror image did not prove to be as strong a stimulus as the other two. The two freely swimming fish tended to stay close together, whereas the freely swimming fish did not remain close to the caged fish. In experiments with olfactory stimuli he mentions a short-term decline in responsiveness, but in his visual stimuli studies he does not discuss whether or not there was a decline in the time the freely swimming fish remained near caged fish or the mirror.

The recent experiments confirmed that vision was foremost in attraction-approach but they did not reveal whether or not additional cues, mediated through other sensory systems, are employed. The major difficulty in interpreting some results is that experienced fish can react to familiar conditions, even when certain stimuli are absent: a fish with plugged nostrils may approach a species mate, even though the olfactory cues used for finer discrimination are lacking. Not being able to see, of course, is a severe deprivation but, even then, other cues may cause a general increase in excitatory level giving rise to an occasional approach and orientation that is not sustained. For example, blinded *Menidia menidia* and

Alosa pseudoharengus placed into appropriate schools swam parallel within the school for several seconds, lost the school, but swam again with it for several seconds when the school returned to their vicinity. Although we cannot rule out a totally random meeting here, we also cannot overlook the possibility that other cues brought on these moments of schooling.

On the other hand, some experiments (for example those carried out by Keenleyside and Hemmings) severely restrict the orientative capability of the enclosed fish. It is apparent to the observer of schooling that fish constantly make spatial adjustments within a school and that feedback from the position of other fish plays a major role in coordinated participation. With continuous "give and take" occurring between and among fish, parallel orientation requires constant adjustment of position. Thus, when fish were enclosed in a box, as in the experiments of Hemmings and Keenleyside, they were limited spatially and unable to make orientative adjustments. Thus, even though the freely swimming fish oriented, it did not sustain the orientation. In point of fact, other experiments with fish separated by a long partition, but left free to swim freely on each side, showed that when parallel orientation was established the fish remained near one another for hours or days. Each fish could adjust to positional changes in the other fish. Shlaifer (1942) found that 2 mackerel *Pneumatophorus grex* showed parallel orientation despite their separation by a glass partition. In some of my studies with carangids, I observed that individual *Caranx hippos* continually maintained parallel orientation despite separation by a distortion-free transparent or translucent partition (Shaw, 1969). These fish responded regularly and interchangeably to each other's changes in direction and in speed, each approaching very closely to the physical barrier and swimming its length almost touching the sides. Moreover, the barrier was only ¼" thick, which meant that the fish swam much more closely than they would under normal schooling conditions. Recently, Cahn (1967) noted that when *Euthynnus affinis* were separated by a long transparent partition, schooling orientation lasted for several days if no major swimming speed adjustments were required. Thus, even though the cues were transmitted only visually, schooling behavior continued as long as parallel orientation could be maintained.

An interesting point emerging from these studies relates to Parr's hypothesis of schooling. He suggested that visual stimuli checked the approach and helped maintain fish spacing. If the size of retinal image were critical, then I believe that the fish would not have oriented so closely to the transparent partition. It appears that the visual system functions in the approach mechanism but that other systems may function in a withdrawal mechanism, keeping fish spacing within maximum and minimum limits.

Visual stimuli must also operate in alerting fish to changes in the velocity of their neighbors and in choosing a particular position, at the head, middle, rear, side, or anywhere in the school. These, however, are probably generalized stimuli rather than highly specific cues.

Shaw and Tucker (1965) noted similar behavior between fish (*Caranx ruber* and *Selar crumenophthalmus*) in a school and fish in an optomotor apparatus. The stimuli in the apparatus were a series of vertical dark-light contrasts. They suggested that the fish's optomotor response was a mechanism that maintained parallel orientation under changing spatial and velocity conditions. In their discussion they pointed out that in an optomotor response, (1) the fish moves in the same direction as and parallel to the moving field of stripes, (2) the fish adjusts its speed to that of the moving field, and (3) the fish can swim along with any part of the moving field. The optomotor response continues while fish are changing velocity and position and operates no matter what the position of the individual within the school. The similarity between this behavior and certain features of schooling is striking: (1) fish in schools move in the same direction, (2) a fish adjusts its speed to that of the surrounding fish, and (3) a fish changes its position relative to other fish, but nonetheless remains parallel to the school. Shaw and Tucker (1965) suggested that fish in a school may be providing, for other fish, a series of successive dark-light contrasts that increase sharply as the fish move. In support of this Baylor and Shaw (1962) found many schooling species had laterally far-sighted eyes that may be well suited for picking up the slightest movement of these dark-light contrasts.

In continued studies on optomotor response I have replicated the results of the *Caranx* studies with *Menidia menidia* and *Atherina mochon*. In addition to the standard stimulus, namely alternating vertical black and white stripes, different patterns were placed horizontally or vertically onto one small area of the rotating drum. The most striking result was the orientation of the fish, swimming most frequently slightly in front of the moving pattern. So consistent was this that when a fish swam too far ahead of the stimulus, it would slow down to allow the stimulus to catch up. (The fish did not appear to be fleeing from the moving stimulus.) This suggests that in a school, fish may be responding to movement coming from behind them. It could also explain why fish at the leading edge continue their forward movement and consequently provide the onward thrust of the school itself—without themselves having another fish to follow.

Before leaving the discussion of the optomotor response I want to mention a few more points. As discussed earlier this response improves in *Phoxinus* when the forebrain is removed (Janzen, 1933) and in *Tilapia* when treated with a tranquilizer. Under normal conditions fish may show erratic behavior, darting around the test container or swimming actively up and down, which some investigators call flight behavior. However, after treatment with Frenquel, the erratic behavior disappears and the fish and drum rotations seem to synchronize. These findings suggest again that schooling is based on two orders in the brain: the optomotor response and parallel orientation on the lower order, mutual attraction and its associated stimuli on a higher one. When the forebrain function is either eliminated or probably depressed under tranquilizer treatment, sensitivity to external cues is reduced. Perhaps there is an increase in the arousal threshold or loss of a regulatory mechanism of the forebrain and, subsequently, responses seated in another part of the brain become dominant. Although an optomotor response is elicited from nonschooling fish, this does not negate its role in schooling behavior after the school is formed. First of all, there must be mutual attraction-approach, based on complex stimuli. Then, once the approaches are made, the optomotor response could be employed to maintain the classic parallel orientation of schooling fish.

Going on now with our discussion of sensory components, we will now take up another important system in schooling, that of the lateral line. Though it is hard to see any connection between the lateral line system and biosocial attraction, it is easy to see how it may function in parallel orientation, that is, in helping to keep the fish polarized and in fish-to-fish spacing in a school (see comments above). Dijkgraaf (1963) particularly stressed the role of the lateral line in perceiving water pressure changes and suggested that it could easily serve in keeping contact and distances between school members. A fish could possibly determine the distance of another fish nearby, because of changes in water displacements caused by the presence and/or movement of the other fish.

In addition, Harris and van Bergeijk (1962) showed that neuromasts within the head canals of the lateral line are sensitive to "near-field" water displacements. This sensitivity may be employed in parallel orientation as well as in achieving the fine spatial adjustments fish make to one another. Unfortunately, there have been no definitive experiments to demonstrate that fish spacing in a school is a direct result of the stimuli received by lateral line end organs. An indication of the complexity of this system is found in its anatomical qualities: the positions of neuromasts along the head and body walls, the pits and canals, enclosed or non-enclosed neuromasts, the distribution of hair cells and kino-cilia within the neuromasts, all influence the fish's sensitivity to the water displacements coming from different directions. A good model of the lateral line, utilizing our present anatomical findings (Wersäll and Flock, 1965) and physiological knowledge (Kuiper, 1956) would give schooling students considerable help in understanding how it functions in fish spacing and orientation.

In other sensory studies, the olfactory system has stimulated interest, but its role in schooling remains unresolved. One very important consideration is the odor gradient, another is previous experience with the odorous substances. It seems logical that odors come into play during attraction-approach; whether or not they are utilized in parallel orientation, however, is highly conjectural. What is called "alarm substance" makes this matter all the more complicated. In some species skin derivatives from the bodies of conspecies are believed to trigger an "alarm reaction" (Pfeiffer, 1962; Thines and Vandenbussche, 1966) in which the group of fish draws away from the skin derivative and subsequently clusters with greater cohesiveness. On the other hand, Tester (1959), Rosenblatt and Losey (1967), and Hemmings (1966) found that these skin derivatives stimulated approach to the substance instead of withdrawal. Unfortunately, the concentration of the odorous substances in the various tests had not been quantitatively determined. Here again we may be dealing with an approach-withdrawal system with odorous stimuli of mild intensity causing approach and that of stronger intensity, withdrawal. Also the fish's previous experiences with the odorous substances cannot be overlooked.

Auditory stimuli have not been shown to operate in biosocial attraction-approach, parallel orientation, or maintenance of the group during resting periods. Moulton (1960) was able to record sounds produced by schooling fish, but these came from swimming movements. Breder (1967) does not believe that auditory communication is a critical aspect of school maintenance and, in fact, points out that schooling fish tend to be quiet, almost noiseless. However, a number of schooling fishes produce sounds at night (Hobson, 1968). These sounds may serve to keep the fish together in the dark when visual references are less accessible.

The recent studies examining the sensory components of schooling behavior demonstrate that vision is the primary sensory modality in attraction-approach. Though it may play a primary role in parallel orientation, it probably plays a lesser one in fish spacing. Other sensory systems, acting in conjunction with the visual one, are integrated to give the fish adequate information for species discrimination, attraction-approach, parallel orientation, fish-to-fish distances and general school maintenance. We have yet to discover the way they are integrated.

THE GEOMETRY OF SCHOOLS
AND ITS MODIFICATIONS

The consistency of spacing between the members of a school created considerable speculation on how fish establish and maintain it. As mentioned earlier, water displacement cues received by the lateral line and visual cues serve in determining the intraschool distances taken by the fish. However, only recently have authors attempted to analyze, quantitatively, fish spacing in a school and, concomitantly, other aspects of school geometry. Each author had to decide what measurements would be most informative, simple to carry out, and applicable to a variety of species.

Breder (1954) proposed a set of equations describing fish-to-fish spacing. He also made measurements of fish-to-fish distances taken from photographs. These photographs were taken from above and the fish appeared to be on the same horizontal plane. Schools are three-dimensional, however, and even a relatively large difference in the height of two fishes from the bottom of a tank or sandy bay is not reflected on a horizontal plane. Keenleyside (1955) measured density of fish in a school within specified areas, taking account of the three-dimensional structure of schools but not spatial orientation, since he did not consider parallel orientation to be a major factor. Williams (1964), John (1964), and Hunter (1966) also measured school structure, but as a two-dimensional system on a horizontal plane. Williams, in measuring space occupied by a school by determining the smallest polygon that encloses a school, did not concern himself with either orientation or fish-to-fish distances. John developed a schooling index by which he considered fish to be schooled if they swam in the same direction and the distance between individuals was less than 4 times their length.

Hunter calculated mean separation distance, mean distance to the nearest neighbor, and mean angular deviation. He pointed out that separation distance to nearest neighbor is a competent parameter and that "for a relative measure of changes in spacing of an entire school, separation distance is preferable, but for a comparison of samples containing different numbers of individuals . . . distance to nearest neighbor must be used." As an example of how these measurements are applied and can give specific and reproducible data, Hunter (1968) measured schools of jack mackerel under various light intensities. He found that mean angular deviation dropped from 60° at 6×10^{-7} foot lamberts to approximately 25° at 6×10^{-6} foot lamberts and that mean separation distance was significantly reduced when the light increased from 6×10^{-6}, as was mean distance to nearest neighbor.

Other students of schooling (Dambach, 1963, and Cullen, Shaw, and Baldwin, 1965) have examined school geometry as a three-dimensional structure. Dambach reported a way of determining the depth of fish in water from vertical photographs. This was essentially the same as the shadow method developed by Cullen, Shaw, and Baldwin, who developed two techniques to study three-dimensional schools, one utilizing stereophotography and the other utilizing the shadow cast by individual fish. From photographs of schools they measured: (1) the distance between the head of each fish and the head of the nearest neighbor in whatever direction, (2) the angular heading of the fish relative to each other, that is, how closely the fish approached parallel orientation, (3) the compactness or dispersion of the school and its shape, that is, whether longer than broad, etc., and (4) the space (volume of water) occupied by the school. The results of their study indicated that fish are packed together with equal density in all planes and that they tend to favor diagonal positions to each other over positions in front, behind, or directly abeam. "It is as if there were a roughly spherical bubble around the head of each fish which is maintained in contact at some point with the bubble of another fish." The diameter of this bubble (for *Harengula*) would be 3–6 cm, so that much of the body and tail of the fish would project behind it (Cullen, Shaw, and Baldwin, 1965). The spatial preferences corresponded to the observations and analyses of Breder (1965) in his study of fish vortices and schools. He proposed that vortices "cast off by each fish" are critical for fish spacing. Probably the most uncomfortable place for a fish to be would be directly behind the tail of another fish, since it would be constantly buffeted by the turbulence from the other fish's tailbeats.

One interesting sidelight is the way these techniques have been used to indicate biological changes during development. Dambach discovered that as *Tilapia macrocephala* grew, their fish-to-fish distances tended to enlarge. However unknown might be the biological factors that contribute to these changes, here at least we have a precise indicator that will make inferences about the sensory system more realistic and testable.

One of the greatest problems faced by Cullen, Shaw, and Baldwin and by Hunter was the large amount of time required to take measurements and the large amount of data that had to be processed. These techniques are slow and unwieldy, and at this time are ineffective in the field. We hope that with scanning devices, which give spatial location of individual fish, and automatic computation, investigators will be able to examine schooling under a variety of environmental and developmental situations and predict natural behavior at sea. For example, exactly how is school organization and geometry altered during development? Do the size of the fish and the number in a school dictate spacing? How consistent is the spacing and general organization among different species and what influence does the external environment and the physiological condition of the fish have on school geometry? Some recent papers, for example, show that the physiological state of fish within a school apparently affects the geometry as well as the movement and swimming direction of the school.

Thines and Vandenbussche (1966) in a paper on activity rhythms suggested that instead of assuming that schooling fish tend to disperse at night only because "visual information progressively diminishes," one should examine the possibility "that dispersion at night may also be partly due to a decline in the responsiveness to external stimuli." They observed schooling among *Rasbora hetermorpha* under the influence of an alarm substance (a skin derivative) and under ordinary aquarium conditions, both day and night. Introduction of the alarm substance into the tank resulted in increased cohesion, a more intense bunching of the fish. The greatest cohesion was obtained during the day under bright lights. However, under the same bright light conditions, at night, school cohesiveness was looser than in the daytime, but was still much greater than that of fish in the dark. A change in the fish reactivity to the alarm substance was apparent. This change suggests that schooling has a diurnal rhythm and that dispersal at night is not entirely dependent on the dark. It may also result from a lessened responsiveness to external cues.

One serious defect of this study is the assumption that increased school cohesiveness, stimulated by the alarm substance, is necessarily related to the biological factors that produce schooling behavior under natural conditions. Also, it may not be possible to bring about schooling formation when a bright light suddenly appears in the middle of a normally dark night; so intense a stimulus might inhibit rather than encourage increased cohesiveness. Despite these shortcomings, the authors have opened an interesting avenue of research in the study of schooling, namely, that fish-to-fish distances and the over-all behavior of a school may have a basic circadian rhythm, reinforced or modified by changes in the external environment.

In another paper on activity rhythms Darnell and Meierotto (1965) investigated 24-hour activity pattern of black bullheads, *Ictalurus melas*. They found that the young swim in schools during daylight with a decrease in activity occurring around midday and at night. The causes for the decrease in midday or night activity are unknown and there is no conclusive evidence that it is due to a change in light intensity.

Other researches into the physiological condition of school members and its effect on school organization have been carried out by Keenleyside (1955), Magnuson (1964), and Hunter (1966, 1968). Keenleyside found that sticklebacks are more densely packed in a school when well fed and tend to disperse when starved. Magnuson (1964) also noted that *Euthynnus yaito* tended to swim more slowly and showed less orientation when hungry than when fed, and Hunter (1966) demonstrated that distances among jack mackerel tended to increase with food deprivation. Hunter's results were particularly informative since he made a quantitative analysis of changes in school structure.

McFarland and Moss (1967) addressed themselves to the problem of changes in total school geometry, rates of movement and swimming direction of the school. Not all directional shifts of a school can be attributed to obvious changes in the external environment: many other factors, such as temperature, salinity, water currents, etc., also have their effect. McFarland and Moss proposed that respiratory gases may bring about certain changes in individual fish and consequently in the behavior of the entire school. They suggested that "reduction of dissolved oxygen and increase of carbon dioxide from school metabolism may be sensed by individual fish." This detection of altered environmental gas concentration could induce modified behavior such as change in direction, spacing, and swimming velocity. They stated that "field data provide positive correlation between oxygen gradients within schools and drastic modifications in school structure." Among *Mugil cephalus,* "the front half and sides of the school were dense and highly polarized, . . . the center . . . less polarized . . . (and the rear) . . . completely unpolarized with fish actively roiling in the surface." Oxygen measurements showed greatest reduction of oxygen at the rear. The authors' reference to small, medium, and large schools is unclear, as is the effect of the observers on the school. If the investigator placed an electrode into a densely packed school, would not the school show "roiling" behavior out of fright, dispersing into several smaller schools, and could not oxygen depletion be the result of increased respiration due to fright? This author believes that the investigative methods influenced the data obtained. If greater controls had been taken to eliminate human interference, the results would be more convincing.

SYNCHRONOUS BEHAVIOR OF SCHOOLS

A most striking feature of fish schools is their apparent synchrony of action. How are changes in direction, speed, positional orientation in one fish communicated to other fish? By what means do the fish maintain the integrity of the school? Many students of schooling have noted that in schools of 50 or more fish, several fish will break away in different directions without in the least altering the course of the school, then within seconds they will reunite. Under certain conditions, however, a similar minority can change the direction of the entire school. Many field investigators, now taking to the depths with SCUBA and other underwater observational devices, have pointed up the variety of behavior shown by schools in reaction to predators. Radakov (1968) reported some behavioral changes occurring in schools, 3–4 meters in length, when approached by a predator. Flurries of activity first appeared around the edge of the school; then if penetrated, the school would form a hole or an open space around the predator. In a recent paper Kühlmann and Karst (1967) report their observations on *Ammodytes,* which form highly cohesive schools during part of the day. They noted that instead of fleeing, these fish, too, form a space around a predator, keeping a certain distance away; and schools that were forcibly divided into several parts attempted to reunite. On the other hand, observations made at Sea Lab II (Clarke, Flechsig and Grigg, 1967) indicated that jack mackerel (*Trachurus symmetricus*) react differently to predators. Only the individual fish reacted when attacked by a scorpion fish or a rockfish, whereas the entire group schooled and dashed off when an individual fish was attacked by sealions or bonitos. White croakers (*Genyonemus liniatus*) near the jack mackerel also moved off as a school for a short distance with the jack mackerel, although they were not being preyed upon. The authors do not say whether the fish returned to the vicinity of Sea Lab II. Von Wahlert (1963) also noted that fish (not identified) tend to draw more closely together in the presence of a predator and leave a hole or an open space around it. Motion pictures projected as a single frame series offer the most direct means for measuring the latency of response in orientative and speed adjustments. Motion pictures taken by this author of a school of *Harengula* sp. in the waters of Bermuda show that the entire school changed to a direction perpendicular to its original course. Single frame analysis showed that this change occurred in pockets forming within the school. Each pocket or group was made up of three or four fish turning 90° as a unit. Three such groups formed before the entire school turned. Interestingly, the groups or pockets were inside the school, not at the edge; they were not contiguous, but disparate. Once the pockets had turned the remainder of the school turned almost simultaneously. The entire action occurred in less than one-half second, i.e., by the 11th frame of a series taken at 24 frames per minute.

Radakov (1966) has been interested in the speed with which a "wave of disturbance" progresses through a school and also the number of fish that change direction before the entire school follows. He reported that when a school of *Atherinomorus stipes* was exposed to frightening stimuli, the fish separated into two zones. The fish in the first zone reacted directly to the stimulus and the fish in the second zone reacted later to the behavioral changes of fish of the first zone. Radakov pointed out that when undisturbed fish come across a "wave of disturbance," they alter their swimming course sharply, and as "the wave increases in size, the speed of fish movement decreases." The "wave of disturbance" moved at about 8 m/sec, greater than the speed of movement of individual fish. Radakov reports that the wave attenuates as it travels farther away from the point of stimulation or as it meets fish going in an opposing direction. Acting in an opposite fashion is the wave that spreads through a school when some of the fish begin to feed. Instead of attenuating, this wave builds up.

We do not yet know the degree to which the behavior of any fish affects the behavior of the entire school. Indeed, we do not even know whether a change in orientation of one or several fish causes a direct change in the others. Several fish may respond independently to an environmental stimulus sooner than others and, as a result, change individual orientation sooner. Though reorientation of the entire school could be dependent on each fish's responding individually to the external stimulus, this is unlikely and would be inefficient for the school as a whole. Each fish is a discrete unit and all the fish around it are external stimuli. It is distinctly advantageous in school living to react to neighboring fish without perceiving the original source of stimulation, such as, for example, a predator swooping down from above.

In all likelihood this is what occurs under some conditions. A small unit of fish (within a large school) reacts to an outside stimulus and the unit changes direction or reorients. The main body becomes alerted to the reorientation and reacts to it, initially by changing direction or reorienting. The outside stimulus may afterwards be perceived by the main body and function in reinforcing the reorientation.

One often mentioned and essential feature of schooling concerns minor changes in orientation and speed adjustment by individual fish and by small groups within a large school. In addition to maintaining the general cohesiveness of the school, these changes may serve in other beneficial ways. Each fish with its constant small spatial adjustments acts as a scanner of its environment, and, conjecturing further, these small movements, in addition to enabling the fish to scan a maximal environment, could very well heighten its sensory perception. With each fish acting as a scanner, then, the entire school becomes a multiple scanning unit, extending the perception of any one of its members. For example, an individual located in a remote part of the school, too distant from a food source or danger to perceive this stimulus itself, would, through its scanning movements, pick up the "message" from the movements of its school members.

ADAPTIVE FEATURES OF SCHOOLING

The adaptive value of schooling has been examined and debated by a number of authors starting with Parr (1927) and continuing, most recently, with Breder (1967). Points of view on this issue have ranged from reproductive facilitation, to protection from predators, to physiological advantages. For example, in goldfish, the white rat of fish studies, Allee (1931) found that poisons were less detrimental to fish in groups than to fish in isolation. Also, he showed that oxygen consumption per fish was lower among fish living in groups than in isolation, though he did not evaluate the effects of stress on the fish from being removed from its normal social environment. Higher oxygen consumption probably was due to increased motor activity of the disturbed fish. In point of fact, Alekseeva (1963) found a higher oxygen consumption rate when individual fish were visually isolated from conspecies but when individual fish were not visually isolated he obtained comparable rates in individuals and groups.

Social facilitation has been suggested as an adaptive advantage to group living and to schooling behavior. Hale (1956) working with the sunfish *Lepomis,* a facultative schooler, found that in a maze they performed better as a group and he suggested that social facilitation "appeared to be primarily a type of mutual stimulation." O'Connell (1960) investigated the conditioning capacity of a school by first conditioning a school of 21 adult sardines, *Sardinopis caerulea,* to approach a food source. Later, over a number of days, he replaced 41 percent of the school with unconditioned fish, which he found had no effect on the latency of the conditioned response. It would have been interesting to carry these experiments further to determine what replacement percentage would finally affect the latency of response.

Hunter and Wisby (1964) found that groups of carp trained to avoid a moving net were more successful than similarly trained isolated carp. Also, untrained individuals placed with trained groups did not show improved performance (lower latencies of escape) until they had spent some time with the group. Their experience with the trained group did not affect the scores of these fish when tested again as individuals. These scores, in fact, revealed higher individual escape latencies. Later, however, when they were replaced together as a group, their escape latencies decreased.

Milanovskii and Rekubratskii (1960) point out that though various schools may have differing adaptive values, they all have in common the feature that the school will respond as an entity "to any change in the relationship with the environment." These authors believe that schools may modify their behavior through learning. This they demonstrated in a study of a single school which ultimately divided into three schools because of a disturbing stimulus. The school (species not named by the authors) consisted of small, medium, and large fish. When mildly startled the largest minnows swam away, and finally only when the observer's hand was immersed, the smallest minnows swam away. Learning, they suggested, occurs as the fish grow, the younger fish not being as receptive to potential danger as the older.

Actually these results can be interpreted differently, i.e., that both biological and experiential changes affecting the sensory thresholds of the larger fish enabled them to respond to minimum cues, whereas the smaller fish could only respond to comparatively intense cues. Evidence, however, is generally to the contrary as learning is known to occur in all fish under natural conditions and, probably, at a faster rate than under laboratory conditions. The fish that survived to the larger size had probably learned to withdraw from even a moderately disturbing stimulus (moderate to the investigator), though at earlier stages of development the stimulus had to be of greater intensity to cause a withdrawal.

Manteifel and Radakov (1960) pointed out that the adaptive value of schooling varies at different ages and stages of the life cycle and at different times of the day. They proposed four adaptive explanations for schooling: protection, food search, migration, and spawning. They gave a number of personal observations on how schooling protects the entire group from predation. Experimentally, they also showed that when one part of the school saw food and rushed to it, this movement attracted the other fish in the school even though the remainder could not see the food. Thus schooling helped all the fish to find food. The authors did not treat the other two adaptive explanations in detail.

Brock and Riffenburgh (1960) postulated that schooling is a mechanism that assures the survival of the school itself even when a number of members are lost to predators. Through manipulation of numbers of prey and the search areas of predators, they showed mathematically that schools above certain numbers could survive an attack that resulted in great losses.

Such a hypothesis is highly plausible, but hardly sufficient to explain the behavior other authors have observed between predator and prey schools. For example, Clarke, Flechsig, and Grigg (1967), making observation from Sea Lab II, found that jack mackerel, *Trachurus symmetricus,* reacted in different ways to different predators. When attacked by sealions and bonitos, common predators, they dashed off as a unit. When attacked by scorpion fish or rockfish, not common predators, only the attacked individuals reacted. The authors noted that the bonito and sealion attacked from above, the others from below. I have noted that mackerel, which school below silver-sides, occasionally swim up, select an individual, and consume it without ruffling the school. Predator-prey relationships and the effects of the predator on the prey population appear to be species-typical: evidently each type of predator has its own mode of interacting with each type of prey. These interactions can be modified for example through experience, general environmental conditions, and by numbers of fish in the prey school, as well as in the predator school.

In evaluating the survival value of fish schools, Breder (1967) considered predation, water visibility, size of groupings, quietness of schools. He concludes that the "whole matter of schooling and aggregating is looked upon as a mechanism of behavioral homeostasis and as such is subject to the influence of selective pressures."

Williams (1964) carried out experiments in which he placed fish in a visually homogeneous environment (which really can't exist if there is more than one fish present) and found that fish formed groups if their own visual reference points were on other fish. In his interesting discussion of the survival values of schooling, he demolished most of the suggestions for it.

As an alternative he proposed that "a school is not an adaptive mechanism itself, but rather an incidental consequence of adaptive individual behavior. . . ." In essence, each fish sought to associate with other fish —"schooling is a form of cover seeking." He emphasized that schooling was conspicuous in pelagic habitats, moderate in intermediate habitats, and undetectable in weedy habitats. Unfortunately, his argument does not withstand the many underwater observations of schooling species inhabiting weedy and rocky areas as well as open water. In fact, the atherinids, very much inshore schooling forms, are found around long rocky jetties, over weedy bottoms, and in great clusters within holes in rocks. *Fundulus heteroclitus,* which fits Williams' definition of a school, is distributed among grasses. Although Williams has provided a most stimulating idea on schooling and its evolution, his theories do not adequately satisfy what is presently known about schools.

Hergenrader and Hasler (1967) incidentally referred to another aspect of schooling: energy conservation. They report that at winter temperatures of 0°–5°, a single yellow perch, *Perca flavescens,* swam 50 percent slower than a school of the same fish. These observations, although not experimentally verified, suggest that schooling behavior may result in energy conservation of individual fish. Shaw (1962) suggested that fish living in a school can utilize the energy produced by the swimming activities of surrounding fish. Breder (1965) stated that the energy of a vortex created by a swimming fish represented "the energy that the fish had expended to propel itself." Neighboring fish, utilizing a vortex, could "coast" for a while and reduce their energy expenditure. Of course, this would be only temporary, but such action distributed among individuals throughout the school could effectively result in the school's ability to swim faster than an individual who had to exert constant energy.

My belief is that we are making a serious error in forcing ourselves to find a single adaptive feature for schooling. Fish that school obviously have survived probably due to the compounding of many adaptive features. The group effect, the confusion effect (Welty, 1934), reproductive facilitation, social facilitation, large numbers, mimicry in appearing as a large animal, energy conservation, etc., may all make schooling highly adaptive.

THE DEVELOPMENT OF SCHOOLING

Contrary to the earlier assumption that fish form schools immediately after hatching (Morrow, 1948), the researches of Shaw (1960, 1961) on

Menidia showed that schooling develops gradually over a period of several weeks during which the fish show characteristic patterns of approach and orientation. The first approaches of the newly hatched fish are immediately followed by withdrawals. Gradually, however, instead of withdrawing, the fish orient parallel to each other and swim in synchrony for several seconds until finally, they approach, orient parallel, and swim in a school continuously. Schooling, i.e., swimming continuously as a group, is clearly established when the fish have reached 10–12 mm in length.

One of the most interesting aspects of early schooling, and one to which Schneirla's approach-withdrawal hypothesis (1965) most aptly applies, is the change that occurs in the direction of approach. During the preschooling stage the approach is head-on, almost as if the fish were incapable of adjusting their swimming direction sufficiently to avoid one another. The fish, however, never collide but instead veer off in what appears to be a fright response. Several possibilities are suggested for these approaches in terms of Schneirla's approach-withdrawal hypothesis. (1) During the early preschooling phase, each fry (young fish) appears to the other as a stimulus of low intensity causing a mutual approach. However, as they come closer and closer the stimulus intensity heightens and withdrawal occurs (Shaw, 1960). Approach to food, on the other hand, is generally head on; little readjustment of position occurs, and there is no veering off. The size of the food particle is tiny compared to the young fish and would therefore be a low level stimulus with attraction continuing as the fish approached more closely. Undoubtedly, in feeding, other sense organs are stimulated and act to reinforce the initial approach stimulus. (2) It is possible that the motor capacity to readjust position is not fully developed in the early preschooling stage. (3) Neural connections may be only partially developed, and although the A-processes (Schneirla, 1965) have begun, they cannot be carried to completion. These neural connections may be a prerequisite for establishing a system of reciprocal stimulation, which we can see later in the ability of young fish to readjust themselves in space and in their tendency to remain near one another during the schooling phase.

In another developmental study, Jorné-Safriel and Shaw (1966) found that in *Atherina mochon* (a species that forms obligate schools with good parallel orientation and close cohesion), approaches of one fry to another were followed by withdrawals during the first week of life. The number of times that an approach was followed by parallel orientation was low, but from the 7th–12th days it increased sharply while the number of withdrawals decreased. In this latter period the fry grew from 8.5–11.0 mm in length. The consistent parallel orientation seen so clearly in *Menidia* was not always apparent in *Atherina mochon* but the observational techniques, radically different in both, might have accounted for some of the differences in the duration of parallel orientation. In contrast to the group-reared *Atherina mochon,* the isolates showed a dramatic tendency to school. Even fry that had been raised in total darkness schooled after they had become adjusted to the light conditions. Their patterns of fish-to-fish orientation soon became indistinguishable from those of isolates reared under normal light conditions.

Dambach (1963) examined the schooling development of three cichlids, *Tilapia macrocephala, T. nilotica,* and *T. tholloni.* The last lays its eggs on a substrate, the first and second are mouth breeders. In these studies he defined schooling as a cohesion based on mutual attraction without including the criterion of parallel orientation. He studied the attraction only of two fish or a "twin-group" as he called it, and found that the readiness or tendency to cohere was well established in all three forms by the time the fry had reached 11 mm in length, although in *T. tholloni* it was weak.

Dambach (1963) also reared the various *Tilapia* in isolation and found, too, that the "twin-group" formed readily when two fry were put together. Placing an inexperienced with an experienced fish did not enhance group cohesion. Dambach experimented with artificial schools, made up of colored wax globules of assorted sizes which the young fish tended to follow when moving. Upon testing several fish in an optomotor apparatus and seeing that they followed the optomotor stimulus, Dambach postulated that this reaction may be related to schooling.

Shaw and Sachs (1967) examined the optomotor reaction and its role in schooling development in *Menidia menidia.* They found that an optomotor response could be elicited from newly hatched *Menidia* as well as all the larger sizes tested. Results led them to believe that the response was not essential to the attraction-approach phase but may be employed in the maintenance of parallel orientation, once established. Shaw (1960) found that a young freely swimming *Menidia* would approach another fish that was enclosed in a glass tube. The freely swimming fish lined up horizontal to the tube and attempted to coordinate its direction of swimming with that of the fish in the tube. But after approximately one minute the freely swimming fish swam away and demonstrated no further approaches. During its orientation to the enclosed fish, the freely swimming fish vibrated almost continuously. Similar vibrations (Jorné-Safriel and Shaw, 1966) were noted in the first approaches of *Atherina mochon* during the schooling phase. Two fish approaching and orienting for the first time vibrated for several seconds before swimming together in parallel formation. These observations stirred an investigation of the lateral line system and the establishment of parallel orientation. Histological changes in the lateral line system were followed during the preschooling and schooling stages. In addition to changes in the flexibility of cupulae, Cahn and Shaw (1963) and Cahn, Shaw, and Atz (1968) found that the innervation of neuromasts occurs at the same time that the fish begin to spend increasingly long periods in parallel orientation. Subsequent stages in the maturation of neuromasts and surrounding structures (as seen by histological technique) is accompanied by the increasing stability of schools. In the developing fish, during the preschooling phase, neuromasts may be responsive to the diffuse stimuli of moving water particles. But as the fish develops, the sense cells become capable of filtering specific stimuli thereby providing the fry with one mechanism for maintenance of parallel orientation and another for regulating fish-to-fish spacing. In *Menidia,* during the early stages, parallel orientation happens only occasionally and for a short time; later it is maintained continuously. Whether or not the lateral line is concerned with the first approaches, however, remains highly conjectural.

The results of a recent study by the author and Madelaine Williams, now being prepared for publication, bring some new and significant findings to the study of behavioral development. In this study fish were reared under three different conditions: as communities, as individuals without experience with species mates, and as individuals with a single experience (one half to two hours long) with species mates. The experience for the latter group occurred anywhere from the 1st to the 35th day after hatching. The significant findings were: (1) when isolated fish were placed together at the 20th day after hatching, schooling appeared and continued to appear in all tested fish until the 55th day after hatching when the experiment was terminated; (2) schooling did not appear in isolates tested together prior to the 20th day; (3) both groups of isolates, unlike the smooth-swimming, community-reared fish, constantly readjusted position, making zig-zag swimming movements and a great many 45° turns; and (4) the isolates with a single experience given from the 1st to the 34th day showed similar behavior to the absolute isolates. But those fish given a single experience at 35 days and tested at 55 days showed smooth swimming and continuous parallel orientation; the time spent in parallel orientation among absolute isolates of comparable age diminished at 55 days and the zig-zag swimming was still present.

The isolated fish spend many short periods in parallel formation interspersed with withdrawal behavior that remains at a very high level throughout all the tests. In contrast, incidence of withdrawal among the community-reared fish is low, almost nonexistent.

I believe that the seemingly ambivalent behavior shown by the isolates gives insight into the effect of isolation on behavioral development. When isolates are placed together, two antithetical behaviors appear: approach-orientation and approach-withdrawal. The approach-orientation results in a social grouping, the approach-withdrawal results in dispersion. That these two behaviors exist simultaneously in isolates but not in community-reared fish suggests that without adequate social experiences, even if not directly related to the specific behavior under analysis, the fish do not develop proper integrative mechanisms. The behaviors remain disparate. The approach-withdrawal, for example, in the preschooling fish, is initially high and tapers off before approach-orientation ensues. As approach-orientation mechanisms develop through behavioral organization, incorporating along the way maturational changes and experiential stimuli, the approach-withdrawal mechanisms may be inhibited as they become integrated into the total behavioral complex. Without appropriate experiential stimuli, perhaps the approach-withdrawal is neither properly integrated nor inhibited. The result is a fish that schools but also withdraws and tends to disrupt schooling.

SUMMATION

In recent times there has been a great deal of disagreement among specialists over what characteristics or behavioral qualities define a school. Breder (1967), by subdividing schools into facultative and obligate, has provided a convenient terminology to describe those fish that school occasionally and only under certain conditions and those that school continuously and virtually under all conditions.

The suggestion, made in the present paper, that schooling should be considered as a two order system, one order, mutual attraction-approach and the other order, parallel orientation, will hopefully assist in analyzing various experiments and reevaluating them in a new light.

The recent gathering of quantitative data on school structure will be an invaluable aid toward assessing the effects of intrinsic and extrinsic stimuli on school geometry and behavior. In future studies on environmental changes and on the physiological state of the fish within the school, it will be possible to say precisely what happens to orientation, spatial distribution, speed, and so forth. Researchers now have the technological means to make truly comparative analyses. They can determine, for example, the absolute amount of light present in a test situation; even schooling behavioral changes can now be measured much more precisely.

Recent interest in the physiological condition of the fish and its effect on schooling has opened new areas of study, such as the nature of diurnal rhythms in schooling and the effects of gas concentration and hunger on school organization.

Several important study areas, however, have not yet been investigated. Much conjecture remains around the speed with which one fish responds to changes in the behavior of others in the school, e.g., reorientation, increased or decreased speed, but as yet no one has made any substantial experimental analyses. Although we know that the visual system is unequivocally connected to schooling, the degree or extent to which other sensory systems operate in schooling is, at present, only vaguely suggested. In fact, all that we know about the visual system is that fish must see in order to school. Also unknown is the nature of sensory stimuli which bring about attraction-approach and orientation.

Studies on the development of schooling behavior have shown that schooling develops gradually and that fish do not require experience with species mates in order to form a school at the appropriate time. Apparently the gradual development of schooling results from a combination of maturational-experiential factors.

The many treatises on the adaptive value of schooling have not provided any one feature that has clear priority over another. Schooling is a multiadaptive phenomenon dependent on the age of the fish (e.g., juvenile schools), the numbers, the reproductive season, the metabolic state, etc. Its evolution in disparate species speaks for its usefulness as a behavioral mechanism.

ADDENDUM

At the 11th International Ethological Congress, Rennes, France, a group of students of schooling behavior discussed the designation of the terms *school* and *aggregation*. They agreed that the term *school* should be used to designate any grouping of fish that is a result of biosocial attraction among the fish. To differentiate the way in which the fish are oriented to each other, the term *polarized school* signifies a group showing parallel orientation and a *nonpolarized school* is a group not showing parallel orientation. There are gradations in the degree of polarization. Orientation within a school varies during a typical day such that schooling fish may not be polarized continuously.

The word *aggregation,* then, should not be used when referring to fish that are socially attracted to each other. It could still be employed to indicate that fish come together as a direct response to extrinsic conditions or when there is reasonable doubt as to whether or not fish are socially attracted to each other.

This author urges that in all ensuing reports on the schooling of fish, investigators distinguish between nonpolarized schools and polarized schools.

ACKNOWLEDGMENTS

The author is a recipient of a Research Scientist Development Award, NIMH, #K3MH5349. Part of the author's research was supported by the National Science Foundation, grant no. GB 351.

REFERENCES

Alekseeva, K. D. 1963. The significance of vision in the "group effect" in a few Black Sea fishes. *VOP Ikhtiol.* 3 (4): 726–733.

Alee, W. C. 1931. *Animal aggregations.* Chicago: Univ. Chicago Press.

Aronson, L. R., and H. Kaplan. 1968. Function of the teleostean forebrain. In D. Ingle, ed., *The central nervous system and fish behavior.* Chicago: Univ. Chicago Press. Pp. 107–125.

Atz, J. W. 1953. Orientation in schooling fishes. In T. C. Schneirla, ed., *Proceedings of a conference on orientation in animals.* Washington, D. C.: Office of Naval Research. Pp. 115–130.

Baylor, E. R., and E. Shaw. 1962. Refractive error and vision in fishes. *Science* 136 (3511): 157–158.

Berwein, M. 1941. Beobachtungen und Versuche uber das gesellige Leben von Elritzen. *Z. Vergleich. Physiol.* 28: 402–420.

Breder, C. M., Jr. 1954. Equations descriptive of fish schools and other animal aggregations. *Ecology* 35 (3): 361–370.

Breder, C. M., Jr. 1959. Studies on social groupings in fishes. *Bull. Am. Mus. Nat. Hist.* 117 (6): 397–481.

Breder, C. M., Jr. 1965. Vortices and fish schools. *Zoologica* 50: 97–114.

Breder, C. M., Jr. 1967. On the survival value of fish schools. *Zoologica* 52: 25–40.

Breder, C. M., Jr., and F. Halpern. 1946. Innate and acquired behavior affecting aggregation of fishes. *Physiol. Zool.* 19: 154–190.

Brock, V. E., and R. H. Riffenburgh. 1960. Fish schooling: a possible factor in reducing predation. *J. Conseil Int. Explor. Mer* 25 (3): 307–317.

Cahn, P. H. 1967. Some observations on the schooling of tunas. *Am. Zool.* 7 (2): abst. no. 64.

Cahn, P. H., and E. Shaw, 1963. Lateral line activity in rheotactic orientation of schooling fishes. *Am. Zool.* 3: 517 (abstr.)

Cahn, P. H., E. Shaw, and E. Atz. 1968. Lateral line nerve maturation related to the development of schooling in the atherinid fish *Menidia. Bull. Marine Sci.* 18: 660–670.

Clarke, T. A., A. O. Flechsig, and R. W. Grigg. 1967. Ecological studies during project Sealab II. *Science* 157 (3795): 1381–1389.

Cullen, J. M., E. Shaw, and H. A. Baldwin. 1965. Methods for measuring

478   EVELYN SHAW

the three-dimensional structure of fish schools. *Anim. Behav.* 13: 534–543.

Dambach, M. 1963. Vergleichende Untersuchungen uber das Schwarmverhalten von *Tilapia*-Jungfischen (Cichlidae, Teleostei). *Z. Tierpsychol.* 20: 267–296.

Darnell, R. M., and R. R. Meierotto. 1965. Diurnal periodicity in the black bullhead, *Ictalurus melas* (Rafinesque). *Trans. Am. Fish. Soc.* 94 (1): 1–8.

Dÿkgraaf, S. 1963. The functioning and significance of the lateral-line organs. *Biol. Rev.* 38 (1): 51.

Hale, E. B. 1956. Social facilitation and forebrain function in maze performance of green sunfish, *Lepomis cyanellus*. *Physiol. Zool.* 29 (2): 93–107.

Harris, G. G., and W. A. van Bergeijk. 1962. Evidence that the lateral-line organ responds to near-field displacements of sound in water. *J. Acoust. Soc. Am.* 34 (12): 1831–1841.

Hemmings, C. C. 1966a. Olfaction and vision in fish schooling. *J. Exp. Biol.* 45: 449–464.

Hemmings, C. C. 1966b. The mechanism of orientation of roach, *Rutilus rutilus* L. in an odour gradient. *J. Exp. Biol.* 45: 465–474.

Hergenrader, G. L., and A. D. Hasler. 1967. Seasonal changes in swimming rates of yellow perch in Lake Mendota as measured by sonar. *Trans. Am. Fish. Soc.* 96 (4): 373–382.

Hobson. E. S. 1968. Predatory behavior of some shore fishes in the Gulf of California. *U.S. Bureau Sport Fish. Wildl. Res. Rep.* 73: 1–92.

Hunter, J. R. 1966. Procedure for analysis of schooling behavior. *J. Fish. Res. Bd. Canada* 23 (4): 547–562.

Hunter, J. R. 1968. Effect of light on schooling and feeding of jack mackerel, *Trachurus symetricus*. *J. Fish. Res. Bd. Canada* 25 (2): 393–407.

Hunter, J. R., and W. J. Wisby. 1964. Net avoidance behavior of carp and other species of fish. *J. Fish. Res. Bd. Canada* 21 (3): 613–633.

Janzen, W. 1933. Untersuchungen uber grosshirnfunktionen des goldfisches (*Carassius auratus*). *Zool. Jahr. Abt. Allg. Zool. u. Physiol. der Tiere.* 52: 591–628.

John, K. R. 1964. Illumination, vision and schooling of *Astyanax mexicanus* (Fillipi). *J. Fish. Res. Bd. Canada* 21 (6): 1453–1473.

Jorné-Safriel, O., and E. Shaw. 1966. The development of schooling in the atherinid fish, *Atherina mochon*. *Pubbl. Staz. Zool. Napoli* 35: 76–88.

Keenleyside, M. H. A. 1955. Some aspects of the schooling behaviour of fish. *Behaviour* 8: 183–248.

Kühlmann, D. H. H., and H. Karst. 1967. Freiwasserbeobachtungen zum verhalten von tobiasfischschwarmen (Ammodytidae) in der westlichen ostsee. *Z. Tierpsychol.* 24(3): 282–297.

Kuiper, J. W. 1956. The microphonic effect of the lateral line organ. *Publ. Biophys. Group, Naturkundig Lab.,* Groningen, 1–159.

McFarland, W. N., and S. A. Moss. 1967. Internal behavior in fish schools. *Science* 156: 260–262.

Magnuson, J. J. 1964. Activity patterns of scombrids. *Proceedings of the Hawaiian Academy of Science,* 39th annual meeting, p. 26, abst. #24.

Manteifel, B. P., and D. V. Radakov. 1960. The adaptive significance of schooling behaviour in fishes. *Progr. Modern Biol.* 50 (3/6): 362–370.

Milanovskii, Yu. E., and V. A. Rekubratskii. 1960. Methods of studying the schooling behavior of fishes. *Nauchn. Dokl. Vysshei Shkoly, Biol. Nauki,* no. 4, pp. 77–81.

Morrow, J. E., Jr. 1948. Schooling behavior in fishes. *Quart. Rev. Biol.* 23 (1): 27–38.

Moulton, J. M. 1960. Swimming sounds and the schooling of fishes. *Biol. Bull.* 119 (2): 210–223.

O'Connell, C. P. 1960. Use of fish school for conditioned response experiments. *Anim. Behav.* 8 (3/4): 225–227.

Parr, A. E. 1927. A contribution to the theoretical analysis of the schooling behaviour of fishes. *Occasional Papers, Bingham Oceanogr. Coll.,* no. 1, pp. 1–32.

Pfeiffer, W. 1962. The fright reaction of fish. *Biol. Rev.* 37: 495–511.

Radakov, D. V. 1966. On a peculiarity of schooling behaviour in fish. *International Council for the Exploration of the Sea,* pp. 1–5.

Radakov, D. V. 1968. The biological and practical importance of schooling behaviour in fishes. FAO Proceedings. (in press)

Rosenblatt, R., and G. S. Losey, Jr. 1967. Alarm reaction of the top smelt, *Atherinops affinis:* Reexamination. *Science,* 158 (3801): 671–672.

Schneirla, T. C. 1965. Aspects of stimulation and organization in approach/ withdrawal processes underlying vertebrate behavioral. In D.S. Lehrman, R. A. Hinde, and E. Shaw, eds., *Advances in the study of behavior,* vol. 1. New York: Academic Press. Pp. 1–74.

Shaw, E. 1960. The development of schooling behavior in fishes. *Psysiol. Zool.* 33: 79–86.

Shaw, E. 1961. The development of schooling behavior in fishes, II. *Physiol. Zool.* 34: 263–272.

Shaw, E. 1962. The schooling of fishes. *Sci. Am.* 206: 128–138.

Shaw, E., and A. Tucker. 1965. The optomotor reaction of schooling carangid fishes. *Anim. Behav.* 13 (2–3): 330–336.

Shaw, E. 1969. The duration of schooling among fish separated and those not separated by barriers. *Am. Mus. Novitates,* no. 2373, pp. 1–13.

Shaw, E., and B. D. Sachs. 1967. Development of the optomotor response in the schooling fish, *Menidia menidia. J. Comp. Physiol. Psychol.* 63 (3): 385–388.

Shlaifer, A. 1942. The schooling behavior of mackerel: A preliminary experimental analysis. *Zoologica* 27: 75–80.

Spooner, G. M. 1931. Some observations on schooling in fish. *J. Marine Biol. Assoc. United Kingdom* 17: 421–448.

Tester, A. L. 1959. Summary of experiments on the response of tuna to stimuli. *Modern Fishing Gear of the World,* sec. 12, pp. 538–542.

Thines, G., and E. Vandenbussche. 1966. The effects of alarm substance on the schooling behaviour of *Rasbora heteromorpha* Duncker in day and night conditions. *Anim. Behav.* 14: 296–302.

Tobach, E., and T. C. Schneirla. 1968. The biopsychology of social behavior in animals. In R. E. Cooke, ed., *The biologic basis of pediatric practice.* New York: McGraw-Hill. Pp. 68–82.

Wahlert, Gerd von. 1963. Observations on schools of fish. *Veroffentl. Inst. Meeresforsch. Bremerhaven* 8: 197–213.

Welty, J. C. 1934. Experiments in group behavior of fishes. *Physiol. Zool.* 7: 85–128.

Wersäll, J., and A. Flock. 1965. Functional anatomy of the vestibular and lateral line organs. In W. D. Neff, ed., *Contributions to sensory physiology,* vol. 1. New York: Academic Press. Pp. 39–61.

Williams, G. C. 1964. Measurement of consociation among fishes and comments on the evolution of schooling. *Publ. Mus. Michigan State Univ.* 2: 351–383.

HOWARD MOLTZ

Department of Psychology
Brooklyn College of the City University of New York
Brooklyn, New York

Effects of Breeding Experience on the Maternal Behavior of the Laboratory Rat

That the primiparous rat is able to act in adequate maternal fashion, that she is able, without specific practice, to engage in those activities necessary for the survival and growth of a litter, indicates that breeding experience is not essential for the expression of maternal behavior. But although obviously unessential, previous parity may still influence maternal behavior. For example, perhaps certain nurtural responses are expressed more efficiently or with greater alacrity by the multiparous rat than by the primiparous rat. Or perhaps there are conditions under which the primiparous rat will fail to react maternally, while the multiparous rat will respond in a perfectly normal manner. In the experiments that follow, we have attempted to study the role of breeding history, comparing systematically and under a variety of conditions the maternal behavior of experienced and inexperienced females.

NORMAL BREEDING CONDITIONS

The most obvious influence previous parity might exert on maternal behavior would be to increase the efficiency with which the mother comes to execute those nurtural responses necessary for the survival and growth of her litter. Our first experiment, therefore, was designed to determine whether the multiparous female is in fact more proficient than the primiparous female in such components of the maternal complex as nursing, nest building, and retrieving (Moltz and Robbins, 1965).

Animals that had reared two litters and animals that had not bred previously were each mated. Beginning 24 hours after parturition and continuing until the pups were 21 days of age, each of the groups just referred to was tested daily for nursing, nest building, and retrieving.

Analysis of the data left no doubt as to the capability of the inexperienced female; in the responses we observed, she proved every bit as proficient as her experienced counterpart. In brief, not only is the primiparous rat able to rear and care for a litter but she can do so in a manner indistinguishable from the multiparous rat.[1]

But what would have occurred had our primiparous and multiparous animals been observed under abnormal instead of normal breeding conditions? More specifically, what if a distinctive set of events characteristic of a particular phase of the breeding episode had been changed or had been prevented from occurring? Would the primiparous female still exhibit nursing, nest building, and retrieving with a proficiency equal to that of the multiparous female, or would she perhaps be less proficient or even altogether incapable of acting maternally? To study some of the possible effects of breeding history and, hopefully, to gain some insight into the development of maternal behavior, we undertook in one experiment to manipulate the experience of parturition itself; in a second experiment, to modify some of the hormonal mechanisms believed to underlie the termination of pregnancy; and in a third, to alter certain stimulus characteristics of the young.

CAESAREAN DELIVERY

One aspect of the breeding episode that would seem critical for the development of certain components of the mother-young relationship involves those experiences centering on parturition. Parturition in the rat, as in most other mammalian forms, is characterized by a complex of responses that includes self-licking—particularly of the vaginal area—the ingestion of birth fluids, and the consumption of fetal and placental membranes. These responses, paralleled by such endogenous phenomena as oxytocin release, uterine contractions, and the emergence of the fetuses, appear by their very nature to be instrumental in orienting the female toward her young and consequently in instigating some of the nurtural responses typically exhibited in the litter situation.

This view of parturition as providing experiences important and perhaps even essential for the occurrence of maternal behavior in the rat has been advanced previously by other investigators. Schneirla (1956), for example,

---

[1] This conclusion, it should be pointed out, confirms the predictions of Beach and Jaynes (1956), Rosenblatt and Lehrman (1963), and Wiesner and Sheard (1933).

speaks of "parturitive behavior" as "facilitating initiation of nursing and other stimulative relations of mother and newborn" (p. 421). Similarly, Rosenblatt and Lehrman (1963), in discussing the dramatic changes in the female's behavior, suggest that these changes might be occasioned, in part, by the experience of parturition itself.

In an effort to determine whether, in fact, the preclusion of parturition does interfere with maternal behavior and, if it does, whether such interference would be mitigated by previous parity, we undertook to deliver females by caesarean section (Moltz, Robbins, and Parks, 1966).[2]

Caesarean section on these females, half of whom were breeding for the first time and half of whom had reared two previous litters, was performed shortly before term. Then, after a period of time sufficient to permit mammary parenchyma to reach a secretory level comparable to that found in normal puerperae, the females were presented with normally delivered foster pups. Thereafter, for 21 days, each female was tested for nest building and retrieving; in addition, all foster litters were weighed daily. It is important to note that these foster litters were between 6 and 15 hours of age at the time they were offered to our caesarean females and, moreover, that they had already been licked and nursed by their own mothers.

The results of our observations indicated that the rat can enter into an effective nurtural relationship despite having been deprived of the experience of parturition and that this is as true for the primiparous as for the multiparous female. Of relevance to these results is the fact that Rosenblatt (1965) also recently observed the occurrence of maternal behavior following caesarean delivery in the primiparous rat. Perhaps caesarean delivery, and with it, of course, the preclusion of "parturitive behavior," does not provide conditions departing sufficiently from the normal to reveal the influence of previous breeding experience.

OVARIECTOMY

The hormonal events underlying the termination of pregnancy are complex and at the present time only incompletely understood. However, there appears to be general agreement that a critical change in ovarian output is associated with the birth process and is partly responsible for the occurrence of that process. This change results in an increase in the level of estrogen and a decrease in the level of progesterone, thereby reversing the prevailing gestational ratio of these steroids (cf. Schofield, 1957; Zarrow, 1961). In our third study, we attempted to determine whether manipulation of this presumably focal endocrine mechanism would influence the

---

[2] See earlier attempts to answer this question (Labriola, 1953; Wiesner and Sheard, 1933).

expression of maternal behavior and, if so, whether the influence thus exerted would be manifested differentially in the primiparous and multiparous female.

Initially, our aim was to ovariectomize female rats approximately 12 to 15 hours from term. We were aware, of course, that the rat, unlike several other mammalian forms, requires the presence of the ovaries to maintain pregnancy. However, we thought that if an ovariectomy were performed sufficiently close to term, the rat might deliver normally and that we might thereby have available for study puerperal females in whom circulating levels of estrogen and progesterone had been markedly reduced. Unfortunately, normal delivery did not occur. Of the six animals on whom we performed this near-term ovariectomy, one died during delivery, while the remaining five either expelled fetuses that evidently had been crushed in the delivery process or, having delivered viable fetuses, had required more than 72 hours to complete parturition.

It is obvious that we could not perform an ovariectomy just prior to parturition without interfering with the normal birth process. Consequently, we decided to perform a caesarean section in conjunction with our ovariectomy. Since, as already discussed, caesarean section in itself does not have a discernible influence on the expression of maternal care in either the primiparous or the multiparous rat, any effect on behavior of this twofold operation could reasonably be attributed to the humoral modifications consequent upon ovariectomy alone.

We found that virtually all the multiparous animals on whom we operated subsequently accepted and reared to weaning the foster litters offered them (foster litters which, like those in our caesarean study, had already been licked and nursed by their own mothers), while, in contrast, only 50 percent of our primiparous animals did so, the remaining 50 percent having either eaten or neglected their young. Evidently, breeding history can influence the expression of maternal behavior, at least under conditions of ovariectomy or, more precisely, under the reduced hormone levels effected through ovariectomy.

It is of interest to note that such an influence parallels one already reported in the literature. Valenstein and Young (1955), for example, found that the expression of sexual behavior in the castrated male guinea pig was influenced significantly by the precastrational history of the animal. Similarly, Rosenblatt and Aronson (1958) reported that sexual behavior declined after castration at an appreciably slower rate in the sexually experienced than in the sexually inexperienced male cat. And finally, Beach (1964) observed that male dogs will maintain a high level of sexual performance after castration if, prior to castration, they had acquired experience with receptive females. These results indicate, as do those of the present experiment, that the effect on behavior of certain types of hormonal manipulation can be influenced significantly by previous experience.

STIMULUS CHARACTERISTICS OF THE YOUNG

It is possible that parturition, or more precisely the experience of parturition, does affect maternal responsiveness but in a way that was simply not evidenced under the conditions of our caesarean study. To be sure, the surgically delivered females of that study were thoroughly adequate mothers. However, it must be remembered that they were so in response to foster young of a particular kind—foster young who, by virtue of being 6–15 hours of age and having been licked and nursed previously by their natural mothers, may well have possessed behavioral and other characteristics optimal for the initiation of such responses as nursing, nest building, and retrieving. Had a different type of pup been proffered, it is conceivable that our caesarean-delivered mothers, in contrast to normal mothers, or our primiparous caesarean-delivered mothers, in contrast to their multiparous counterparts, would not have entered into a successful nurtural relationship.

There is evidence in the literature (Denenberg, Grota, and Zarrow, 1963) indicating that the presence or absence of the placenta is a significant variable in determining whether or not normally delivered lactating females will accept foster young. This finding has been confirmed in the author's laboratory: specifically, we observed, as did Denenberg, Grota, and Zarrow, that normally delivered mothers, whether primiparous or multiparous, were more likely to accept and rear to weaning "noncleaned" foster pups as compared with "cleaned" foster pups. The purpose of the present experiment was to determine whether the same "preference" would be shown by caesarean-sectioned females and, moreover, whether their breeding history would prove to be a significant variable.

"Donor" females approximately 12–15 hours from term were killed by cervical fracture and their pups taken. One half of these pups then had the placentas and amniotic sacs removed after which they were also washed with distilled water; the remaining half were not cleaned. These two classes of foster pups were offered, respectively, to caesarean-delivered primiparous and multiparous females, and behavioral comparisons were also made with females after normal delivery.

The incidence of litter acceptance was, in general, significantly lower among caesarean-delivered than among normally delivered females. Whereas, on the average, only 39 percent of the surgically delivered pups proffered to caesarean females were accepted, as many as 66 percent of those proffered to normal females were accepted. Moreover, the behavior of the caesarean-delivered female, in contrast to that of the normally delivered females, was affected significantly by previous parity. Specifically, the multiparous caesarean-female showed a preference for cleaned as compared with noncleaned foster young, with 75 percent of the cleaned pups surviving to weaning as compared with 30 percent of the noncleaned pups; the primiparous caesarean-female, on the other hand, did not show a preference for either class of foster pup, accepting, in fact, only 25 percent of the cleaned and 26 percent of the noncleaned pups respectively.

The results of the present experiment are noteworthy in two respects. First, they indicate that the experience of parturition does have an effect on maternal behavior, although, to be sure, an effect more subtle than we had ventured to anticipate. Thus, it now appears that caesarean-delivered and normally delivered females, although comparable with regard to their acceptance of foster pups that have been licked and fed by the natural mother, are strikingly different in their acceptance of foster pups delivered by surgical intervention. Moreover, it also appears that caesarean and normal females have different preferences—the normal mother, when offered surgically delivered young, prefers the noncleaned pup, while the caesarean mother, when she does exhibit a preference, prefers the cleaned pup. And second, under the conditions described, breeding history proved to exert an important influence on maternal behavior as evidenced by the greater success of the multiparous caesarean-female in the nurturance of cleaned young as compared with that of the primiparous caesarean-female.

We must confess that we cannot offer a consistent explanation either of the differences between caesarean and normal females or of why previous parity acted to affect the behavior of caesarean females in the direction of increasing their acceptance of cleaned young. Hopefully, elucidation will come through additional observation and experimentation.

It is apparent now that the nurtural responsiveness of the multiparous rat is not equivalent in all respects to that of the primiparous rat; previous parity, under some conditions, evidently does influence the expression of maternal behavior. However, merely to have shown that breeding history can exert an effect is only a preliminary step. The more important and complex task is to discover the precise nature of the multiform relationships that our research thus far has intimated may exist between breeding history and maternal responsiveness. Experiments designed to this end are now in progress in the author's laboratory.

ACKNOWLEDGMENTS

The author's research was supported by Research Grant MH–07967 and by Public Health Service research career program award K3–MH–21, 781, both from the National Institutes of Health.

REFERENCES

Beach, F. A. 1964. Biological bases for reproductive behavior. In W. Etkin, ed., *Social behavior and organization among vertebrates*. Chicago: Univ. Chicago Press. Pp. 117–142.

Beach, F. A., and J. Jaynes. 1956. Studies on maternal retrieving in rats: II. Effects of practice and previous parturitions. *Am. Naturalist* 90: 103–109.

Denenberg, V. H., L. J. Grota, and M. X. Zarrow. 1963. Maternal Behaviour in the rat: analysis of cross-fostering. *J. Reprod. Fertility* 5: 133–141.

Labriola, J. 1953. Effects of caesarean delivery upon maternal behavior in rats. *Proc. Soc. Exp. Biol. Med.* 83: 556–557.

Moltz, H., and D. Robbins. 1965. Maternal behavior of primiparous and multiparous rats. *J. Comp. Physiol. Psychol.* 60: 417–421.

Moltz, H., D. Robbins, and M. Parks. 1966. Caesarean delivery and the maternal behavior of primiparous and multiparous rats. *J. Comp. Physiol. Psychol.* 61: 455–460.

Rosenblatt, J. S. 1965. The basis of synchrony in the behavioural interaction between the mother and her offspring in the laboratory rat. In B. M. Foss, ed., *Determinants of infant behaviour*, vol. III. London: Methuen.

Rosenblatt, J. S., and L. R. Aronson. 1958. The decline of sexual behavior in male cats after castration with special reference to the role of prior sexual experience. *Behaviour* 12: 285–338.

Rosenblatt, J. S., and D. S. Lehrman. 1963. Maternal behavior of the laboratory rat. In H. L. Rheingold, ed., *Maternal behavior in mammals*. New York: Wiley. Pp. 8–57.

Schneirla, T. C. 1956. Interrelationships of the "innate" and the "acquired" in instinctive behavior. In P.-P. Grassé ed., *L'Instinct dans le comportement des animaux et de l'homme*. Paris: Masson. Pp. 387–452.

Schofield, B. M. 1957. The hormonal control of myometrial function during pregnancy. *J. Physiol.* 138: 1–10.

Valenstein, E. S., and W. C. Young. 1955. An experiential factor influencing the effectiveness of testosterone propionate in eliciting sexual behavior in male guinea pigs. *Endocrinology* 56: 173–177.

Wiesner, B. P., and N. M. Sheard. 1933. *Maternal behaviour in the rat.* London: Oliver and Boyd.

Zarrow, M. X. 1961. Gestation. In W. C. Young, ed., *Sex and internal secretions.* Baltimore: Williams & Wilkins. Pp. 958–1031.

JAY S. ROSENBLATT
Institute of Animal Behavior
Rutgers University
Newark, New Jersey

Views on the Onset and Maintenance of Maternal Behavior in the Rat

In the laboratory rat, maternal behavior begins at parturition and is maintained for 3 or 4 weeks before it declines and the mother resumes estrous cycling in preparation for her next pregnancy. Considering the entire process of reproduction, including the 22 days of pregnancy, maternal behavior appears about midway between conception and weaning of the young. Under semifield conditions (Calhoun, 1963) the young remain with the mother for at least 5 weeks after birth, but since descriptions are lacking, it is difficult to estimate the mother's actual role in the relationship with the young at this time and to compare the duration of maternal behavior under these conditions with that under laboratory conditions.

During parturition the mother is mainly concerned with delivery of the young, licking and cleaning the pups, licking birth fluids from her body and wherever they have spread to, and eating the placentas. Nevertheless elements of maternal behavior appear as she retrieves and licks the cleaned pups, lies down over them, and, on occasion, nurses them. During the first day the full pattern of maternal care appears, as nestbuilding and ano-genital licking is added to the mother's activity, soon after the young are born.

Descriptions of maternal behavior thus far have been dictated by the mother's behavior in relation to the needs of the young. The descriptions therefore emphasize the *functions* of maternal activities. How maternal behavior arises from the behavioral capacities of the female, from her perceptual abilities, and from her motivational state, conditioned by hormones, is only beginning to be studied. Lazar's (1967) study of retrieving in the mouse, which traced the many components of retrieving from the 6th day to its decline 1 or 2 weeks later, suggests the kind of analysis, free of functional preconceptions, which can be done also on the rat.

The aim of the present article is to discuss primarily *motivational aspects of maternal behavior,* particularly the physiological conditions underlying *maternal responsiveness* to young. In the first section I shall discuss the onset of maternal behavior at parturition in relation to the development of maternal responsiveness. The second section will deal with the maintenance of maternal behavior following its onset and will attempt to deal with the question: how is maternal responsiveness maintained following the disappearance of the conditions during pregnancy and parturition that led to its onset. This will lead naturally to a discussion of the role of the young that will introduce the concept of behavioral synchrony in relation to the interaction of the mother and her young during the course of maternal behavior.

ONSET OF MATERNAL BEHAVIOR

The first appearance of maternal behavior is quite sudden, following in most cases shortly after the female has completed parturition. Wiesner and Sheard (1933) demonstrated the dramatic change by showing that females, while pregnant, retrieved a food pellet in preference to a pup but immediately after parturition they retrieved the pup in preference to the food pellet. This has been confirmed in tests with females during the last half of pregnancy (Rosenblatt and Lehrman, 1963; Rosenblatt, 1965; Labriola, 1953). A recent report (Plume et al., 1969) indicates however that nonpregnant females retrieve pups as readily as small objects (i.e., plastic toys). We shall discuss this report later. In our study 5- to 10-day old pups were presented to pregnant females and elicited neither retrieving nor any other maternal behavior. When pups of the same age were presented to the same females shortly after they had given birth, retrieving and other maternal responses were readily elicited. The conclusion drawn from these observations is that the female's motivational state undergoes a change that is related to the physiological and behavioral events of parturition.

Physiological Basis

What are the physiological events leading up to parturition and how are they related to the onset of maternal behavior? The more notable physiological events of interest to us are the increase in uterine motility and contractility, leading to the onset of labor, and the initiation of lactation, which provides the basis of postpartum nursing. These physiological changes appear late in pregnancy but there is no evidence that in themselves either the uterine changes or activity of the mammary glands are the basis of maternal behavior postpartum. The removal of the uterus of late-pregnant females (including the fetuses, of course) does not prevent the onset of maternal behavior shortly afterward, nor does removal of the mammary glands prior to the beginning of pregnancy significantly alter the development of maternal behavior or its onset (Wiesner and Sheard, 1933; Labriola, 1953; Lott and Rosenblatt, 1969; Moltz, Robbins, and Parks, 1966; Moltz, Geller, and Levin, 1967). Rather the late-pregnancy changes in uterus and mammary glands are the result of changes in hormonal conditions which in turn reflect altered activity of various endocrine glands including the anterior pituitary, the ovaries, and, to some extent, the placentas. The effects of hormones secreted by these glands upon neural centers is what is believed to underly the onset of maternal behavior (see excellent reviews by Lehrman, 1961 and Richards, 1967).

Until recently only indirect methods of assessing the hormonal condition of the late-pregnant female were available; it has not been possible to determine the precise nature of the hormonal changes which produce late-pregnancy conditions in the uterus and mammary glands with these methods. Common to the changes at both sites is an altered ratio of ovarian hormones (i.e., estrogen and progesterone) secreted by the ovary and the placenta (Cowie and Folley, 1961; Zarrow, 1961; Carsten, 1968; Vorherr, 1968). The shift, in favor of estrogen, results, presumably, from a decrease in progesterone secretion or from a combination of increased estrogen secretion and decreased progesterone secretion (Cowie and Folley, 1961; Meites, 1966). At the uterus the shift in hormonal balance induces spontaneous motility and increased responsiveness to the stimulating action of oxytocin. Uterine activity begins as spontaneous, weak contractions but it develops into vigorous spasms of the uterus when oxytocin, released in response to cervical stimulation during labor, reaches the uterus. At the mammary glands the altered ratio of circulating estrogen and progesterone permits prolactin, one of the lactogenic hormones secreted by the anterior pituitary gland, to exert its influence on lactation. Estrogen may also be responsible for the increased secretion of prolactin at parturition.

It has therefore been proposed that an *increase* in circulating estrogen and a *decrease* in progesterone trigger the onset of maternal behavior. Prolactin is considered by some as also eliciting or contributing to the stimulation of maternal behavior.

Recently Terkel and Rosenblatt (1968) showed that blood plasma taken from mothers within 48 hours after parturition was capable of inducing maternal behavior when injected into virgins that were at the same time given 5- to 10-day old pups. The latencies for the onset of retrieving and other items of maternal care were considerably shorter than the latencies of virgins injected with nonmaternal blood plasma or saline solution. However the hormonal composition of the "maternal blood plasma" has not been analyzed. We are forced therefore to depend upon estimates of plasma estrogen and progesterone made by gas chromatography and immunoassay during the course of pregnancy, up to parturition, in order to evaluate the thories of hormonal induction of maternal behavior presented above. Estimates of prolactin secretion have been obtained directly from prolactin assay of the anterior pituitary gland and plasma at various times during pregnancy and shortly after parturition.

The proposed decrease in plasma progesterone at the end of pregnancy has been confirmed by several investigators who found a sharp decrease beginning about a week before parturition and reaching a low point at parturition (Grota and Eik-Nes, 1967; Hashimoto et al., 1968). On the other hand, plasma estrogen is also at a low level at parturition. This however contradicts other evidence indicating that the secretion of estrogen increases towards the end of pregnancy. Greenwald (1966) reported an increase in ovarian weight and in the number of larger-sized follicles during the last week of pregnancy; pituitary concentration of follicle stimulating hormone also increased. The changes in the ovary were accompanied by renewed responsiveness to exogenous luteinizing hormone, administration of which resulted in ovulation for the first time during pregnancy. The most recent report confirms these histological indications of increased estrogen secretion at the end of pregnancy. Yoshinaga, and Hawkins (1969) using a bioassay method (i.e., intravaginal) found a generally low rate of secretion of estrogen by the ovary during most of pregnancy and a rapid increase near term. At term the secretion rate was equal to the maximal rate found during the estrous cycle, at proestrus; following this, most females enter the phase of postpartum estrus.

It has long been assumed that prolactin secretion increases towards the end of pregnancy to account for the initiation of lactation at this time. Grosvenor and Turner (1960) found depletion of pituitary prolactin during the last half of pregnancy and on this basis assumed that circulating levels of this hormone rose. However a recent report by Grindeland, McCulloch, and Ellis (1969) provides evidence obtained by radio-immunoassay that the prolactin level remains the same and quite low throughout pregnancy (i.e., 7th, 14th, 20th days), comparable to the level of circulating prolactin during diestrus. Whether it rises sharply after the 20th day until parturition was not studied.

Behavioral Basis

The behavioral events of parturition are also believed to play a role in the onset of maternal behavior. These events are numerous and could include the various stimuli from the fetuses, placentas, birth fluids, delivery process, etc. Attention has mainly been focused however on the licking of pups. Birch (1956) suggested that licking of the pups establishes a bond between the mother and her young for which the mother was prepared by the licking of her own body during pregnancy (Roth and Rosenblatt, 1967). While Birch's own study supported this hypothesis in that females that wore collars from weaning through the end of pregnancy and therefore were prevented from licking themselves when giving birth (i.e., without the collars) failed to lick and take care of their newborn. Nearly all the newborn died. Other investigators, however, have not been able to substantiate these findings using the procedures described by Birch (Coomans, 1955; Friedlich, 1962; Christopherson and Wagman, 1965).

Doubt has been cast on the role of parturitive behavior in the onset of maternal behavior by several studies in which delivery has been accomplished by caesarean section, thus eliminating the mother's behavior entirely (Wiesner and Sheard, 1933; Labriola, 1953; Moltz, Robbins, and Parks, 1966). The appearance of maternal behavior within 24 hours after delivery, without prior contact with pups during birth and in the absence of eating of the placenta, emphasizes the importance of the nonbehavioral (i.e., physiological) processes in the onset of maternal behavior.

Before leaving the question of the importance of behavioral events during parturition in relation to the onset of maternal behavior I would like to suggest that issue has not yet been adequately settled. It has been shown that maternal behavior can proceed after caesarean-section delivery and to that extent parturitive behavior is not crucial. However the high level of maternal responsiveness which exists at parturition may allow for alternative paths to the onset of maternal behavior. One path, that normally taken, is through contact with young during parturition. Another, shown to be equally effective, is through postparturition contact with the young. It would be important to study three separate groups of new mothers that have had contact with young either during parturition, for an equivalent period after parturition, or during both periods and to compare them with respect to the onset of maternal behavior 24 hours after parturition.

Maternal Responsiveness During Pregnancy

The onset of maternal behavior has been seriously interfered with by removing the ovaries between the 19th and 21st day of pregnancy; such removal is followed either by a difficult natural delivery or a caesarean section (Jost, 1959; Moltz and Wiener, 1966). Fifty percent of the mothers fail to exhibit maternal behavior and their litters die shortly after birth while the remaining mothers are adequate and rear their litters to weaning. While it is apparent therefore that ovarian hormones and possibly pituitary hormones whose release is stimulated by the ovarian hormones play a role in the onset of maternal behavior it has not been easy to find what hormones are involved. An earlier report of the effectiveness of prolactin in inducing maternal behavior in virgins (Riddle, Lahr, and Bates, 1942) has not been confirmed in recent studies in which this pituitary hormone, and estrogen and progesterone, given singly and in combinations, to virgins and experienced mothers, have not induced maternal behavior (Beach and Wilson, 1963; Lott, 1962; Lott and Fuchs, 1962).

It is not clear how ovariectomy a day or two before delivery interferes with the onset of maternal behavior. Among the possibilities are that plasma estrogen is prevented from increasing, and plasma progesterone, which is already low, is decreased further; perhaps pituitary secretion of prolactin is also affected. In a recent study an attempt was made to determine whether preventing the fall in progesterone level would produce any interference with the onset of maternal behavior in primiparous females (Moltz, Levin, and Leon, 1969). Progesterone was injected into pregnant females during the last four days before parturition (day 19 to 23) and caesarean sections were performed (i.e., on day 21) to deliver the young. Fifty percent of these mothers, like the ovariectomized ones, did not show maternal behavior when they were given newborn young. Since these females retained their ovaries, the injection of progesterone may have had the effect not only of maintaining a high level of progesterone but also of preventing an increase in estrogen secretion by preventing the release of FSH (Rothchild, 1965). Moltz et al. (1969) have chosen however to interpret these findings as indicating that the injected progesterone blocked the stimulating effect of prolactin, which, they believe, is the hormone responsible for the onset of maternal behavior. According to this view the reduction of circulating progesterone which normally occurs at parturition removes the block of prolactin effects, enabling prolactin to exert its effect on maternal behavior. The rise in estrogen secretion also normally plays a role presumably in that it stimulates the release of prolactin from the anterior pituitary gland.

This theory is similar to the theory that has been proposed to explain the initiation of lactation in the rat which also occurs at parturition and involves progesterone blocking the action of prolactin at the mammary gland (see Cowie and Folley, 1961). It has the virtue, therefore, of proposing a common hormonal basis for the onset of maternal behavior and lactation, two maternal functions that parallel one another in several respects and diverge in others.

An important assumption underlying attempts to find the hormonal basis of maternal behavior is that these events take place around parturition. In reality, however, we know very little about when maternal responsiveness first arises. Wiesner and Sheard (1933) suggested that maternal responsiveness develops gradually during pregnancy in synchrony with gestational processes, but independently of them. They provided no substantial evidence for this view, however. On the other hand, studies on the rabbit have shown that terminating pregnancy about two-thirds of the way to term, either by caesarean section or ovariectomy, is followed by the appearance of nestbuilding in a significant percentage of the females (Zarrow et al., 1962).

Dr. Dale F. Lott and I undertook preliminary studies to determine the earliest time during pregnancy when we might find a significant increase in the maternal responsiveness of pregnant females over what we believed at that time was the nonresponsiveness of virgins (Lott and Rosenblatt, 1969). We felt that if we could find a period during pregnancy when the first increase in maternal responsiveness could be measured then we could study that period intensively for the hormonal basis.

Since our studies up until that point indicated that pregnant females do not show maternal behavior, we used the procedure of terminating pregnancy by caesarean-section delivery or total hysterectomy to stimulate maternal behavior. Pups, 5 to 10 days of age, were given to the females 24 hours after the operation and remained with them continuously, being replaced daily with fresh pups because the females did not lactate. Maternal responsiveness was measured by the latency (in days) for the appearance of retrieving in tests carried out each morning when the pups were exchanged. After the retrieving test the females were observed for 2 hours using 1-minute spotchecks every 20 minutes during which the appearance of crouching over the young (as in nursing), nestbuilding, licking and, again, retrieving was recorded. Each morning the presence or absence of a nest was also recorded.

Pregnancies were terminated in different groups of females at the 8th, 10th, 13th, 16th, and 19th day, and tests were begun 24 hours later. Differences between groups in the average latency for the onset of retrieving (and the other items of maternal behavior) were our measures of differences in the maternal responsiveness due to the intervening period of pregnancy.

On the basis of the results, which are given below, we studied the maternal behavior of pregnant females: two groups, 10-day and 16-day pregnant females, were studied. They were given pups on the 11th and 17th days of pregnancy, to match the earlier procedure, and latencies for retrieving (and other items of maternal behavior) were measured. In a second study, the same two groups (i.e., 10-day and 16-day pregnant females) were hysterectomized but in addition the ovaries were removed at the same time; this procedure eliminated postoperative secretion of ovarian hormones. They were then observed in the manner described above.

Our final group was used as a control but it proved to be perhaps more interesting than the main experimental groups, as we shall show later; this group consisted of virgins who were simply given young and observed like the experimental females.

The results of our first experiment (Table 1, column 4) were surprising

TABLE 1

Latencies (days) for the onset of retrieving in pregnant rats exposed to pups after various treatments. Latencies (mean + SD) are given from day of operation or the equivalent day of pregnancy.

				Treatment				
(1)	(2)	(3) Day pups	(4)	(5)	(6)	(7)	(8) Hysterectomy +	(9) (4–8)
Group	N	given	Hysterectomy[a]	N	Pregnant[b]	N	Ovariectomy	p-value
19-days pregnant	8	20	2.25 ± 0.28					
16-days pregnant	16	17	2.31 ± 0.67	8	4.63 ± 0.99[c]	14	4.57 ± 2.02[d]	< .05
13-days pregnant	9	14	3.67 ± 1.35					
10-days pregnant	11	11	4.18 ± 2.01	9	7.33 ± 1.89	7	6.42 ± 3.02	< .005
8-days	8	9	6.37 ± 1.05					
Virgins	14	—	6.78 ± 2.80					
			$F = 14.23$ $df = 5/60$ $p < .01$		$p < .005$		$p < .05$	

[a] Some animals were hysterectomized and others had only their fetuses and the placentas removed by Caesarean-section. Thus far these two procedures have given the same behavioral results.

[b] Comparison between means of columns 6 and 8 were not significant (t test).

[c] Mean latency is significantly shorter than the latency of virgins at the .01 level of confidence.

[d] Mean latency is significantly shorter than the latency of virgins at the .02 to .05 levels of confidence.

since at no time following the different terminations of pregnancy was it not possible to stimulate the onset of retrieving and other items of maternal behavior, and even the virgins showed these behaviors, eventually. Although retrieving could be elicited from all of the groups, nevertheless there were important differences in the latencies and therefore in maternal responsiveness. Starting at the 10th day of pregnancy, the latencies for the onset of retrieving became progressively shorter (compared to the virgins) as pregnancies were allowed to continue longer before the females were hysterectomized. By the 16th to 19th day, termination of pregnancy was followed by the onset of retrieving in an average of 2 days.

Since other studies have shown that on the 21st day termination of pregnancy (i.e., by caesarean-section delivery) is followed by maternal behavior within 24 hours (Labriola, 1953; Moltz, Robbins, and Parks, 1966), the increase in maternal responsiveness (i.e., shortened latency for the onset of retrieving) seen during the last half of pregnancy, from about the 10th day up until the 19th day, evidently continues until what would be the time of parturition.

Evidence of an increase in maternal responsiveness during pregnancy was also obtained in the study in which 10- and 16-day pregnant females were given pups (on the 11th and 17th day) without terminating the pregnancies. Maternal responsiveness increases some time between the 11th and 17th day (Table 1, column 6) but at the 17th day it was not as high in the pregnant females as it had been in the hysterectomized animals. Evidently terminating pregnancy has an effect on maternal responsiveness over and above that caused by pregnancy itself, and at the 10th day of pregnancy, only the effect of terminating pregnancy could be measured in the hysterectomized females.

One further point emerged from our study of 10- and 16-day pregnant females that were both hysterectomized and ovariectomized. Removal of the ovaries eliminated the effect of the hysterectomy; the females were no more responsive to young than they were as pregnant animals. Thus, in effect, ovariectomy delayed the onset of retrieving in the hysterectomized females (Table 1, column 8). It will be recalled that Jost (1959) and Moltz and Wiener (1966) found that 50 per cent of the females that had been ovariectomized just prior to parturition (or caesarean-section delivery) did not show maternal behavior. These females may have been delayed in the onset of maternal behavior. After their litters died, however, they were not given new litters and therefore they may not have had the opportunity to initiate maternal behavior after the delay.

These studies show that there is considerable development of maternal responsiveness during pregnancy prior to the beginning of parturition. By the time the female is about to give birth, her maternal responsiveness has already developed to a high level and needs only to be increased a relatively small amount for the onset of maternal behavior to occur. Termination of pregnancy by normal delivery provides the *increment* required for the onset of maternal behavior, a fact which has been the basis, we believe, for the impression that the onset of maternal behavior is based entirely upon changes occurring around parturition. According to this hypothesis, females ovariectomized just prior to normal parturition (Jost, 1959; Moltz and Wiener, 1966), having passed through most of pregnancy, are likely to have already developed a high level of maternal responsiveness, and, therefore, the onset of maternal behavior is probably delayed but not prevented by the removal of the ovaries. Our findings suggest also that hormonal changes throughout the last half of pregnancy are probably the basis for the onset of maternal behavior insofar as they influence the level of maternal responsiveness during this period. We are not yet in a position, however, to identify these hormonal changes.

Nonhormonal Basis of Maternal Responsiveness

Maternal behavior has its origins even earlier than the last half of pregnancy: virgins are also capable of displaying maternal behavior in response to prolonged stimulation by pups (Table 1, column 4). The maternal behavior exhibited by virgins includes not only retrieving pups to a nest, but nestbuilding, licking the young, and crouching over them in a nursing position, without lactating, of course. Moreover, once aroused by pup stimulation, virgins can maintain their maternal responsiveness over an interval of at least two weeks during which they receive no further stimulation by pups (Rosenblatt, 1967).

The maternal behavior exhibited by virgins does not appear to depend upon pituitary or ovarian hormones (Cosnier and Couturier, 1966; Rosenblatt, 1967). Neither hypophysectomy nor ovariectomy prevents virgins from exhibiting maternal behavior in response to pup stimulation. Furthermore, males also can be induced to show all elements of maternal behavior except in some cases, nestbuilding, and removal of the testes does not prevent this. Maternal behavior is therefore a basic characteristic of the rat which is not dependent upon sex or hormones for its appearance, but which may be induced more rapidly as a result of stimulation by hormones.

During the period of induction, the virgin is initially attracted to the young pups and makes repeated contacts with them. The stimulation received during these contacts apparently mounts to the point where licking, retrieving, etc., are gradually aroused, all four items of maternal behavior appearing within a short time of one another.

It has recently been reported, however, that retrieving alone may occur in virgins without any prior exposure to pups and that pups as well as small plastic toys will be retrieved equally (Plume et al., 1968). Virgin females, males, and pregnant and lactating females were presented with a pup and a small toy for seven daily retrieving tests and the proportion of tests in which the pup or the toy was retrieved was compared and found to be equal for all groups except the lactating mothers. Lactating mothers retrieved the toy as frequently as the virgins but their retrieving of pups was three times more frequent.

It is difficult to determine from this report how many virgins retrieved pups while it is clear that all of the lactating mothers retrieved them and many of the pregnant females also retrieved pups. We have therefore repeated this study following as closely as possible the procedure used but with closer attention to the behavior of the virgins in response to the pup and the toy. We recorded instances of sniffing the pup and toy, licking and mouthing, and picking up, carrying, and depositing them.

We find that retrieving of pups is rare among the virgins but that they sniff the pup a good deal and then frequently lick it but almost never mouth it or pick it up. When retrieving of the pup does occur the complete sequence is performed rapidly: the virgin sniffs, licks, picks up the pup, carries it to a corner and deposits it, then usually leaves.

The virgin's response to the toy, almost from the beginning, is quite different from its response to the pup. The toy is sniffed as frequently as the pup but it is never licked; the virgin mouths the toy and begins to bite it, carrying it a short distance in the process, dropping it, picking it up again, mouthing it, etc., until the toy is brought to a corner. She remains with the toy and may often carry it away from the corner and return again. The toy continues to elicit "play" throughout the test, while the pup frequently elicits sniffing but no further actions in the retrieving sequence, except occasionally.

It appears to us that while the virgin's reactions to the pup and toy share many movements in common, and responses to both are more frequent in more active animals, they are different behaviors based upon the virgin's ability to perceive differences in the two objects. Towards the pup the virgin displays only actions preliminary to retrieving (i.e., sniffing and licking) but not being "maternal" as yet, she does not retrieve it, except occasionally. The toy elicits play activity and this is not based upon the maternal condition. In support of this view is the finding that while pup retrieving was three times more frequent in the lactating mother than in the virgin, retrieving of the toy by mothers did not increase or decrease proportionately but remained at the same frequency as in the virgin (Plume et al., 1968).

Such tendencies as to differentiate between toy and pup, to mouth the first and lick the second, may indicate the perceptual and response basis of maternal behavior in females who until then have not been exposed to pups nor stimulated by the hormones related to maternal care. That a significant basis already exists in females prior to exposure to young is indicated in a recent study by Thoman and Arnold (1968) in which females reared in isolation, and fed by hand from birth, became adequate mothers and reared litters that showed only slight effects of their rearing by these mothers as compared to normally reared mothers. Our studies have not yet reached the point where we can formulate ideas about the developmental origins of maternal behavior in the rat in the way that such ideas are being studied in other rodents (see review by Richards, 1967).

With maternal behavior now shown to be present in the virgin, the problem arises as to what action hormones have on maternal behavior during pregnancy. There is no direct evidence on this point but the study described earlier (Terkel and Rosenblatt, 1968), which in fact was undertaken following a suggestion of Stone (1925), provides a partial answer to this question. The latency for the onset of retrieving was shortened considerably by the injected blood plasma from the mothers; latencies averaged a little more than 2 days. In some cases maternal behavior (e.g., licking pups and attending to them) appeared within 4 hours. The various control groups that were injected with blood plasma taken from estrous-cycling females, or with saline solution had average latencies ranging from 4 to 7 days. Clearly then the blood plasma of a recent mother engaged in maternal behavior contains a substance or substances (i.e., hormones) capable of accelerating the effect of pup stimulation in eliciting maternal behavior.

MAINTENANCE OF MATERNAL RESPONSIVENESS

Hormonal or Nonhormonal Maintenance of Maternal Responsiveness

As was indicated previously, following parturition there is a period of 2 to 3 weeks during which the maternal behavior established at parturition is maintained at what appears to be a fairly constant or slightly declining level (Moltz and Robbins, 1965; Grota and Ader, in press). Nestbuilding, nursing, and licking the young appear regularly and retrieving occurs whenever the young are outside the nest. At the end of this period, nestbuilding declines as the young no longer remain at the nest site, and whatever nesting material is gathered by the mother is soon scattered by the young. Also, few opportunities for retrieving and licking occur since the young run around the cage and frequently avoid the mother or engage in play among themselves. Nursing, now initiated by the young, is seen only infrequently and it lasts only for a brief period as the young become capable of independent feeding from sources other than the mother (Rosenblatt and Lehrman, 1963; Rosenblatt, 1965; Moltz and Robbins, 1965). The close correlation between changes in the behavior of the young and the maternal responsiveness of the female, leading to a decline in maternal behavior during the 3rd and 4th weeks, suggests that the maintenance of maternal responsiveness during the preceding weeks is based upon the relatively unchanging character of the stimuli from the young and their effect upon the mother, a suggestion that has been partially confirmed by several studies (Rosenblatt and Lehrman, 1963; Rosenblatt, 1965).

There are, however, several indications that the physiological (i.e., hormonal) basis for the development of maternal responsiveness during pregnancy and parturition, discussed in the previous section, is no longer operative once maternal behavior has been established. Obias (1957) hypophysectomized females on the 13th day of pregnancy; pregnancy was maintained and although many animals had difficulty in giving birth, a number of females dying in the process, the remaining mothers exhibited maternal behavior (i.e., nestbuilding, crouching over young, and retrieving) after parturition, and, except for their failure to lactate, they were similar to intact mothers. The maintenance of pregnancy by these hypophysectomized females can be explained by the placental secretion of hormones with luteotropic effects which reaches a maximum by the 11th day of pregnancy. This maintains the secretion of progesterone by the ovary (Ray et al., 1955; Kinzey, 1968). However, after parturition the hypophysectomized mother does not have any source of prolactin, yet she is able to maintain her maternal responsiveness.

I have found that prolactin and oxytocin do not maintain the maternal responsiveness of mothers who have given birth to young and have begun to exhibit maternal behavior (Rosenblatt, unpublished). The young were removed from the mothers 12 hours postpartum, the mothers were ovariectomized and prolactin (2× daily) and oxytocin (4× daily) injections were given for the next 10 days. At the end of this time, young were presented to the mothers and only 30 percent exhibited maternal behavior (i.e., retrieving), a percentage that was not greater than was found in uninjected or saline-injected mothers used as controls. The results were no different when the injections of prolactin and oxytocin were begun after caesarean-section delivery of 19-day pregnant females that were also ovariectomized. Moreover, when the young were allowed to remain with them continuously, with daily exchanges of the litter for fresh pups, the rates at which the mothers resumed maternal behavior were also not different among the experimental and control groups and none differed significantly from untreated virgins that were made maternal by pup stimulation alone. It appears therefore that while prolactin is necessary for lactation postpartum, it is not necessary for the display of maternal behavior.

The maintenance of maternal behavior does not appear to depend upon ovarian hormones either, although the development of maternal responsiveness, as we have seen, is prevented or seriously delayed by ovariectomy before parturition. The development of maternal behavior was delayed or prevented in 50 percent of the females that were ovariectomized between the 19th and 21st day of pregnancy (Jost, 1959; Moltz and Wiener, 1966). I have found that ovariectomy within 12 hours after parturition does not affect the maintenance of maternal behavior and mothers go on to rear their litters until weaning (Rosenblatt, unpublished). Progesterone administered to females in late pregnancy (i.e., 19th to 23rd day) also causes a delay or inhibition of maternal behavior, but after maternal behavior has been established, it has no apparent effect (Moltz, Levin, and Leon, 1969).

Recently it has been reported that maternal care is improved by either estrogen, progesterone, or prolactin. These hormones increase the success with which 10-day lactating mothers rear to weaning cleaned newborn that are given to them in exchange for their own litters (Denenberg, Grota, and Zarrow, 1963; Zarrow, Grota, and Denenberg, 1967). In the next section we shall discuss these studies in detail, but for now we can say that it is not clear that the improved success in rearing the pups caused by these hormones results from an improvement in maternal behavior (Grota, 1968).

There is no clear evidence, therefore, that the mother requires either ovarian or pituitary hormones for the maintenance of maternal behavior after parturition. Since removal of the pituitary also prevents the release of adrenocortical and thyroid hormones, maternal behavior after parturition does not appear to be dependent upon these hormones either. The question arises then: what does maintain the mother's responsiveness to her young after parturition?

Stimuli from the young, as we have already shown, are capable of inducing maternal behavior in virgins and of maintaining it for an indefinite time (Rosenblatt, 1967). The maternal behavior shown by the virgins, in response to the 5- to 10-day old pups that are used, resembles in nearly all respects, except actual nursing, the maternal behavior of lactating mothers that are rearing their own 5- to 10-day old pups (Fleming and Rosenblatt, unpublished). It is likely that the hormones necessary for the onset of maternal behavior at parturition and the stimuli from the young which induce maternal behavior in the virgin act upon common sites in the nervous system, though this idea has yet to be experimentally tested. Evidence suggests however that maternal behavior can be induced more rapidly (i.e., about 2 days) by confining virgins in narrow cages where they are forced to remain in contact with pups continuously rather than only sporadically as in our larger, standard cages (Terkel and Rosenblatt, unpublished). By the same token, the induction of maternal behavior is delayed when pups are presented in wire baskets attached to the inside of the female's cage where the female cannot lick or contact them in other ways (Roth, 1967). The stimulating effect of young upon the maternal responsiveness of the female varies, therefore, with the amount of contact she has with them which permits the various stimuli from the young to exert their effects. The onset of maternal behavior at parturition, and its onset during exposure to pups, in virgins, ensure that from then on the female will have close and continuous contact with pups. The most favorable conditions for the young to play an important role in the maintenance of maternal behavior are therefore established under these two differing circumstances, that of hormonal induction and that of pup stimulation of maternal behavior.

The onset of maternal behavior in virgins (i.e., with pup stimulation) is gradual but in the postpartum mother it is sudden, beginning shortly after the pups have been cleaned following their delivery. It is possible to remove pups from parturient mothers therefore before she has had any opportunity to "mother" them, apart from cleaning them during parturition. Normally with pups present from birth onward, the mother begins to exhibit maternal behavior and continues until the 3rd to 4th week. However, if the young are removed at birth and mothers receive no further stimulation for the next 4 days, on the beginning of the 5th day, they are unresponsive to pups and neither retrieve nor crouch over them, or build nests (Rosenblatt, and Lehrman, 1963; Rosenblatt, 1965). If they are permitted to remain with the pups for several days, the mothers gradually begin to show all the items of maternal behavior but these fall off again as the young become weak and gradually die, since after a 4-day period without nursing, mothers no longer lactate.

Parturition therefore serves as the transition period during which hormonal processes that contribute to the development of maternal behavior at delivery come to a more or less abrupt end and new processes begin, associated with stimulation from the pups. These serve to maintain the female's maternal behavior.

Removal of the litter for a period of 4 days, after maternal behavior has been in progress for 3 days or longer, does not have as great an effect on the maintenance of maternal responsiveness as the removal of pups at parturition (Rosenblatt and Lehrman, 1963; Rosenblatt, 1965). Maternal behavior is resumed by 40 to 60 percent of the mothers immediately after the 4-day interruption and the remaining mothers resume their care of pups within 1 or 2 days. The mother's dependence upon the young is thus greater at parturition than several days later when her maternal behavior is more firmly based upon postparturitional processes.

Maternal behavior in virgins, induced by exposure to pups, is also maintained for some time after the pups have been removed. Maintenance is not dependent upon hormones since both hypophysectomized and ovariectomized virgins immediately resume maternal care of pups 2 weeks after the original removal (Rosenblatt, 1967). Cosnier and Couturier (1966) reported that mothers who had weaned their young some 2 months earlier and were ovariectomized shortly afterward, nevertheless showed maternal behavior towards newborn pups almost immediately. Intact females with the same histories behaved similarly.

SYNCHRONY IN THE BEHAVIORAL INTERACTION
OF THE MOTHER AND HER YOUNG

The concept of synchrony, referring to the fact that the mother's behavior is adapted to the needs and behavioral capacities of the young and that it changes as these develop in the young, has proven useful. As a descriptive concept it raises such questions as: how is the behavior of the mother and her young continually adapted to each other, what sensory stimuli are provided by each, and what are the changes that occur from birth to weaning that maintain the synchrony between mother and young? As a causal concept it suggests that the relationship between the mother and her young, in particular the motivational condition of the mother, and the developing socialization of the young underlying this relationship, are established and maintained by the reciprocal exchange of stimulation (i.e., trophallaxis). Moreover, this view suggests that weaning is a natural outgrowth of processes already present in the preweaning relationship and represents a weakening of this relationship as a result of changes in both the mother and her young. Furthermore, synchrony between mother and young is the outcome of natural selection in which each has exerted selective pressure upon the evolution of behavior in the other; this is an aspect of social selection to which Scott (1967) has recently referred. Questions about the evolution of maternal behavior and of early behavioral development can therefore be raised in terms of the concept of synchrony. Finally, this concept makes it necessary in studying early development to consider the social context, to be aware that any influences upon the long term development of the young that act during the suckling period are likely to affect the synchronous relationship with the mother first and through this the behavioral development of the young.

Synchrony is established early and mainly through the initiative of the mother as indicated in nursing and retrieving behavior. In nursing her newborn the mother enters the nest, stands above the young then lowers herself on top of them, settling into a nursing position in this manner. In preliminary studies it has been noted that young pups suckle only while on their backs, reaching upward towards the mother's nipples, nuzzling and attaching to a nipple. To test whether pups could initiate suckling an anesthetized mother was placed on her side near the pups. The pups did not crawl the short distance necessary to reach her; when she was placed in contact with them while lying on her side, they did not nuzzle or attach to her nipples. The mother's behavior is therefore adjusted to the suckling response of the newborn which is normally elicited by contact from above (and can also be elicited in tests with an anesthetized mother held above the pups) and proceeds only when the pups are in supine position. In this position also the young are easily stimulated to eliminate by the mother's licking of their ano-genital region a maternal function made necessary by the inability of the pups to eliminate voluntarily.

Retrieving improves rapidly during the first week after birth: the mother initiates it sooner when pups are out of the nest, fewer extraneous activities interrupt the completion of retrieving, and what is most significant, the pups are dropped much less often while being carried back to the nest. Carlier and Noirot (1965) have shown that around the 4th day the mother rapidly adjusts her mouth grip to reduce the amount of squeaking by the pups. Squeaking occurs when the mother grasps the pups by the legs, tail, head or neck. It is very much reduced when the pups are grasped centrally on the body, that is at the back, flank, or abdomen. The mother's adjustment to the pups during retrieving, is, very likely, in response to the reduced vocalization which occurs when she grasps them centrally rather than at the extremities or head region. In addition to provoking less squeaking, the mother achieves a more balanced mouth grip for carrying the pups back to the nest. Once this adjustment has been made by a primiparous mother, it is retained when her second litter is born.

In the following two sections several recent studies on maternal care of young in the rat will be discussed in terms of the concept of synchrony with the aim of examining the value of this concept for the understanding of a wide range of phenomena in the relationship of the mother and her young.

Synchrony in the Nursing-Suckling Relationship

Bruce (1961) reported some years ago that there was a high rate of pup mortality shortly after birth among mothers that had become pregnant during postpartum estrus and had carried the second litter while suckling and rearing the first. She attributed the infant deaths to nutritional depletion of the mother during gestation and to inadequate lactation after the second parturition. Recently, however, while studying the maternal behavior of late-pregnant females, we found a similar high rate of mortality among the newborn. They died of nutritional deficiency which resulted from their inability either to grasp the mother's nipples or to express milk from them. Dr. Benjamin Sachs and I were struck by the similarity between our findings and those of Bruce although the two situations were quite different: the prospective mothers in our study were exposed to pups during late-pregnancy in order to induce maternal behavior, but the pups, that were 5- to 10-days of age, did not elicit milk secretion during their suckling as did the pups reared by the mothers in Bruce's study (Masson, 1948). Thus, nutritional depletion of the mother could not explain our findings. We set about investigating the basis for the high rate of mortality of pups born to these mothers with the possibility in mind that our findings might help us to understand Bruce's results also.

Study of the mothers indicated that the exposure to pups had not interfered with pregnancy nor with lactation: normal vaginal smears of pregnancy were recorded, parturitions took place at the normal time, the pups were alive and well at birth, the females showed good maternal behavior, and the mammary glands contained adequate amounts of milk. We noticed however that in the first hours after parturition the nipples were often bleeding, lacerated, and had been stretched thin, a condition not found in mothers that had not been suckled by pups before parturition. Our attention therefore turned to the suckling behavior of the older pups (i.e., 5- to 10-day old pups) that were used to induce maternal behavior during pregnancy.

Two possible explanations for the condition of the mother's nipples came to our minds as a result of examining our procedures: the first was that the older pups had suckled from the pregnant female and had perhaps damaged the nipples and the second concerned the time when the suckling could have produced such damage to the nipples. We therefore covered the mouths of the older pups with collodion before placing them with the late-pregnant female; this procedure prevented them from suckling but it had no effect on their efficacy as stimuli to induce maternal behavior. The females became maternal towards the older young but were not suckled and when their own litters were born hardly any of the young succumbed from starvation; collodion placed elsewhere as a control (i.e., on the pups' noses) did not prevent postpartum death of the mothers' own young. Thus, suckling was shown to be crucial in the deaths of the mothers' litters.

Our procedure had been to allow the older pups to remain with the mothers until a few hours *after* the litters were born. The older pups were then removed and the newborn had uncontested access to the mothers' nipples. In a second study the pups were removed *before* parturition, after they had already induced maternal behavior in the pregnant females. With this change in our procedure pup deaths also declined to nearly zero. We examined the nipples of females that were suckled for the extra 12 hours during which parturition occurred and compared them with the nipples of females that were suckled earlier but were not suckled during this period. The former as we indicated above, were often bleeding, while the latter looked red from suckling but were not damaged.

Nursing the newborn must have been painful for the mothers that had been suckled by the older pups during the previous 12 hours that included parturition. As a result the mothers either did not allow the newborn access to the nipples although she crouched over them or the newborn were unable to express milk from the nipples.

The vigorous sucking of the older pups during the last 12 hours before the mothers gave birth to their young damaged the still small and tender nipples. Among human mothers the nipples become larger and more protractible during pregnancy enabling the infant more easily to draw the nipple into the mouth and suck (Gunther, 1961; Hytten, 1963). It is likely that similar changes occur in the nipples of the rat under the hormonal stimulation of pregnancy. At delivery the nipples are adapted to the sucking of newborn but not to the more vigorous sucking of 5- to 10-day old pups (Bruce, 1961). Yet older pups are of course able to suckle from their own mothers without damaging her nipples. Apparently the nipples normally undergo a gradual change which enables them to withstand the more vigorous sucking of the young as they grow. It is more than likely that this change is produced by the young themselves during suckling.

The relationship between suckling and maternal lactation appears to be synchronized throughout the entire period of maternal care. The synchrony is progressive with respect to the amount of suckling stimulation the mother receives and the amount of milk she produces, and with regard to the stimuli from the young which are capable of promoting lactation. Early in lactation only suckling is capable of causing the release of pro-lactin from the anterior pituitary gland but later, after a week and a half of nursing, prolactin is released without actual suckling when the mother can receive exteroceptive stimulation of various sorts from the young near by (Ingelbrecht, 1935; Eayrs and Baddeley, 1957; Grosvenor, 1965). While the neuroendocrine mechanisms underlying the synchrony between mother and young during active lactation are fairly well understood the nature of weaning and of the termination of lactation are not yet well understood (Meites, 1966). Some recent results have been reported when newborn are fostered to mothers at various states of lactation which can contribute to our understanding of the feeding synchrony.

It has been reported that there was a high rate of mortality (i.e., 33 percent) among newborn (i.e., less than 1 hour old) that were given to 10-day lactating foster mothers in place of their own litters (Denenberg, Grota, and Zarrow, 1963). The rate of mortality was slightly lower with 5-day lactating foster mothers (i.e., 29 percent) and it approached the normal rate when 1-day lactating females were used as foster mothers (i.e., 6 percent). Among all groups, deaths occurred soon after the exchange but they were not due to cannibalism: cannibalism was low among all the groups and was not higher among the 10-day than the 1-day lactating foster mothers (Grota, 1968). Where the discrepancy between the status of the newborn and of the foster mother was small, as for example, when newborn were given to mothers in their first day of lactation, the young gained weight at a normal rate from the start. Newborn given to 10-day lactating mothers, on the other hand, gained very little weight up to the 5th day, and during this period many of them died from starvation. Furthermore, fewer pups survived in large litters (i.e., 10 pups) than in small litters (i.e., 4 pups) with 10-day lactating mothers; those that did survive until the 5th day began to increase their weight at the normal daily rate from then on. Taken together these findings suggest that the feeding relationship between the mother and the newborn was disturbed by the exchange of newborn to mothers, further along in the feeding sequence (i.e., feeding of 10-day old pups). Many of the young were unable to survive the period required for establishment of a synchronized feeding relationship between the new mother and the developing young. Those that did, however, eventually were able to establish such a relationship and from then on feeding progressed normally.

The initial failure of feeding was based on either behavioral incompatibility between the foster mother and the newborn (reflected in the poorly established feeding relationship) or a failure of lactation; perhaps both were involved. Observations of feeding are lacking in the reports cited, but the 10-day lactating mothers appeared no less maternal towards the newborn than the 1-day lactating mothers and in fact when tested, they retrieved their newborn more rapidly than did the 1-day lactating mothers.

Failure of lactation as the cause of the high rate of pup mortality is suggested by Bruce (1961). She found that the transfer of newborn pups to mothers that were nursing older pups (i.e., 6 days or older) caused almost complete cessation of lactation; the newborn showed little weight gain for several days until they reached the age of 6 days when normal weight gains began. Bruce suggested that after maximum lactation is established about the 6th day the suckling by younger pups is not enough to maintain it and there is an abrupt reduction of milk production. These findings so closely parallel those reported by Denenberg, Grota, and Zarrow (1963), Zarrow, Grota, and Denenberg (1967), and Grota (1968) as to provide an explanation for the latter.

In addition to the findings discussed above, these investigators also reported that eating of placentas attached to the newborn (i.e., less than 12 hours old) given to the 10-day lactating mothers improved the survival rate of the newborn. Pups with placentas had a survival rate of about 70 percent but those without placentas survived in only 26 percent of the cases. No behavioral observations were made to indicate the basis for the difference in survival: the difference might reside in the mothers or it might be due to the young, since after they were 12 hours old this difference in survival based upon the presence or absence of placentas no longer appeared.

Some indication that mothers are influenced by hormones present in the placentas was obtained indirectly by substituting injected hormones. The difference in the rate of survival of pups with and without placentas disappeared when estrogen or prolactin were administered to the mothers at the same time they received the newly born pups. Progesterone had a similar but less pronounced effect. In all instances however the basic low rate of survival of newborn given to 10-day lactating mothers was not reduced: what was reduced was the very much lower rate of survival of the pups without placentas. The latter rate appears to be affected by both placenta-eating and the hormones cited above if they accompany the transfer of young to the foster mothers; the former rate however is un-affected by placenta eating or the hormone treatments. These studies re-main inconclusive therefore with regard to the role of ovarian hormones and prolactin in maternal behavior. They do point to the importance of maintaining the synchronized relationship between the mother and her young for optimal survival of the young.

Maternal Care and Behavioral Development of the Young

The synchronous character of the mother-young relationship during the lactation period ensures that the physical and behavioral development of the young will occur under optimal ecological and social conditions insofar as these are under the control of the mother (Sturman-Hulbe and Stone, 1929; Nolen and Alexander, 1966). The mother, responding to environ-mental disturbances, in most cases provides protection against them not only for herself but for her litter as well. The young, on the other hand, who are also directly affected by various disturbances (e.g., separation from the mother and littermates), react in ways (e.g., vocalize) that elicit the appropriate maternal care from the mother.

An attempt to evaluate the stabilizing effect of the mother-young rela-tionship in mice and rats on the behavioral development of the young was made by Tobach and Schneirla (1962) and Tobach (1966, 1968). They compared the effects of transferring litters from one mother to another, daily and transferring litters with their mothers to new cages, daily. Young that remained with their own mothers throughout the daily transfers to new cages were emotionally more stable after weaning than young that were moved from mother to mother each day. The measures used were defecation and activity in an open field. A second group of mothers with their young were transferred to new cages each day but, unlike the group described above, the new cages differed in many features from the cages to which the animals had been accustomed. The mother herself was therefore markedly disturbed by the transfer, and this persisted from some time, being renewed each day when another transfer occurred. The mother's disturbance was transmitted to the young and despite the constancy of the social group from day to day, the young developed a fearfulness that was similar to that of the young that had been transferred from one mother to another each day.

The mother's response to her young is highly discriminating and may include the ways in which her young differ from similarly aged young of another mother. Beach and Jaynes (1956) found that rat mothers retrieved their own young in preference to alien young of the same age (i.e., 6 days of age). Among mice Meier and Schutzman (1968) reported maternal discrimination between male and female young with preferred retrieving of the females first. The basis exists therefore for the mother to be responsive to subtle characteristics of her young—certainly she responds to those features which change with age—and to alter her behavior in response to any changes they undergo.

On the basis of this capacity of the mother, Richards (1966) proposed that many of the developmental effects of early handling and other types of special treatment of the young, reported to produce long lasting effects on emotionality and other behavior characteristics, exert their influence through changes in the mother's behavior towards the treated young. Thus a brief period of treatment (i.e., 3 minutes), early in the litter period (i.e., 2nd to 5th day), may in fact result in an altered relationship between the mother and her young which lasts for the entire suckling period.

It is in regard to the reported effects of brief separation of the young from the mother and exposure to cold, handling, rotation, and other special stimulation during early development that Richards' proposal assumes considerable importance. These treatments are reported to have short-term effects indicated by the earlier maturation of various reflexes (e.g., startle reflex), and of eye opening, and of adrenocortical hormone release to cold and other stresses. They also have long-term effects which include earlier vaginal opening, more effective resistence to stress, infection and other adverse conditions, and less fearfulness as well as more efficient reproduction (Levine, 1962; Barnett and Burn, 1967). Richards has suggested, in effect, that rather than being a product solely of the brief period (i.e., the "critical period") of special stimulation during the experimental treatment, these behavioral effects are the result of a change in the mother's relationship to the treated young, and as such, the young are exposed to an altered treatment over a prolonged period. Levine (1967) and Meier and Schutzman (1968) have subscribed to the same view.

There is evidence that mothers respond to the effects of such treatment of their young and in some instances this does result in a different kind of relationship than the normal one between the mother and her young. Young (1965) exposed individual members of several litters to either a brief period of rotation or cooling and found that mothers retrieved these pups sooner than they did pups of the same litters that were not treated. Among mice, Barnett and Burn (1967) found that those pups of several litters that had been exposed to cooling outside the nest for 1½ hours per day were visited by both parents more frequently than their littermates that had not been cooled. This in effect increased the amount of parental care they received since during these visits the parents licked, nuzzled, grasped them in the mouth, and carried them. Ear-punching, which clearly alters the stimulating properties of the young, yet is common practice in many laboratories, had the same effect upon parental care as cooling.

While different mothers provide a degree of equivalent maternal care, they also differ and the difference is greater when the mother herself has been subjected to various kinds of treatments during her life. Mothers that have been handled as infants transmit an effect to their young. Their young are more fearful at weaning and remain so a month later; their behavior as well as their adrenocortical hormone response in novel situations distinguish them from young reared by mothers that were not stressed as infants (Denenberg and Whimbey, 1963; Levine, 1967). Extensive handling of young including feeding in the absence of a mother but with a warm, moist, pulsating tube as a substitute, has effects which are discernible even at the end of a year. The young, reared under these minimal conditions of equivalence with their own mother, exhibit marked stress (i.e., increased adrenocortical hormone release) in response to disturbance (Thoman, Levine, and Arnold, 1968).

ACKNOWLEDGMENTS

The research reported in this article was supported by Research Grant MH-08604. Portions of the research was done in collaboration with Drs. Dale F. Lott and Benjamin Sachs, and with graduate students Joseph Terkel, and Alison Fleming. Alice Trattner and Donna Chruslinski have been my assistants. The research has benefitted from discussions with Dr. Daniel S. Lehrman and my other colleagues. Contribution number 50 of the Institute of Animal Behavior.

REFERENCES

Barnett, S. A., and J. Burn. 1967. Early stimulation and maternal behaviour. *Nature* 213: 150–152.

Beach, F. A., and J. Jaynes. 1956. Studies on maternal retrieving in rats. I. Recognition of young. *J. Mammal.* 37: 177–180.

Beach, F. A., and J. R. Wilson. 1963. Effects of prolactin, progesterone, and estrogen on reactions of nonpregnant rats to foster young. *Psychol. Rep.* 13: 231–239.

Birch, H. G. 1956. Sources of order in the maternal behavior of animals. *Am. J. Orthopsychiat.* 26: 279–284.

Bruce, H. M. 1961. Observations on the suckling stimulus and lactation in the rat. *J. Reprod. Fertility* 2: 17–34.

Calhoun, J. B. 1963. *The ecology and sociology of the Norway rat.* U.S. Public Health Service Publication No. 1008. P. 288.

Carlier, C., and E. Noirot. 1965. Effects of previous experience on maternal retrieving by rats. *Anim. Behav.* 13: 423–426.

Carsten, M. E. 1968. Regulation of myometrial composition, growth, and activity. In N. S. Asali, ed., *Biology of gestation*, vol. I. New York: Academic Press. Pp. 356–425.

Christophersen, E. R., and W. Wagman. 1965. Maternal behavior in the albino rat as a function of self-licking deprivation. *J. Comp. Physiol. Psychol.* 60: 142–144.

Coomans, H. E. 1955. Experiments concerning the maternal behavior of hooded rats. Lecture given at Amer. Mus. Nat. Hist.

Cosnier, J., and C. Couturier. 1966. Comportement maternal provoque chez les rattes adultes castrées. *Compt. Rend. Soc. Biol.* 160: 789–791.

Cowie, A. T., and S. J. Folley. 1961. The mammary gland and lactation. In W. C. Young, ed., *Sex and internal secretions*. Baltimore: Williams & Wilkins. Pp. 590–642.

Denenberg, V. H., L. J. Grota, and M. X. Zarrow. 1963. Maternal behaviour in the rat: analysis of cross-fostering. *J. Reprod. Fertility* 5: 131–41.

Denenberg, V. H., and A. E. Whimbey. 1963. Behavior of adult rats modified by experiences their mothers had as infants. *Science* 142: 1192–1193.

Eayrs, J. J., and R. M. Baddeley. 1956. Neural pathways in lactation. *J. Anat.* 90: 161–171.

Friedlich, O. B. 1962. A study of maternal behavior in the albino rat as a function of self-licking deprivation. Master's thesis, Southern Illinois University.

Greenwald, G. S. 1966. Ovarian follicular development and pituitary FSH and LH content in the pregnant rat. *Endocrinology* 79: 572–578.

Grindeland, R. E., W. A. McCulloch, and S. Ellis. 1969. Radio-immunoassay of rat prolactin. Paper presented at meeting of the Endocrine Society, New York, June, 1969.

Grosvenor, C. E. 1965. Evidence that exteroceptive stimuli can release prolactin from the pituitary gland of the lactating rat. *Endocrinology* 76: 340–342.

Grosvenor, C. E., and C. W. Turner. 1969. Pituitary lactogenic hormone concentration during pregnancy in the rat. *Endocrinology* 66: 96–99.

Grota, L. J. 1968. Factors influencing the acceptance of Cesarean delivered offspring by foster mothers. *Physiol. and Behav.* 3: 265–269.

Grota, L. J., and R. Ader. 1970. Continuous recording of maternal behaviour in the rat. (In press.)

Grota, L. J., and K. B. Eik-Nes. 1967. Plasma progesterone concentrations during pregnancy and lactation in the rat. *J. Reprod. Fertility* 13: 83–91.

Gunther, M. 1961. Infant behaviour at the breast. In B. M. Foss, ed., *Determinants of infant behaviour*. London: Methuen. Pp. 37–44.

Hashimoto, I., D. M. Henricks, L. L. Anderson, and R. M. Melampy. 1968. Progesterone and pregn-4-en-20α-o1-3-one in ovarian venous blood during various reproductive states in the rat. *Endocrinology* 82: 333–341.

Hytten, F. E. 1963. *Changes in the breasts and nipples in pregnancy.* World Health Organization, Geneva, December, 1963. Pp. 1–7.

Ingelbrecht, P. 1935. Influence du système nerveaux central sur la mammelle lactante chez le rat blanc. *Compt. Rend. Soc. Biol.* 120: 1369–1371.

Jost, A. 1959. Développement des foetus, accouchement et allaitement chez des rattes castrées en fin de gestation. *Arch. Anat. Micros.* 48: 133–40.

Kinzey, W. G. 1968. Hormonal activity of the rat placenta in the absence of dietary protein. *Endocrinology* 83: 266–270.

Labriola, J. 1953. Effects of caesarian delivery upon maternal behavior in rats. *Proc. Soc. Exp. Biol.*, N. Y., 83: 556–557.

Lazar, J. W. 1967. A response analysis of retrieving behavior in the mouse. Doctoral thesis, University of Tennessee.

Lehrman, D. S. 1961. Hormonal regulation of parental behavior in birds and infra-human mammals. In W. C. Young, ed., *Sex and internal secretions*. Baltimore: Williams & Wilkins. Pp. 1268–1382.

Levine, S. 1962. The effects of infantile experience on adult behavior. In A. J. Bachrach, ed., *Experimental foundations of clinical psychology*. New York: Basic Books. Pp. 139–169.

Levine, S. 1967. Maternal and environmental influences on the adrenocortical response to stress in weanling rats. *Science* 157: 258–260.

Lott, D. F. 1962. The role of progesterone in the maternal behavior of rodents. *J. Comp. Physiol. Psychol.* 55: 610–613.

Lott, D. F., and S. S. Fuchs. 1962. Failure to induce retrieving by sensitization or the injection of prolactin. *J. Comp. Physiol. Psychol.* 55: 1111–1113.

Lott, D. F., and J. S. Rosenblatt. 1969. Development of maternal responsiveness during pregnancy in the rat. In B. M. Foss, ed., *Determinants of infant behaviour*, vol. IV. London: Methuen. Pp. 61–67.

Masson, G. M. C. 1948. Effects of estradiol and progesterone on lactation. *Anat. Rec.* 102: 513–521.

Meier, G. W., and L. H. Schutzman. 1968. Mother-infant interactions and experimental manipulation: confounding or misidentification? *Devel. Psychobiol.* 1: 141–145.

Meites, J. 1966, Control of mammary growth and lactation. In L. Martini and W. F. Ganong, eds., *Neuroendocrinology,* vol. I. New York: Academic Press. Pp. 669–707.

Moltz, H., D. Geller, and R. Levin. 1967. Maternal behavior in the totally mammectomized rat. *J. Comp. Psychol.* 64: 225–229.

Moltz, H., R. Levin, and M. Leon. 1969. Differential effects of progesterone on the maternal behavior of primiparous and multiparous rats. *J. Comp. Physiol. Psychol.* 67: 36–40.

Moltz, H., and E. Wiener. 1966. Ovariectomy: effects on the maternal behavior of the primiparous and multiparous rat. *J. Comp. Physiol. Psychol.* 62: 382–387.

Moltz, H., and D. Robbins. 1965. Maternal behavior of primiparous and multiparous rats. *J. Comp. Physiol. Psychol.* 60: 417–421.

Moltz, H., D. Robbins, and M. Parks. 1966. Caesarian delivery and the maternal behavior of primiparous and multiparous rats. *J. Comp. Physiol. Psychol.* 61: 455–460.

Nolen, G. A., and J. C. Alexander. 1966. Effects of diet and type of nesting material on the reproduction and lactation of the rat. *Lab. Anim. Care* 16: 327–366.

Obias, M. D. 1957. Maternal behavior of hypophysectomized gravid albino rats and development and performance of their progeny. *J. Comp. Physiol. Psychol.* 50: 120–124.

Plume, S., C. Fogarty, L. J. Grota, and R. Ader. 1968. Is retrieving a measure of maternal behavior in the rat? *Psychol. Rep.* 23: 627–630.

Ray, E. W., S. C. Averill, W. R. Lyons, and R. E. Johnson. 1955. Rat placental hormonal activities corresponding to those of pituitary mammotropin. *Endocrinology* 56: 359–373.

Richards, M. P. M. 1966. Infantile handling in rodents: a reassessment in the light of recent studies of maternal behaviour. *Anim. Behav.* 14: 582.

Richards, M. P. M. 1967. Maternal behaviour in rodents and lagomorphs. In A. McLaren, ed., *Advances in reproductive physiology,* vol. II. London: Lagos Press. Pp. 53–110.

Riddle, O., E. L. Lahr, and R. W. Bates. 1942. The role of hormones in the initiation of maternal behavior in rats. *Am. J. Physiol.* 137: 299–317.

Rosenblatt, J. S. 1965. The basis of synchrony in the behavioural interaction between the mother and her offspring in the laboratory rat. In B. M. Foss, ed., *Determinants of infant behaviour,* vol. III. Methuen, London, Pp. 3–45.

Rosenblatt, J. S. 1967. Nonhormonal basis of maternal behavior in the rat. *Science* 156: 1512–1514.

Rosenblatt, J. S., and D. S. Lehrman. 1963. Maternal behavior of the laboratory rat. In H. L. Rheingold, ed., *Maternal behavior in mammals.* New York: Wiley. Pp. 8–57.

Roth, L. L. 1967. Effects of young and of social isolation on maternal behavior in the virgin rat. *Am. Zool.* 7: 800 (abst).

Roth, L. L., and J. S. Rosenblatt. 1967. Changes in self-licking during pregnancy in the rat. *J. Comp. Physiol. Psychol.* 63: 397–400.

Rothchild, I. 1965. Interrelations between progesterone and the ovary, pituitary, and central nervous system in the control of ovulation and the regulation of progesterone secretions. *Vitamins and Hormones* 23: 209–327.

Scott, J. P. 1967. The evolution of social behavior in dogs and wolves. *Am. Zool.* 7: 373–381.

Stone, C. P. 1925. Preliminary note on maternal behavior of rats living in parabiosis. *Endocrinology* 9: 505–512.

Sturman-Hulbe, M., and C. P. Stone. 1929. Maternal behavior in the albino rat. *J. Comp. Physiol. Psychol.* 65: 203–237.

Terkel, J., and J. S. Rosenblatt. 1968. Maternal behavior induced by maternal blood plasma injected into virgin rats. *J. Comp. Physiol. Psychol.* 65: 479–482.

Thoman, E. B., and W. J. Arnold. 1968. Effects of early social deprivation on maternal behavior in the rat. *J. Comp. Physiol. Psychol.* 65: 55–59.

Thoman, E. B., S. Levine, and W. J. Arnold. 1968. Effects of maternal deprivation and incubation rearing on adrenocortical activity in the adult rat. *Devel. Psychobiol.* 1: 21–23.

Tobach, E. 1966. The objective study of emotional behavior in the rat and mouse. *Proc. XVIII Inter. Psychol. Cong.,* 3: 1966, 130–135.

Tobach, E. 1968. Long term effects of differential pre-weaning experience in the Wistar Rat. *Proc. XXIV Inter. Union of Physiol. Sci.,* 7.

Tobach, E., and T. C. Schneirla. 1962. Eliminative responses in mice and rats, and the problem of "emotionality". In E. L. Bliss, ed., *Roots of behavior*. New York: Harper. Pp. 211–231.

Vorherr, H. 1968. The pregnant uterus: process of labor puerperium, and lactation. In N. S. Asali, ed., *Biology of gestation,* vol. I. New York: Academic Press. Pp. 426–448.

Wiesner, B. P., and N. M. Sheard. 1933. *Maternal behaviour in the rat.* London: Oliver and Boyd.

Yoshinaga, K., and R. A. Hawkins. 1969. Estrogen secretion by the rat ovary during the estrous cycle and pregnancy. Paper presented at meeting of The Endocrine Society, New York, June, 1969.

Young, R. D. 1965. Influence of neonatal treatment on maternal behavior: a confounding variable. *Psychonom. Sci.* 3: 295–296.

Zarrow, M. X. 1961. Gestation. In W. C. Young, ed., *Sex and internal secretions,* 3rd ed. Baltimore: Williams & Wilkins. Pp. 958–1031.

Zarrow, M. X., P. B. Sawin, S. Ross, and V. H. Denenberg. 1962. Maternal behavior and its endocrine bases in the rat. In E. L. Bliss, ed., *Roots of behavior,* New York: Harper. Pp. 187–197.

Zarrow, M. X., L. J. Grota, and V. H. Denenberg. 1967. Maternal behavior in the rat: survival of newborn fostered young after hormonal treatment of the foster mother. *Anat. Rec.* 157: 13–18.

V

HUMAN BEHAVIOR

HEINI P. HEDIGER
Zoological Gardens
Zurich, Switzerland

The Development of the Presentation and the Viewing of Animals in Zoological Gardens

Like all things in this world, the keeping of animals in the zoo, especially the manner of presenting them and of viewing them by visitors, has undergone changes. The viewing of animals in the zoo and their presentation are related to one another as the positive print is to the negative. One has to examine both, therefore, and their mutual effects upon one another in order to illustrate the evolution of one or another aspect of either. It must be understood that we are not attempting to illustrate the historical development of zoological gardens in general, but only the development of "presentation," i.e., the exhibition of animals, on the one hand, and the way of viewing them, on the other.

This development has taken a parallel course in the zoo and the museum. I am suggesting a close link between this development and the work of T. C. Schneirla, surprising as this may sound. We need only look back a few decades, as one does in the study of evolution, to discover that the relationship between museum and zoo are really quite intimate and that they both have common roots. In these roots are interlaced also the roots of much more mundane institutions, e.g., the circus and its precursors, the county fair and traveling menageries. I am thinking, in this context, not only of the stream of interesting dead animals that have flowed into museums all over the world during the years, and have kept on flowing, even until today. Even more, I have in mind the common roots of origin of the presentation and viewing of animals in the zoo and the museum by man who, in this respect, as in his scientific activity in general, has also undergone evolution.

There can be no harm on the occasion of honoring an outstanding scientist, in taking an opportunity to pause briefly during the unswerving rush of progress in scientific research for a retrospective view. A straightforward approach can lead to new insights and directions. It may appear to be risky, however, if I, a representative of show-business, or, I would not even care if called, of monkey-business, undertake such an attempt.

Schneirla would forgive me if I chose as a starting point for my exposition the first exhibition of an exotic large animal in North America. According to a publication of the American Antiquarian Society, which even deals with the subject of the circus (R. W. G. Vail, 1934), this was the appearance of an African lion in the New World in the year 1720. The newspapers referred to the animal not only as "the most noble, but the tamest and most beautiful creature." At the first occasion of exhibiting exotic animals, it was the rule—in Europe as well as in America—to approach them with true respect and esteem and to stress their harmlessness, nobility, and docility. This was true also for the camel, i.e., the dromedary, which was the second exotic large animal shown in America, in the year 1721. It was described as, among other things, "a very wonderful and surprising creature." Adults had to pay 1 shilling to see it, children sixpence. Also, when the first pair of camels was displayed in Wall street, 66 years later in 1787, it was stressed that they were "remarkably harmless and docile." The lion and camel were followed by the polar bear in 1733; it was only in 1796 that the first Indian elephant set foot upon American soil. Advertisements proclaimed that he possessed the "adroitness of the beaver, the intelligence of the ape, and the fidelity of the dog." Thus, the positive qualities were primarily emphasized and admired. However, by 1816, exhibited elephants were not accepted with respect everywhere; one was shot in broad daylight by a brute. Later the skeleton of this unlucky elephant was exhibited and its skin is said (according to Vail) to have gone to the American Museum. The elephant was a female.

At the time, no one could foresee that the first importation of a male elephant in 1818 would, so to speak, automatically bring on a catastrophe. I have shown (1965, p. 226) with numerous examples from the old world that bull elephants, as a rule, cause deaths amongst their keepers. This macabre list was started in America by the bull mentioned. One day, Columbus—this was the name of the animal that had become dangerous —killed his keeper, William Kelly.

However, it cannot be my task here to trace the history of American importations of exotic animals. I mention, therefore, only one more arrival in the year 1789, that of the first ape—an orang-utan. R. W. G. Vail (1934, p. 21) claims that the second orang-utan came, apparently alive, to the American Museum in New York in 1821. There existed several "American Museums" in New York, at that time, e.g., the von Scudder (1811), and The Grand Museum of Natural History (1819). The American Museum of Natural History, where T. C. Schneirla did his work, was only founded in 1869. Its precursors, mentioned above, housed not only preserved animals but also living ones. The point was to show curiosities of nature to the curious public in return for jingling coins, regardless of whether these curiosities were dead or alive.

The idea of presenting separately dead animal material in museums, and live animals in zoos was absolutely secondary; as a consequence, incidentally, the separation is not carried out everywhere even today. For example, the two Paris zoos, that is, the old Jardin des Plantes and the new Parc Zoologique du Bois de Vincennes, together form a unit within the Museum National d'Histoire Naturelle. In Antwerp, also, the zoo and museum form a unit and are under the same direction. We could give many more examples which obviously are determined by their historical development.

Seen as a whole, there has been the definite tendency in the last century not only to separate living and dead animal material sharply, into zoos on the one hand and museums on the other, but also to remove the display of exhibits from the realm of business ventures and to consider it a function of a cultural institution in the service of education and scientific research. Thus, natural history museums are no longer shops for the products of nature, and zoos are no longer pet shops; the commercial element, in the sense of making profit, has been eliminated in favor of the educational and scientific function. It is set forth in the regulations of the international association of zoo directors that zoos operating on a purely commercial basis cannot be admitted. Money-making is not prohibited by this but it should not be an end in itself; instead the money should function to develop a cultural institution, scientific activity, and conservation of natural resources.

The separation of living and dead material into zoos and museums, which has been effected, with exceptions, does not always include the manner of presentation. With regard to the technique of display, the development has not been completed, i.e., the tasks of museums and zoos are not separated clearly enough. This is shown probably most clearly in the keeping of birds which differs from that of other groups of animals where the development of display techniques took place previously.

The old style cages were a kind of solid box in which the animal was housed as a living specimen until, with its death, it became ready for the museum. The death chambers of the menageries were, in a way, the anterooms or waiting rooms of the museums. Animal lovers and zoo enemies did not refrain from using such expressions. There was really not much difference in the manner of presentation: the living animal in its narrow cage was provided with food, the stuffed one with preservative.

As mentioned above, the museum-like exhibition of living creatures

FIGURE 1
An enormous collection of single birds—a misconception of the zoo's function; such collections are more appropriate as the task of the museum.

persisted longest with regard to birds, where representatives of all possible species were housed singly in cages arranged in several long rows on top of one another.

The ambition to have near-complete collections is justifiable only in museums, not in zoos. In zoos it is not the completeness of the collection that matters but much more the naturalness of the habitat, the food, and the social structure, i.e., the success of breeding. The latter is the keystone of biological animal keeping.

Since at least a million different species exist, no zoo can hope to arrive at anything near a complete collection. Each zoo should limit itself to a particular selection. This necessary selection should be based upon a plan of stocking animals which takes into consideration the special circumstances of the particular zoo, i.e., the history of its supporting town, its geographical and topographical situation, its special climate, and other special features. Each town should make an effort to have its own specialized zoo; the worst thing would be the uniform zoo with which we are threatened today in which many species are represented by a single individual (Hediger, 1965).

The display of single individuals belonging to many different species must be rejected because it is based on a confusion with the methods of the museum. For that matter modern museums need not, and do not usually, exhibit complete collections. Most often, the major parts of collections are kept in the scientific departments while the visitor is presented only with a selection of impressive samples of nature.

An additional reason for being against the isolation of animals, which has persisted longest with regard to birds in zoos that go to extremes, is that such isolation is essentially unbiological and irresponsible. It is of greater importance today to show each species, selected for exhibition in an overall plan, in a reproductive society, i.e., in the natural family group or larger social association.

As dealt with here, an essential feature of the development of the presentation of living animals in the zoo is to show the visitor, in the best way possible, a careful selection of species in near-natural spatial areas, as a natural social group with near-natural feeding and treatment, rather than to present as many individuals as possible, singly, in more or less abstract jail-like rooms.

A discussion of feeding will not be given here; it was described by me in the Zoo Yearbook of 1966. The use of space is more important in our context, because it has significance not only for the exhibited animal but also for the man looking at the animal.

Contemporary man, particularly the inhabitant of the large city, visits the zoo for different reasons than man 50 or 100 years ago visited its precursors, the menageries and the cases of objects of nature. Formerly it was only a matter of admiring curiosities of the animal kingdom, whether these were alive behind bars or dead in glass cages.

Nowadays the zoo visitor no longer wants to observe only a living *specimen* but rather a happy animal, family groups or herds, in near-natural surroundings, i.e., surroundings which provide a natural environment for the animal and which provide for its viewer, in need of recreation, the beneficial stimulus of exotic scenery that lets him forget the un-naturalness of his own highly technologized shop and home. The most perfect example of this kind of display that I have found so far is the aquatic bird house in the Bronx Zoo.

On closer examination it can be seen that man, the man of the large city, actually lives in cages, i.e., in a chaos of iron structures, electrical and other cables, and in an artificial climate with artificial light—far from any contact with nature, away from animals and plants. One should not forget that our parents or grandparents still had much more contact with animals and plants, i.e., with nature, and that we are primarily not created for a termite-like life in sterile complexes of skyscrapers.

This adaptation cannot be accomplished easily in one or two generations. Most human beings of our generation still feel a fairly intense hunger for nature which finds expression in different symptoms: the continuously growing number of zoos between Peking and Los Angeles and between Johannesburg and Toronto is only one of the many symptoms. The growing number of visitors to each zoo is another one. Never before have so many pets, aquariums and terrariums been kept in private apartments; never before did the pet market in large cities flourish as it does right now.

The achievements of modern technology, the conquests of space with ever newer ships, the improbable achievements of cybernetics and of electronic brains, by themselves, do not manage to fully satisfy man of our times. There remains a compensating hunger for living plants and animals, even for dead parts of plants. Old pieces of wood and roots that one would have thrown into the fire just a few decades ago, now suddenly have nearly invaluable decorative value. They are put into lobbies of elegant hotels and into the windows of exclusive jewelers: the city dweller starves for each little piece of nature. The little piece of artificial virgin forest in the building of the Ford Foundation (42nd Street, New York City) is a striking and charming example.

The development of the exhibition of living animals in the zoo that has begun to appear over the last five decades may be characterized by two features: the kennel and the territory. By kennel we mean a jail-like narrow cage in which an animal, most frequently a single one, is kept by force and viewed in this situation as a living specimen. "Territory" on the other hand means the natural division of space, with species-specific habitat organization for the use of an animal or of a social unit of animals (e.g., family, herd, pack, etc.).

The illusion that the so-called free-living animals live really free, that they are so to speak, in possession of the much exalted "golden freedom" has been shattered as a biological illusion based on a human projection. The modern science of territoriality knows as much about the territory of many animals species as morphology knows about the number of toes or the structure of teeth: the size, the structure and the markings of the territory, are well known in several cases.

Only in its territory, in this interesting system of fixed reference points and connecting lines with the corresponding time program (space-time system), can an animal feel comfortable. Therefore, in the zoo one must make the utmost effort to place at the animal's disposal all the necessary details in a smaller territory in which the greatest importance is attached to the home as the most important fixed reference point because only at this place of optimal security can the animal find rest.

Before we go further into a consideration of the territory, it should be remembered again that the separation of zoo and museum, of living and dead animal material, has not been carried out everywhere, speedily and completely. Also the separation from another root, that of show business, the circus and its precursors, has not been effected everywhere.

The circus has never pursued the aim of showing animals in near-natural surroundings but it seeks to demonstrate animals in their maximum accomplishments and mainly human accomplishments: horses and dogs walking on two legs, elephants in grotesque costume, apes eating at the set table, etc. The point here is maximum anthropomorphizing whilst the zoo aims at optimal animal living conditions. We are dealing with completely opposite tendencies. In the "Official Illustrations of the St. Louis Zoological Park" for the year 1922, there are pictures of an orangutan with a rifle on "drill parade" and "at attention" (Figure 2).

A strong circus component has remained typical of the St. Louis zoo up until today. However, no zoo can afford to present orangutans in this way anymore, partly because this kind of anthropomorphizing is in poor taste, but mainly because this species of ape has become so rare that one can no longer dispose of it at will. According to the careful census of Barbara Harrison (1961) there remain only about 2800 individuals in the limited refuge areas of the orangutan in Borneo and Sumatra. The species was therefore just placed under protection, and it is expected of those zoos, which today, on the basis of their qualifications, obtain permission to purchase this precious ape, that they give it optimal care in order to bring it to reproduction.

Today the species has become valuable to the culture and must no longer be abused merely for the entertainment of a certain kind of public. This example illustrates clearly the disappearance of the former circus mentality in favor of a sense of responsibility for the conservation of nature and of science: today animal training in zoos is only permissible for occupational therapy, compensating for lack of activity due to the conditions of captivity (Hediger, 1964).

FIGURE 2
Demonstration of an orangutan in the 1920s: an example of extreme anthropomorphizing.

The increasing significance of the apes and other primates for medical research (Conway, 1966) will not be discussed here since it is outside the scope of our topic. What has been said thus far may be summarized once more: the basic tendency in the development of presentation as well as of viewing may be characterized as a decrease of anthropomorphizing in favor of a more biological treatment and view.

This progressive "biologizing" is expressed mainly in animal housing: the old jail-like cage has given way to the territory-like area in which the animal finds all the elements that are important to it, such as fixed points for security, for comfort, for rest, for feeding and digestion, for marking, etc. The precursor of the territory idea was the "abundant room idea" as it was propagated in the preface to the official guide to the New York Zoological Park by William T. Hornaday in 1907. It was, at first, just a typical reaction to the narrow kennel, but in addition this idea was based upon the misconception of the free-living animal living really free. This idea was not yet recognized then as human wishful thinking, i.e., as real anthropomorphizing, but at least it led to relieving the animal from the confinement of the jail-like kennels. However, today we know that the well-being of the zoo animal is not in the least proportional to the *quantity* of room at its disposal; it is much more dependent upon the *quality* of the room, i.e., on the similarity of the area landscaping to its territory but also on the naturalness of the structure of family and social group, on the positive relationship to humans, and finally, on the adequacy of the food.

As in animal behavior, so also in animal keeping in the zoo, the deepseated tendency for anthropomorphizing turned out to be one of the main stumbling blocks to progress.

It took a long time to discover that real "free-living" for an animal is actually "nonfreedom," i.e. the animal often has extremely strong ties to space, time, and social structure.

The analysis of the very tragic tendency to anthropomorphize led to the inclusion within the sphere of a scientific analysis of its counterpart—its parallel in the animal kingdom—the tendency to zoomorphize (Hediger, 1964). The tendency to identify creatures of different species with an animal's own species and, as a consequence, to build them into the animal's own social structure is by no means a human monopoly but it is also found in the animal kingdom, mainly in mammals and birds, and, perhaps, even in certain fishes.

This means that the keeper is regarded by many of his charges as a conspecific and therefore has to adapt himself skillfully to his role, i.e., to the role of the alpha animal, namely, the top individual of the society. This has the advantage, amongst others, that the human keeper is not taken as a disturbing stranger but as the aloof dominant animal, the orders of which one naturally has to obey. I have treated this fact more exhaustively elsewhere (1964, 1965); there is no space here for repeating myself.

In summary, it should be emphasized only that in our time the zoo forms a piece of secondary nature, even a piece of man's biotope, for the human city inhabitant who is otherwise almost completely isolated from nature. In this very fascinating overlapping zone of the human city habitat and the neo-habitat of the zoo animals, encounters between animal and human beings today are entirely different in character from those of half a century ago in the old-fashioned menageries.

Modern man does not want to see simply curiosities of the animal kingdom, in the sense of living specimens, but animals in near-natural artificial territories, in natural family- and herd-associations, in an appealing setting, in which man also, starving for nature, can relax and be refreshed. The presentation of the animal must be done in so natural a way that man can also make observations about its natural behavior. The scientific analysis of observations made possible in the biological zoo is a self evident requirement today. The caging of animals merely for the satisfaction of a primitive "voyeurism" is not enough today; animals living in zoos under secondarily natural conditions must serve education and research to the utmost extent. Where this elementary condition of animal keeping is not fulfilled, we deal not with a zoo but with a menagerie of the old style. The need to analyze exhibited animals scientifically, particularly their behavior, which can only be studied in the living animal—and under near natural conditions—has also led to the conception of zoos as behavior museums and simultaneously to the conception of their function as valuable introductions to animal behavior studies in free nature.

Many field workers, e.g., H. Kummer, F. Walther, agree that certain behaviors of animals in the wild, such as several details of facial expressions and gestures in the behavior of primates and the marking behavior of gazelles, would not have been noted if it had not been possible to observe them first in the zoo animal. The zoo animal is at one's disposal nearby at any time, and moreover is recognizable as an individual and observable in the group.

Compared with comparative anatomy and comparative physiology, which are unquestionably significant to medical research, comparative psychology is still extremely young and, accordingly, is in need of further development. This need is so pertinent that it affects even the museums which, for a long time, occupied themselves only with dead animals.

If I am not mistaken, it was G. K. Noble who introduced the observation of the living animal in museums as a scientific method, first in the American Museum of Natural History in New York. One no longer wanted to restrict oneself merely to systematic and morphological studies in the museum. T. C. Schneirla continued to develop this method of studying animal behavior in the museum most fruitfully, as shown by many of his studies and those of his collaborators. In this way the living creature, although with respect to different and loftier aspects, returns to the museum in a very promising way. The museum, however, can only deal with small animal species for reasons of space and technical facilities. The study of the behavior of large animals must be conducted in the wild.

Translated by
MONICA IMPEKOVEN and J. S. ROSENBLATT;
BRIGITTE CAPPELLI and E. TOBACH

REFERENCES

Conway, W. G. 1966. The availability and long term supply of primates for medical research. *International Union for the Conservation of Nature Bulletin,* n. s., no. 18.

Harrisson, B. 1961. Orang-utan: what chances of survival? *The Sarawak Museum J.,* n. s., 10 (17–18): 238–261.

Hediger, H. 1961. The evolution of territorial behavior. In S. L. Washburn, ed., *Social life of early man.* New York: Viking Fund.

Hediger, H. 1964. *Wild animals in captivity: An outline of the biology of zoological gardens.* New York: Dover.

Hediger, H. 1965. *Mensch und Tier im Zoo—Tiergartenbiologie.* Zürich: Albert Müller Verlag AG.

Hediger, H. 1966. Diet of animals in captivity. *International Zoo Yearbook* 6: 37–58.

Kummer, H., and F. Kurt. 1965. A comparison of social behavior in captive and wild hamadryas baboons. In H. Vagtborg, ed., *The baboon in medical research.* Austin: Univ. Texas Press.

Maier, N. R. F., and T. C. Schneirla. 1963. *Principles of animal psychology,* enlarged edition. Gloucester, Mass: Peter Smith.

Vail, R. W. G. 1934. *Random notes on the history of the early American circus.* Worcester, Mass.

Walther, F. 1965. Verhaltensstudien an der Grantgazelle (Gazella granti) im Ngorongoro-Krater. *Z. Tierpsychol.* 22 (2): 167–208.

ALEXANDER THOMAS and STELLA CHESS
Department of Psychiatry
New York University School of Medicine
New York

Behavioral Individuality
in Childhood

THE INTERACTIONIST APPROACH

Two main opposing views have in the past characterized the approach to the study of behavioral individuality in childhood. These views stemmed from the attempt to identify the sources from which such individuality develops. The *preformist* or *constitutionalist* approach suggested that individual differences in development derived primarily from prenatal influences which were presumably largely genetic in origin. The developmental process was viewed as a succession of stages in which new layers of hereditarily determined patterns of organization were exhibited. The *epigenetic* or *environmentalist* approach proceeded from the assumption that the new-born child and young infant had little patterned organization of individuality at birth and was psychologically a *tabula rasa*. In this view, individuality in behavioral patterning and personality was achieved in response to the formative influences of the surrounding environment.

The preformist approach, which flourished in the nineteenth and early twentieth centuries, was discredited as it became evident that life experience had a profound influence in shaping the complex patterns of behavior and personality, which had been labelled inborn and constitutional. In recent decades the environmentalist view has been dominant, and most professional workers have concentrated on the study of environmental influences on the child's development—first and foremost the mother, then the father and other intrafamilial influences, and finally larger sociocultural factors. These studies have delineated constellations of parental attitudes and practices, sibling and other family relationships, social values, and cultural norms that significantly affect the course of psychological development. However, as with the constitutionalist view, recent studies have made it increasingly evident that individuality in behavioral development cannot be adequately explained on the basis of environmental factors alone, just as such individual differences could not be derived exclusively from constitutional-hereditary sources.

Strict adherence to such a "heredity" versus "environment" dichotomy only serves to obscure the basic interaction of these two categories. Such an interactionist approach, in which behavioral phenomena are considered to be the expression of a continuous organism-environment interaction from the very first days of life onward, has been formulated in various ways in the work of Pavlov (1927), Stern (1927) and Lewin (1935), among others, and has been precisely stated by Schneirla and Rosenblatt (1961):

> Behavior is typified by reciprocal stimulative relationships. . . . Behavioral development[,] because it centers on and depends upon reciprocal stimulative processes between female and young, is essentially social from the start. Mammalian behavioral development is best conceived as a unitary system of processes changing progressively under the influence of an intimate interrelationship of factors of maturation and of experience —with maturation defined as the developmental contributions of tissue growth and differentiation and their secondary processes, experience as the effects of stimulation and its organic traces on behavior.

An interactionist approach to individual development, therefore, involves the simultaneous scrutiny and analysis of responses of the organism and of the objective circumstances in which these responses occur. It also necessitates the distinction between objective situation and effective environment. Lewin (1935) and Koffka (1924) have each considered the objective situation to be divisible into geographic environment, reflecting its general attributes, and behavioral environment, reflecting the selected aspects that are effective in influencing function. In other words, while polarized light is present in the geographic environment for both man and bee, it is a feature of the effective visual environment only for the bee. Similarly, antigens, which may be ubiquitously present, represent effective environment only for the allergic person. In addition to an all-or-nothing relationship between the organism and its surroundings, there is abundant evidence to suggest that there are individual differences in the response to the effective environment; these differences may be either in the degree of responsiveness or even in the qualities of the response. Thus, to a constant stimulus a given direction of action may vary from individual to individual in intensity, in direction, or in both. It is clear from these considerations that determination of the effective environment is dependent upon the characteristics of the organism. Viewed obversely, our knowledge of the characteristics of the organism defines its effective environment. To say that an organism is color-blind indicates that its effective environment does not include certain frequency differences in the visual spectrum. The identification of a child as shy signifies that strangers are specifically definable features of his effective environment.

The simultaneous consideration of organismic and environmental factors in interaction with each other appeared to offer an approach to the study of behavioral individuality in childhood more promising than either the preformist or epigenetic views. We and our coworker, Dr. Herbert Birch, felt that such an approach might deal with a number of unanswered basic questions in child development. Why do youngsters exposed to the same kind of parental influences so often show markedly different directions of personality and development? Why is there such variability in the effect that mentally ill parents have on their children? Some of these children develop serious psychological disturbances, others develop mild problems, and still others mature as if immune to the stresses produced by their parents' mental illness. Why do some parents who show no evidence of any significant psychiatric disturbances and who provide a good home for their children sometimes have a child with serious psychological disturbances? Why do the rules for childcare in feeding, weaning, toilet training and so on never seem to work equally well for all children, even when applied by intelligent and conscientious mothers?

When we became interested, some 15 years ago, in dealing with these questions it was evident that very little systematic knowledge existed as to the contribution made by the child's own reactive characteristics as an organism to his individual development. It was common knowledge to experienced pediatricians and baby nurses and to many mothers that newborn and very young infants vary markedly in their behavior and responses to various types of stimuli. But few specific studies of these individual differences existed and these were in the main restricted to short-term investigations of a limited number of behavioral attitudes, such as general activity level or sensory thresholds. Most workers in the child development field were concentrating on the analysis of environmental influences, with organismic influences receiving at most ad hoc consideration.

Our own study of behavioral individuality had, therefore, to first concentrate on the systematic delineation of characteristics of individuality in the very young infant without neglecting the simultaneous consideration of environmental factors. The next concern was to trace the persistence and stability of these characteristics as the child grew older, and their significance for psychological development. A consistent interactionist approach also required the consideration at each level of development of both the environment's influence on the child and the influence of the child's own reactive characteristics on the parents and other significant figures in his environment.

THE NEW YORK LONGITUDINAL STUDY

Only an anterospective longitudinal study can provide the necessary detailed and accurate data for such an analysis of behavioral development. Information obtained by retrospective recall, whether from parents or the subjects themselves, is subject to various distortions because of both idiosyncratic and cultural influences (Robbins, 1963; Wenar, 1963). Our New York longitudinal study, in progress since 1956, has followed the behavioral development of 136 children from early infancy onward. Information has been gathered longitudinally and anterospectively at sequential age-levels on the nature of the child's own individual characteristics of functioning at home, in school, and in standard test situations; on parental attitudes and child-care practices; on special environmental events and the child's reactions to such events; and by psychological testing. The data have been gathered by parent and teacher interviews and by direct observational techniques. The staff child psychiatrist has done a psychiatric evaluation of each child presenting symptoms. Wherever necessary, neurological examination or special testing, such as perceptual tests, has been done.

The collection of data has been guided by the following principles:

1. Behavior is described in objective terms. Strict avoidance of interpretations of behavior by parents, interviewer, or observer is maintained throughout. Thus, a statement "the baby hated his cereal" or "he loved his bath" is considered unsatisfactory for primary data. Instead the question is always asked "what did he do that made you think he loved or hated it" and a detailed description of the child's actual behavior is recorded. In this way contamination of the raw data by interpretations based on preconceptions, evident in a number of studies, is avoided. Since interpretations of a specific item of a child's behavior by several simultaneous observers may vary greatly, it is difficult if not impossible to obtain accurate and reduplicable data on an interpretative basis.

2. The basic data are obtained from the details of the child's behavior in the natural daily activities of his life. These include, among others, sleeping, feeding, dressing, bathing, nail-cutting, and hair-brushing in the young infant. As the child grows older, other activities such as involvement with individual people, play, toileting, vaccinations, discipline, etc., are added. Detailed information on the child's behavior in his daily life insures that the data will reflect the child's characteristic modes of functioning. This procedure is in contrast to those methods which, by relying primarily on

observations made in unfamiliar testing situations, raise questions of atypical and unrepresentative behavior in such artificial settings.

3. In comparing the characteristics of responses among infants, obtained differences may be due to two factors. On one hand, the stimulus may vary in intensity or in quality. On the other, the stimulus may be constant but differences derive from individual differences in reactivity. This latter factor is the principal focus of interest in this study. However, any attempt to define the variations in the individual response of different infants by presenting them with the same constant stimulus would be limited and even undesirable. Such constancy involves artificial test situations, which introduce such new and uncontrolled variables as strange environment, special manipulations, and testing devices. Furthermore, a controlled, delineated set of stimuli may elicit responses that represent only a narrow segment of the child's pattern of reactivity. The present study, therefore, started with the assumption that no special advantage derives from constancy of stimuli. By acquiring information on the infant's behavioral responses in as many types of situations as possible, the data sample becomes representative of the child's functioning. This range of information has been obtained by determining the specific responses that a baby makes to the various features of day-by-day living, such as the bath, the taste of different foods, periods of food deprivation, loud noises, the crib, bright lights, and so on. The data thus comprise a record of the responses to different stimuli of varying intensity and quality. In contrast with the artificiality of the limited experiment, this method is based on the classic approach used in biological field studies of function in living organisms, namely, the delineation of behavior within the environmental context in which it occurs. The major goal is to obtain detailed and accurate data on a consistent type of reaction pattern in the individual infant, whether responding to a hunger stimulus, the taste of new food, the temperature of the bath water, or an attitude of the mother.

In data collection special emphasis is placed on the recording of the details of the child's first response to a new stimulus and his subsequent reactions on exposure to the same stimulus until a consistent, long-term response has been established. Such stimuli may be simple, as the first bath or the introduction of a new food; they may be complex, as the move to a new home or the introduction of a new person into the household. In either case the sequence of responses to new stimuli can give very valuable information as to the individual pattern of reactivity of the child.

We have found that the use of structured interviews that focus on gathering information on a descriptive factual level has made it possible to utilize the parents as a rich source of meaningful and accurate data on the child's behavior. The parental reports have correlated closely with descriptions of behavior obtained by independent, direct observations. They have also been broad and detailed enough to permit content analysis (Thomas, et al., 1963).

NINE CATEGORIES OF REACTIVITY

Starting from the position that at present no adequate theoretical framework exists that can serve as the basis for a deductive approach to individual child development, the delineation of categories of individuality in reaction patterns was approached through an inductive content analysis of the behavior protocols. The following 9 categories have been found to be present and item scorable in the parental interview protocols:

1. Activity Level: the motor component present in a given child's functioning and the diurnal proportion of active and inactive periods. Protocol data on motility during bathing, eating, playing, dressing and handling, as well as information concerning the sleep-wake cycle, reaching, crawling, and walking are used in scoring this category.
2. Rhythmicity: the predictability or the unpredictability in time of any function. It can be analyzed in relation to the sleep-wake cycle, hunger, feeding pattern, and elimination schedule.
3. Approach or Withdrawal: the nature of the response to a new stimulus, be it a new food, new toy, or new person.
4. Adaptability: responses to new or altered situations. Our concern is not with the nature of the initial responses, but with the ease with which they are modified in desired directions.
5. Intensity of Reaction: the energy level of response, irrespective of its quality or direction.
6. Threshold of Responsiveness: the intensity level of stimulation that is necessary to evoke a discernible response, irrespective of the specific form that the response may take, or the sensory modality affected. The behaviors utilized are those concerning reactions to sensory stimuli, environmental objects, and social contacts.
7. Quality of Mood: the amount of pleasant, joyful, and friendly behavior, as contrasted with unpleasant, crying, and unfriendly behavior.
8. Distractibility: the effectiveness of extraneous environmental stimuli in interfering with or in altering the direction of the ongoing behavior.
9. Attention Span and Persistence: two categories that are related. Attention span is the length of time a particular activity is pursued by the

child. Persistence refers to the continuation of an activity in the face of obstacles to the maintenance of the activity direction.

Each category is scored on a 3-point scale, in terms of two polar extremes and one intermediate level. Details of the criteria for each category, methods of scoring, and levels of scoring reliability have been reported elsewhere (Thomas et al., 1963).

It has been possible to score each parental interview of every child for the 9 categories, starting at two months of age. Teacher interviews and direct observations at home, in school, and in test situations have also been scorable to the same criteria. The direct observations cannot be scored for the categories of rhythmicity and adaptability, which require behavioral data over a significant period of time. The specific criteria for the scoring of the 9 categories vary, of course, with the age-level of the child. Thus, distractibility at age 6 months may be expressed by the ease with which bottle feeding is interfered with by an extraneous sound, and at age 8 years by the ready disturbance of concentration on ongoing academic work by the appearance of a stranger in the room. Approach-withdrawal responses in early infancy are concerned with the first bath and new foods, at 3 years of age with a new baby sitter and the first day at nursery school, and at 8 years of age with a new bicycle and academic subject. Comparable scorable protocols have been obtained in the New York longitudinal study from a relatively homogeneous native-born, highly educated middle-class group of parents and from a group of 70 working-class parents of Puerto Rican background.

TEMPERAMENTAL INDIVIDUALITY

We have considered that a child's characteristic mode of functioning at any age-level as determined by these 9 categories of reactivity represents his pattern of *temperamental individuality*. The term temperament, in our usage, contains no inferences as to genetic, endocrine, somatologic, or environmental etiologies. It refers to the *how* rather than to the *what* or the *why* of behavior. No implication of permanence or immutability is attached to this conception.

Comparison of the temperament scores of the group of children as a whole show significant degrees of consistency from one age period to another, but also shifts that are slight in some children, marked in others. Temperament resembles the I.Q. score in this combination of consistency and mutability over a period of time. In many children the consistency of expression of temperament at different ages has been very striking. A child who adapted very quickly to the bath at 1 month and to new foods at 3 months adapted just as quickly to a new baby sitter at 18 months and to nursery school at 4 years. One child who was easily distracted from feeding at 2 months by anybody walking into the room, was just as easily distracted when playing at 3 years; another child was as consistently non-distractible at both ages. In other children, modification of temperamental characteristics that occurred in time appeared definitely related to parental handling or to special environmental events.

TEMPERAMENT-ENVIRONMENT INTERACTION

The anterospective longitudinal nature of our data has made it possible to assess the influence of temperament and environment on both normal psychological development and on the ontogenesis of behavior problems. In no instance has a consideration of temperament alone or environment alone been sufficient to understand the child's functioning, whether in his response to child-care practices, the evolution of specific personality characteristics, functioning in school, or the development of behavior problems. Rather, it appears that behavioral normality as well as behavioral disturbance is the result of the interaction between a child with a given patterning of temperament and significant features of his developmental environment. Influential environmental factors include intrafamilial as well as extra-familial circumstances such as school and peer group. In several cases, additional special factors such as brain damage or physical abnormality also interacted with temperament and environment to produce deviations in behavioral development.

PARENTAL CHILD-CARE PRACTICES

Our data indicate that individuality in the responses of different children to parental child-care practices is related to temperamental characteristics as well as to the manner in which the parents apply these practices (Chess, Thomas, and Birch, 1959; Thomas et al., 1961a). The infant with regular sleep-wake and hunger cycles usually does very well with self-demand feedings, in which the infant's cry determines the time of feeding. Such regular infants are in the great majority, but the minority with irregular sleep-wake and hunger cycles often do not develop regular feeding schedules with a self-demand approach, and do better with scheduled feedings. A quickly adaptive child with predominant approach responses usually takes quickly to new foods or to weaning; the opposite temperamental type does not. Persuading a child to sit quietly for 10 minutes on a toilet seat for training purposes is much easier with a low activity level child than one with high activity. These findings suggest that no one rule of child-care may be equally applicable to all children and that individualization of child-care practices requires a consideration of the specific temperamental characteristics of the child (Chess, Thomas, and Birch, 1965).

It has also been evident that no formula that assumes that all children will react similarly to the same parental approach is valid. In addition, interaction is a reciprocal process—infants with differing temperaments may influence the response of the mother in varying ways, even mothers with the same basic maternal attitudes, and children with similar temperamental patterns may evoke dissimilar responses in different parents (Thomas et al., 1961b).

TEMPERAMENTAL TYPES AND STRESS

Of special interest has been the finding that a stressful environmental situation or demand is related to the temperamental responses of the child as well as to the objective features of the environment (Thomas et al., 1968). In other words, the effective environment is the result of the child-environment interaction and not of the objective situation as such. We have been able to delineate a number of temperamental constellations by quantitative and qualitative analysis of the item scores and ratings for the 9 categories and to identify those situations that are most likely to be stressful for the child with a particular combination of reactive characteristics.

One such temperamental constellation comprises the combination of irregularity in biological functions, negative (withdrawal) reactions to many new situations, slow adaptability to change, predominantly intense responses and frequent negative mood expressions. A child with such a pattern is aptly called a *difficult child,* because routines in sleeping and feeding are established with difficulty and much crying. New places, strangers, new rules may all produce an initial reaction of withdrawal and loud crying. The difficult children constitute approximately 10 percent of our study population but a disproportionately high percentage develop behavior problems (Rutter et al., 1964). The stressful demands for these children are most typically those involved in the socialization process both within the family and outside the home—the demands for modification and change in behavior to adapt to the rules of social living of the family, the peer group, and the school. If such demands are presented to the child in a friendly, patient, and consistent fashion, the stress will be minimized and usually only transient. If, on the contrary, rules and expectations are presented impatiently, angrily, and inconsistently, significant behavioral disturbance is likely to ensue. Characteristically, once such a child adapts to a new situation his functioning may then show the same intensity and consistency but in a positive direction.

At the opposite end of the temperamental spectrum from the difficult child is the *easy child,* with the reactive characteristics of regularity in biological functions, positive (approach) reactions to most new situations, quick and easy adaptability to change and predominantly positive mood expressions of mild or moderate intensity. These children, who comprise a much larger proportion of the study population, develop regular sleep and feeding schedules quickly, take to most new foods easily, are usually easy to wean and toilet-train, and generally adapt quickly to a new place, a stranger, or a new school situation. In contrast to the difficult child the easy child usually does not find the demands of socialization stressful and most frequently easily adapts to family expectations and standards that are appropriate for his age-level. The typical stress situation for the easy child occurs where there is a severe dissonance between the standards and demands of the intra- and extrafamilial environments. The child learns patterns of behavior expected and demanded at home in the first few years of life. If these patterns then conflict sharply with the norms and demands of peer groups and school, stress and malfunctioning may develop as the child moves more and more into activities outside the home.

Conflict between intrafamilial and extrafamilial standards and demands may produce stress and maladaptive behavioral development in many types of children, but in our study population it has been most specifically pathogenic for the easy child.

The demands of the new are also stressful for another type of child—the one with initial negative responses of mild intensity to new situations and people. If the child can reexperience the new a number of times without pressure, he typically adapts gradually and slowly until finally a stable positive response is established. We have called the child with this pattern the *slow to warm up child.* He differs from the difficult child in that the negative responses are mild rather than intense and that biological functions are usually regular rather than irregular. Stress typically develops for the slow to warm up child when parents, teachers, or peers demand and insist on the kind of quick adaptation to the new that is very difficult or impossible for this child.

In contrast to the difficult and slow to warm up children in whom stressful situations are most likely to develop in relationship to the new, there are other children in whom stress occurs most typically not with the new but during the course of ongoing activity after the initial positive adaptation has occurred. One such temperamental type is the very *persistent* and *nondistractible* child who resists the demand that he terminate an activity in which he is selectively and intensively involved. If such a demand is made insistently or forcibly, tension and frustration will mount and may finally express itself explosively. The *nonpersistent* and *distractible child* with short attention span is at the other extreme. He finds it difficult to sustain an involvement with a difficult task for any length of time and the stressful situation for him is represented by the demand for such sustained attention and activity.

A *high activity child* will typically have difficulty in situations in which he is expected to sit quietly with restraint of motor activity for long periods of time. The *low activity child,* on the other hand, will experience stress when exposed to the demand for liveliness and speed of movement of which he is not capable.

The one category of reactivity for which we have not as yet been able to identify typical stressful child-environment interaction patterns is level of sensory threshold. It is possible that further data analyses now in progress may delineate such patterns.

There has been no evidence to indicate that any of the temperamental constellations described above are the result of specific patterns of parental attitudes or practices. Parental functioning is indeed influential in the child's development, but not in producing the child's initial temperamental characteristics. Parental influence is mediated through the child-environmental interactional process, as is the child's influence on the parent.

It is clear that our analysis of the factors involved in the young child's behavioral development has not utilized various currently influential hypotheses about theoretical subjective states in the infant and young child— instinctual drive states and their vicissitudes, anxiety, intrapsychic conflict and psychodynamic defenses. Our data indicate that theoretical concepts of instinctual drive states are not required to explain the course of psychological development. As for anxiety, intrapsychic conflict and psychodynamic defenses, these phenomena appear to develop as secondary phenomena resulting from the stress of a maladaptive temperament-environment interaction. Once any or all of these secondary factors appear they can add a new dimension to the child-environment interaction and influence the subsequent course of development of behavioral disturbances.

CONCLUSION

The findings of the New York longitudinal study indicate that an interactionist approach that considers behavioral development in terms of a continuously evolving "reciprocal stimulative process" between organism and environment is as pertinent for the study of the human child as for other organisms. By contrast, currently prevalent concepts in child psychology and psychiatry assume that a child's behavior is a direct one-to-one reaction to maternal influences. Such a view all too often leads to a study of the mother as a substitute for a study of the child's temperament, other organismic characteristics, the level of the child's functioning, other intrafamilial and extrafamilial influences, and the continuously evolving interaction process among all these factors. We have elsewhere characterized this preoccupation of psychologists and psychiatrists with the role of the mother as the "Mal de Mère" syndrome (Chess, 1964). This preoccupation has caused innumerable mothers who have been incorrectly held exclusively responsible for their children's problems to be unjustly burdened with intense guilt feelings. Diagnostic procedures have tended to be focused primarily on the mother's assumed noxious attitudes and practices, and treatment plans have concentrated on changing these presumed pathogenic maternal characteristics and ignored other significant etiological factors. An approach that gives the same serious attention to defining individuality in temperament and other organismic characteristics as well as to identifying maternal and other environmental influences is the best treatment for this "Mal de Mère" syndrome.

ACKNOWLEDGMENTS

This research was supported in part by a grant from the National Institute of Mental Health, MH-3614.

REFERENCES

Chess, S. 1964. Editorial: Mal de Mère. *Am. J. Orthopsychiat.* 34: 613–614.
Chess, S., A. Thomas, and H. G. Birch. 1959. Characteristics of the individual child's behavioral responses to the environment. *Am. J. Orthopsychiat.* 29: 791–802.
Chess, S., A. Thomas, and H. G. Birch. 1965. *Your child is a person.* New York: Viking.
Koffka, K. 1924. *The growth of the mind.* New York: Harcourt, Brace.
Lewin, K. 1935. *Dynamic theory of personality.* New York: McGraw-Hill.
Pavlov, I. P. 1927. *Conditioned reflexes.* Trans. and ed., G. V. Anrep. London: Oxford University Press.
Robbins, L. 1963. The accuracy of parental recall of aspects of child development and child-rearing practices. *J. Abnorm. Soc. Psychol.* 66: 261–270.

Rutter, M., H. G. Birch, A. Thomas, and S. Chess. 1964. Temperamental characteristics in infancy and later development of behavioral disorders. *Brit. J. Psychiat.* 110: 651–661.

Schneirla, T. C., and J. S. Rosenblatt. 1961. Behavioral organization and genesis of the social bond in insects and mammals. *Am. J. Orthopsychiat.* 31: 223–253.

Stern, W. 1927. *Psychologie der frühen kindheit, bis zum sechsten lebensjahre.* Leipzig: Quelle and Meyer.

Thomas, A., H. G. Birch, S. Chess, and M. E. Hertzig. 1961a. The developmental dynamics of primary reaction characteristics in children. *Proceedings Third World Congress Psychiat.* Toronto: Univ. Toronto Press. Pp. 721–726.

Thomas, A., H. G. Birch, S. Chess, and L. Robbins. 1961b. Individuality in responses of children to similar environmental situations. *Am. J. Psychiat.* 117: 798–803.

Thomas, A., S. Chess, and H. G. Birch. 1968. *Temperament and behavior disorders in children.* New York: New York Univ. Press.

Thomas, A., S. Chess, H. G. Birch, M. E. Hertzig, and S. Korn. 1963. *Behavioral individuality in early childhood.* New York: New York Univ. Press.

Wenar, C. 1963. The reliability of developmental histories. *Psychosom. Med.* 25: 505–509.

HERMAN A. WITKIN, JUDITH BIRNBAUM, SALVATORE
LOMONACO, SUZANNE LEHR, and JUDITH L. HERMAN
State University of New York Downstate Medical Center
Brooklyn, New York

Cognitive Patterning in Congenitally Totally Blind Children[1]

It has been shown in a number of studies that the development of cognitive styles is influenced by experiences the child has had while growing up (Barclay and Cusumano, 1967; Berry, 1966; Corah, 1965; Dawson, 1967a, 1967b; Dershowitz, 1966; Dyk and Witkin, 1965; Seder, 1957; Witkin et al., 1962). A great deal remains to be done, however, in identifying more specifically the kinds of life experiences that are influential in the development of different cognitive styles and the precise ways in which these experiences exert their influence. Our program of research on the role of experiential factors in the development of cognitive style has been exploiting both naturally occurring variations in life experiences and "accidents in nature." On the side of naturally occurring variations, we have been investigating cognitive functioning in children who have grown up under special life circumstances, as identical twins (Winestine, 1964) and boys from highly orthodox Jewish families (Dershowitz, 1966). On the side of "accidents in nature," we have been studying two kinds of children growing up under the influence of a severe deficit: retarded children (Witkin et al., 1966) and congenitally totally blind children.

The study of the blind was undertaken with specific expectations as to how the absence of vision, and the inevitably greater reliance of the blind child on other persons, would be likely to affect cognitive development.

One expectation was that the blind would be more global in their cognitive functioning as compared to the sighted, who would be more articulated. The global-articulated cognitive dimension is the cognitive style with which we have been particularly concerned in our past studies (Witkin et al., 1954; Witkin et al., 1962). Experience is articulated, rather than

---

[1] Reprinted from *Child Development*, 1968, 39: 767–786, by permission of the Society for Research in Child Development, Inc.

global, if the person is able to experience parts of a field as discrete from background when the field is structured and to impose structure on a field, and therefore experience it as organized, when the field has relatively little inherent structure. Articulation thus has two aspects—analysis and structuring. The concept of articulation may be applied to experience of an immediately present stimulus configuration (perception) and to experience which is more in the realm of symbolic representation (intellectual activity). Results of our earlier cross-sectional and longitudinal studies have made it abundantly clear that the ordinary progression in cognitive development in the sighted is from relatively global to relatively articulated (Witkin, Goodenough, and Karp, 1967).

In congenitally totally blind children, the development of articulation is, in a variety of ways, likely to be hindered. Vision, lacking in the blind child, is the sensory channel most useful in the development of articulation. Because through vision we are able to apprehend parts of the field and the field as a whole in one view, experiencing items as discrete from background and the field as structured—in other words, articulated perception —is facilitated. Further, drawing on his past visual experiences, the sighted person can visualize the articulation of a field even when the field is apprehended through other sense modalities alone.

Studies of the sighted have shown that children who experience the world about them in articulated fashion are also likely to have an articulated body concept—that is, to have an impression of the body as having definite limits or boundaries and the parts within as discrete yet interrelated and formed into a definite structure (Witkin et al., 1962; Witkin, 1965a). Vision facilitates apprehension of the body as an entity apart from the field as well as apprehension of the specific parts of the body and their interrelation. Moreover, through vision it is possible to form an impression of how the bodies of others are "put together," and this in turn also contributes to development of an articulated body concept. To form an impression of objects as discrete and of field or body as structured through the senses that remain when vision is lacking is, of course, possible, though surely much more difficult.

In still another way is the development of articulation likely to be hindered in the blind child. Because he cannot see, dependence on others in all kinds of activities is intensified. Our studies of sighted children have shown that children who interact with their mothers in ways that interfere with the opportunity for separation from mother are likely to show a global style of cognitive functioning and to give evidence of limited differentiation in other psychological areas.

In all these ways, the absence of vision may be expected to slow the pace of the usual progression in cognitive development from global to articulated.

These speculations led us to expect that blind children, as a group, would give evidence of less developed articulation in their cognitive functioning when compared to sighted children of the same age. But again drawing on our past studies, we expected to find self-consistency in mode of cognitive functioning—that is, cognitive styles—among the blind as we have among the sighted. To the extent that blind children grow up in widely varying family and social contexts and are surely different in constitutional characteristics, we may also expect them to show individual differences in cognitive style.

Results of past factor analytic studies with the sighted (Goodenough and Karp, 1961; Karp, 1963) led us to expect still another characteristic of cognitive patterning in the blind. Loading one factor in these studies, called an "analytical" factor, were the Block Design, Picture Completion, and Object Assembly subtests of the Wechsler intelligence scales and tests of perceptual field dependence. Performance on these Wechsler subtests in fact provides a measure of the global-articulated cognitive style. A second factor was defined by the Vocabulary, Information, and Comprehension subtests of the Wechsler scales. It has been labeled a "verbal-comprehension" factor. A third factor, called an "attention-concentration" factor, was defined by the Wechsler Digit Span, Arithmetic, and Coding subtests. In the sighted, verbal-comprehension factor scores have been found to relate at a low level or not at all to measures of the global-articulated dimension. Since verbal competence and competence at articulation are thus more or less distinct abilities, we expected measures of these two kinds of abilities also to show a low relation in the blind.

This study of the blind was undertaken, then, with three hypotheses: (1) The global-articulated cognitive style identified in the sighted will be evident in the blind as well. With respect to this dimension, blind children will be self-consistent in performance across cognitive tasks and will show individual differences. (2) Blind children will be less articulated in their cognitive functioning than sighted children. (3) Measures of articulation and measures of verbal comprehension will show little or no relation to each other.

TESTS

Articulation, as noted, implies both analysis and structuring of experience. In the present study, the analytical aspect only has been considered. The tasks used evaluated this aspect of articulation in both perceptual and intellectual activities. The perceptual component of the analytical aspect of articulation is in fact the field-dependence-independence dimension. It now seems clear that individual differences along the field-dependence-independence dimension represent differences in capacity to overcome an embedding context or to perceive parts of a field as discrete from organized background. The factor analytic studies cited have shown that both perceptual and intellectual tasks having the requirement of overcoming an embedding context load the same factor (Goodenough and Karp, 1961; Karp, 1963).

To evaluate analytical functioning in the blind, the tests we ordinarily employ, all of which require vision, obviously could not be used. When this study was undertaken, several nonvisual tests of analytical functioning were available from work by other investigators. In addition, we adapted several visual tests from those of our original battery which could easily be translated into nonvisual form. Of the tests used, some are specifically perceptual in nature and are in effect tests of field dependence, while others are of a kind commonly classified as problem-solving or intellectual tasks, although there is, of course, no assumption of a hard and fast division between them.

All tests were given in the same form to the blind subjects and to the control group of sighted subjects. In the tactile tests, we prevented the sighted subjects from seeing the test materials by placing the materials behind a rectangular screen with a curtained cut, through which the subject placed his hands.

Tests of Analytical Ability (Field Dependence) in Perception

THE TACTILE EMBEDDED-FIGURES TEST

This test was developed by Axelrod and Cohen (1961) and is a direct tactile counterpart of the visual embedded-figures test we ordinarily use to evaluate field dependence (Witkin, 1950). The subject first feels a simple figure with his fingers and then must find this figure in a complex, organized design in which it is embedded. Axelrod and Cohen found a very high correlation (.78 for a group of young subjects and .68 for a group of elderly subjects) between this tactile embedded-figures test and Thurstone's version of the Gottschaldt figures, which is a visual embedded-figures test.

Each of the simple and complex figures of the tactile embedded-figures test is in the form of a raised design on a thin, 4-inch-square sheet of plastic. The test consists of 14 trials, preceded by three practice trials. Our method of administering the test was essentially the same as that of Axelrod and Cohen, differing only in a few details. On each trial, the subject is first presented with the complex figure for 15 seconds and asked to describe it. This figure is then removed and the simple figure embedded in it is presented for 10 seconds. Finally, the simple figure is removed and the complex figure is given once again to the subject, with the task of locating the simple figure within it. The subject is told that in every case the simple figure is present in the larger complex design and always in an upright position. He is asked to work as quickly as possible, since he is being timed, but he is also told not to rush and to make sure that when he identifies the simple figure it is exactly the right one. He is also informed that should he ever forget the simple figure while searching for it in the complex one, he may ask for time out to examine the simple figure again. If the figure the subject picks out as the simple one is incorrect, he is so informed and asked to continue his search. A maximum of 300 seconds is allowed per trial. If the subject fails to find the simple figure within that time, his score for the trial is 300 seconds. The subject's total score for the test is the sum of his time scores for the 14 trials of the test. Axelrod and Cohen did not report the reliability of the test. In the data of the present study, the uncorrected odd-even reliability was .68 for the sighted and .71 for the blind.

THE AUDITORY EMBEDDED-FIGURES TEST

This test, developed by White (1953), is an auditory translation of the visual embedded-figures test. A short three-, four-, or five-note tune is played followed by a longer, more complex tune, which may or may not contain the simple tune. White reports a significant correlation of .63 between scores for the auditory embedded-figures test and scores for the Thurstone Gottschaldt figures test.

The auditory embedded-figures test consists of 50 pairs of simple and complex tunes recorded on tape. The method used in administering the test followed White's, except for a few small features. For practice purposes, the subject is first given four trials in which the tunes are played by the experimenter on a xylophone, and then three additional trials in which the tunes are played on a tape recorder. Before the test, the subject is informed that he will be presented with a series of pairs of notes, each consisting of a short set of notes followed by a longer set. Sometimes the shorter series of notes will be present in the longer, complex tune, in identical sequence and pitch, at the beginning, middle, or end of the complex tune; sometimes it will not be present at all. After each pair of simple and complex notes, he is to indicate whether the short series of notes occurred in the complex melody. If he is not sure, he is to guess. Finally, the subject is informed that after every 10 pairs of tunes, there is a 10-second rest period, but he may ask that the tape be stopped at any time if he wants to take a break (no subject in the study asked for a break). The test takes about 30 minutes. The subject's score for the test is the percentage of correct answers for the 50 problems of the test. White has reported a Kuder-Richardson reliability coefficient of .83 for the test with his sighted subjects.

Tests of Analytical Ability in Problem Solving

THE TACTILE BLOCK DESIGN TEST

In this tactile form of the block design test, as in the original visual form, the subject must reproduce a reference design by the appropriate arrangement of a set of blocks. The reference design has an organization which must be "broken up" into component blocks if it is to be reproduced. Numerous studies have shown a high relation between the visual form of the block design test and tests of field dependence (see, e.g., Goodenough and Karp, 1961; Karp, 1963; Witkin et al., 1962).

The blocks and reference designs making up the test are the same as in the Wechsler Block Design subtest, except that the patterns are represented by different textures (smooth oaktag and wire screening) rather than by color, as in the original. There are four cube-shaped blocks. Each has two surfaces covered with wire screening, two covered with oaktag, and two divided in halves diagonally, one-half covered with oaktag and the other with screening. On each of eight trials, the subject is given a design made up of oaktag and screening, and he is required to compose the design by appropriate combination of the four blocks. The first trial is a practice one and is not counted. On each of the other eight trials, the subject is given three points if he makes the correct design within 30 seconds, two points if he makes it in from 31 to 60 seconds, and one point if he makes it in from 61 to 180 seconds. The maximum score possible is thus 24. The uncorrected odd-even reliability of the test was .61 for the sighted subjects and .76 for the blind subjects of the present study.

THE TACTILE MATCHSTICKS TEST

In the visual version of this test, as used by Guilford, Frick, Christensen, and Merrifield (1957) in their studies of adaptive flexibility, the subject is shown a set of matches arranged in the form of a lattice and asked to reduce the total number of squares in the lattice by removing a specified number of matchsticks. In order to solve the problem the subject must overcome the organization of the presented latticework and discover within it a subsidiary organization which meets particular requirements. It has been demonstrated repeatedly that performance in this test, in its visual form, relates highly to performance in tests of field dependence (Guilford et al., 1957; Karp, 1963; Witkin et al., 1962). This test was easily changed to a nonvisual form by having the subject feel the latticework with his fingertips instead of seeing it. The test consists of 10 problems, each scored right or wrong, so that 10 is the maximum score possible. Preceding the test are three practice trials. In the data of the present study, the uncorrected odd-even reliability of the test was .71 for the sighted subjects and .81 for the blind subjects.

Test of Body Concept

As already noted, sighted children and adults who experience the world about them in articulated fashion are also likely to have an articulated concept of their bodies. One of the techniques we have used with sighted subjects to evaluate articulation of body concept is the figure drawing. In the drawings of children with an articulated cognitive style, the human body is likely to be portrayed in realistic proportions; parts of the body are represented in appropriate relation to each other and to body outline; and sexual differentiation is clearly indicated.

Underlying the figure-drawing technique is the idea that, in sketching the body on paper, the person is dealing with the body vicariously and, in doing so, projects something of his concept of what the body is like, a concept compounded out of the experiences he has had with his own body and the bodies of others. Following this view, we sought a nonvisual medium in which our blind subjects might represent the body and thereby perhaps express their concept of the body. Clay modeling seemed to provide such a medium. Obviously, in this translation of a primarily visual technique into a nonvisual form, we took a far greater leap than was involved in the translation of the tests considered thus far.

The procedure used followed the method usually employed in administering the figure-drawing test. The subject is given a ball-shaped piece of self-hardening clay weighing about ¾ pound to 1 pound and asked to make a person. He is told he may use as much clay as he wants and that he may have more clay if he wishes. If the subject asks whether he should make a man or woman, sitting or standing, etc., he is told it is up to him. If he asks whether to make a head or whole person, he is told to make a whole person.

The clay models were rated for articulation on a five-point scale. The scale used was patterned closely after the sophistication-of-body-concept scale developed in our earlier studies to evaluate figure drawings along the articulation dimension (Witkin et al., 1962). The five categories used in rating clay models are: (1) amoebic masses, lumps, or blobs bearing no recognizable resemblance to the human shape; (2) primitive attempt at human form, grossly misshapen and/or gross size distortion; (3) intentional but only gross approximation of the human figure; (4) intentional shaping, fewer distortions of size, shape, and attachments; (5) clearly represented human form without major distortions.

Two blind subjects and two sighted subjects did not make clay models, because they were no longer available for testing.

Intelligence Tests

All subjects were given the verbal section of the Wechsler; the younger subjects received the Wechsler Intelligence Scale for Children (WISC) and the older subjects the Wechsler Adult Intelligence Scale (WAIS) or the Wechsler-Bellevue (W-B). The sighted subjects were given the performance section in addition.

SUBJECTS

Twenty-five subjects, 13 boys and 12 girls, totally blind from birth, constituted the experimental group. None had pattern vision or light vision. In 20 cases the cause of blindness was retrolental fibroplasia, and there was one case each of optic atrophy, congenital glaucoma, phtesis bulbi, bilateral congenital anophthalmia, and agenesis corticalis. The subjects ranged in age from 12–7 to 18–9, with a mean age of 14–6. (Partialing out age from the correlations in the tables presented in the Results section had only a negligible effect.) In no case did the subject's record show any evidence of brain damage, although neurological examinations had not been conducted in all instances. The IQ's of the group, computed of course for the verbal section alone, ranged from 92 to 153, with a mean of 114.8. All were thus at least within the normal range, and many were superior. The mean school grade of the group was 8.7.

The control group consisted of 28 sighted subjects—15 boys and 13 girls—matched to the blind group, by group means, for age and school grade. The age range of the control group was from 12–3 to 19–9, with a mean age of 14–11. Their mean school grade was 8.8. The blind and sighted groups also turned out to be equated in verbal-comprehension factor IQ (computed from the Vocabulary, Information, and Comprehension Wechsler subtests according to the procedure described in the Results section), although this characteristic was not used in the selection of subjects. The mean values for the two groups were 115.5 and 113.7. It was originally planned to equate the blind and sighted groups for socioeconomic status as well. However, a statistical check made when the two groups were almost completed showed no significant relation between socioeconomic status and measures for our various tests of articulation. This criterion for matching was therefore abandoned. In other studies with the sighted, socioeconomic status also typically failed to relate to articulation measures.

RESULTS

Self-Consistency in Cognitive Functioning of the Blind

Table 1 shows the intercorrelations among the scores for the five tests given the blind. To compute the values of this table, correlations for boys

TABLE 1
Intercorrelations for blind subjects.[a]

	Matchsticks	Tactile block design	Auditory embedded figures	Clay models
Tactile embedded figures.........	−.76[b]	−.73[b]	−.30	−.56[c]
Matchsticks....................	· · ·	.60[b]	.30	.54[c]
Tactile block design.............	· · ·	· · ·	.24	.56[c]
Auditory embedded figures.......	· · ·	· · ·	· · ·	.36

[a] $N = 25$, except for clay models, where $N = 23$.
[b] $p < .01$.
[c] $p < .05$ (one-tail tests have been used throughout).

and girls (who in past studies have shown differences in performances on tests of the global-articulated dimension) were computed separately and then combined by the method of r to z transformation. All the correlations in the table are quite high and significant, with the exception of those with the auditory embedded-figures test. Correlations of scores for the latter test with scores for the other tests are all in the expected direction, but none reaches significance. Except for results for the auditory embedded-figures test, the outcome is in keeping with expectations.

The result for the auditory test was unexpected. Several hypotheses suggested themselves to account for it. An obvious one is based on the fact that all the tests which are significantly intercorrelated are tactile in nature, whereas the auditory test is not. It is possible that competence at tactile discrimination, rather than competence at overcoming embedding contexts, as we had assumed, may be primarily responsible for individual differences in performance of the tactile tests.

To check this hypothesis, we gave our subjects a test which requires discrimination but not disembedding. This is a haptic-forms test developed by Gollin. In this test the subject is allowed to feel an irregularly shaped flat piece of plexiglas in order to become familiar with its form. He is then given this same piece mixed in with three other pieces of similar shape and asked to pick out the original piece. The test consists of 20 such plexiglas sets, and the subject's score is the total number of correct identifications. In a personal communication, Gollin has reported for nine groups of adults odd-even reliability coefficients ranging from .85 to .96. Performance of this task clearly involves form discrimination, but does not involve overcoming an embedding context in the sense of breaking up an organized structure in order to identify a sought-after component part. We may expect form discrimination, as tapped by this haptic-forms test, to figure in performance of our tactile disembedding tests and in the clay-models test as well. At the same time, we would not expect this factor to account entirely for performance in these situations if they do in fact involve disembedding.

As anticipated, when our blind subjects were given the haptic-forms test, we found that scores for the test did relate to measures for the various other tactile tests. However, they typically related at a lower level than these tactile tests related to each other. Moreover, partialing out the haptic-forms test from the correlations in Table 1, though lowering the correlations, did not affect the pattern of interrelations; all the significant correlations remained significant and all the nonsignificant correlations remained nonsignificant.

These results suggest that performance in the various tactile tests involves more than tactile form discrimination alone. They argue against the possibility that the involvement of this factor in the tactile tasks is responsible for their failure to relate to the auditory embedded-figures test.

A second hypothesis we considered, to account for the absence of a relation between the auditory test and the other tests, was that the auditory test demands prolonged sustained attention to a much higher degree than do the other tests. This hypothesis was suggested by our own experience in taking the test and the experience reported by some of our subjects. Within a trial, each note of the simple melody must be "held onto" as the other notes are played, and then the simple melody as a whole must be "held onto" as each note of the complex melody is played. This task is repeated 50 times during the approximately 30-minute period that the test requires. The record on which the test material is recorded plays on regardless of lapses in attention or forgotten notes. It is obvious that the format of the test places considerable strain upon attention, so that ability to maintain prolonged focused attention would be of considerable help in its performance.

To check on the possibility that this kind of competence is in fact important in the auditory embedded-figures test, we examined the performance of our subjects on two Wechsler subtests which in past factor analytic studies of the Wechsler scales loaded an attention-concentration factor (Cohen, 1957, 1959; Goodenough and Karp, 1961; Karp, 1963). These are the Digit Span and Arithmetic subtests. To represent competence on this factor, an attention-concentration factor IQ was computed for each subject. The sums of the age-corrected scaled scores for the Digit Span and Arithmetic subtests were converted so that the underlying distribution had a mean of 100 and a standard deviation of 15. The conversion procedure utilized the tables of means, standard deviations, and intercorrelations provided in the Wechsler manual. Because such tables are not available for the Wechsler-Bellevue, factor IQ's could not be computed for the five subjects who received this form of the Wechsler; these five subjects are therefore not included in analyses employing factor IQ's.

Four blind children who were not available to complete the entire test series were not given the Digit Span and Arithmetic subtests.

Table 2 shows, for our blind subjects, the correlations of attention-

TABLE 2

Correlations for blind children between test scores and attention-concentration factor IQ's.[a]

	Attention-concentration factor IQ
Tactile embedded figures...........	−.40
Matchsticks......................	.20
Tactile block design..............	.03
Auditory embedded figures........	.65[b]
Clay models.....................	.01

[a] $N = 17$, except for clay models, where $N = 15$.
[b] Significant at $<.05$ level.

concentration factor IQ's with scores for the various disembedding tests and the clay-models test. The attention score does in fact show a particularly high relation to auditory embedded-figures scores, and the correlation between these two measures is the only significant one in Table 2. In the blind, the auditory embedded-figures test does then seem to measure—perhaps most of all—capacity for sustained attention to a much greater degree than the other tests. As will be discussed later, it may be this difference which contributes particularly to the lack of a relation between the auditory test and the other tests in the blind.

Going back to our starting hypothesis, we may say that, as expected, the global-articulated dimension is to be found in the cognitive functioning of the congenitally blind. The individual blind subject was highly consistent in his performance across a variety of tactile tests of articulation and a test of articulation of body concept; and marked individual differences were evident in the extent to which the cognitive functioning of the blind subject was relatively articulated or relatively global.

Comparison of the Blind and Sighted

Our hypothesis here was that congenitally totally blind children would tend to be more global in their cognitive functioning than sighted children. Table 3 presents the mean scores for the blind and sighted groups. The

TABLE 3
Means for blind and sighted groups.[a]

Tests	Blind	Sighted
Tactile embedded figures............	134.5	64.5[b]
Tactile matchsticks................	4.8	5.3
Tactile block design..............	10.9	12.3
Clay models.....................	2.6	3.4[c]
Auditory embedded figures.........	84.9	66.7[b]

[a] N for blind: 13 boys and 12 girls, except for clay models, where N for boys is 12 and N for girls is 11. N for sighted: 15 boys and 13 girls, except for clay models, where N for girls is 11.
[b] $p < .01$ for difference between blind and sighted.
[c] $p < .05$ for difference between blind and sighted.

differences between the two groups on the tactile embedded-figures test, the tactile matchsticks test, the tactile block design test, and the clay-models test are all in the expected direction, performance of the sighted showing greater articulation than performance of the blind. The significance of these differences was tested by a two-way analysis of variance, with a check made on the effects of both blindness and sex. Sex effects were checked, since, as already noted, among the sighted, sex differences have commonly been found in tests of articulation. The analysis of variance showed the difference between blind and sighted to be significant for the tactile embedded-figures test ($F = 19.36$, $p < .01$) and the clay-models test ($F = 6.77$, $p < .05$).

There was a significant sex effect for the tactile matchsticks test only, but in a direction opposite to that expected on the basis of past results for the sighted. It may be noted, however, that the number of cases in each sex group is quite small.

The auditory embedded-figures test again behaved differently from each of the other tests in the battery. The blind as a group did significantly *better* than the sighted on this test ($F = 30.52$, $p < .01$).

To check the effect of musical experience on performance in the auditory embedded-figures test, scores of subjects with no musical training were compared with scores of subjects who had musical training of any kind and duration. In both the blind and sighted groups, musically experienced subjects did significantly better than nonexperienced subjects. However, when blind and sighted subjects with no experience were compared, the blind were significantly better; and a significant difference in favor of the blind was also found when musically experienced blind and sighted subjects were compared.

It is noteworthy that on the haptic-forms test the mean scores for the blind and sighted were not significantly different and, in fact, were almost identical. This test, which does not involve disembedding, thus shows a different picture from the tactile tests which have such a requirement. Taking the results together, our expectation that articulation will be more developed in sighted children than in blind children is, on the whole, supported.

Relation between Measures of Articulation and Verbal Ability

Our third hypothesis was that measures of articulation and measures of verbal comprehension would show little or no relation in the blind. The measure of verbal ability we used was the verbal-comprehension factor IQ, based on the Vocabulary, Information, and Comprehension subtests of the Wechsler scales. As already mentioned, in past factor analytic studies of the Wechsler scales, these three subtests showed the highest loading on what has been called a verbal-comprehension factor. Factor IQ's were used rather than scaled scores, since, as noted earlier, the factor IQ's are age corrected; they were computed by the same method already described for computing attention-concentration factor IQ's.

In keeping with expectations, correlations of verbal-comprehension factor IQ's with measures of each of the five tests given the blind were all low (ranging from .10 to −.21), and none was significant. Thus, in still another way, we find a parallel between the cognitive picture in the blind and the cognitive picture previously observed in the sighted.

DISCUSSION

The finding that the blind show less articulation in their cognitive activity than the sighted provides another demonstration that differences in developmental history make for differences in cognitive functioning. The particular aspects of developmental history implicated in this study are the absence of the sense modality most useful in articulation of experience and enforced dependence on others, although what has been done thus far does not permit us to assay the specific and separate contributions of these factors to the less developed articulation found in the blind.

Noteworthy as it is that congenitally totally blind subjects show less developed articulation than thc sighted, of equal interest is the finding that the difference is not as great as might be expected in view of how severely handicapped the blind are. Some individual blind subjects, in fact, show highly developed articulation in their cognitive functioning and, from the suggestive findings for the clay-models test, in body concept as well. How do they achieve this high level of differentiation? The possibility must be considered that, in at least some children, blindness may actually serve as an impetus to the development of differentiation. The special effort the blind child must make to achieve an articulated concept of the world, precisely because vision is lacking, may conceivably encourage great investment in articulation and so actually foster the development of differentiation. Further, parents, teachers, and others may go to great lengths to help the blind child achieve an articulated impression of his surroundings and of himself, thereby again giving impetus to the development of differentiation. What is being suggested is that in addition to the negative influences of blindness upon differentiation, there may also be positive influences. These positive influences may, in some cases, be sufficiently great to compensate for and even exceed the inevitably present negative ones. Special characteristics of cognitive functioning may thus themselves contribute to the overall development of psychological differentiation.

Many basic similarities are evident in the cognitive patterning of the blind and the sighted, and there are important differences as well.

With regard to similarities, intraindividual consistencies—in other words, cognitive styles—and individual differences are evident in the blind and sighted. In both, also, measures of the global-articulated dimension show independence of measures of verbal-comprehension ability. With the finding of cognitive styles of the same basic kind in the blind and sighted, and in other recent studies in the deaf (Fiebert, 1967) and retarded (Witkin et al., 1966) as well, the cognitive-style concept acquires greater generality.

With regard to differences in cognitive patterning, one major difference is suggested by the results for the auditory embedded-figures test. To summarize these results: (1) In the blind, the auditory embedded-figures test behaved differently from all the other tests of articulation. Scores for these other tests were all highly interrelated, but none related significantly to scores for the auditory embedded-figures test; and auditory test scores showed a particularly high relation to measures of capacity for concentrated attention. (2) The auditory embedded-figures test showed a different pattern of relation to the other tests in the blind than in the sighted. In contrast to the picture in the blind, just reviewed, in the sighted, the auditory embedded-figures tests did relate to other tests of articulation. White (1953), as already reported, found a high relation between the auditory test and a visual embedded-figures test. In the sighted control group of the present study, the auditory test again tended to relate to other tests of articulation in their visual form, though at a lower level than in the White study; and it also tended to relate to the various tactile tests of articulation. Moreover, it did not relate to the attention-concentration tests. (3) The relation between mean scores of the blind and sighted was different for the auditory embedded-figures test than for the various other tests. The blind did significantly better than the sighted on the auditory embedded-figures test, whereas the sighted did better on all the other tests.

One way in which we may make sense of all these results is by hypothesizing that the auditory embedded-figures test, as presently constituted, measures different kinds of ability in the sighted than in the blind. In the sighted it is primarily a test of capacity to overcome an embedding context; in the blind it is mainly a test of capacity for sustained attention in the auditory realm. For future studies we are seeking to change the present format of the test so as to reduce the demand for sustained attention, thereby permitting greater expression of individual differences in capacity for disembedding.

If the auditory embedded-figures test in fact now measures in the blind capacity for sustained auditory attention, it is not at all surprising that the blind should do better than the sighted. Deprived of the opportunity to experience the world through vision, the blind rely greatly on audition, the next best sense available to them. There is clear evidence that in their orientation, blind persons make greater use of auditory cues than do the sighted (see, for example, Burlingham, 1964; Supa, Cotzin, and Dallenbach, 1944). With their greater alertness to auditory cues, the blind seem to have greater ability to maintain prolonged attention to auditory material. Evidence that the blind are outstanding in this ability comes from the findings of the present study for the Digit Span subtest of the Wechsler. This test requires the subject to repeat a series of numbers presented auditorially, and so provides a particularly direct measure of ability to "hold onto" elements in the auditory realm. The blind were markedly superior on Digit Span to both the sighted control group and the general population. The mean scaled score for the blind was 15.1, in contrast to a mean scaled score for the sighted of only 10.8 ($p < .01$). (A mean scaled score of 10 is, of course, average for the general population.) Thus, the blind seem to have highly developed ability in maintaining prolonged attention to auditory material, gained undoubtedly through long enforced reliance on that modality. This special competence very likely contributes to their superior performance in the auditory embedded-figures test.

There is another hypothesis to be considered in interpreting the network of findings for the auditory embedded-figures test. Because of their special investment in auditory experience, among the blind, development of articulation in the auditory realm need not be congruent with the development of articulation in other realms. Though they may show stylistic tendencies within one modality, they need not show the consistency across modalities repeatedly observed in the sighted, where discrepancies in investment from one sense modality to another are not as marked. The hypothesis that in some groups with special developmental histories the global-articulated cognitive style may be sensory specific cannot be adequately evaluated on the basis of the results of the present study. Not only is it unclear whether the auditory embedded-figures test does in fact provide a measure of articulation in the blind, but we did not obtain data for several different auditory disembedding tasks for the same subjects, as we would need to do in order to assess stylistic tendencies along the global-articulated dimension in the auditory realm.

From the data this study has provided on cognitive abilities in the blind and sighted, the following picture emerges: The blind and sighted groups are about *equivalent* in verbal-comprehension ability (with verbal-comprehension factor IQ's of 115.5 and 113.7, respectively). On the other hand, the blind are *superior* to the sighted in tasks requiring sustained auditory attention (with mean attention-concentration factor IQ's of 124.0 and 110.9, respectively, and with percentage correct on the auditory embedded-figures test of 84.9 and 66.7, respectively); and they are *inferior* to the sighted in tasks featuring competence in articulation (see Table 3). The difference between blind and sighted in capacity for auditory attention and in articulation ability may be attributed specifically to differences in the kinds of experiences they have had while growing up. The overall outcome of the special experiences the blind have had is marked unevenness in level of functioning from one cognitive area to another.

There are two problems of assessment of the blind on which the results of the present study have a bearing.

The first is assessment of intelligence and of cognitive functioning in general. Obviously, only the verbal parts of the usual standard intelligence tests can be used with the blind, so that when we refer to a blind child's IQ we are in fact speaking of his IQ for the verbal section of the test. (See Witkin, 1965b, for a discussion of the consequences, especially marked in the blind, of an emphasis on verbal competence in "routing" children through life.) Included among the performance subtests which cannot be used are several which in the factor analytic studies cited loaded what we have called an analytical factor (Cohen, 1957, 1959; Goodenough and Karp, 1961; Karp, 1963). One of the Wechsler subtests which shows a high loading on this factor is the Block Design subtest. Also loading this same factor are our various tests of perceptual field dependence, including the visual form of the embedded-figures test; the matchsticks test in its visual form; and figure drawings, scored in much the same way as the clay-models productions. In the sighted control group of the present study, a pattern of consistently significant correlations is found between the usual tests (in visual form) loading this analytical factor and the various tactile tests of articulation as well as the clay-models test. It seems fair to assume that the tactile tests used in this study are good representatives of the analytical factor and, more specifically, of the analytical subtests of the Wechsler. It is noteworthy, for example, that in our control group, scores for the tactile block design test correlated .78 with scores for the visual Block Design subtest of the Wechsler.

The tactile tests of this study may be used to assess an ability—analytical competence—tapped by the usual standard intelligence tests but not measurable in the blind when they are given these standard tests. That the nonvisual articulation tests do in fact contribute something not now represented in intelligence testing of the blind is indicated by the consistent picture of very low, nonsignificant correlations found in our blind group between total IQ—essentially, of course, verbal IQ—and our various measures of articulation.

It is most encouraging that a performance intelligence test for the blind (the Haptic Intelligence Scale for Adult Blind, or HIS-AB) has become available (Shurrager and Shurrager, 1964). This test includes tactile versions of two Wechsler subtests loading the analytical factor—Object Assembly and Block Design. The HIS Block Design test, like the block design test used in the present study, is based on the Wechsler Block Design subtest, so that the two tests, though developed independently, are essentially identical in form.

The second issue of assessment of the blind for which our findings are relevant is personality assessment. Obviously, projective tests in common use with sighted persons cannot be used with the blind. If it is true for the blind, as we have found it to be for the sighted, that cognitive style carries a message about broad dimensions of personal functioning, then the nonvisual cognitive tests used in this study may serve in personality assessment of the blind. In addition, the clay-models procedure may potentially have broad value in personality assessment as its counterpart, the figure-drawing technique, has proved to have for the sighted. A better sense of the clay-model technique may perhaps be obtained by examining some of the clay figures produced by our blind children. Figure 1 gives samples of productions rated "1," that is, showing very global representations of the human body. Figure 2 shows a clay figure rated "5" (the only one in the group so rated), that is, showing a highly articulated representation of the human body. We have used the clay-model procedure for the first time in this study, so the findings are altogether preliminary, but the results seem promising. Particularly encouraging is the observation that, in the blind, measures of articulation of clay models relate to other measures of articulation in a way that may be expected from our past studies with figure drawings of the sighted. Further, in the sighted, clay-model ratings and articulation measures also tend to relate as expected. However, the correlation between clay-model and figure-drawing ratings in the sighted, though again in the expected direction, is only .26 (not significant). This suggests a need for caution in considering the clay-model technique, in its present format, as directly comparable to the figure-drawing technique. However,

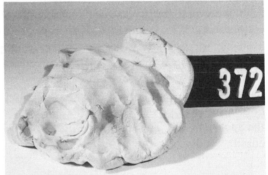

FIGURE 1
Clay models rated "1."

FIGURE 2
Clay model rated "5."

the results for the clay-model technique to this point seem sufficiently promising to warrant further work with it.

ACKNOWLEDGMENTS

The studies reported here were supported by a grant (M-628) from the U.S. Public Health Service, National Institutes of Health. This paper was presented in abbreviated form at a symposium on Cognitive Structure and Personality at the American Psychological Association, Chicago, September 7, 1965. We are deeply grateful to Miss Josephine Taylor of the New Jersey Commission for the Blind for her invaluable help in recruiting the blind subjects for this study and in obtaining and interpreting medical information about them. A debt of gratitude is also owed to the staffs of the Lighthouse and the Industrial Home for the Blind for their aid in subject recruitment. We are indebted to the following colleagues for making available for this study the tests they developed: Drs. Seymour Axelrod and Louis Cohen for the tactile embedded-figures test; Dr. Benjamin White for the auditory embedded-figures test; and Dr. Eugene Gollin for the haptic-forms test. We are also very grateful to Dr. Hanna Marlens for devising and applying the articulation scale for assessing clay models of the human figure. The assistance of Anne McKinnon is gratefully acknowledged.

REFERENCES

Axelrod, S., and L. D. Cohen. 1961. Senescence and embedded-figures performance in vision and touch. *Percept. Mot. Skills* 12: 283–288.
Barclay, A. G., and D. Cusumano. 1967. Father absence, cross-sex identity, and field dependent behavior in male adolescents. *Child Devel.* 38: 243–250.
Berry, J. W. 1966. Temne and Eskimo perceptual skills. *Int. J. Psychol.* 1 (3): 207–229.
Burlingham, D. 1964. Hearing and its role in development of the blind. In R. Eissler, A. Freud, H. Hartmann, & M. Kriss, eds., *Psychoanalytic study of the child.* New York: International Universities Press. Pp. 95–112.
Cohen, J. 1957. The factorial structure of the WAIS between early adulthood and old age. *J. Consulting Psychol.* 21: 283–290.

Cohen, J. 1959. The factorial structure of the WISC at ages 7–6, 10–6, and 13–6. *J. Consulting Psychol.* 23: 285–299.

Corah, M. L. 1965. Differentiation in children and their parents. *J. Personality* 33: 300–308.

Dawson, J. L. M. 1967a. Cultural and physiological influences upon spatial-perceptual processes in West Africa, Part I. *Int. J. Psychol.* 2: 115–128.

Dawson, J. L. M. 1967b. Cultural and physiological influences upon spatial-perceptual processes in West Africa, Part II. *Int. J. Psychol.* 2: 171–185.

Dershowitz, Z. 1966. Influences of cultural patterns on the thinking of children in certain ethnic groups: a study of the effect of Jewish subcultures on the field-dependence-independence dimension of cognition. Doctoral dissertation, New York University.

Dyk, R. B., and H. A. Witkin. 1965. Family experiences related to the development of differentiation in children. *Child Devel.* 30(1): 22–55.

Fiebert, M. 1967. Cognitive styles in the deaf. *Percept. and Mot. Skills* 24: 319–329.

Goodenough, D. R., and S. A. Karp. 1961. Field dependence and intellectual functioning. *J. Abnorm. Soc. Psychol.* 63: 241–246.

Guilford, J. P., J. W. Frick, P. R. Christensen, and P. R. Merrifield. 1957. A factor-analytic study of flexibility in thinking. *Report Psychology Laboratory,* no. 18. Los Angeles: University of Southern California.

Karp, S. A. 1963. Field dependence and overcoming embeddedness. *J. Consulting Psychol.* 27: 294–302.

Seder, J. A. 1957. The origin of differences in extent of independence in children: developmental factors in perceptual field dependence. Bachelor's thesis, Radcliffe College.

Shurrager, H. C., and P. S. Shurrager. 1964. *Manual for the haptic intelligence scale for adult blind.* Chicago: Psychology Research.

Supa, M., M. Cotzin, and K. M. Dallenbach. 1944. "Facial vision": the perception of obstacles by the blind. *Am. J. Psychol.* 57: 133–183.

White, B. W. 1953. Visual and auditory closure. *J. Exp. Psychol.* 48: 234–240.

Winestine, M. C. 1964. Twinship and psychological differentiation. Doctoral dissertation, New York University.

Witkin, H. A. 1950. Individual differences in ease of perception of embedded figures. *J. Personality* 19: 1–15.

Witkin, H. A. 1965a. Development of the body concept and psychological differentiation. In H. Werner and S. Wapner, eds., *The body percept.* New York: Random House.

Witkin, H. A. 1965b. Some implications of research on cognitive style for problems of education. *Archivio di Psicologia, Neurologia e Psichiatria* 26(1): 28–55.

Witkin, H. A., R. B. Dyk, H. F. Faterson, D. R. Goodenough, and S. A. Karp. 1962. *Psychological differentiation.* New York: Wiley.

Witkin, H. A., H. F. Faterson, D. R. Goodenough, and J. Birnbaum. 1966. Cognitive patterning in mildly retarded boys. *Child Devel.* 37: 301–316.

Witkin, H. A., D. R. Goodenough, and S. A. Karp. 1967. Stability of cognitive style from childhood to young adulthood. *J. Personality Soc. Psychol.* 7: 291–300.

Witkin, H. A., H. B. Lewis, M. Hertzman, K. Machover, P. B. Meissner, and S. Wapner. 1954. *Personality through perception.* New York: Harper.

MUZAFER SHERIF and CAROLYN W. SHERIF
Pennsylvania State University
University Park, Pennsylvania

Motivation and Intergroup Aggression: A Persistent Problem in Levels of Analysis

The age-old problem of man's inhumanity to man has been analyzed by philosophers, commentators, moralists, and legal experts as well as social scientists and psychologists. Most accounts have recognized that intense emotions and motivational states are associated with aggressive deeds.

This discussion concerns the levels of analysis required for specifying the essential variables that generate aggressive behavior between members of different human groups. Many psychological accounts of aggression have failed to differentiate between interpersonal aggression, on the one hand, and intergroup aggression, on the other. Conversely, some writers who deal with aggression in intergroup and international affairs have contended that psychological variables are irrelevant to the analysis.

The present thesis is that motivations to aggression toward members of a different human group involve variables at a different level of analysis than those in strictly interpersonal aggression, as well as psychological variables. By recognizing the levels of analysis required, an integrated account of aggressive motives comes within our grasp. In an age when technological tools provide an aggressor with the power to destroy much of mankind and render the environment unsafe for those remaining, a proper grasp of these problems becomes essential.

AGGRESSIVE MOTIVES AND THE SEARCH FOR MAN'S NATURE

In accounting for aggression or cooperation in human relations, traditional theorizing fell into two main camps, each with variations on its theme: the instinctivist camp and the environmentalist camp.

The instinctivist camp explained the state of human relations in terms of primordial dispositions inherent in human nature. These dispositions, it was suggested, seek releasers that have no function other than to trigger an eruption. Thus, Hobbes depicted a human nature that is selfish, sneaky, and aggressive. Rousseau, disgusted by human relations in the hands of a degenerate and decaying aristocracy, depicted human nature as essentially innocent and good, but distorted by the shackles of the corrupt social order of his age. The environmentalist camp, on the other hand, put the praise or blame for aggression entirely on the cultural, social, political and economic conditions surrounding man, discounting the significance of his motivations and emotions.

If our aim is to understand human motives and ambitions as men transact in group units, we fail utterly if we limit the analysis entirely to environmental conditions. On the other hand, the instinctivist position amounts to leaving out the environmental conditions, hence gives a mutilated picture of the problem.

Until very recently, the most influential psychological theories tenaciously put the blame for destruction and murder committed by human groups on destructive and aggressive dispositions inherent in human nature. For example, one well-known list of innate dispositions was posited by William McDougall. McDougall included in his list the instincts of ascendance and submission, along with instincts of pugnacity and acquisition (McDougall, 1923).

The "instinctive" terms included in such lists varied according to the social philosophy and pessimistic or optimistic outlook of the author himself. The presence or absence of aggressive dispositions in such lists is one index of the conception of human relations held by particular authors. According to its social arrangements, each society has its "myth" about the Nature of Man, as cogently noted by Tolman. Being a peace-loving man in a war-torn world, Tolman included "loyalty to the group" and "sharing" in his own list of social drives (Tolman, 1942).

First, we shall examine the diverse means by which aggression has been sought as a fundamental part of man's nature, starting with the unabashed positing of a "death instinct." The pitfalls of limiting the analysis to psychological variables will be indicated. Finally, the levels of analysis will be specified that are required for adequate understanding of man's inhumanity to those he sets apart invidiously as beyond the pale of his moral standards.

"Death Instinct"

The instinctivist position explains the vindictiveness of group toward group on the basis of blind aggressive forces erupting from the depths of human nature. Perhaps its most influential proponent was Sigmund Freud, the towering figure of the psychoanalytic movement. In his earlier work, Freud saw aggression as a response to the frustration of impulses that he then conceived as more basic. In his later writings, however, he posited an "innate, independent, instinctual disposition in man" towards aggression (Freud, 1930, p. 102). Specifically, as problems of group relations became of greater concern to him, he developed the notion of two classes of instincts: "Eros or the sexual instincts" and a "death instinct, the task of which is to lead organic matter back into the inorganic state" (1927, p. 55).

For Freud, it was Eros that held human beings together in groups, but Eros was easily ravaged by the death instinct's aggressive impulses. Freud was very explicit on this point (1930, pp. 85–86):

> This aggressive cruelty usually lies in wait of some provocation, or else it steps into the service of some other purpose, the aim of which might as well have been achieved by milder measures. . . . It also manifests itself spontaneously and reveals men as savage beasts to whom the thought of sparing their own kind is alien. . . . The existence of this tendency for aggression . . . makes it necessary for culture to institute its high demands. Civilized society is perpetually menaced with disintegration through this primary hostility of men towards one another.

Thus, the ultimate reason for the existence of human culture as well as the greatest threat to that culture were seen in man's innate destructiveness. Society, in Freud's eyes, was an inevitable enemy of the individual. The individual conscience was essentially "dread of society" (1922, p. 10).

For Freud, interaction in groups and collective encounters did not produce creative outcomes. On the contrary, they served only to release man's instinctive impulses.

> From our point of view we need not attribute so much importance to the appearance of new characteristics. For us it would be enough to say that in a group the individual is brought under conditions which allow him to throw off the repressions of his unconscious instincts. The apparently new characteristics which he then displays are in fact the manifestation of his unconscious, in which all that is evil in the human mind is contained as a pre-disposition. (1922, pp. 9–10.)

Freud's contention is, of course, in sharp opposition to results obtained in research on social interaction by social scientists during the last thirty years.

Doctrine and Fact on Aggressive Instincts

All of us, psychologists and laymen alike, seem irresistibly fascinated with musing over ourselves, over our desires, and over the attachments and fantasies we had when we were children and adolescents with kaleido-scopic dreams. There is fascination in abandoning the usual restrictions and revealing ourselves freely to a trusted friend, to a beloved person, or in the permissive atmosphere of a confessional or therapeutic session.

The range of conscious awareness is very limited, as many studies have established. The happenings of our past are inseparable parts of our self-identity, whether they are horrid or pleasant, whether the source of our guilt, shame, inner conflict, or gratification. Beset with tedious preoccupa-tions and role conflicts in a modern life that is casually, even contradictorily patterned, we develop a craving to unearth those parts of ourselves hidden by the limited awareness of the living present. This fascination may be partially responsible for the appeal of dramatic accounts of unbridled im-pulses lying beneath conscious awareness, such as presented in the aggres-sion doctrine under consideration.

The doctrine of all-powerful aggressive impulses inhering in a "death instinct" had its critics even in the heyday of the psychoanalytic movement, both within the fold of the movement and outside it. Today, experimental, developmental and other research evidence on intergroup attitudes and deeds renders the explanation of aggression in human relations on the basis of a death instinct completely untenable.

In his comprehensive survey on aggression, Berkowitz (1962) sum-marized theories of aggression and critically evaluated them in the light of available research findings from several disciplines. He concluded that re-search evidence does not support Freud's conception of a "death instinct" whose energy must be released and whose inherent tendency is toward return to the "quiescence of inorganic matter." In fact he noted, very few psychoanalysts today accept the "death instinct." "Dramatic though it may be, the concept of an innate drive for destruction, as Freud posited it, is scientifically unwarranted. . . . There are a number of bases upon which the hypothesis can be attacked, some logical, others factual" (ibid., p. 8).

The Error of Attributing Man's Aggression to Biological Nature

Another lingering misconception is the postulation of a fighting impulse inherent in biological organisms. This misconception seems to arise from the undeniable fact that frequent fights occur among some species of animals. A fighting instinct has been postulated without detailed examination of the causes of these fights.

The entire concept of animal "instinct" is being subjected to searching examination today in the light of recent research (see Schneirla, 1964). Much animal behavior that was called "instinctive" because it occurred with relative invariance among members of a species turns out to involve a complex learning process by organisms developing in particular environments and with species-specific capacities for perception, locomotion, etc.

With respect to a fighting instinct, it is important to note that the occurrence of fighting among animals is not without reference to environmental events. Indeed, Scott (1958) has observed in his survey of studies on aggression that the impetus for fighting is closely related to environmental conditions. Contrasting the sequence of events associated with the physiology of hunger, he concluded that there was "no physiological evidence of any spontaneous stimulation for fighting arising within the body" and that "the chain of causation in every case eventually traces back to the outside" (ibid., p. 62).

As Schneirla has demonstrated in a series of penetrating analyses, the criteria for drawing valid analogies between animal behavior and human behavior cannot rest solely on similarity of outcomes (e.g., injury to another), or complexity of performance (witness the army ant's maneuvers), or degree of adaptation to the environment. The criteria for proper analogies between behavior of different species must be found in the comparability of the *processes* underlying the behavior (Schneirla, 1946; 1951; 1953). According to the level of evolutionary development and capacities of the species, the variables that affect the animal's behavior differ. Thus, man's dominance-aggression patterns, while superficially comparable to the pecking order of fowls in some conventionalized situations, are by no means governed by the same variables, physiological or environmental (Schneirla, 1953).

Accounts of human aggressiveness must include the process, specific to the human species, by which man learns and responds to external stimulation in accordance with verbally formulated rules or generalizations. Because of this process, man's aggression to man does not vary systematically according to the same conditions as it does in lower species, such as food shortages and ecological arrangements. On the contrary, the most violent acts of human aggression seem to involve individuals and groups with greater food supplies, more space, and greater facilities for adaptation without aggression.

Error of Generalizing from Atypical Populations

Probably one reason that psychoanalysts postulated a self-generated instinct of aggression was that they based their theorizing on a select sample of individuals—disturbed, deranged, at odds with their fellow men. We borrow this inference from a leading researcher on human development, Robert R. Sears, who spent years studying children in their friendly and hostile interpersonal relations. Sears (1960, p. 96) drew attention to the pitfall of making wholesale inferences from data on a select and unrepresentative sample:

Much of what we know—or at least hypothesize—about the moral properties of human behavior comes from the clinic. Only recently have more experimental and more replicable investigations begun to examine these matters. Quite naturally, such studies have begun with the clinically obtained hypotheses. And clearly, the clinic historically has drawn its main clientele from the ill, the disturbed, the badly socialized. These are the ones who resist temptation too little or too much, and who are in trouble one way or another with their feelings of guilt. A clinic population does not draw so heavily from those who are achieving their own ideals. Even among the ill, the therapeutic focus seems to fall more intensively on lapses from grace than on failures to achieve the ideal.

A select sample of clinical cases, who admittedly are at odds with their fellow men, is certainly an inadequate basis for generalizations concerning the nature of human motivation. There is a related habit, hard to overcome because it is so strongly ingrained. The tendency is to jump to conclusions and to evaluate human beings everywhere on the basis of the standards of value inculcated in us within the confines of human arrangements prevailing in our own culture.

Although modern anthropology has documented the wide variations in competition, submission, and aggression in different cultures, some psychological research on these topics proceeds uncritically from the assumption that the pattern prevailing in the researcher's society is typical. The use of any human culture as a prototype for generalizations about man's social motives is risky, as we shall see.

The Error in Seeking the Roots of Aggressiveness in Primitive Cultures

In theorizing about human nature, it has been almost an article of faith that the pristine traits of humanity would be revealed more clearly among peoples in more primitive cultures. Technically developed societies should be further removed from original human impulses, according to this reasoning. Therefore, if one would know whether human nature is innately aggressive, look to man in a less developed society.

Especially since the latter half of the 19th century, treatises marshaling evidence from various primitive cultures were written in efforts to prove the particular author's assumptions about the traits of original human nature. The selectivity of the cases included by various authors, in line with their particular assumptions, is a fascinating example of the intrusion of subjective factors in shaping discrimination and decision, which has also been found in a number of laboratory experiments.

Psychological theorizing on influences contributing to friendship or enmity between human groups gains greater perspective if it starts from the vantage point provided by cultural anthropology as to the occurrence, frequency and scope of friendly or hostile intergroup events. Especially helpful for the student of intergroup relations is the orientation pursued for years by Leslie White and his associates. White and his associates do not present psychological analyses; on the contrary, they provide a notion of culture at its own *level* of organization. Their conception of culture is not reductionistic; i.e., culture is not reduced to unrelated, meaningless bits of behavior with no context. Culture is conceived as a meaningful structure. From a psychological viewpoint, relevant aspects of this pattern are parts of the stimulus world the individual confronts. His perceptions can be assessed more effectively within this cultural context.

A distinctive feature of White's orientation makes it particularly pertinent for assessing the human and physical arrangements contributing to positive or negative relations between human groupings. Instead of considering each cultural group as a more or less closed "culture pattern," each a law virtually unto itself, this orientation puts cultures in a *time* perspective. Lawful recurrences over time enable comparisons between cultures at different times, in terms of the actualities of prevailing conditions, techniques, modes of life, and premises regulating human arrangements (e.g., White, 1959).

Working from the viewpoint of cultural evolution, Marshall D. Sahlins (1960) in a paper on the "Origin of Society" reached a conclusion (pp. 81–82) about the search for the roots of aggression in primitive societies. It should serve as a corrective to statements about original human impulses formulated without regard to the prevailing mode of life and human arrangements:

> Territorial relations among neighboring human hunting-and-gathering bands (a term used technically to refer to the cohesive local group) offer an instructive contrast. The band territory is never exclusive. . . . Warfare is limited among hunters and gatherers. Indeed, many are reported to find the idea of war incomprehensible. A massive military effort would be difficult to sustain for technical and logistic reasons. But war is even further inhibited by the spread of a social relation—kinship—which in primitive society is often a synonym for "peace." Thomas Hobbes' famous fantasy of a war of "all against all" in the natural state could not be further from the truth. War increases in intensity, bloodiness, duration and significance for social survival through the evolution of culture, reaching its culmination in modern civilization. Paradoxically the cruel belligerence that is popularly considered the epitome of human nature reaches its zenith in the human condition most removed from the pristine. By contrast, it has been remarked of the Bushmen that "it is not in their nature to fight."

Human Motives and Their Social Context

Early in this century, theory in academic psychology was primarily intellectualistic. Reacting against this heritage, the emphasis shifted toward the psychodynamics of instinctual impulses conceived as the driving force of man's deeds, even though he is not aware of them. Reaction against the over-intellectualistic tradition was carried too far. In some quarters, emphasis on the vicissitudes of instincts is still very much alive. In others, motivational forces, even acquired motives, are taken as absolute determinants of behavior. An impressive series of assertions by well-known therapists and other psychodynamicists could be quoted to document the excessive emphasis on motives.

As Eysenck (1950, p. 64) stated:

> In their excitement about the discovery of the powers of "emotion" over "intellect," many psychologists have gone to extremes, portraying the "man in the street" as the mere plaything of uncontrollable unconscious forces which cannot in any way be influenced by reason. Such a view is no less contrary to fact than the previous over-estimation of rationalistic influences; what is needed is a more realistic appraisal of the relative importance of these two factors in each individual case.

The psychodynamicists have tended to view the individual as a self-contained powerhouse, generating his own impulses without regard to transactions with his cultural setting; to view the development of human conscience as merely the "dread of society"; to view society itself solely as the agent of oppression to the individual's impulses and passions. If we have pretense of a scientific approach, we must heed the indications of converging lines from psychiatry, studies of human development, animal studies, social psychology, and cultural anthropology. A balanced view must consider society as more than a system of suppressive prohibitions and conscience as more than "dread of society," but also composed of the values and imperatives of social groups of which the individual is an integral part (cf. Pear, 1950, p. 134).

Because of past emphases, a social scientist is still expected to be an apologist for the group or culture, and the psychiatrist and psychologist partial to psychodynamic explanations, ignoring group and social influences. Developments in recent theory and research are eliminating the one-sided emphasis in both social science and psychology. The adequate study of man and his relations requires taking into account his intimately felt motives, yearnings, aspirations, and cognitive processes in the context of his affiliations with other people and the sociocultural setting of which he is a part (Sherif and Koslin, 1960; Sherif and Sherif, 1964). No man feels, yearns, thinks, hates or fights altogether in isolation.

For the individual, the social context constitutes a stimulus situation that must be included in analysis of his experience and behavior. It is also true that man can affect the social context itself by his actions. However, it is erroneous and misleading to conclude that events occurring in the social context are simply manifestations of psychological processes. On the contrary, a different level of analysis is required when we move from the individual as a unit of analysis to aspects of his social context: other persons, groups, institutions, social values, technology, and so forth. This does not mean that the two levels of analysis—psychological and sociocultural—are contradictory. Instead, each is necessary for a rounded account of individual behavior and sociocultural events.

What is uniquely individual and what is sociocultural must be integrated in the study of man's relation to man, including his aggressive confrontations. This view has gained substantial support in recent years, factually and theoretically (see Cloutier, 1965). A recent expression is found in a monograph on *Psychiatric Aspects of the Prevention of Nuclear War* prepared by the Group for the Advancement of Psychiatry (1964). After reviewing a considerable body of relevant material the authors recapitulate their conclusion (p. 118):

> War is a social institution; it is not inevitably rooted in the nature of man. Although war has traditionally served as an outlet for many basic human psychological needs, both aggressive and socially cohesive ones, the increasing mechanization and automation of modern warfare has rendered it less and less relevant to these needs. There are other social institutions and other means of conducting conflict between groups of people, or between nations, that can serve these psychological needs more adaptively in our modern world.

A Second Look at the Springs of Aggressive Acts

In the light of accumulating research, psychologists and social scientists have felt the necessity in recent years of examining the springs of aggressive deeds more carefully. After World War II, there was a crop of post-mortem verdicts in professional journals (including the *Journal of Abnormal and Social Psychology* published by the American Psychological Association) analyzing the horrors of the recent war in such terms as "collective guilt," "early childhood frustrations" of Axis leaders, and "unleashing of hidden aggressive tendencies" accumulated during the critical period preceding the war. Shortly thereafter, psychologists, social scientists and psychiatrists turned to more serious study of the causes of aggressive tendencies and deeds. Necessarily, such study required considerations outside of the provincialism of their particular specialty. They gave more receptive consideration to evidence from various disciplines.

They found, for example, studies of the American soldier conducted during the war (Stouffer et al., 1949). Included were answers to questions about reasons for fighting. The soldiers were the men actually doing the fighting, in situations where the expression of aggressive feelings toward the enemy was socially permissible. Yet the majority of the soldiers said they fought to "get the job done," or because they did not want to let their outfits down. Only 2 percent said they fought out of anger, revenge, or "fighting spirit." Another 3 percent gave replies that might be interpreted as aggressive, such as "making a better world," "crushing the aggressor," "belief in what I am fighting for" (p. 109). Certainly aggressive and vindictive feelings were not the most salient reported by these American fighting men.

Such reports led to a closer look at the locus of aggressive and destructive emotions, since quite clearly they were not universal. Pear (1950) commented on this point, asking whether wars were the outcome of aggressive impulses of the general populace or whether such impulses had to be fanned in the name of things sacred and just to a people. He reached the following conclusion about a theory of war based on individual impulses toward aggression:

> It fails to distinguish between the aggressiveness of the warmakers, which can be very real indeed (though frequently personal greed, still socially disapproved if found out, masquerades as socially approved aggressiveness) and the attitudes of the general population, many of whom may not know of the impending war, of the combatant, the semi-combatant soldiers, and of the victims. In a war involving more than half the population of the world, a vast number of people who had nothing to do with declaring war suffered passively. Often aggressiveness had to be stirred up and intensified even in the fighters (we have recently read about the experimental army "hate school" abolished as a result of psychiatrists' reports)[;] in the uniformed sections many people of both sexes lived an unaggressive life and yet helped to win the war[;] "backroom boys" and scientists are unlikely to have done their best thinking if viscerally stirred: "beating the enemy" cannot have been a constant day-and-night goal giving incentive to all non-combatants, as the excellent book *War Factory,* among others, showed.

At about the same time, another eminent psychologist known for his continued interest in "personal factors" wrote on the "role of expectancy" of war in bringing about conflict. Gordon Allport of Harvard University concluded (in Cantril, 1950, p. 43):

The people of the world—the common people themselves—never make war. They are led into war, they fight wars, and they suffer the consequences; but they do not actually make war. Hence when we say that "wars begin in the minds of men" we can mean only that *under certain circumstances leaders can provoke and organize the people of a nation to fight*. Left alone people themselves could not make war.

Foundations and organizations have arranged several joint committees of experts to pool their findings on the causes of aggressive passions and deeds. For one of these, the sessions and comments by psychologists, psychiatrists, and social scientists of various scientific positions and ideologies were published for UNESCO with Hadley Cantril as editor (1950). (The participants were Gordon W. Allport, Gilberto Freyre, Georges Gurvitch, Max Horkheimer, Arne Naess, John Richman, Harry Stack Sullivan, and Alexander Szalai.)

Despite their differences on several points, all participants agreed on several fundamental issues. Their common statement included agreement on the following conclusion: "To the best of our knowledge, there is no evidence to indicate that wars are necessary and inevitable consequences of 'human nature' as such" (p. 17).

Neither the psychodynamicist nor the newspaper commentator who concludes that wars are inevitable because of "human nature" has provided adequate evidence for his theory. Their arguments begin with the observation that men make wars, then state that men make wars because they are aggressive by nature. The proof of men's aggressive nature lies in the fact that they make wars. Nothing has been added to the first observation but words to confuse the unwary.

Frustration-Aggression in Its Social Context

As mentioned earlier, Freud's position before he posited the "death instinct" as the source of hostile and destructive dispositions was that aggression is the result of frustration of basic drives, especially in early childhood. This earlier formulation by Freud found its most systematic expression in a highly influential book published over a quarter of a century ago (Dollard et al., 1939). It advanced the theory that all aggressive behavior (excluding only instrumental acts) was the outcome of frustration, and that frustration invariably led to aggressive tendencies.

In the face of criticism from various quarters, the formulation was subsequently qualified to indicate that every frustration does not necessarily lead to aggression. In the words of one of the principal proponents of the formulation, frustration "produces instigators to a number of different types of responses, one of which is an instigation to some form of aggression" (Miller, 1941, p. 338).

Of course, frustration suffered when activities directed toward attainment of goal objects are blocked is a state of arousal and tension with some consequences. But, in a systematic examination of the possible consequences, the following questions have to be faced:

1. (a) Is every act of aggression the outcome of frustration?
   (b) Is aggression the invariable response to frustration?
2. Within what sets of circumstances and within what framework of interpersonal and group ties is frustration conducive or *not* conducive to aggressive deeds?

In relation to the first question, Berkowitz' critical survey of research (1962) led him to conclude unequivocally that "there *are* some aggressive acts. . . . that are not necessarily instigated by frustrations" (p. 30). Among such acts, he includes a number with important consequences, such as wholesale killing and destruction initiated as policies during wartime. He also cites the interesting research by Bandura and his co-workers demonstrating that children can acquire hostile modes of behavior merely by observing the aggressive actions by an adult, even when the adult is "nurturant" to the children.

J. P. Scott (1958) cites evidence from experiments on animal behavior in his own and other laboratories that there are important causes of aggressive behavior other than frustration. "For example, we have seen that the best way to train a mouse to be highly aggressive is not to frustrate him but to give him success in fighting" (p. 33). *"Frustration leads to aggression only in a situation where the individual has a habit of being aggressive"* (p. 35).

The further question of whether frustration invariably leads to aggressive behavior in animals is answered by Scott in these words: "In short, while frustration is highly likely to produce aggression, the result may also be other kinds of behavior" (p. 34).

Himmelweit (1950) reached a similar conclusion in her survey of studies of human frustration. She listed the following responses to frustration observed in different studies of human subjects: aggression, regression (lowering the level of performance), evading the situation by leaving it or daydreaming, apathy and resignation (especially for prolonged frustration), repression or "forgetting." Frequently several of these responses were observed in the same frustrating situation.

When applied to problems of hostility and prejudice between human groups the frustration-aggression hypothesis is an inadequate and even misleading explanation of aggressive behavior and of reaction to frustration. It is misleading because it directs investigation exclusively toward an aroused psychological state and a class of actions (aggressive). Any adequate hypothesis about human behavior must include specification of *the environmental circumstances in which a psychological state occurs and in which a class of actions occurs.* As applied to frustration and aggression in group relations, this means that an adequate hypothesis must include statements about the culture and organizational context of both tension arousal and behavior. The second question raised above is pertinent here.

If the frustrating experience is strictly an individual affair within the context of day-to-day interpersonal relations, how can it become the basis for prejudice, injustice, and aggressive actions that follow existing dominance-subordination arrangements in society? Why is there not more variation in the targets of hostility? It is necessary to ask whether a frustration is *individually* experienced or is seen by other members of a group as a *common frustration.* If the frustration is seen as shared, does it have any bearing on the direction of the greatest prejudice and the most hostile actions toward other groups? According to an adapted version of the frustration-aggression hypothesis, the most vehement aggressive impulses should be generated among those who share the greatest frustrations.

Is the direction of prejudice and aggressive actions primarily from the most downtrodden and deprived groups toward dominant groups who maintain their mighty position, or are these not frequently directed from the mighty and powerful *toward* the underprivileged groups, who one may assume are more frustrated? Such questions direct serious consideration to the organizational or group context of frustration and of aggressive actions. Answers to such questions are necessary before we can conclude that frustration suffered by single individuals without becoming a common concern have a major role in intergroup aggression.

It should be clear that human societies vary a great deal in *whom* they select as proper targets of hatred and aggressive acts. These social variations have profound significance for the role of frustration in generating aggressive deeds. The individual who develops in a society where discrimination is directed against a particular group and who is not obviously a member of that group learns to conceive of himself as belonging to those groups that consider themselves to be superior to the discriminated group. If members of the discriminated group challenge his superior position, he may feel very frustrated indeed, but frustrated as a member of the dominant and superior kind. Thus, the social context is essential for analyzing both his experience of frustration and its consequences for his behavior.

Perhaps the inadequacies of the frustration-aggression hypothesis for predicting violence within an organizational context are most striking when it is applied to lynchings or murders of dominated groups. Such deeds have been attributed to frustrations or deprivations suffered by the lynchers, who frequently, though not always, have included "poor whites."

There may be a grain of truth that personal frustrations have something to do with the degree and extent of brutality by individual lynchers. But the presence or absence of lynching *as an institution,* to be engaged in without serious fear of punishment, is not adequately explained in these terms. When this brutal institution is outside the bounds of the organizational and normative orientation of the group, when it is forbidden and punishable, the mere thought of such deeds in concert with one's fellows is personally appalling, no matter what the individual's frustrations may be. The point was made well in a comparison reported by Klineberg (1950, p. 198):

> White Brazilians are, on the whole, much more economically frustrated than white Americans. The economic standards of the former are definitely much lower, and relatively many more of them live near or at a bare subsistence level. There are fluctuations in economic conditions in Brazil, just as in the United States. However, there are no lynchings of Brazilian Negroes. . . . This fact makes it clearly inadequate to explain aggression against Negro or, in more general terms, hostility against other groups (which may take the form of war in extreme cases) entirely in terms of the aggressive impulses developed within the individual as a result of his frustrations.

It would be difficult to contend that there are no white Americans frustrated for reasons of poverty or lack of status. But despite this, it may be safe to predict that lynching and publicly condoned murder will be driven out-of-bounds in the United States, including the South. If so, it will be because of a changing context for intergroup behavior. This changing context will reflect the determined efforts of many Americans toward equal civil rights for all citizens.

Relationships within and between human groups, which form the context for frustration and associated aggression toward others because of their group membership, set limits for the degree and targets of aggression and chart the direction of what is desirable, or even ideal, in intergroup action. Frustration and aggression are undeniable human phenomena. Unless they are assessed within the context of group relationships, researchers can scarcely hope to have more than fragments of data, useful for valid predictions only in the restricted interpersonal relationships of the laboratory.

SOCIAL-PSYCHOLOGICAL PERSPECTIVE ON INTERGROUP AGGRESSION

The individual is the unit of psychological analysis, but any psychological analysis necessarily includes variables from the stimulational background and the immediate stimulus context of the individual. Attempts to deal with human aggression exclusively through motivational concepts amount to ignoring the stimulational background and context.

When aggression is directed toward members of other groups, the relevant stimulational background and context are organizational and sociocultural. As the student of vision specifies the properties of visual stimuli in nonpsychological units, so too must the student of human aggression attempt to specify social stimuli at a different level of analysis. The relevant units of analysis are not merely other individuals, but groups, social and economic organizations, institutions, cultures and societies. The properties of these units, of which the individual is a part, are as crucial to analysis of his behavior as any emotion or motive he may experience. Necessarily, then, assessment of the etiology of man's inhumanity to men in other groups requires juncture of the psychological and sociocultural levels of analysis.

This perspective permits specification of the conditions in which psychological factors, such as frustration, are and are not conducive to aggression toward another group or its members. In studies guided by this perspective, we have found that the sufficient condition for the rise of aggressive tendencies toward another group is conflict between the goals of the groups in question (Sherif and Sherif, 1953; 1969). To the extent that the group constitutes a stable pattern of interpersonal ties with binding standards of conduct for the members, the individual member moves to further group goals, and experiences pleasant emotions at its successes and personal frustration at its reverses. When, over time, the actions of another group appear to threaten a cherished goal of his group, the individual joins with enthusiasm into the formulation and execution of aggressive plans.

The research subjects in these experiments were normal, well-adjusted, and healthy in their usual environments, not warped or perverted. The crucial importance of the properties of the groups and their relationships was demonstrated by changing the conditions of their interaction. At a time when intergroup antagonism was so intense that each group refrained from contact with the other, they faced a series of experimentally introduced goals that had high appeal value for both groups, but could be attained only if both pooled their resources and efforts (superordinate goals). Over time, the erstwhile aggressors began to cooperate, then to change their attitudes toward one another. Friendship was extended across group lines (Sherif, 1966).

The experiments demonstrate the necessity of including variables at the sociological level of analysis (in this case, the properties of groups and intergroup relations) in dealing with events at the psychological level (in this case, the aggressive or friendly attitudes and behavior of the individual). The events that ensued cannot be accounted for by eliminating their social context and its change. Nor is the picture complete without considering the significant consequences of individual experience and action, for these were the medium by which the social context was established and then transformed from intense conflict to harmonious interplay between groups.

The traditional dichotomy between psychological and sociological levels of analysis dies hard. As long as its remnants influence theory and research, accounts of man's social behavior will be filled with contradictions and apparent paradox. These can be resolved by recognition that both levels of analysis are necessary and must be integrated in a comprehensive conception.

REFERENCES

Berkowitz, L. 1962. *Aggression: A social psychological analysis.* New York: McGraw-Hill.
Cantril H., ed. 1950. *Tensions that cause wars.* Urbana: Univ. Illinois Press.
Cloutier, F. 1965. International tensions and mental Health. In M. Schwebel, ed., *Behavioral science and human survival.* Palo Alto, Calif.: Science and Behavior Books, Inc. Pp. 96–102.
Dollard, J., N. Miller, L. W. Doob, O. H. Mowrer, R. R. Sears. 1939. *Frustration and aggression.* New Haven: Yale Univ. Press.
Eysenck, H. J. 1950. War and aggressiveness: a survey of social attitude studies. In T. H. Pear, ed., *Psychological factors of peace and war.* New York: Philosophical Library.
Freud, S. 1922. *Group psychology and the analysis of the ego.* London: Hogarth.
Freud, S. 1927. *The ego and the id.* London: Hogarth.
Freud, S. 1930. *Civilization and its discontents.* London: Hogarth.
Group for the Advancement of Psychiatry. 1964. *Psychiatric aspects of the prevention of nuclear war.* New York.
Himmelweit, Hilda. 1950. Frustration and aggression, a review of recent experimental work. In T. H. Pear, ed., *Psychological factors of peace and war.* New York: Philosophical Library.
Klineberg, O. 1950. *Tensions affecting international understanding.* New York: Social Sciences Research Council, Bull. 62.
McDougal, W. 1923. *Outline of psychology.* New York: Scribner.
Miller, N. 1941. The frustration-aggression hypothesis. *Psychol. Rev.* 48: 337–342.
Pear, T. H. 1950. Peace, war and culture patterns. In T. H. Pear, ed., *Psy-*

chological factors of peace and war. New York: Philosophical Library.

Sahlins, M. D. 1960. Origin of Society. *Sci. Am.* 203 (Sept. 1960): 76–87.

Schneirla, T. C. 1946. Problems in the biopsychology of social organization. *J. Abnorm. Soc. Psychol.* 41: 385–402.

Schneirla, T. C. 1951. The "levels" concept in the study of social organization. In J. H. Rohrer and M. Sherif, eds., *Social psychology at the crossroads.* New York: Harper.

Schneirla, T. C. 1953. The concept of levels in the study of social phenomena. In M. Sherif and C. W. Sherif, *Groups in harmony and tension.* New York: Harper. (1966 edition, Octagon Books, New York.)

Schneirla, T. C. 1964. Instinctive behavior, maturation-experience or development. In N. R. F. Maier and T. C. Schneirla, *Principles of animal psychology.* New York: Dover.

Scott, J. P. 1958. *Aggression.* Chicago: Univ. Chicago Press.

Sears, R. R. 1960. The growth of conscience. In I. Iscoe and H. Stevenson, eds., *Personality development in children.* Austin: Univ. of Texas Press.

Sherif, M. 1966. *In common predicament. Social psychology of intergroup conflict and cooperation.* Boston: Houghton-Mifflin.

Sherif, M., and B. Koslin, 1960. The "institutional" vs. "behavioral" controversy in social science with special reference to political science. In M. Sherif, *Social interactions: Process and products.* Chicago: Aldine, 1967. Chapter 4.

Sherif, M., and C. W. Sherif. 1953. *Groups in harmony and tension.* New York: Harper. (1966 edition, Octagon Books, New York.)

Sherif, M., and C. Sherif. 1964. *Reference groups: Exploration into conformity and deviation of adolescents.* New York: Harper and Row.

Sherif, M., and C. W. Sherif. 1969. *Social psychology.* 3rd ed. New York: Harper and Row. Chapters 11–12.

Stouffer, S. A., A. A. Lumsdaine, M. H. Lumsdaine, R. M. Williams, Jr., M. B. Smith, I. L. Janis, S. A. Star, L. S., Cottrell. 1949. *Studies in social psychology in World War II,* vol. II. *The American soldier.* Princeton N. J.: Princeton Univ. Press.

Tolman, E. C., 1942. *Drives toward war.* New York: Appleton-Century.

White, L. A. 1959. *The evolution of culture.* New York: McGraw-Hill.

NORMAN R. F. MAIER

Department of Psychology
The University of Michigan
Ann Arbor, Michigan

The Integrative Function in Group Problem-solving

PROBLEM-SOLVING MECHANISMS IN THE INDIVIDUAL

Levels of Problem-solving

With changing environmental conditions, survival depends upon the organism's ability to cope with problems. A number of problem-solving mechanisms can be identified in the evolutionary development of animals. The process of *natural selection* achieves problem-solving in that only organisms that have adaptive responses survive to perpetuate their own kind. Changes brought about through mutation tend to be retained if they have survival value, whereas unadaptive changes tend to be lost. Thus, the experience of previous generations builds up innate behaviors or behavior tendencies that serve to solve problems by individuals or groups.

The second mechanism that makes an early appearance in the life history of animals is that of *variability*. Whenever an organism is confronted with an obstacle, it displays a variety of behaviors. The paramecium tends to move in several directions—back and forth and right and left—when it reaches a barrier; the human organism in the same situation entertains a variety of alternatives. Commonly called "trial and error" behavior, variability solves problems in the sense that some of the varied movements may take the organism around the obstacle. Thus even blind variability may have some value, whereas unvarying behavior or repetition of an ineffective behavior would trap an organism.

The ability of an animal to *form associations and to build habits* is a third mechanism. This mechanism permits an organism to sort out the most adaptive responses developed in its own lifetime and thereby to profit from its own experience. Responses that were effective in one situation thus become more available in subsequent situations. This learning mechanism has been improved through evolutionary development in that susceptibility to modification of behavior through previous experience has greatly increased.

A learned response is expressed not only when the situation reappears but when similar situations occur. This type of problem-solving may be described as generalization or the transfer of training by means of similarity. Thus, an animal that has discovered that it can escape by digging tends to repeat its newly acquired response to a barrier when confined on a later occasion.

Like innately determined behaviors, habits may become a disservice if the nature of the problems encountered change too rapidly. Acquired responses as well as instincts serve their purpose best if future problems tend to duplicate past problems.

The disadvantages of overgeneralization can be offset if the organism is capable of *differentiation*. Through the process of differentiation the organism learns the conditions under which the response is adaptive and the conditions under which it is not. This ability to differentiate becomes a fourth mechanism in problem-solving. Differentiation restricts the learned responses to specific aspects of the stimulus conditions and thus permits a variety of responses to different but related stimuli. Through differentiation a dog can be trained when to sit up and when to lie down to his master's cues.

Differentiation represents a refinement in learning and, when sufficiently developed, becomes the mechanism for concept formation. Concept formation, therefore, may be regarded as the highest degree to which the fourth type of problem-solving mechanism may evolve. It enables the organism to learn to respond to classes of stimuli by recognizing the essential class characteristics.

Finally, we came to productive problem-solving (Koehler, 1926), a process by which past experiences are reorganized so as to solve a unique problem. Reorganization not only implies new groupings of past learning but a fragmentation of previously learned combinations (Maier, 1931, 1960). In order for a chimpanzee to stack boxes to reach a banana overhead, the relationship between boxes lying in different parts of the room must be perceived in a new configuration, and when this occurs the boxes take on new meanings (insight). The same functional changes occur when a person uses a dime as a screwdriver. The dime must be taken out of the context of money before it may be perceived as merely a sturdy, but thin piece of metal. The fact that ice picks will be tried out as screwdrivers before coins indicates how previous learning by similarity can stand in the way of problem-solving (Maier, 1930).

The process of productive problem-solving requires more than a chain or pattern of associative bonds derived from past experiences in which the elements have been contiguous. Experiences gained in a variety of contexts must become available in the reorganization process. The many interconnecting nerve fibers and nuclei of the brain become integrative centers for diversified incoming information.

It has been shown that rats (see summary in Maier and Schneirla, 1935), given a variety of disconnected past experiences can use various segments of the experience and recombine them to solve a new problem. Thus, given the problem of how to reach food placed behind a barrier, rats are able to construct a detour around the barrier by taking relevant items of information from various past experiences and combining them in a new and meaningful manner. Since the route (a combination of elements) is entirely new for the animal, it cannot be reduced to associations built up during previous experience. Furthermore, brain lesions (Maier, 1932) reduce this ability, while at the same time leaving the ability to build associations intact.

The ability of children to solve a similar problem matures only after the age of four years (Maier, 1936), considerably after their ability to learn. It is for reasons such as these that Maier and Schneirla have emphasized the importance of levels in development.

Finding original and spontaneous solutions in individual problem-solving represents an end product that depends upon a high degree of communication between various brain centers. This processing of information is essential to obtaining a single organized solution or decision rather than a series of disconnected responses. Thus, a creative solution to a new problem represents a unified reorganization of data received from the various brain centers in which memory traces have been stored.

Group Problem-solving as a Fifth Mechanism in Problem-solving

When a group of people engage in problem-solving, the essential integrative function is lacking. Each individual responds as a separate entity and a single unified solution to a problem is difficult to achieve. A majority decision does not represent unity, since the minority pulls in a different direction. Often the minority is forced into submission by the majority (social pressure). When members do not resolve their differences, the full potential of a group is lacking. Groups that do not utilize all of their resources are deficient in their group behavior in a manner similar to that of individuals with damaged integrative centers.

From this analysis it follows that unless some integrative function is introduced into group problem-solving, the product of group-thinking becomes a function of dominance rather than the integration of the resources that are potentially present in the group.

Research evidence supporting this deficiency in group problem-solving is available (Hoffman and Maier, 1964). When four-person groups were asked to solve a problem for which solutions ranged from the naive to the creative, it was found that a rather limited number of ideas were discussed. These ideas were generated by some and evaluated by others. Participants tended to react negatively or positively to the alternatives discussed. Since these discussions were taped, it was possible to code the various ideas and count the supportive and critical comments. The difference between supportive and critical comments for each idea discussed yields a value that has been described as a solution's valence. It was found that the solution that first achieved a valence in the vicinity of +15 was the solution that was adopted in 80% of the cases, regardless of its quality. Occasionally, valence values of +30 were needed before a solution was adopted, but for the particular problem studied the value +15 seemed to be a critical one. Because a solution achieves high valence does not mean that it is representative of the group's opinion. It was found that one or two talkative members could raise a solution's valence and cause its adoption.

Satisfaction of members with the solution they adopted was generally present; in some instances, even for individuals who did not initially support it. The correlation of .59 between an individual's valence value for a solution and his satisfaction with it, however, indicates that members, in general, were inclined to support a popular solution.

Solutions of high quality, which often were suggested after a lesser quality solution had reached a valence of +15, had little chance of being adopted. Once the trend toward a solution occurred, critical evaluation evidently ceased to operate. In contrast, member satisfaction with the objective quality of a solution showed a positive but low correlation revealing the group's emphasis on reaching agreement rather than achieving a high quality product.

In another study (Hoffman, Burke, and Maier, 1965), it was found that satisfaction was more a function of the influence a member exerts in determining which solution is reached than on his degree of participation or the quality of the solution adopted. Even though the selected solution usually had general support, minority opinions tended to be suppressed by group pressure. To prevent the suppression of minority opinions and the explorations of differences, a group leader seems to be needed. The mere presence of a leader, who conducts the discussion, but is not allowed to contribute ideas, upgrades a group's product (Maier and Solem, 1952). This finding suggests that groups may profit from a person who can function in a manner analogous to that played by the central nervous system of an individual. If a discussion leader serves the purposes of facilitating communication and preventing interpersonal conflict, the product of group thinking theoretically should surpass that of the individual, since a group's resources of knowledge and variety of thinking exceed that of any member in the group. If, however, these resources are not used and differences result in conflict, the end product may be inferior to that of an individual because popularity and compromises would invariably be involved. The results of researches comparing group with individual problem-solving point to conflicting conclusions regarding the merits of individual versus group thinking (Lorge, Fox, Davitz, and Brenner, 1958). Although the nature of the problem and the level of the group may be factors influencing the varied outcomes, the function of the leadership also must be taken into account. Merely comparing leaderless groups with leader-led groups is not valid since leaderless groups permit dominant individuals to emerge; and leader-led groups may have varied outcomes because of differences in leadership styles.

If leaders were trained to function in the role of increasing a group's ability to communicate, and if interpersonal conflict could be turned into interstimulation, group problem-solving might emerge as the fifth and highest type of problem-solving. Our task is to explore the functions a leader might play in facilitating group problem-solving when he serves as a discussion leader rather than a resource person, persuader, or decision maker.

RESEARCH FINDINGS ON GROUP DISCUSSION LEADERSHIP

We have carried out nine different studies, using the same basic problem, in order to test the various factors that influence the solution. It was our desire to simulate a real-life situation in which the leader's authority and various interpersonal conflicts would become issues because of conflicting objectives.

The Problem

The *Changing Work Procedures* (Maier, 1952) problem has been found to be very sensitive for measuring leadership skills and permits a variety of solutions, ranging from choices between obvious alternatives to innovative resolutions of conflict. Four-person groups are needed to role-play the case. Each person in a group was assigned one of the following roles: Gus, the foreman; Jack, a good worker; Walt, the fastest worker; and Steve, the slowest worker. The three men work as a team assembling fuel pumps, each man having a work position and a supply of parts to add to the casting. General instructions inform them that the work is monotonous and that the men decided several years ago to change positions hourly. Pay is based on a group piece-rate.

The foreman's role requires him to call a meeting to discuss a change in work methods. He is supplied with time-study data that suggest that production (and pay) could be increased if each of the men worked in the position in which he showed the best times: Jack on Position 1, Steve on Position 2, and Walt on Position 3. This would mean abandoning the rotation between positions.

Jack's role shows that monotony bothers him and that changing positions is important in reducing it. He dislikes speed-ups and time studies and is satisfied with his pay. When he is on his best position and has time, he helps Steve a little.

Walt's role shows him to be a team worker who protects Steve and helps him out. He handles monotony by talking, daydreaming, and changing his pace. His main gripe is the time-study man.

Steve's role shows his appreciation of the help he gets from Jack and Walt and his desire to please them. He recognizes that he performs better on one of the positions than on the others and he likes this position best. Since his preference coincides with the time-study data, there is no conflict in the information supplied.

The decisions reached may be divided into three types:

1) Old Method: Continue the rotation method, including minor variations, such as, helping each other, additional training, improved tools, etc.

2) New Method: Each work his best position, including minor variations, such as, rest pauses, music, etc.

3) Integrative Method: Includes solutions such as (a) two men rotate; (b) all rotate between their two best positions; and (c) all rotate but spend more time on best position. Variations and combinations of these three work patterns also are included under this method.

The first of the above types of decisions represents a failure to cope with the facts supplied by the time study, i.e., that each man works better at one position than in the other two. Sometimes the foreman is convinced by the workers that the time-study data are unreliable, while in other cases the foreman feels that a change that is not accepted by the men will result in lower productivity.

The second type of solution indicates that the men either have been persuaded or forced to give up rotation and to accept increased earnings while tolerating the increased monotony.

The third type of solution indicates a resolution of the conflicting views in that the fact of individual differences and the fact of monotony are recognized and dealt with. In addition, the Integrative solutions have to be generated. They are the product of the discussion and are not present in the minds of the participants. In contrast, the Old solutions, except for minor modifications, are known to all since they continue the work method in progress; while the New solutions are suggested by the leader and obviously follow from an examination of the time-study figures supplied. The conflict between these two obvious types of solutions can lead to any one of three outcomes: the Integrative-type of solution, the winning of one over the other, or hard feelings. The Integrative solutions represent the most innovative alternative.

The Acceptance of the Decisions by Type

How do the workers feel about the three types of decisions? This was experimentally tested by Maier and Hoffman (1965). Each participant was asked whether production would "go up," "remain the same," or "go down" as a result of the discussion and decision reached. The opinions of the three workers (plus, zero, or minus) were algebraically totaled to obtain a group score which could range between +3 and −3. It was found that the Integrative solutions received the most support, with a mean acceptance score of +2.6; while the New solutions received the least support, with an acceptance score of +1.1. The acceptance score for the Old solutions was +2.0. Thus, the men are most inclined to have confidence in a solution that was generated during the discussion (i.e., the Integrative solution). The foreman tended to have faith in all solutions; the percentage of his favorable estimates for the Old, the New, and the Integrative solutions being 78.4, 89.9, and 97.9, respectively. He rightly judged the value of the Integrative solution, but often was not aware of the men's resistance to the New solutions, which frequently were associated with negative acceptance scores.

The Leader's Perception of his Subordinates

Maier and Hoffman (1965) also found that when a subordinate responds negatively to ideas suggested by the leader, he may be seen either as an "idea man" or as a "problem employee." Foremen were asked to indicate which of their workers fell into these classifications. Integrative solutions were least often associated with groups in which the foreman perceived one or more of his workers as problem employees (4.5%) and most often with groups in which foremen perceived one or more idea men (43.2%).

This finding is related to an earlier study by Solem (1958). He demonstrated that persons playing the role of foreman do a better job of utilizing participation when the group members are placed in the role of peers than when they are subordinates. Thus, the outcome of a discussion is influenced by the attitude and perceptions of the leader.

Organizational Background

The population sample also seems to influence the outcome. One variable is the extent to which the subjects used are oriented toward the organization. To test this Maier and Hoffman (1961) compared the results of these different groups of subjects: middle management supervisors and executives, students in the school of business, and literary college students. The first of these groups have experienced the organizational influences, the second are studying and anticipating it, while the third have little knowledge of or interest in business organizations. Organizational orientation may be expected to influence both the behavior of the leaders and the workers because the problem used involves superior-subordinate relationships.

The striking result was the progressive increase in Integrative solutions as organizational distance declined it being 11.6% for the management population; 21.4% for the business students and 46.9% for the literary college students. It appears that pressures against innovation are most apparent where organizational orientation was most pronounced.

Comparison of the frequency of Old solutions for the business executives and business administration students also is of interest. The executive group more frequently rebelled against the New solutions and more often caused the foreman to settle for the Old solutions. Thus, 28.3% of the executive groups settled for the Old solutions as compared to 7.1% of the student group. Since the predominant behavior of Gus, the foreman, is persuading the workers to adopt the New solution, it seems that executives who have experienced organizational pressures are less easily persuaded by their supervisors than are business administration students.

Comparison of First and Second Solutions

The predominant behavior in problem situations involving a conflict of interests is that of selling or persuading. Each person attempts to get others to see things his way. What will happen if, after groups have reached a decision, they are asked to find a second solution? Maier and Hoffman (1960) used management groups to examine the differences between first and second solutions.

It is apparent that persons who dominated the discussion and had their solutions adopted on the first attempt will find themselves in a different situation when a second solution is required. Those who have already won, in that their solutions were adopted, now have no solutions to promote. This applies particularly to Gus, the foreman, who becomes more inclined to entertain ideas suggested by the workers because, having won his point, he feels less threatened by their resistance.

Since there is a variety of solutions of each type, it is not necessary for a group to shift solution types in order to find a second solution. The major difference between the first and second solutions was the sharp increase in Integrative solutions from 11.1% to 42.6% and these were achieved at the expense of New solutions. This shift reflects a lowered persistence on the part of Gus thereby improving the communication process.

Groups tend to reach decisions before fully exploring the alternatives they are able to generate. Requiring these groups to return to problem-solving resulted in their generating the less obvious integrative solutions, solutions which frequently were not discovered on the first attempt. The time requirements for reaching the second solution were also less; the mean being 11 minutes as compared to 17 for the first solution.

The results of this study indicate that the leader reduces the level of a group's thinking by having a solution in mind himself. When he reduces his persuasive activities, additional ideas are generated and expressed.

Delaying the Process of Reaching a Decision

In another experiment (Maier and Solem, 1952) an attempt was made to reduce the leader's preoccupation with his solution. In 96 groups, made up of student and supervisory personnel, the leader was asked to go through three preliminary steps before attempting to reach a decision. The purpose of the steps was to get the leader to explore the feelings of the workers regarding job problems and to make a list of factors that concerned them.

As a control, 50 other groups selected from the same population (student and supervisory personnel) followed the standard procedure. The results of this comparison showed an increase in Integrative solutions, from 12.0% for the standard procedure to 41.7% for the experimental 3-step. The preliminary steps required in exploring the problem and de-layed concern with the solution thus led to the generation of the more innovative solutions. (Both supervisory and student groups showed signifi-cant benefits from the 3-step delay.)

Generally speaking, conditions favorable to generating alternatives in the early stages of a discussion seem to upgrade the final product of problem-solving discussions. The discussion leader, being in a position to facilitate or to inhibit this process, plays a unique and influential role.

Utilization of Disagreement

One of the important aspects of a group discussion is the leader's ability to generate disagreement when it is lacking and to prevent it from leading to interpersonal conflict when it is excessive (Maier, 1963). Experimental evidence of the value of disagreement has been obtained in two ways. One approach (Hoffman, 1959) compared the product of homogeneous groups with that of heterogeneous groups. The homogeneous groups were made up of persons who had similar scores on a personality test (Guilford-Zimmerman Temperament Survey, 1949), whereas the heterogeneous groups had dissimilar scores. It was reasoned that disagreement among members would be greater in the heterogeneous groups, and hence these groups should show the greater frequency of Integrative solutions. Students from the literary college were used as subjects in this study.

The trend in results was in the expected direction, but the number of groups was too small to yield a significant difference. However, additional problems were also used, and when the products of these are also considered, the superiority of the heterogeneous groups is significant. In a subsequent study (Hoffman and Maier, 1961) a significant difference between the performance of homogeneous and heterogeneous groups on the *Changing Work Procedures* problem was obtained; the percentage of Integrative solutions being 30.8 and 58.8, respectively. However, in this experiment the scoring of solution types was somewhat modified so that the absolute values of these should not be compared to those in the other experiments.

The second approach to the study of the value of disagreement (Hoffman, Harburg, and Maier, 1962) was that of altering the content of the roles. In one set of groups the roles of the workers were modified to reduce (1) the fear of management, (2) the hostility toward the time-study man, and (3) the magnitude of boredom, whereas in the other set of groups, the roles were modified to increase these negative factors. This study also used college students, a population for which Integrative solutions tend to be relatively high.

In this study Integrative solutions occurred 2½ times as frequently in the strong opposition condition, as in the weak opposition groups, demonstrating the value of disagreement in generating additional alternatives. However, strong opposition also was associated with greater resistance. This was indicated by the greater number of Old solutions when the opposition was strong (22.9%) than when it was weak (14.6%). Thus, the leader was less prone to have his way with strong opposition, and, unless the discussion led to generating new alternatives, groups were reluctant to change their work method.

Increasing Communication

When groups enter into discussion, they sometimes supply relevant information in their possession, while on other occasions they select from it or color it so as to support a position they have taken. It is not uncommon for participants to make statements that have no factual support at all. Thus, information, misinformation, and varied interpretations become part of the discussion process. It will be granted that when individuals go to extremes to protect their positions, they are not solving the problem under consideration, but instead are attempting to achieve a personal goal. These behaviors on the part of participants, as well as the leader, prevent the efficient processing of information. How much does this failure to communicate accurately interfere with the efficiency of group problem-solving?

To experimentally test this question a variation in procedure was introduced by Colgrove (1966). The role information previously distributed among the members of a group was made available to individual problem-solvers. They were asked to act as consultants and, after studying the roles, to recommend the best solutions for Gus to follow.

The next step was to combine these individuals into groups of four and assign each person one of the roles. These groups then role-played the case in accordance with the standard procedure. For this experimental test, therefore, each role-player was in possession of all of the information, a condition which might be expected if groups were highly effective and accurate in their communication.

Groups taken from the same population followed the standard procedure and were used as controls. If Integrative solutions represent evidence of good communication, then the experimental groups should reach more of them than the control group.

The results of this investigation (Colgrove, 1966) showed that both Integrative and Old solutions to be higher in the experimental than in the control situation: the percentage of Integrative solutions being 49.1 and 33.3 respectively and that for Old solutions being 26.3 and 14.7 respectively. The large reduction in the New solutions accounts for the significant difference obtained.

It is apparent, therefore, that Integrative solutions can be increased by improved communication. The condition described represents optimum communication since each individual knows everything. However, conflict of interests has been retained so that this conflict sets a limit on what can be accomplished by full communication.

Problem-solving and Motivation for Change

Resistance to change is a common concept used to describe the opposition encountered when attempts are made to improve a situation. The problem used in our studies is a typical instance of this phenomenon. Is this resistance to change an opposition to change *as such* or is it directed toward the change agent? In the standard version of the problem, the leader is able to offer an incentive for change in the form of increased earnings that would result from improved working procedures. We have seen that even with this incentive, resistance to change is evident by the appearance of Old-type solutions. What would happen if the leader could offer no incentive for change?

In order to answer this question Maier and Hoffman altered the problem in one respect only: instead of being based on a group piece-rate, the pay was based on an hourly rate. A control group from the same population was used for comparison.

The striking finding from this experimental variation was that the frequency of Integrative solutions was almost identical, being 43.0% for the piece-rate condition and 40.0% for the hourly condition. Thus, change in work methods, which the group can influence, occurs with equal frequency, regardless of whether the financial incentive is present or not. This nicely demonstrates that change, as such, is not the source of resistance. However, when the leader goes to the group with a New solution, and persists in his attempts to sell it, he has an advantage if he can offer an incentive. This point is indicated by the difference in the frequency of Old solutions, which was greater for the nonincentive than in the incentive groups (25.0% versus 13.9%; Maier and Hoffman, 1964).

Resistance to change, therefore, must be regarded as something that occurs when the change is initiated from the outside. Changes growing out of group discussion or initiated inside the group do not generate this resistance. There is no need to fear such changes. The more the leader is inclined to go to a group with a problem and the less he is inclined to promote a solution, the more motivation for change is gained from a discussion. It appears that the activity involved in problem-solving builds up its own motivation for change.

Effects of Training in Group Decision

Can a leader be trained to conduct better discussions? In order to test this possibility, supervisory personnel participating in a management training program were used as subjects.

The experimental variable was training in the group-decision concept, which was given to the experimental groups and omitted from the control groups. Thus, in 44 groups the persons were exposed to a presentation on the nature of the group-decision method and how it could be adapted to dealing with job problems. The presentation emphasized the point that if employees are allowed to participate in solving a problem, they will be able to get together on a solution that they are willing to accept. The foremen's functions in such discussions are to present the facts and conditions bearing on the problem and to conduct a permissive discussion.

Opportunities for discussion also were included in this supervisory training. These discussions permitted (a) an expression of attitudes of the participants, (b) opportunities to ask questions, and (c) opportunities to air personal views. The presentation and discussion periods were combined in three training periods of nearly three hours each. The role-playing problem was introduced during the fourth period and was given as a part of the training program.

A total of 36 control groups were given no training except for a half-hour lecture that preceded the introduction to the problem for role-playing. The lecture was on the subject of resistance to change and emphasis was placed on (a) the importance of recognizing differences in attitudes in employees, (b) the importance of employees accepting changes in jobs if these are to be made successfully, (c) the need for recognizing that all employees do not have the same motivations, and (d) the value of listening to what employees have to say in order to understand them.

The results of this experiment showed that the groups without training in group-decision restricted their considerations primarily to Old and New solutions. The trained leaders, in contrast, achieved 34.4% Integrative solutions and these solutions primarily replaced the solutions that were classified under the Old-type category for the untrained groups. The number of New solutions was roughly the same for the two groups.

This pronounced difference in results does not mean that a few hours of training is all that is needed to change a leader's style in conducting a discussion. The trained group leaders followed the typical procedure and invariably tried to sell the New-type solution. However, when they met resistance they were less inclined to resort to threats, and instead were more prone to listen. The most noticeable change, therefore, was that of increased permissiveness. The training objective, which is to approach the group with a problem and avoid even the suggestion of a solution, requires much more training.

It is of interest to compare the results of the leaders who received training in group decision-making with those who were asked to follow the 3-step procedures (see page 558). If we compare the comparable populations in these two studies we find that these two leadership conditions yield about the same frequency of Integrative solutions: 32.7% for the

3-step and 36.4% for the trained leaders. The difference is in the frequency of Old solutions; the percentage for the 3-step method was 25.5 and for the group-decision training method it was only 4.5. Since this difference is significant at the .02 level of confidence, it seems safe to conclude that although the two methods accomplished positive results, they did it in different ways. The increased permissiveness attained through the group-decision training increased the Integrative-type solutions primarily by reducing the frequency of Old-type solutions, while the 3-step delay procedures increased the Integrative-type solutions largely by reducing the number of New-type solutions.

Changing Patterns of Managerial Behavior

During the 10-year period between 1952 and 1961, management personnel in large companies have been exposed to a great deal of executive training. Programs range in length from one day to three months. Furthermore, results from employee opinion surveys have made industries sensitive to morale and interpersonal problems. Has this change in management climate exerted a measurable influence?

Since we have gathered data from similar management groups during this period, it is of interest to make some comparisons between groups in which the same procedure has been followed with comparable management levels. The differences over the nine-year period are not significant, although the trend suggests that the emphasis in human relations aspects of executive training may be changing leadership styles in the direction of control more by persuasive methods than by punitive methods. However, the discussion leadership skills, which we have shown to yield significant differences in the outcomes of group problem-solving, have not revealed themselves in industrial practice. The type of leader who can utilize the resources of his group members is still pretty much a laboratory product. Thus, our society is far from achieving the potential of group cooperation and group innovation.

DEMONSTRATED LEADER FUNCTIONS

Let us now reexamine the leadership functions that have been found to facilitate group problem-solving. First of all, it is clear from our analysis that the leader's function is to pose the problem, not to solve it. If he goes to the group with a solution, he shuts out their opportunity to generate ideas. We have found this short-circuiting approach most pronounced in people who have an organizational orientation. It is customary in organizations for ideas to be initiated at the top. A weakness in this approach, even if used, is that modern problems are becoming so complex and organizations so large that the individual leader cannot supply all of the needed initiative and knowledge.

Another reason why the leader should pose rather than solve a problem is that solutions are effectively carried out only if they gain the support and acceptance of those who must execute them. A solution, no matter how elegant, does not contain the motive power to put it into highly efficient action. People who are strongly motivated are needed to do justice to the execution of a decision. We have found that this motivation is best achieved when the individuals, who must execute a decision, have had a hand in formulating it. This means that solutions imposed from above are destined to be weak in their implementation.

A second function of the leader is to conduct the discussion so as to maximize the communication of information and reduce the need to contribute misinformation in order to defend a position. Without leadership, groups are poor problem-solvers. They tend to terminate problem-solving when a particular solution gains a certain amount of valence. The tendency for groups to confuse reaching agreement with finding a solution lowers the quality of group thinking. The leader must protect persons holding minority views (Maier and Solem, 1952), and he must delay the evaluation of preliminary solutions so as to permit less obvious ideas to be generated. The fact that second solutions are more innovative than first solutions demonstrates that, by nature, people are too prone to terminate their problem-solving activity in order to complete their task.

One of the best ways to delay solutions and to remove pressures to conform is to make disagreement an asset rather than a fault. This does not mean that disagreement is necessarily a virtue, but disagreement can delay a solution and, in the process of trying to understand conflicting viewpoints, one finds that additional viewpoints are generated. When individual problem-solvers take stands and each tries to win, the problem-solving process turns into interpersonal conflict. It is the leader's function to collect facts possessed by participants, assist in the communication of a variety of viewpoints, encourage disagreement, and protect individuals from group pressure. A leader can turn a participant, who is opposing another's ideas, into a contributor by asking for his ideas. If the various contributions are entertained (writing them briefly on a chalkboard is most helpful), the conflicts are turned away from personalities and toward the situation.

After ideas are collected, they can be evaluated in terms of their strengths and weaknesses. This process again makes it clear that evaluation is not a matter of choosing between black and white, but of selecting an alternative that has optimum merit in terms of the desired objective; be it a sound theory, an economical method, a new product, or an artistic work.

Our research up to this point has only begun to measure some of the things a leader can do. If we examine the potential resources present in groups and locate potential hazards, additional leadership principles may be generated.

RESOURCES IN A GROUP

What are the potential advantages of group-thinking versus those of individual-thinking? To assess the advantages one must explore the resources. Unless the resources are exploited and weaknesses are avoided, the group, as an organism, is not functioning efficiently.

The Resource of Increased Knowledge

First of all, the information in a group exceeds that of any individual in the group. Often the most informed individual is the leader. If he conducts the discussion instead of supplying the solution, may we not sacrifice his wisdom? The answer to this is in the negative because information can be supplied by the leader without his suggesting a solution. In the problem discussed in this study, Gus, the foreman, can show the time-study data to the workers, not for the purpose of supporting his solution, but for the purpose of seeing whether any good use can be made of them. It is of interest to observe that less than half of the foremen show the workers the data. Rather, they present an interpretation of the data, so slanted that the New-type solution is the only reasonable action. Raw data permit a variety of interpretations, and this type of information can be supplied by a leader without thereby implying any particular action or solution. When solutions are not implied, facts cease to be threatening.

Thus, the leader's information need not be lost when he becomes a discussion leader, rather he should become aware of the differences between facts and the meanings that are added to facts when he has a solution in mind. Further, he not only should be concerned with his own information, but also with that of members of his group. The fact of monotony in our experimental problem, for example, is something Gus and the time-study man may have overlooked. How to use the time-study data without increasing monotony becomes a challenging problem.

The Resource of Varied Problem Obstacles

There are many ways to view a problem. Once an individual sees an obstacle, he is inclined to pursue it. Persistence in habitual approaches (Maier, 1931) prevents an individual from exploring other approaches. In a group situation the differences in outlook become potential advantages because they tend to help individuals out of their habitual lines of thinking. The leader can maximize the difference by exploring the obstacles seen by various group members, and then stimulate the search for others.

The Resource of Interstimulation of Ideas

A group also has the potential of generating more ideas than the same individuals working alone because one person's suggestion may be a stimulant for ideas in others (Osborn, 1953; Clark, 1958). Although this interstimulation may not always occur in leaderless groups (Dunnette, 1964), the fact remains that the potential for variety is increased. We found women to be inferior to men in solving problems, yet a group made up of one woman and three men tended to be superior to a group of four men (Hoffman and Maier, 1961). Apparently even a difference in sex is a factor favorable to innovation.

Increased Number of Value Standards

The process of evaluation also is potentially greater in the group than in the individual because a greater variety of value systems is brought to bear when solutions are appraised. The tendency for the individual to perseverate reduces the factors considered, and hence increases the opportunity to overlook essential considerations.

Higher Frustration Threshold

A group's tendency to become discouraged by a problem doubtless is less than that of the same individuals working separately. If a group is more prone to remain objective and problem oriented, a solution's quality is more assured, provided the disadvantages of group processes are avoided.

POTENTIAL DISADVANTAGES IN GROUP-THINKING

Conformity Pressures

Perhaps the greatest negative factor in group-thinking is the pressure to conform and thus be a "good" group member. We have already shown that a "problem employee" and an "idea man" greatly depend on the leader's perception. This means that the leader must be trained, not only to avoid his own perceptions that make for conformity, but also to minimize the tendency of dominant participants to use group pressure to reject strange or unusual ideas.

Time Requirements

Time is another consideration that must be respected. The group process requires more man-hours to solve a problem than does the individual process. This may be an advantage in that it can prevent the reaching of premature decisions, but it must be granted that much time often is wasted in the group process. This waste obviously is something the leader can influence. However, he must use his influence carefully, because the presence of time pressure is not consistent with innovation in problem-solving.

Reduced Recognition

Inventive and creative persons sometimes feel that opportunities to gain individual recognition are lost when groups solve a problem. They may fear that the leader will take the credit or that he may patent the product. These are individual motivations that cannot be ignored. Interpersonal trust thus becomes a factor for consideration under such conditions.

However, it should be pointed out that status in the group and stimulation from the exchange of ideas can lead to other satisfactions, which in the long run may be greater. The opportunities for individual recognition and individual status are rapidly dwindling, so that the search for group satisfactions and achievement in groups should receive greater attention.

Group Immaturity

A fourth disadvantage in group-thinking is present if the participants are inexperienced or immature. For example, if the leader of a technical group is a scientist and the participants are graduate students, it would be inefficient for the leader to conduct himself in a way similar to that of a leader with a group of subordinates made up of scientists. In the former case, the discussion would have value as a training device, and it would be unwise to discourage participation; but the leader's role would have to be adapted to the group's qualifications. The same reasoning would apply to all training situations; pooling inexperience and naiveté neither makes for reliable information nor assures a good product. This means that the role of a problem-solving conference leader and that of a trainer should be differentiated, and the gap between them should be narrowed as the group matures in growth and responsibility.

SUMMARY

Five problem-solving mechanisms in living organisms have been described. Four of these represent individual development, whereas the fifth represents a possibility to be achieved only if the resources latent in a group are utilized. The achievement of group problem-solving, as the highest level of problem-solving, requires that one member serve as a leader trained in group discussion. The major portion of the paper deals with the role of the leader in making groups good problem-solvers.

In order to demonstrate the leadership function, the same problem situation was presented to groups under a variety of conditions. Altering the climate but leaving the facts the same is found to produce sharp differences in outcome.

The final section of the paper deals with leadership potentials that still require experimental verification. These potentialities hinge upon the full utilization of the latent resources in the group and the reduction of the latent deficiencies.

ACKNOWLEDGMENTS

The research reported was supported by USPHS Grant No. MH-02704, United States Public Health Service.

REFERENCES

Clark, C. H. 1958. *Brainstorming*. Garden City, New York: Doubleday.
Colgrove, M. 1966. Increasing innovative solutions by the manipulation of individual mental sets. Doctoral dissertation, University of Michigan.
Dunnette, M. D. 1964. Are meetings any good for solving problems? *Personnel Admin.* 27: 12–16, 29.
Guilford, J. P., and W. S. Zimmerman. 1949. *The Guilford-Zimmerman temperament survey*. Beverly Hills: Sheridan Supply.
Hoffman, L. R. 1959. Homogeneity of member personality and its effect on group problem-solving. *J. Abnorm. Soc. Psychol.* 58: 27–32.
Hoffman, L. R., R. J. Burke, and N. R. F. Maier. 1965. Participation, influence, and satisfaction among members of problem-solving groups. *Psychol. Rep.* 16: 661–667.
Hoffman, L. R., E. Harburg, and N. R. F. Maier. 1962. Differences and disagreements as factors in creative group problem solving. *J. Abnorm. Soc. Psychol.* 64: 206–214.
Hoffman, L. R., and N. R. F. Maier. 1961. Quality and acceptance of problem solutions by members of homogeneous and heterogeneous groups. *J. Abnorm. Soc. Psychol.* 62: 401–407.

Hoffman, L. R., and N. R. F. Maier. 1964. Valence in the adoption of solutions by problem solving groups: concept, method, and results. *J. Abnorm. Soc. Psychol.* 69: 264–271.

Koehler, W. 1926. *The mentality of apes.* New York: Harcourt, Brace.

Lorge, I., D. Fox, J. Davitz, and M. Brenner. 1958. A survey of studies contrasting the quality of group performance and individual performance, 1920–1957. *Psychol. Bull.* 55: 337–372.

Maier, N. R. F. 1930. Reasoning in humans. I. On direction. *J. Comp. Psychol.* 10: 115–143.

Maier, N. R. F. 1931. Reasoning and learning. *Psychol. Rev.* 38: 332–346.

Maier, N. R. F. 1932. The effect of cerebral destruction on reasoning and learning in rats. *J. Comp. Neurol.* 54: 45–75.

Maier, N. R. F. 1936. Reasoning in children. *J. Comp. Psychol.* 21: 357–366.

Maier, N. R. F. 1952. *Principles of human relations.* New York: Wiley.

Maier, N. R. F. 1953. An experimental test of the effect of training on discussion leadership. *Human Relations* 6: 161–173.

Maier, N. R. F. 1960. Selector-integrator mechanisms in behavior. In B. Kaplan and S. Wapner, eds., *Perspectives in psychological theory.* New York: International Universities Press.

Maier, N. R. F. 1963. *Problem solving discussions and conferences.* New York: McGraw-Hill.

Maier, N. R. F., and L. R. Hoffman. 1960. Quality of first and second solutions in group problem solving. *J. Appl. Psychol.* 44: 278–283.

Maier, N. R. F., and L. R. Hoffman. 1961. Organization and creative problem solving. *J. Appl. Psychol.* 45: 277–280.

Maier, N. R. F., and L. R. Hoffman. 1964. Financial incentives and group decision in motivating change. *J. Soc. Psychol.* 64: 369–378.

Maier, N. R. F., and L. R. Hoffman. 1965. Acceptance and quality of solutions as related to leaders' attitudes toward disagreement in group problem solving. *J. Appl. Behav. Sci.* 1(4): 373–386.

Maier, N. R. F., and T. C. Schneirla. 1935. *Principles of animal psychology.* Chapter 20. New York: McGraw-Hill. (1964 enlarged edition, Dover Publ., New York.)

Maier, N. R. F., and A. R. Solem. 1952. The contribution of the discussion leader to the quality of group thinking: the effective use of minority opinions. *Human Relations* 5: 277–288.

Maier, N. R. F., and A. R. Solem. 1962. Improving solutions by turning choice situations into problems. *Personnel Psychol.* 15: 151–157.

Osborn, A. F. 1953. *Creative imagination: principles and procedures of creative thinking.* New York: Scribner.

Solem, A. R. 1958. An evaluation of two attitudinal approaches to delegation. *J. Appl. Psychol.* 22: 36–39.

MAX SCHUR

Department of Psychiatry
State University of New York
Downstate Medical Center
Brooklyn, New York

LUCILLE B. RITVO

Department of the History of Science
  and Medicine
Yale University
New Haven, Connecticut

The Concept of Development and Evolution in Psychoanalysis

ΔΑΙΜΩΝ, *Dämon*
> *Wie an dem Tag, der dich der Welt verliehen,*
> *Die Sonne stand zum Grusse der Planeten,*
> *Bist alsobald und fort und fort gediehen*
> *Nach dem Gesetz, wonach du angetreten. . . .*

ΤΥΧΗ, *das Zufällige*
> *Die strenge Grenze doch umgeht gefällig*
> *Ein Wandelndes, das mit und um uns wandelt;*
> *Nicht einsam bleibst du, bildest dich gesellig,*
> *Und handelst wohl so, wie ein andrer handelt: . . .*

Goethe, URWORTE. ORPHISCHE

The work of T. C. Schneirla bares some of "the underlying bedrock" that Freud, as biologist-turned-therapist, subsumed for psychoanalytic psychology. Freud always assumed that "for the psychical field, the biological field does in fact play the part of the underlying bedrock" (1937, p. 252). In his "Project for a Scientific Psychology" (1895) he attempted to explain psychic phenomena in terms of contemporary neurophysiology. Subsequently, he tried to restrict himself to the explanation of psychical phenomena in psychological terms. However, he never ceased to apply evolutionary concepts to psychic phenomena, e.g., development, conflict, continuum ("the necessary acquirement of each mental power and capacity by gradation," Darwin, 1859, p. 413), adaptation and survival value. In his early biological researches on the eel, crayfish, and *Ammocoetes planerii* in the laboratories of Claus, Brücke, and Meynert, Freud had successfully pursued both ontogenetic and phylogenetic investigations and contributed supportive evidence for evolutionary theory (L. Ritvo, 1963). As early as his high-school days "the theories of Darwin, which were then of topical interest, strongly attracted me, for they held out hopes of an extraordinary advance in our understanding of the world" (Freud, 1925, p. 8). When for economic reasons he had to renounce the career in biology on which he was well launched, Freud brought into the therapy room the training and attitudes of a research worker in evolutionary biology. It is not surprising that forced by the exigencies of life to turn his attention to the problems of the human personality he should eventually try the ontogenetic and phylogenetic approaches which had served him so well in biology. The ontogenetic approach proved richly rewarding in Freud's writings and continues to do so in today's psychoanalytic research. The phylogenetic approach, however, is limited by the fact that psychoanalysis studies only a single species. Freud was aware that the answers to questions of phylogeny lie in biology. However, he permitted himself to raise questions and to speculate in terms of the biological thought of his day such as Haeckel's "ontogeny repeats phylogeny" and Darwin's neo-Lamarckism (Ritvo, 1965). Freud sometimes took big leaps in his evolutionary speculations that he did not permit himself in his scientific formulations on ontogeny. His evolutionary statements, therefore, may cause those whose knowledge of his work is restricted to his later speculative writings to question the quality of his scientific thinking. Schneirla's studies of animal behavior and his writings on the subject provide for modern analysts some of the phylogenetic data and insights lacking in Freud's time and they corroborate at lower phyletic levels some of Freud's findings at the human level. Working at lower phyletic levels Schneirla has encountered some of the same problems as Freud and dealt with them in a remarkably similar fashion, most notably the nature-nurture dilemma. Freud recognized the need to study more than just the mature stages of man's development; Schneirla points to the same need in understanding the development of animal behavior. Schneirla has found that even at lower phyletic levels the "instinct" concept is an oversimplification leading to unnecessary confusion. Schneirla's biphasic theory of approach and withdrawal has implications for psychoanalytic metapsychology that are beginning to be explored (Schur, 1962, 1966).

ΔΑΙΜΩΝ AND ΤΥΧΗ—FATE AND CHANCE

The traditional heredity-environment dilemma stands out more clearly as a pseudo-problem as further evidence indicates that in all animals intrinsic and extrinsic factors are closely related throughout ontogeny. . . . The question is how development occurs in the particular animal under prevailing conditions, not what heredity specifically contributes or environment specifically contributes, or how much either contributes proportionally, to the process. (Schneirla, 1957.)

When Freud entered practice the etiology of hysteria was explained by hereditary taint, a view supported by the great Charcot. Freud did not discard this view as rapidly as he did Erb's popular electrotherapy. In "Fragment of an Analysis of a Case of Hysteria" (1905a) Freud still mentioned the syphilitic history of the patient's father as he had in his earlier hysteria papers. But that same year (1905b, p. 173) he pointed out that

> writers who concern themselves with explaining the characteristics and reactions of the adult have devoted much more attention to the primaeval period which is comprised in the life of the individual's ancestors—have, that is, ascribed much more influence to heredity—than to the other primaeval period, which falls within the lifetime of the individual himself—that is, to childhood. One would surely have supposed that the influence of this latter period would be easier to understand and could claim to be considered before that of heredity.

One view of the etiology of neurotic illness did emphasize the experiential and can be likened to the presence of a foreign body that needs to be excised. The cathartic method which Freud learned from Josef Breuer, co-discoverer of the Hering-Breuer reflex, attributed the source of the symptoms to a forgotten trauma, i.e., a repressed memory, which Freud found was always of a sexual nature and always, when traced to its source, an infantile sexual trauma. Freud's realization of the importance of experiential factors was derived from the tales of early childhood seduction of his hysteria patients. In 1897 Freud had to recognize that his patients were not telling him the truth. Turning this defeat into victory is one of Freud's greatest triumphs. He (1897) confided to Fliess:

> Were I depressed, jaded, unclear in my mind, such doubts might be taken for signs of weakness. But as I am in just the opposite state, I must acknowledge them to be the result of honest and effective intellectual labour, and I am proud that after penetrating so far I am still capable of such criticism. Can these doubts be only an episode on the way to further knowledge?

It did indeed lead to knowledge of the importance of the early fantasy life in mental development—to the discovery of "psychic reality." From here on Freud was able to put hereditary disposition and experiential factors in their proper perspective, at least ontogenetically. Benjamin (1961, p. 23) points out that Freud did not completely relinquish what he had regarded as his "great discovery" of the importance of the experiential, but he displaced it from the ontogenetic to the phylogenetic, i.e., from the history of the individual to the history of the race. Freud's displacement of the prime importance of the experiential from the ontogenetic to the phylogenetic, plus the fact that psychoanalytic observation is restricted to a single species, may explain the very marked difference in the fate of Freud's ontogenetic and phylogenetic formulations. The phylogenetic were for the most part speculative and controversial; some, for instance, the death instinct, contributed to a great deal of confusion (Schur, 1966). The ontogenetic led to important discoveries.

Freud in his lifetime experienced the entire gamut of the nature-nurture controversy: from the emphasis on heredity, then on experience, to the recognition of the interplay between them and finally to the study of the intricate nature of that interplay.

> We find in psychoanalysis that we are dealing not with one *anlage* but with an infinite number of *anlagen* which are developed and fixed by accidental fate. The *anlage* is so to speak polymorphous. . . . Now, it seems that in single cases all these possibilities of variation are realized in such a way that in each individual sometimes this, sometimes that part of the inherited disposition becomes so dominant that it selects some experiences and rejects others, whereas accidental influences work here and there so powerfully that this or that part of the originally dormant *anlage* will be aroused and fixated. (Freud, 1911b, p. 284.)

We have only to substitute the word "genes" to have a formulation in up-to-date genetic terms. Freud saw the nature-nurture controversy as "a case in which scientifically thinking people distort a cooperation into an antithesis. . . . Δαιμων και Τυχη [fate and chance] and not one or the other are decisive" (ibid.). For the interplay of the accidental and the constitutional factors Freud developed the etiological or complemental series which he added to the 1915 edition of the *Three Essays* (1905b, pp. 239–240):

> To cover the majority of cases we can picture what has been described as a "complemental series," in which the diminishing intensity of one factor is balanced by the increasing intensity of the other.

At this time and later in *Analysis Terminable and Interminable* (1937) he explained the importance of the concept of the complemental series for the etiology of mental illness: In some cases the *anlage* is so "lethal" that even given a normal (average-expectable, Hartmann, 1939a) environment, mental illness develops, whereas in other cases severe environmental noxae can be mainly responsible for pathology. He applied the series not only to differences between individuals but to ontogeny. "We shall be in even

closer harmony with psycho-analytic research if we give a place of preference among the accidental factors to the experiences of early childhood"
(1915 addition to 1905b, p. 240). Near the end of his life Freud (1940,
p. 185) used an analogy from embryology:

> The damage inflicted on the ego by its first experiences gives us the
> appearance of being disproportionately great; but we have only to take
> as an analogy the differences in the results produced by the prick of a
> needle into a mass of cells in the act of cell-division (as in Roux's experi-
> ments) and into the fully grown animal which eventually develops out
> of them.

Schneirla cautions against studying the behavior of only mature animals
and overlooking early stages of development. Psychoanalysis from its very
beginning, with the discovery that "hysterics suffer mainly from reminis-
cences" (Breuer and Freud, 1893), has been more concerned with the era
of growth and development than with maturity. Freud sought to corroborate
the insights he gained from reconstruction of his patients' pasts with observations of his own children and those of his psychoanalytic colleagues. In
1909 he published the case history of "Little Hans" (1909a). Since the
twenties a new field, child analysis, has attracted many workers and
made accessible to observation "a host of intimate data about the child's
life, his fantasies as well as his daily experiences" (Kris, 1950). More
recently psychoanalytically oriented observations of children in nonanalytic
situations, e.g., nursery school and well-baby clinic, have become available
through such service-oriented research projects as that of Anna Freud and
her associates at the Hampstead Clinic in London (A. Freud, 1959;
A. Freud, Burlingham, and de Monchaux, 1958) and the Longitudinal
Study initiated in 1950 at the Yale Child Study Center by Dr. Milton Senn
and the late Ernst Kris (S. Ritvo et al., 1963). In the Yale Longitudinal
Study observations of constitutional differences commenced at the time of
delivery; environmental factors were investigated even earlier in prenatal
studies of the expectant parents. This longitudinal study sought among
other things the "interaction between the concrete environment and the
development of the child's capacities" (Kris, 1950, p. 28). Longitudinal
studies offer the opportunity to observe development in the preverbal stage,
particularly the development of object-relationships (the mother-child
interactions, etc.), whereas the hereditary or genetic factors may be more
explicitly sought in studies of twins, such as those of Hartmann (1934–
1935) and Kallman (1959). But even in his twin study Hartmann points
out (p. 427):

> The possibilities it [the twin-study method] opens can be of particular
> use . . . if we do not limit ourselves to determining whether or not a

given trait (in the phenotype) appears in one or both twins—which would, of course, be sufficient for a biogenetic analysis of certain diseases—but also observe the growth of the personality by means of the interaction between heredity and environment.

The most recently published extension of Freud's developmental or genetic approach is Anna Freud's *Normality and Pathology in Childhood: Assessments of Development* (1965). Half a century after her father she still finds it necessary to reiterate that "our task is not to isolate the two factors [innate and environmental] and to ascribe to each a separate field of influence but to trace their interactions" (p. 85).

A problem arises in the practice of psychoanalysis because of the great range of man's development from the prevalent dependence on external stimulation which we encounter in childhood and in lower animals to the highest level of relative independence of external stimuli:

> While the analysts of adults have to remind themselves of the frustrating, external, precipitating causes of the disorder of their patients, so as not to be blinded by the powers of the inner world, the child analysts have to remember that the detrimental external factors which crowd their view achieve their pathological significance by way of interaction with the innate disposition and acquired, internalized libidinal and ego attitudes. (A. Freud, 1965, pp. 51–52.)

When Freud enunciated his complemental series he saw "no reason to deny the existence of extreme cases at the two ends of the series" (1905b, p. 240). These extremes have been investigated in child analysis by the selection of patients with either innate or environmental damage of a massive nature. Major constitutional factors militating against normal development may be such severe sensory or physical handicaps as blindness, deafness, and deformities; at the other end are children traumatized by the lack of the outward setting for normal development, e.g., raised by psychotic parents or orphaned and institutionalized. These investigations may have been attempted in order "to probe whether there are any quantitative limits beyond which the pathogenic influences can be seen as unilateral" (A. Freud, 1965, p. 52). Anna Freud concludes that the "range of [pathological formations] and the detailed characteristics of the personalities of the children depend, as in the less severe cases, on the interaction between the two sides, i.e., on the manner in which a particular constitution reacts to a particular set of external circumstances" (p. 53).

Psychoanalysis has grown since Freud first recognized that neither heredity alone nor the accidental but their interaction was the basis for the etiology of the neuroses. But his conceptualization continues to be valid.

THE DEVELOPMENT OF THE MENTAL APPARATUS

Freud expressed his theories about the development of the mental apparatus in evolutional and developmental terms. According to Freud's theory, the activity of a hypothetical primitive psychic apparatus is "regulated by an effort to avoid an accumulation of excitation and to maintain itself so far as possible without excitation. For that reason it is built upon the plan of a reflex apparatus" (1900, p. 598). According to this theory "the accumulation of excitation . . . is felt as unpleasure and . . . sets the apparatus in action with a view to repeating the experience of satisfaction, which involved a diminution of excitation and was felt as pleasure" (ibid.). He calls "a current of this kind in the apparatus, starting from unpleasure and aiming at pleasure . . . a 'wish' " (ibid.). Both external and internal (maturational) stimuli cause the development of the mental apparatus from a simple reflex plan to what Freud calls "its present perfection" (p. 565). Freud (1915a, p. 120) expressed this theory in evolutional terms when he described the development of instinctual drives:

> External stimuli impose only the single task of withdrawing from them; this is accomplished by muscular movements, one of which eventually achieves that aim and thereafter . . . becomes a hereditary disposition. Instinctual stimuli, which originate from within the organism, cannot be dealt with by this mechanism. Thus they make for higher demands on the nervous system and cause it to undertake involved . . . activities by which the external world is so changed as to afford satisfaction to the internal source of stimulation. . . . We may therefore well conclude that instinct[ual drive]s and not external stimuli are the true motive forces behind the advances that have led the nervous system, with its unlimited capacities, to its present high level of development. There is naturally nothing to prevent our supposing that the instinct[ual drive]s themselves are, at least in part, precipitates of the effects of external stimulation, which in the course of phylogenesis have brought about modifications in the living substance.

Trieb—Instincts or Instinctual Drives?

In *Instincts and Their Vicissitudes* (1915a) Freud formulated his theory of instinctual drives within both a biological and a psychological framework:

> If now we apply ourselves to considering mental life from a *biological* point of view, an 'instinct[ual drive]' appears to us as a concept on the frontier between the mental and the somatic, as the *psychical representa-*

tive of the stimuli originating from within the organism and reaching the mind, as a measure of the demand made upon the mind for work in consequence of its connection with the body. (Pp. 121–122, second italics ours.)

The term *Trieb* was used by Freud to connote not behavior patterns but the inner forces that motivate them. He did not, however, apply it to his controversial "death instinct" (1920). The instinctual drives have energies at their disposal and these energies have their *source* and their *force*. This force exerts *pressure* on the executive apparatus and thereby becomes a motivational agent of behavior. An instinctual drive has an *aim* which can be somewhat variable, but the most variable factor in an instinctual drive and "not originally connected with it" is its *object,* "the thing in regard to which or through which the instinct[ual drive] is able to achieve its aim" (1915a, p. 122). These assumptions Freud conceptualized, in contradistinction to the word *Instinkt* used in biology, as *Trieb,* a term best rendered into English as "instinctual drive." The instinctual drives, as *mental representations* of stimuli originating within the organism, are seen as the motivational forces contributing to the development of ego functions which mediate between the demands of these drives and the requirements of reality. Unfortunately, the term "instinct" with its implications of a given behavior pattern has been used throughout *The Standard Edition of the Complete Psychological Works of Sigmund Freud.* Apart from the valid objection that Schneirla raised against the usage of the "instinct" concept, the choice of the word "instinct" for *Trieb* in the otherwise outstanding English translation of Freud's work has introduced additional confusion (Hartmann, 1939a; Hartmann, Kris, and Loewenstein, 1946; Schur, 1960, 1961). Schneirla (1956, p. 430) suggested that "the term 'instinct' be retired from scientific usage, except to designate a developmental process resulting in species-typical behavior." Therefore, whenever the term "instinct" appears in a quotation from the *Standard Edition,* the authors of this paper have modified it to read "instinct[ual drive]."

Freud first discerned maturational factors in the libidinal or sexual drive, which has its somatic sources in specific organs of the body. Accordingly Freud's developmental approach is best known for the psychosexual stages of development—the oral, anal, phallic, genital, and latency phases. As there is not space in this paper to discuss Freud's devotion to a continuum in his ontogenetic formulations in contrast to the big leaps in some of his phylogenetic speculations, it is well to point out here his statement "that we are picturing every such [sexual] trend as a current which has been continuous since the beginning of life but which we have divided up, to some extent artificially, into separate successive advances" (1916–1917, p. 340). Although Freud looked upon the libidinal drive as existing from "the beginning of life" in the individual, he speculated (1920, pp. 56–57) in non-Haeckelian fashion that phylogenetically it was a late-comer.

> We need more information on the origin of sexual reproduction and of the sexual instinct[ual drive]s in general. . . . The origin of reproduction by sexually differentiated germ-cells might be pictured along sober Darwinian lines by supposing that the advantage of amphimixis, arrived at on some occasion by the chance conjugation of two protista, was retained and further exploited in later development. On this view "sex" would not be anything very ancient; and the extraordinarily violent instinctual drives whose aim it is to bring about sexual union would be repeating something that had once occurred by chance and had since become established as being advantageous.

We will not deal in this paper with Freud's concept of the death instinct but only with the aspect conceptualized by him as the destructive instinctual drive. (For discussion see Hartmann, Kris, and Loewenstein, 1946; Schur, 1966.) Freud realized the difficulty of following the developmental vicissitudes of the destructive instinctual drive because it is rarely encountered in a pure form. The aggressive drive sometimes combines with and sometimes conflicts with the libidinal. Such a relationship is indicated in the names that Freud gave to the oral-aggressive and the anal-sadistic stages. During the Oedipal phase the aggression turns against the parent of the same sex. Freud attributed the Oedipal conflict in typical neo-Lamarckian fashion to the hereditary precipitates of parricide at the period of the primal horde described by Darwin in *The Descent of Man*. Dissolution of the Oedipal conflict normally leads to internalization of the parental figure and the formation of the superego. The strictness of the superego is related to the strength of the aggressive drive. The aggressive drive also reinforces the ego in controlling the id; it operates against the ego as a "punishing force," often felt as guilt or exhibited by self-destructive tendencies. With the formation of the superego, the ego not only has to mediate between the instinctual drives and the environment but also between the drives and the demands of the superego. Thus the ego, in addition to its self-preserving function, has a synthesizing and harmonizing function as organizer of the three systems of personality, the ego, id and superego, a function which "has rightly been compared with Cannon's concept of homeostasis, or described as one level of it" (Hartmann, 1956, p. 291).

Darwin had related man's morality to his lengthy dependency. Freud (1940, p. 147) expressed a similar view in terms of the superego and indicated a problem for animal psychology:

> This general schematic picture of a psychical apparatus may be supposed to apply as well to the higher animals which resemble man men-

tally. A super-ego must be presumed to be present wherever, as is the case with man, there is a long period of dependence in childhood. A distinction between ego and id is an unavoidable assumption. Animal psychology has not yet taken in hand the interesting problem which is here presented.

To Freud it seemed that analogous structures of the psychical apparatus could have developed in higher animals before man. Although Freud occasionally referred to the Oedipal fantasy as having biological origins (1923, p. 35, n. 1), he more often attributed it to actual historical events (see above). Another of his earliest findings, infantile sexuality, Freud placed at the period of hominization.[1] His theory of diphasic sexuality

is confirmed by the anatomical investigation of the growth of the internal genitalia [see Hutchinson, 1930]; it leads us to suppose that the human race is descended from a species of animal which reached sexual maturity in five years and rouses a suspicion that the postponement of sexual life and its diphasic onset [in two waves] are intimately connected with the history of hominization. Human beings appear to be the only animal organisms with a latency period and sexual retardation of this kind. Investigations on the primates (which, so far as I know, are not available) would be indispensable for testing this theory. (Freud, 1939, p. 75.)

Man's assumption of the upright gait "made his genitals, which were previously concealed, visible" and consequently

the organic periodicity of the sexual process has persisted, it is true, but its effect on psychical sexual excitation has rather been reversed. This change seems most likely to be connected with the diminution of the olfactory stimuli by means of which the menstrual process produced an effect on the male psyche. Their role was taken over by visual excitations, which, in contrast to the intermittent olfactory stimuli, were able to maintain a permanent effect. The taboo on menstruation is derived from this "organic repression", as a defence against a phase of development that has been surmounted. (Freud, 1930, p. 99, n. 1.)

With the more persistent effect of visual stimulation on his need for genital satisfaction "the male acquired a motive for keeping the female, or speaking more generally, his sexual objects near him" (ibid.), thus founding the family. The exposure of the genitals by the upright stance made protection necessary and "provoked feelings of shame." The development of the feeling of shame in the individual is one of the dams against libido which leads to the civilizing of the individual. Freud related it to civilization itself.

---

[1] This is Strachey's translation of Freud's word "Menschwerdung"—the process of becoming human.

Freud was aware that "this is only theoretical speculation, but it is important enough to deserve careful checking with reference to the conditions of life which obtain among animals closely related to man" (1930, p. 100, n. 1). Such checking into the nature of other animals lies outside the field of psychoanalysis. On evolution Freud could only speculate and offer hypotheses from his psychoanalytic findings. The verifications and corrections necessary for a theory he realized he had to leave to the biologists. Nevertheless, he could not avoid touching in this fashion on a field originally his own.

> We may be struck by the fact that we have so often been obliged to venture beyond the frontiers of the science of psychology. The phenomena with which we are dealing do not belong to psychology alone; they have an organic and biological side as well, and accordingly in the course of our efforts at building up psycho-analysis we have also made some important biological discoveries and have not been able to avoid framing new biological hypotheses. (1940, p. 195.)

Schneirla's Biphasic Theory and Freud's Pleasure-Unpleasure Principles

> Much evidence shows that in *all* animals the species-typical pattern of behavior is based upon biphasic, functionally opposed mechanisms insuring approach or withdrawal reaction according to whether stimuli of low or high intensity, respectively, are in effect. . . . Through evolution higher psychological levels have arisen in which through ontogeny such mechanisms can produce new and qualitatively advanced types of adjustment to environmental conditions. (Schneirla, 1959, p. 4.)

The concepts "pleasure and unpleasure principles" Freud formulated first in the context of his genetic formulations of the development of the mental apparatus. He used two main models for both the development of the mental apparatus and the concepts unpleasure and pleasure principles. Both models had evolutional and ontogenetic implications. One of the models is generally assumed to follow Fechner's "constancy principle." Freud (1920, pp. 8–9) quoted Fechner (1873):

> "every psycho-physical motion rising above the threshold of consciousness is attended by pleasure in proportion as, beyond a certain limit, it approximates to complete stability, and is attended by unpleasure in proportion as, beyond a certain limit, it deviates from complete stability; while between the two limits, which may be described as qualitative thresholds of pleasure and unpleasure, there is a certain margin of aesthetic indifference."

Fechner's formulation was concerned with the experiences of pleasure and unpleasure but Freud used it as the basis for a biophysiological regulating principle. It is essential for an understanding of these "principles" to distinguish "economic" from experiential formulations. The economic point of view, according to Freud, "endeavours to follow out the vicissitudes of amounts of excitation and to arrive at least at some *relative* estimate of their magnitude" (1915b, p. 181).

THE UNPLEASURE PRINCIPLE—WITHDRAWAL

The steps by which Freud arrived at his formulation of the unpleasure principle are: the model of a reflex apparatus which tries to maintain itself free of stimuli; "the fiction of a primitive psychical apparatus" whose activities are regulated by an effort to avoid an accumulation of tension, which is equated with unpleasure; and finally the model of withdrawal from excessive *external* stimulation whether external to the organism or to the mental apparatus.

> Let us examine the antithesis to the primary experience of satisfaction— namely, the experience of an external fright . . . when the primitive apparatus is impinged upon by a perceptual stimulus which is a source of painful excitation. Uncoordinated motor manifestations will follow until one of them withdraws the apparatus from the perception and at the same time from the pain. . . . There will be an inclination in the primitive apparatus to drop the distressing memory-picture immediately, if anything happens to revive it, because if its excitation were to overflow into perception it would provoke unpleasure. . . . This avoidance by the psychical process of the memory of anything that had once been distressing is the prototype . . . of *psychical repression*. . . . As a result of the unpleasure principle, the first Ψ-system is totally incapable of bringing anything disagreeable into the context of its thoughts. (Freud, 1900, p. 600.)

It is evident that the necessity for withdrawal—physical withdrawal if the apparatus for this is available, withdrawal of the cathexis of the percept and later also of the memory trace of a "painful" excitation—is the model for Freud's unpleasure principle. The need to withdraw from the source of pain and danger has not achieved in the course of evolution the degree of internalization that is the basis for an instinctual drive. There is no motivational force to seek an object in order to withdraw from it. We therefore cannot speak of an "instinctual drive" to withdraw from pain and danger. There is only a propensity to respond to certain stimuli with withdrawal and certain species-specific discharge patterns. The response regulated by the unpleasure principle is the model both for the anxiety response and for defense. In Schneirla's terms the need for withdrawal, a physiological concept, develops into the tendency of avoidance, which is a psychological concept. Avoidance of danger is essential for survival and adaptation. In man it plays a great role in normal and abnormal development.

In *Beyond the Pleasure Principle* (1920, pp. 26–27) Freud speculated further about the response to intense, potentially traumatic external stimulation:

> Let us picture a living organism in its most simplified form as an undifferentiated vesicle of a substance that is susceptible to stimulation. . . . Suspended in an external world charged with the most powerful energies; it would be killed by the stimulation if it were not provided with a protective shield against stimuli. . . . *Protection against* stimuli is an almost more important function for the living organism than *reception* of stimuli.

Both the function of withdrawal and the added safeguard of the protective barrier against stimuli have eminent survival value and must have played an important role in evolution.

THE PLEASURE PRINCIPLE—APPROACH

Freud contended that in the course of evolution "the external protective cortical layer had migrated inwards." Ontogenetically the central nervous system develops from the ectoderm. The evolutional development of somatic apparatuses with a gradual shift away from primary dependence on external stimulation has led to a progressive internalization and to the development of the instinctual drives. As mentioned earlier Freud attributed to the instinctual drives the development of the mental apparatus from the primitive reflex plan. "The exigencies of life [the somatic needs] interfere with this simple function [of reflexive withdrawal] and it is to them, too, that the apparatus owes the impetus to further development" (Freud, 1900, p. 565). Freud distinguished clearly between the demands of the instinctual drives and the response to external danger. The response to external danger is withdrawal or avoidance. Instinctual drives press for discharge and initiate (motivate) alloplastic action. The accumulation of drive tension preceding gratification of instinctual demands is equated with unpleasure. Drive discharge through need (drive) satisfaction is equated with pleasure. The tendency to achieve such gratification is regulated by the pleasure principle.

What follows is the crucial developmental formulation and establishes the model of an internalized motivational force:

> An essential component of this experience of satisfaction is a particular perception (that of nourishment, in our example) the mnemic image of which remains associated thenceforward with the memory trace of the

excitation produced by the need. As a result of the link that has thus been established, next time this need arises a psychical impulse will at once emerge which will seek to recathect the mnemic image of the perception and to reevoke the perception itself, that is to say, to reestablish the situation of the original satisfaction. An impulse of this kind we call a wish; the reappearance of the perception is the fulfillment of the wish. . . . Thus the aim of this first psychical activity was to produce a "perceptual identity"—a repetition of the perception which was linked with the satisfaction of the need. (Freud, 1900, pp. 565–566.)

The emergence of a wish marks the phylogenetic as well as the ontogenetic beginnings of psychic structure. The transition is from "somatic needs," a physiological concept, to the psychological concept of instinctual drives as mental representations of stimuli arising within the soma (Schur, 1961, 1966). In Schneirla's terminology the emergence of a wish marks the transition from "approach" to "seeking," from a biosocial to a psychosocial pattern.

Early in life the mnemic image extends not only to the experience of satisfaction but to the total situation that brings it about. The wish must extend not only to the recathexis of the memory of satisfaction, which Freud equates with the disappearance of unpleasure, but to the recathexis of more and more details of the situation of satisfaction. With the emphasis on the wish to recathect the memory of the whole situation, including certain stimuli, it becomes clear that the "constancy" and unpleasure principles, based on a withdrawal response, cannot apply. The need to recreate any situation of satisfaction, or at least to recathect its memory trace, is regulated by the pleasure principle. The pleasure principle regulates the need to seek the object for gratification through approach response. The energy behind the wish is represented by the instinctual drives. We usually assume that both the gratification of a wish and the disappearance of pain-danger result in tension reduction. However, differences in the tension reduction, neurophysiologically and psychologically, are evident in normal development.

In some of his later work Freud came close to an explicit expression of the distinction between the pleasure and unpleasure principles implicit in his models and genetic formulations. However, he obscured his distinction by expressing it in terms of the antithesis between the death-instinct and libido (Schur, 1966). Schneirla's biphasic theory provides a more satisfactory principle than Fechner's "constancy principle" as the *biological substratum* Freud was seeking for psychoanalytic psychology.

THE EGO AND ITS ROLE IN ADAPTATION

After his attempt in the "Project" (1895) Freud had had to recognize the impossibility of basing a general psychology on a neurophysiological model. His "magnum opus," *The Interpretation of Dreams* (1900), and his *Psychopathology of Everyday Life* (1901) reveal a hope, which he did not have to abandon, that psychoanalysis might provide the framework for a general theory valid for both normal and abnormal behavior. This goal could be fully reached only with the development of ego psychology. Freud had recognized the importance of the ego in its role of "the repressing agency" as early as his "Neuropsychoses of Defense" (1894). Freud's letter to Jung, December 19, 1909 (Schur, 1970), explains why Freud refrained for many years from further study of the ego: "Thus far I have only described the repressed, which is the novel, the unknown, as Cato did when he sided with the *causa victa*. I hope I have not forgotten that *there also exists a victimizer*." [2]

In 1911 Freud took an important step toward reaching his goal of a general developmental psychology of human behavior. The work was continued by Anna Freud, Hartmann and others. In "Two Principles of Mental Functioning" (1911a) Freud describes the transition from the "pleasure ego" which seeks nothing but *immediate* gratification of its wishes to the "reality ego" which can both tolerate delay of satisfaction and initiate alloplastic changes. The transition from the pleasure ego to the reality ego is achieved through maturation and development. It also represents the gradually decreasing dependency of behavior on external stimulation traceable in the evolutionary series. Reality testing requires the development of such functions (later designated as ego functions) as attention, anticipation, memory, and judgment. Simple motor discharge in response to internal or external stimulation is converted into purposive action. Restraint upon immediate action

> was provided by means of the process of *thinking*, which was developed from the presentation of ideas. Thinking was endowed with characteristics which made it possible for the mental apparatus to tolerate an increased tension of stimulus while the process of discharge was postponed. It is essentially an experimental kind of acting, accompanied by displacement of relatively small quantities of cathexis together with less expenditure (discharge) of them. (Freud, 1911a, p. 221.)

In 1926 Freud took another important step toward a general developmental psychology. He had always recognized the problem of anxiety as "a nodal point at which the most various and important questions converge, a riddle whose solution would be bound to throw a flood of light on our whole mental existence" (1916–1917, p. 393). Freud never lost interest in the biophysiological manifestations of anxiety but he recognized that the vicissitudes of anxiety and the place of anxiety in normal and abnormal

---

[2] We are indebted to the International Universities Press for permission to quote this letter.

development must be formulated mainly in psychological terms. By applying the genetic point of view in *Inhibitions, Symptoms and Anxiety* (1926) Freud established the link between biological and psychological concepts. He recognized the phylogenetic origin of anxiety in a biological response analogous to those in animals and an ontogenetic origin in the *anlage* present in the newborn for the development of certain responses. He established a hierarchy of danger situations which correspond to maturation and development in close interaction with the environment. Man at birth is not equipped with preformed responses to certain stimuli indicating "danger." The appraisal of danger and the response to it become an ego faculty of highly sophisticated abstraction. The normal development of this faculty is of crucial importance for adaptation. The pathology of this development is of equal importance for psychopathology and also somatic illness (Schur, 1953, 1958, 1960).

In the last decade of his life Freud frequently stressed his "partiality" for the primacy of the intellect—his interest in ego psychology. In 1937 Freud also stated that one must attribute variations in the functioning of the ego to innate factors. Although seemingly obvious, this was an unusual statement for Freud; partly under the sway of the Darwinian concept of the struggle for existence, Freud remained preoccupied with the idea that the ego develops "out of the id" as a result of inner conflict with instinctual demands and outer conflict with the environment. It therefore remained for Hartmann to establish the concept of a "conflict-free ego sphere." To the *anlage* for the development of the instinctual drives and the set of wishes motivated by them which Freud postulated, Hartmann added a set of inborn apparatuses present at birth or soon after which, interacting with an "average expectable environment," develop functions independently of conflict. The ego, conceptualized as an assembly of functions held together by its synthetic function, may therefore be said to include functions of "primary autonomy," e.g., perception, memory, motor control, which develop from an "undifferentiated ego-id" (Freud, 1940, p. 149) independently of conflict. We can also assume that certain autonomous apparatuses serve the development of the id as well as the ego (Schur, 1966). In addition, according to Hartmann, many attitudes, interests, and aims of the ego that have originated in situations of conflict (defense) may through a "change of function" eventually gain "secondary autonomy," i.e., become part of the conflict-free ego sphere, and serve adaptation. Hartmann distinguishes throughout his work, as does Schneirla, between *anlage,* growth, maturation, and development.

Freud tentatively postulated a self-preservative instinct. He was never fully satisfied with it and later ascribed the self-preservative functions to the ego. The functions of the ego have a much closer relationship to external reality than do the instinctual drives. However, as Hartmann points out, there is survival value in "the whole ensemble of instinctual drives, ego functions, ego apparatuses, and the principles of regulation as they meet the average expectable environmental conditions" (1939a, p. 46). Hartmann (1939b, p. 13) ascribed a role in man's adaptedness to the instinctual drives as well as the ego.

> Probably the true position is that in his ego, especially as expressed in rational thought and action, in its synthetic and differentiating function, man is equipped with a very highly differentiated organ of adaption but this highly differentiated organ is evidently by itself incapable of guaranteeing an optimum of adaptation. A system of regulation operating at the highest level of development is not sufficient to maintain a stable equilibrium; a more primitive system is needed to supplement it.

Hartmann also called attention to the fact that even before the *intentional* processes of adaptation of the ego begin, a state of adaptedness exists. "Strictly speaking, the normal newborn human and his average expectable environment are adapted to each other from the very first moment" (Hartmann, 1939a, p. 51). Schneirla has pointed out that all species from the lowest to the highest by virtue of their current existence have demonstrated the survival value of their equipment. Hartmann's formulations include the interrelationships of man's initial adaptedness and his further adaptational processes with social as well as biological reality.

The development through internalization of the instinctual drives with their variability of aims and objects and especially the development of an ego with its primary and secondary autonomous functions serving intentional processes of adaptation account for the wide range of human behavior and for man's faculty of learning, which exceeds by far that of any other animal. The study of the ego as the organ of adaptation complements the work of Freud, which during the first decades of psychoanalysis had to be concerned mainly with the substratum of the human personality, the id. Psychoanalysis has now reached the stage where it can claim to provide the basis for a general psychology of human behavior. Because psychoanalysis has since its beginning been a developmental psychology, its students have been vitally interested in comparative psychology, to which the work of Schneirla has contributed so much.

REFERENCES

Benjamin, J. D. 1961. The innate and the experiential. In H. W. Brosin, ed., *Lectures in experimental psychiatry*. Pittsburgh: Univ. Pittsburgh Press. Pp. 19–42.

Breuer, J., and S. Freud. 1893. On the psychical mechanism of hysterical phenomena: preliminary communication. In J. Strachey, ed., *The standard edition of the complete psychological works of Sigmund Freud* (hereafter designated *Standard Edition*) vol. 2. London: Hogarth, 1955. Pp. 3–17.

Darwin, C. 1859. *On the origin of species.* London: Watts, 1950.

Darwin, C. 1871. *The descent of man and selection in relation to sex.* London: Murray.

Fechner, G. T. 1873. *Einige Ideen zur Schöpfungs- und Entwicklungsgeschichte der Organismen.* Leipzig.

Freud, A. 1936. *The ego and the mechanisms of defense.* New York: International Universities Press, 1946.

Freud, A. 1959. Clinical studies in psychoanalysis: research project of the Hampstead Child Therapy Clinic. *The psychoanalytic study of the child,* vol. 14. New York: International Universities Press. Pp. 122–131.

Freud, A. 1965. *Normality and pathology in childhood: Assessments of development.* New York: International Universities Press.

Freud, A., D. Burlingham, and C. de Monchaux. 1958. Clinical studies in psychoanalysis. *Proc. Roy. Soc. Med.* 51: 938–947.

Freud, S. 1894. The neuro-psychoses of defense. In J. Strachey, ed., *Standard Edition,* vol. 3. London: Hogarth, 1962. Pp. 45–61.

Freud, S. 1895. Project for a scientific psychology. In M. Bonaparte, A. Freud, and E. Kris, eds., *The origins of psychoanalysis.* New York: Basic Books, 1954. Pp. 347–455.

Freud, S. 1897. Letter to Wilhelm Fliess. In M. Bonaparte, A. Freud, and E. Kris, eds., *The origins of psychoanalysis.* New York: Basic Books, 1954. Pp. 216–217.

Freud, S. 1900. The interpretation of dreams. In J. Strachey, ed., *Standard Edition,* vol. 5. London: Hogarth, 1953.

Freud, S. 1901. The psychopathology of everyday life. In J. Strachey, ed., *Standard Edition,* vol. 6. London: Hogarth, 1960.

Freud, S. 1905a. Fragment of an analysis of a case of hysteria. In J. Strachey, ed., *Standard Edition,* vol. 7. London: Hogarth, 1953. Pp. 3–22.

Freud, S. 1905b. Three essays on the theory of sexuality. In J. Strachey, ed., *Standard Edition,* vol. 7. London: Hogarth, 1953. Pp. 125–245.

Freud, S. 1909a. Analysis of a phobia in a five-year-old boy. In J. Strachey, ed., *Standard Edition,* vol. 10. London: Hogarth, 1955. Pp. 3–149.

Freud, S. 1911a. Formulations on the two principles of mental functioning. In J. Strachey, ed., *Standard Edition,* vol. 12. London: Hogarth, 1960. Pp. 213–226.

Freud, S. 1911b. Letter 149 to Else Voigtländer. In E. L. Freud, ed., *Letters of Sigmund Freud.* New York: Basic Books, 1960. Pp. 283–285.

Freud, S. 1915a. Instincts and their vicissitudes. In J. Strachey, ed., *Standard Edition,* vol. 14. London: Hogarth, 1957. Pp. 111–140.

Freud, S. 1915b. Repression. In J. Strachey, ed., *Standard Edition,* vol. 14. London: Hogarth, 1957. Pp. 159–215.

Freud, S. 1916–1917. Introductory lectures on psychoanalysis. In J. Strachey, ed., *Standard Edition,* vol. 16. London: Hogarth, 1963.

Freud, S. 1920. Beyond the pleasure principle. In J. Strachey, ed., *Standard Edition*, vol. 18. London: Hogarth, 1955. Pp. 7–64.

Freud, S. 1923. The ego and the id. In J. Strachey, ed., *Standard Edition*, vol. 19. London: Hogarth, 1961. Pp. 3–66.

Freud, S. 1925. An autobiographical study. In J. Strachey, ed., *Standard Edition*, vol. 20. London: Hogarth, 1959. Pp. 3–74.

Freud, S. 1926. Inhibitions, symptoms and anxiety. In J. Strachey, ed., *Standard Edition*, vol. 20. London: Hogarth, 1959. Pp. 77–175.

Freud, S. 1930. Civilization and its discontents. In J. Strachey, ed., *Standard Edition*, vol. 21. London: Hogarth, 1961. Pp. 57–145.

Freud, S. 1937. Analysis terminable and interminable. In J. Strachey, ed., *Standard Edition*, vol. 23. London: Hogarth, 1964. Pp. 209–253.

Freud, S. 1939. Moses and monotheism. In J. Strachey, ed., *Standard Edition*, vol. 23. London: Hogarth, 1964. Pp. 1–137.

Freud, S. 1940. An outline of psycho-analysis. In J. Strachey, ed., *Standard Edition*, vol. 23. London: Hogarth, 1964. Pp. 141–207.

Gill, M. M. 1963. Topography and systems in psychoanalytic theory. *Psychological Issues Monograph 12*. New York: International Universities Press.

Goethe, J. W. 1815–1817. Urworte. Orphische. In L. Forster, ed., *The Penguin book of german verse*. Middlesex, England: Penguin Books Ltd., 1957. Pp. 229–230.

Hartmann, H. 1934–1935. Psychiatric studies of twins. In his *Essays on ego psychology*. New York: International Universities Press, 1964. Pp. 419–445.

Hartmann, H. 1939a. *Ego psychology and the problem of adaptation*. New York: International Universities Press, 1958.

Hartmann, H. 1939b. Psychoanalysis and the concept of health. In his *Essays on ego psychology*. New York: International Universities Press, 1964. Pp. 3–18.

Hartmann, H. 1950. Comments on the psychoanalytic theory of the ego. In his *Essays on ego psychology*. New York: International Universities Press, 1964. Pp. 113–141.

Hartmann, H. 1956. The development of the ego concept in Freud's work. In his *Essays on ego psychology*. New York: International Universities Press, 1964. Pp. 268–296.

Hartmann, H. 1964. *Essays on ego psychology*. New York: International Universities Press.

Hartmann, H., E. Kris, and R. Loewenstein. 1946. Comments on the formation of psychic structure. In "Papers on Psychoanalytic Psychology." *Psychological Issues Monograph 14*. New York: International Universities Press. 1964. Pp. 27–55.

Hutchinson, G. E. 1930. Two biological aspects of psychoanalytic theory. *Int. J. Psycho-Anal.* 9(1): 83–86.

Kallman, F. J. 1959. The genetics of mental illness. In S. Arieti, ed., *The American handbook of psychiatry*, vol. 1. New York: Basic Books. Pp. 175–196.

Kris, E. 1950. On preconscious mental processes. In *Psychoanalytic explorations in art.* New York: International Universities Press, 1952. Pp. 303–318.

Ritvo, L. B. 1963. Darwin of the mind. Master's thesis, Yale Medical Historical Library.

Ritvo, L. B. 1965. Darwin as the source of Freud's neo-Lamarckianism. *J. Am. Psychoanal. Assn.* 13: 499–517.

Ritvo, S., A. T. McCollum, E. Omwake, S. Provence, and A. Solnit. 1963. Some relation of constitution, environment and personality as observed in a longitudinal study of child development: case report. In A. Solnit and S. Provence, eds., *Modern perspectives in child development.* New York: International Universities Press. Pp. 107–143.

Schneirla, T. C. 1956. Interrelationships of the "innate" and the "acquired" in instinctive behavior. In P.-P. Grassé, ed., *L'Instinct dans le comportement des animaux et de l'homme.* Paris: Masson. Pp. 387–452.

Schneirla, T. C. 1957. The concept of development in comparative psychology. In C. B. Harms, ed., *The concept of development: An issue in the study of human behavior.* Minneapolis: Univ. of Minnesota Press. Pp. 78–108.

Schneirla, T. C. 1959. An evolutionary and developmental theory of biphasic processes underlying approach and withdrawal. In M. R. Jones, ed., *Nebraska symposium on motivation.* Lincoln: Univ. Nebraska Press. Pp. 1–42.

Schur, M. 1953. The ego in anxiety. In R. M. Loewenstein, ed., *Drives, affects, behavior,* vol. 1. New York: International Universities Press. Pp. 67–103.

Schur, M. 1955. Comments on the metapsychology of somatization. *The psychoanalytic study of the child,* vol. 10. New York: International Universities Press. Pp. 119–164.

Schur, M. 1958. The ego and the id in anxiety. *The psychoanalytic study of the child,* vol. 13. New York: International Universities Press. Pp. 190–220.

Schur, M. 1960. Phylogenesis and ontogenesis of affect- and structure-formation and the phenomenon of repetition compulsion. *Int. J. Psycho-Anal.* 41: 275–287.

Schur, M. 1961. Animal research panel 1960: a psychoanalyst's comments. *Am. J. Orthopsychiat.* 31: 276–291.

Schur, M. 1962. The theory of the parent-infant relationship. *Int. J. Psycho-Anal.* 43: 243–245.

Schur, M. 1966. The id and the regulatory principles of mental functioning. *J. Am. Psychoanal. Assn. Monograph Series 4.* New York: International Universities Press.

Schur, M. 1970. *Freud: living and dying.* New York: International Universities Press (in press).

Author Index

Species Index

Subject Index